Nelson Mindtap | Your Learning amplified

*I love that everything is interconnected, relevant
and that there is a clear learning sequence.
I have the tools to create a learning experience
that meets the needs of all my students and
can easily see how they're progressing.*

NELSON VICscience

Meredith McKague
(Consultant)
Kenna Bradley
(Lead author)
Andrea Blunden
Michael Diamond
Alysha Harkins
Nicole Henry
Leigh Park
Adam Scanlon
Adina Wolters
Contributing authors
Vicki Couzens
Graham Gee
Lynne Kelly
Tyson Yunkaporta

psychology

VCE UNITS 3 + 4

VICscience Psychology VCE Units 3 & 4 Student Book
4th Edition
Meredith McKague (series consultant)
Kenna Bradley (lead author)
Andrea Blunden
Michael Diamond
Alysha Harkins
Nicole Henry
Leigh Park
Adam Scanlon
Adina Wolters
Vicki Couzens
Graham Gee
Lynne Kelly
Tyson Yunkaporta

ISBN 9780170465076

Publisher: Eleanor Gregory
Content developer: Katherine Roan
Project editors: Felicity Clissold, Alex Chambers, Alana Faigen, Alan Stewart
Copyeditors: Jennifer Butler, Holly Proctor, Marcia Bascombe
Proofreader: Vanessa Lanaway
Indexer: Max McMaster
Series text design: Ruth Comey (Flint Design)
Chapter Maps: Leigh Ashforth (Watershed Design)
Series cover design: Emilie Pfitzner (Everyday Ambitions)
Series designer: Cengage Creative Studio
Permissions researcher: Helen Mammides
Production controllers: Karen Young, Alex Chambers
Typeset by: Lumina Datamatics

Any URLs contained in this publication were checked for currency during the production process. Note, however, that the publisher cannot vouch for the ongoing currency of URLs.

Acknowledgements
Selected VCE Examination questions and extracts from the VCE Study Designs are copyright Victorian Curriculum and Assessment Authority (VCAA), reproduced by permission. VCE® is a registered trademark of the VCAA. The VCAA does not endorse this product and makes no warranties regarding the correctness or accuracy of this study resource. To the extent permitted by law, the VCAA excludes all liability for any loss or damage suffered or incurred as a result of accessing, using or relying on the content. Current VCE Study Designs, past VCE exams and related content can be accessed directly at www.vcaa.vic.edu.au

© 2022 Cengage Learning Australia Pty Limited

Copyright Notice
This Work is copyright. No part of this Work may be reproduced, stored in a retrieval system, or transmitted in any form or by any means without prior written permission of the Publisher. Except as permitted under the *Copyright Act 1968,* for example any fair dealing for the purposes of private study, research, criticism or review, subject to certain limitations. These limitations include: Restricting the copying to a maximum of one chapter or 10% of this book, whichever is greater; providing an appropriate notice and warning with the copies of the Work disseminated; taking all reasonable steps to limit access to these copies to people authorised to receive these copies; ensuring you hold the appropriate Licences issued by the Copyright Agency Limited ("CAL"), supply a remuneration notice to CAL and pay any required fees. For details of CAL licences and remuneration notices please contact CAL at Level 11, 66 Goulburn Street, Sydney NSW 2000, Tel: (02) 9394 7600, Fax: (02) 9394 7601
Email: info@copyright.com.au
Website: www.copyright.com.au

For product information and technology assistance,
in Australia call **1300 790 853**;
in New Zealand call **0800 449 725**

For permission to use material from this text or product, please email
aust.permissions@cengage.com

ISBN 978 0 17 046507 6

Cengage Learning Australia
Level 5, 80 Dorcas Street
Southbank VIC 3006 Australia

Cengage Learning New Zealand
Unit 4B Rosedale Office Park
331 Rosedale Road, Albany, North Shore 0632, NZ

For learning solutions, visit **cengage.com.au**

Printed in China by 1010 Printing International Limited.
1 2 3 4 5 6 7 26 25 24 23

Contents

Introduction	ix	To the student	xiv
Author team	xi	To the teacher	xviii
Author acknowledgements	xiii	Nelson MindTap	xxii
Publisher acknowledgements	xiii		

1 Scientific research methods — 2

Know your key terms		6
1.1	The process of psychological research investigations	8
	The logbook	8
	Developing your research question, aim and hypothesis	9
	Designing a research investigation	10
	Collecting data	11
	Analysing data	13
	Drawing conclusions	13
	Limitations and recommendations	13
	Key concepts 1.1	14
1.2	Scientific investigation methodologies	14
	Controlled experiments in brief	14
	Correlational studies	15
	Case studies	17
	Classification and identification	17
	Fieldwork	17
	Product, process or system development	18
	Case study 1.1 SEMA3: an example of a product, process or system development	19
	Literature reviews	20
	Key concepts 1.2	21
1.3	The controlled experiment in detail	21
	Variables in controlled experiments	22
	Kinds of designs for controlled experiments	25
	Key concepts 1.3	28
1.4	Analysing and evaluating research	29
	Processing quantitative data	29
	Data analysis concepts	33
	Evaluating data and investigation methods	37
	Ethics	38
	Key concepts 1.4	41
	Chapter summary	42

UNIT 3 How does experience affect behaviour and mental processes?

Area of study 1 How does the nervous system enable psychological functioning?

2 Nervous system functioning — 47

Know your key terms		50
2.1	The nervous system: roles and subdivisions	51
	The central nervous system (CNS)	51
	Key concepts 2.1a	53
	Concept questions 2.1a	53
	The peripheral nervous system (PNS)	54
	Key concepts 2.1b	56
	Concept questions 2.1b	56
	Investigation 2.1 How does the human nervous system enable a person to interact with the external world?	59
	Key concepts 2.1c	60
	Concept questions 2.1c	60
2.2	Conscious and unconscious responses	61
	Conscious responses	61
	Unconscious responses	61
	Activity 2.1 Conscious and unconscious responses	63
	Analysing research 2.1 Reflex control could improve walking after incomplete spinal injuries	64
	Key concepts 2.2	65
	Concept questions 2.2	65
2.3	The transmission of neural information	65
	The structure of neurons	65

	Neurotransmitters: chemical messengers at the synapse	66
	Key concepts 2.3a	67
	Concept questions 2.3a	67
	The excitatory and inhibitory effects of neurotransmitters	68
	Activity 2.2 Glutamate and GABA balance	69
	Analysing research 2.2 New medication offers hope to patients with frequent, uncontrollable seizures	70
	Neuromodulators: chemical messengers beyond the synapse	71
	Analysing research 2.3 Levodopa better than newer drugs for long-term treatment of Parkinson's, largest ever trial shows	72
	Key concepts 2.3b	73
	Concept questions 2.3b	73
2.4	Neural basis of learning and memory	73
	Effects of experience on neural development	74
	Synaptic changes as a mechanism for learning and memory	75
	Synaptic plasticity	75
	Activity 2.3 Synaptic plasticity	78
	Case study 2.1 How video games can affect your brain	78
	Key concepts 2.4	81
	Concept questions 2.4	81
	Chapter summary	82
	End-of-chapter exam	84

3 Stress as an example of a psychobiological process 87

	Know your key terms	90
3.1	Stress and stressors	91
	Internal and external stressors	91
	Acute and chronic stress	92
	Key concepts 3.1a	93
	Concept questions 3.1a	93
	Stress responses	93
	Stress as a biological process	94
	Activity 3.1 The stress response	98
	Analysing research 3.1 Can meditation reduce chronic stress?	99
	Key concepts 3.1b	100
	Concept questions 3.1b	100
3.2	Selye's General Adaptation Syndrome	101
	Alarm-reaction stage	101
	Resistance stage	101
	Exhaustion stage	102
	Activity 3.2 Explaining Selye's GAS	103
	Strengths and limitations of the GAS	103
	Key concepts 3.2	104
	Concept questions 3.2	104
3.3	Stress as a psychological process	104
	Analysing research 3.2 How stress can lead to inequality	105
	Lazarus and Folkman's Transactional Model of Stress and Coping	106
	Activity 3.3 Lazarus and Folkman's Transactional Model	107
	Key concepts 3.3	108
	Concept questions 3.3	109
3.4	The gut–brain axis	109
	The interaction of the microbiota and the nervous system	111
	Activity 3.4 GBA flowchart	111
	The impact on psychological processes and behaviour	112
	Stress and the microbiota	113
	Key concepts 3.4	114
	Concept questions 3.4	114
3.5	Coping strategies	115
	Context-specific effectiveness	115
	Coping flexibility	115
	Approach strategies	116
	Investigation 3.1 The effectiveness of strategies for coping withstress	117
	Avoidance strategies	118
	Activity 3.5 SAC preparation audit	118
	Key concepts 3.5	120
	Concept questions 3.5	120
	Chapter summary	122
	End-of-chapter exam	124

Unit 3 Area of study 1 review   127

Area of study 2 How do people learn and remember?

4 Approaches to understand learning 133

	Know your key terms	136
4.1	Approaches to understand learning	137
4.2	Behaviourist approaches to learning	137
	Ivan Pavlov and the discovery of classical conditioning	138
	Classical conditioning terminology	140
	Pavlov's three-phases of classical conditioning	141

What is learned in classical conditioning?		142	Case study 4.3 Dadirri: deep listening with Dr Miriam-Rose Ungunmerr-Baumann	172
Key concepts 4.2a		143	Ways of knowing: Yarning	173
Concept questions 4.2a		143	Case study 4.4 Sand talk symbols for five ways of knowing	174
From Pavlov to Watson and the 'Little Albert' experiment		144	Final reflections	175
Analysing research 4.1 Conditioned emotional reactions in a human infant (Watson and Rayner, 1920)		144	Key concepts 4.4	175
			Concept questions 4.4	176
Investigation 4.1 Evaluation of a case study		145	Chapter summary	177
			End-of-chapter exam	179

5 The psychobiological process of memory — 183

The survival value of classical conditioning	146	
Analysing research 4.2 From Little Albert to fMRI: contemporary approaches to fear conditioning in humans	146	
Operant conditioning	148	
Measuring the effects of reinforcement and punishment	149	
Activity 4.1 How to improve study habits using operant conditioning	150	
Key concepts 4.2b	151	
Concept questions 4.2b	151	
The three-phase process of operant conditioning – learning our ABCs	152	
Key concepts 4.2c	154	
Concept questions 4.2c	154	
Case study 4.1 Teaching Groucho the Australian fur seal to brush his teeth	155	
Case study 4.2 Using behaviour modification strategies to change social media habits	156	
Key concepts 4.2d	159	
Concept questions 4.2d	159	

Know your key terms		186
5.1	What is memory?	188
	The multi-store model of memory	188
	Activity 5.1 Short-term memory	191
	Analysing research 5.1 Practising retrieval improves long-term retention	192
	Investigation 5.1 Serial position effect	196
	Key concepts 5.1	197
	Concept questions 5.1	197
5.2	The structure of long-term memory	198
	Explicit memory	198
	Implicit memory	201
	Key concepts 5.2	202
	Concept questions 5.2	202
5.3	The neural basis of explicit and implicit memories	203
	The roles of the neocortex, hippocampus and amygdala in explicit memory	203
	The neural basis of episodic memory	204
	The neural basis of semantic memory	205
	Analysing research 5.2 Where words are stored: the brain's semantic map	206
	The role of the basal ganglia and cerebellum in implicit memory	207
	Activity 5.2 Two truths and a lie	207
	Key concepts 5.3	208
	Concept questions 5.3	209
5.4	Autobiographical memory	209
	The roles of semantic and episodic memory in remembering the past and imagining the future	210
	Key concepts 5.4a	211
	Concept questions 5.4a	211
	The role of mental imagery in ABM and episodic future-thinking	212
	Where are you on the mental imagery/ABM spectrum?	213

4.3	The social-cognitive approach to learning	160
	Albert Bandura and social-cognitive learning	160
	Activity 4.2 Observational learning	161
	Analysing research 4.3 Learned aggression in children through observation of an adult model	162
	Current applications of social-cognitive learning	164
	Key concepts 4.3	166
	Concept questions 4.3	166
4.4	Aboriginal and Torres Strait Islander peoples' approaches to learning	166
	Situating ourselves as learners	166
	Aboriginal and Torres Strait Islander peoples' ways of knowing	168
	Ways of knowing: Dadirri (Deep listening)	171

		Key concepts 5.4b	213
		Concept questions 5.4b	213
	ABM in Alzheimer's disease		214
	Activity 5.3 Autobiographical memory, aphantasia and Alzheimer's disease		216
		Key concepts 5.4c	217
		Concept questions 5.4c	217
5.5	Mnemonics		218
	Mnemonics based on written language		218
	The method of loci		219

Activity 5.4 Method of loci	221
The sung narratives of oral cultures	222
Case study 5.1 The Seven Sisters Songline	223
Final reflections	225
Key concepts 5.5	225
Concept questions 5.5	225
Chapter summary	227
End-of-chapter exam	230
Unit 3 Area of study 2 review	**233**

UNIT 4 — How is mental wellbeing supported and maintained?

Area of study 1 How does sleep affect mental processes and behaviour?

6 The demand for sleep — 239

Know your key terms		242
6.1	What is consciousness?	243
	Normal waking consciousness	244
	Altered states of consciousness	244
	Sleep as a psychological construct	244
	Activity 6.1 Consciousness flowchart	248
	Key concepts 6.1a	249
	Concept questions 6.1a	249
	Techniques used to measure sleep	250
	Activity 6.2 Sleep monitoring	256
	Key concepts 6.1b	257
	Concept questions 6.1b	257
6.2	Regulation of sleep–wake patterns	258
	Circadian rhythms	258
	Analysing research 6.1 Even dim light before bedtime may disrupt a preschooler's sleep	260
	Ultradian rhythms	261
	Activity 6.3 Light and melatonin	262
	Analysing research 6.2 Disruption of circadian rhythms due to chronic constant light leads to depressive and anxiety-like behaviours in rats	263
	Key concepts 6.2	264
	Concept questions 6.2	264
6.3	The changes in sleep over the life span	265
	Analysing research 6.3 Screen time and sleep among school-aged children and adolescents	265

Demand for sleep: how much sleep is enough?	266
Investigation 6.1 Sleep patterns across the life span	269
Activity 6.4 Sleep across the life span	270
Analysing research 6.4 A sound night's sleep grows more elusive as people get older. But what some call insomnia may actually be an age-old survival mechanism, researchers report	270
Key concepts 6.3	272
Concept questions 6.3	272
Chapter summary	273
End-of-chapter exam	275

7 Importance of sleep to mental wellbeing — 279

Know your key terms		282
7.1	Partial sleep deprivation	283
	Effects of partial sleep deprivation	284
	Analysing research 7.1 Chronically sleep deprived? You can't make up for lost sleep	287
	Analysing research 7.2 Fatigue trial test a wake-up to drowsy drivers	288
	Comparison between the effects of sleep deprivation and alcohol consumption	289
	Activity 7.1 Sleepiness Scale	291

	Analysing research 7.3 At what point does fatigue start to affect performance?	291
	Key concepts 7.1	293
	Concept questions 7.1	293
7.2	Sleep disorders	293
	Circadian rhythm sleep disorders	294
	Sleep–wake shifts in adolescence	296
	Treatments for circadian rhythm sleep disorder	299
	Activity 7.2 Sleep disorders	301
	Key concepts 7.2	302
	Concept questions 7.2	302
7.3	Improving sleep–wake patterns and mental wellbeing	303
	Daylight and blue light	303
	Temperature	304
	Activity 7.3 Zeitgebers mind map	306
	Analysing research 7.4 Working night shift burns less energy, increases risk of weight gain	307
	Investigation 7.1 Investigating circadian rhythms	308
	Key concepts 7.3	309
	Concept questions 7.3	309
	Chapter summary	310
	End-of-chapter exam	311

Unit 4 Area of study 1 review 314

Area of study 2 What influences mental wellbeing?

8 Defining mental wellbeing 321

Know your key terms		324
8.1	Ways of considering mental wellbeing	325
	A dual-continuum model of mental health and wellbeing	325
	Activity 8.1 The dual-continuum model	331
	Resilience	331
	Analysing research 8.1 Three hours is enough to help prevent mental health problems in Australian teens	333
	Investigation 8.1 What has been the influence of the COVID-19 pandemic on mental wellbeing?	334
	Key concepts 8.1a	334
	Concept questions 8.1a	335

	Aboriginal and Torres Strait Islander peoples' understandings of social and emotional wellbeing	335
	Social and emotional wellbeing (SEWB)	336
	Activity 8.2 SEWB model contexts	340
	Activity 8.3 Uluru Statement from the Heart	341
	Case study 8.1 Cloaked in strength: possum-skin cloaks in social and emotional wellbeing	342
	Key concepts 8.1b	343
	Concept questions 8.1b	343
8.2	Mental wellbeing as a continuum	343
	The influence of internal and external factors	344
	Case study 8.2 Finding yourself	345
	Stress, anxiety and phobia	346
	Activity 8.4 Internal and external factors influencing wellbeing and phobia	351
	Analysing research 8.2 Writing about exam worries boosts exam performance	351
	Key concepts 8.2	353
	Concept questions 8.2	353
	Chapter summary	354
	End-of-chapter exam	356

9 Application of a biopsychosocial approach to explain specific phobia 359

Know your key terms		362
9.1	Development of specific phobia	363
	Biological factors contributing to phobia	364
	Key concepts 9.1a	365
	Concept questions 9.1a	366
	Psychological factors contributing to phobia	366
	Social factors contributing to phobia	369
	Activity 9.1 Biological, psychological and social contributing factors to specific phobia	370
	Analysing research 9.1 Explainer: why are we afraid of spiders?	371
	Key concepts 9.1b	372
	Concept questions 9.1b	372
9.2	Evidence-based interventions	373
	Biological interventions	373
	Key concepts 9.2a	375
	Concept questions 9.2a	375
	Psychological interventions	376

Case study 9.1 How virtual reality spiders are helping people face their arachnophobia ... 379
Activity 9.2 Developing a fear hierarchy ... 380
Social interventions ... 381
Analysing research 9.2 If you're afraid of spiders, they seem bigger: phobia's effect on perception of feared objects allows fear to persist ... 383
Investigation 9.1 Investigation of a specific phobia ... 383
    Key concepts 9.2b ... 384
    Concept questions 9.2b ... 384
    Chapter summary ... 385
    End-of-chapter exam ... 387

10 Maintenance of mental wellbeing ... 389

Know your key terms ... 392
10.1 The application of a biopsychosocial approach to maintaining mental wellbeing ... 393
    Biological protective factors that maintain mental wellbeing ... 393
    Analysing research 10.1 Reduced sleep quality can aggravate pre-existing psychological conditions ... 397
    Psychological protective factors that maintain mental wellbeing ... 397
    Analysing research 10.2 Focusing awareness on the present moment can enhance academic performance and lower stress levels ... 399
    Investigation 10.1 The influence of mindfulness meditation on wellbeing ... 401
    Social protective factors that maintain mental wellbeing ... 402
    Case study 10.1 How online mindfulness training can help students thrive during the pandemic ... 403
    Activity 10.1 Protective factors ... 404
        Key concepts 10.1 ... 405
        Concept questions 10.1 ... 405
10.2 Cultural determinants of social and emotional wellbeing ... 406
    Cultural continuity ... 406
    Case study 10.2 Tanderrum ... 407
    Self-determination ... 408
        Key concepts 10.2 ... 409
        Concept questions 10.2 ... 410
        Chapter summary ... 411
        End-of-chapter exam ... 412

Unit 4 Area of study 2 review ... 415

Area of study 3 How is scientific inquiry used to investigate mental processes and psychological functioning?

11 Using scientific inquiry ... 421

Know your key term ... 424
11.1 Designing an investigation ... 425
    Step 1: Get your logbook ready ... 425
    Step 2: Determine your research question ... 425
    Step 3: Write an aim ... 426
    Step 4: Write a hypothesis ... 426
    Step 5: Choose a methodology ... 428
    Step 6: Write a method ... 428
    Step 7: Consider safety and ethics ... 429
    Before you begin: things to consider ... 430
        Key concepts 11.1 ... 430
        Concept questions 11.1 ... 430
11.2 Conducting an investigation ... 431
    Step 8: Generate primary quantitative data ... 431
    Step 9: Organise and summarise your data ... 431
    Step 10: Analyse and evaluate your data ... 432
    Step 11: Determine uncertainty and limitations of your data ... 432
    Step 12: Interpret your data ... 433
    Step 13: Draw evidence-based conclusions ... 433
        Key concepts 11.2 ... 433
        Concept questions 11.2 ... 434
11.3 Science communication ... 434
    Step 14: Create your poster ... 434
    Poster sections ... 435
        Key concepts 11.3 ... 438
        Concept questions 11.3 ... 438
    Investigation 11.1 The process of learning based on performance on a puzzle maze ... 438
        Chapter summary ... 440

Glossary ... 441
Selected answers ... 453
References ... 468
Index ... 478

Introduction

VICscience Psychology VCE Units 3 & 4 (fourth edition) has been written to meet the specifications of the VCAA VCE Psychology Study Design 2023–2027. Our author team has been chosen for their secondary or tertiary teaching experience and their comprehensive knowledge of the psychology discipline.

Our author team has produced a resource that is comprehensive, engaging and easily accessible to all students of VCE Psychology. The book map presented on page xxii provides a bird's eye view of the content covered in Units 3 & 4, and its interconnectedness. From this map, the content covered by each chapter is further expanded upon in the chapter maps that sit at the start of each chapter. Each chapter map is an easy-to-use visual navigational tool to gently guide students through the story and connections within each chapter.

VCE can be an exciting but sometimes overwhelming time. Student wellbeing has been a major priority in the construction of this resource. The role of colour and its ability to influence feelings, attention and behaviour when learning has been carefully considered in the designs of the cover and chapters. Authors have carefully constructed chapters, taking into account choice of words, sentence length and use of diagrams to help explain concepts.

At the core of this approach is the inclusion of cognitive learning strategies within each chapter. These include pre-testing at the beginning of each chapter, promoting retrieval of prerequisite material from long-term memory. Key terms are front-loaded so that students can learn definitions early and retrieve them as they read through each chapter. Five evidence-based principles of effective learning drawn from research in cognitive and educational psychology have been embedded within the Key concept questions and higher-order thinking challenges that accompany the text (Weinstein et al., 2018). These include concrete examples **c**, retrieval practice **r**, interleaving of concepts **i**, elaboration **e** and dual-coding **d** of concepts through linking imagery with language-based definitions. Cognitive learning strategies promote the active construction of knowledge and assist students to make connections between concepts – the effective use of cognitive learning strategies deepens students' understanding and enhances their sense of agency and self-efficacy. The cognitive learning strategies can be easily extended to classroom activities. You can learn more and download freely available materials at http://www.learningscientists.org.

Activities in each chapter have been carefully designed to assist students in moving from lower-order thinking in the *Try it* section to higher-order thinking in the *Apply it* and *Exam ready* sections. Students can also practise and build upon their examination skills by attempting graded VCAA exam-style questions in the end-of-chapter exams and Area of Study reviews.

The study of Psychology is investigative. Key science skills associated with the scientific research that underpin the study of Psychology have been brought together in Chapter 1 Scientific research methods. These skills have been woven through all the chapters and students can practise applying them in the investigations and analysing research sections.

A dedicated chapter has been provided to assist students with successfully experiencing psychological research. Chapter 11 provides guidelines on how to complete Unit 4 Outcome 3.

VICscience Psychology VCE Units 3 & 4 (fourth edition) gives students a thorough grounding in VCE Psychology and sets them up for an enjoyable and successful year.

Reference: Weinstein, Y., Sumeracki, M., & Caviglioli, O. (2018). *Understanding how we learn: A visual guide*. Routledge.

INTRODUCTION

The study of Psychology has been traditionally viewed from a Western science perspective, but study designers have become increasingly aware of the need to challenge assumptions and include diverse psychological knowledge and experience. Aboriginal and Torres Strait Islander knowledge and perspectives have been provided through engagement with First Nations authors. Special thanks go to

Dr Vicki Couzens

Dr Couzens is a Gunditjmara, Keerray Woorroong citizen whose lands are located in what is now known as western districts of Victoria. She is a Senior Knowledge Holder of Gunditjmara Language and the Possum Cloak Story. She is also an artist, and holds a PhD from RMIT University where she is a Research Fellow.

Dr Graham Gee

Dr Gee is an Aboriginal-Chinese man, also with Celtic heritage, originally from Darwin. His Aboriginal-Chinese grandfather was born near Belyuen on Larrakia Country. Graham is a clinical psychologist and worked at the Victorian Aboriginal Health Service for 11 years before taking up a senior research fellow position at the Murdoch Children's Research Institute. Graham's early career research has focused on Aboriginal and Torres Strait Islander mental health, social and emotional wellbeing, and healing and recovery from complex trauma. In 2022, he received a Fellowship from the Eisen Family Private Fund that has supported his team to commence work with Aboriginal services dedicated to healing child sexual abuse. Graham recently joined the National Clinical Reference Group for the Prime Minister and Cabinet National Office for Child Safety, and the Research Advisory Committee for the National Centre for Action on Child Sexual Abuse.

Dr Tyson Yunkaporta

Dr Yunkaporta is an Aboriginal scholar, founder of the Indigenous Knowledge Systems Lab at Deakin University in Melbourne, and author of *Sand Talk*. His work focuses on applying Indigenous methods of inquiry to resolve complex issues and explore global crises.

Dr Lynne Kelly AM

Dr Kelly AM is an Adjunct Research Fellow at LaTrobe University researching Indigenous memory methods. She is a teacher and the author of many books, including four on memory.

Author team

Meredith McKague (author and consultant)

Meredith is a cognitive psychologist in the Melbourne School of Psychological Sciences at the University of Melbourne and was the Chief Assessor for the VCE psychology exam between 2013 and 2020. Meredith is also a non-Indigenous member of the Reference Group and Community of Practice with the Australian Indigenous Psychology Education Project (indigenouspsyched.org.au). Meredith conducts research and teaching in the areas of learning, memory and language.

Kenna Bradley (lead author)

Kenna is an experienced Psychology educator, having worked in schools and other education settings for twenty plus years teaching units 1 to 4 Psychology. She is currently a Learning Specialist in a state secondary school where she passionately teaches Psychology. Kenna has co-authored previous textbooks and has held many roles in examination and assessment.

Andrea Blunden

Andrea has taught Psychology since its inception in the 1990s in both government and non-government schools throughout Victoria. She has also worked as a VCE Psychology exam marker. Andrea enjoys teaching Psychology as she believes that it is a subject that is useful and relevant to everybody.

Michael Diamond

Michael Diamond works at the intersection of psychology and data science. His areas of specialty within psychology are research methods, statistics, utilising technology, and creating teaching content at both the VCE and the tertiary level. Michael has worked as a Research Consultant for Unforgettable Research Services (a platform for the collection and analysis of data from smartphones, wearable devices, Internet of Things (IoT), and social media) and as the Hub Coordinator for the Complex Human Data Hub within the School of Psychological Sciences at the University of Melbourne. Michael is a recipient of the Australian Psychological Society APS prize, and holds a Bachelor of Psychological Sciences (Honours), a Graduate Diploma in Psychology, and a Bachelor of Medical and Health Science.

Alysha Harkins

Alysha studied Psychology at Edith Cowan University and subsequently completed a Graduate Diploma in International Health at Curtin University, both in Perth, Western Australia. She discovered a love for education while working with young people teaching in Barcelona and as a Community Outreach Officer for the WA Aids Council. After working as a Research Officer in Dhaka, Bangladesh, she returned to Australian shores to complete a Graduate Diploma of Teaching (Secondary) at the University of Melbourne. She is currently a passionate teacher of English and Psychology.

Nicole Henry

Nicole completed a Bachelor of Science and Diploma of Education at Monash University and a Bachelor of Arts at Deakin University. She has taught in country and regional schools in both government and private sectors over 30 years. During that time, Nicole has primarily taught VCE Psychology and Biology. This has seen her develop a range of curriculum resources for her students. In recent years, Nicole has written and reviewed for Cengage VCE Biology. Nicole is one of the foundation facilitators of the Nature Stewards Program, an adult environmental education program under the umbrella of Outdoors Victoria.

Leigh Park

Leigh has been involved in Psychology education for the past 30 years. He has presented at local, national and international conferences with a focus on the connection between Psychology and the outdoors. He is passionate about involving students in their learning to discover more about themselves, to explore the world around them and to develop an empathy for others.

Adam Scanlon

Adam has worked as a VCE Psychology teacher for almost a decade in the state education system. He has also worked in school leadership and student management. He is extremely passionate about the subject and encourages students to think deeply about how the concepts of Psychology shape their experiences. In his work as a secondary teacher, he has supported students to achieve consistently high results, including perfect study scores, by focusing on writing quality short-answer responses.

Adina Wolters

Adina has a B.A, B.Sc., Dip. Ed. (Psychology and Maths) and is a current teacher of Psychology and Head of Psychology for Years 10–12 at Mount Scopus Memorial College. She has been teaching Psychology for 25 years with a particular passion for experiential learning. She is also a long-term presenter at Psychology teachers' conferences.

Author acknowledgements

Leigh would like to thank Joanna, Jazmin and Brooke for their boundless support and fun.
Nicole would like to thank my friend and colleague Cheryl Watson for your advice. Thank you to Jim, Simone and Nina for all your belief and encouragement.

Adina says thank you to my students for motivating me to continue finding new ways to engage with Psychology course content. You make the journey a fun adventure. Thank you to my husband Ian for his constant support and understanding when I spend countless hours making new activities and to my sons Dylan and Mitchell, who let me test the activities out on them while they were studying Psychology.

Publisher acknowledgements

Eleanor Gregory sincerely thanks the following VCE Psychology teachers for their assistance in reviewing and commenting on the authors' manuscripts during the development of this product:

- David Anderton, St Margaret's Berwick Grammar
- Elisa Baldwin, Wheelers Hill Secondary College
- Amelia Brear, Eltham High School
- Cheryl Watson, The Geelong College

And special thanks to:
Chalsea Chappel, Victoria University Secondary College – St Albans campus for compiling the Area of Study reviews.

To the student

The VCE Psychology course comprises both key knowledge and key science skills components, which will be assessed throughout your studies. We understand that undertaking VCE Psychology, especially at Units 3 & 4, can be an exciting but sometimes overwhelming time. You will learn a lot of content and develop scientific skills throughout very busy years which will culminate at the end of Units 3 & 4 in an external assessment. We have taken these stressors into account when designing the *VICscience Psychology* suite of products. You will not need to go beyond these learning materials to study VCE Psychology; they have been designed to work in unison so you can achieve at your very best level.

10 steps to study success

Ensure you take time to read the 10 ways we have organised your VCE Psychology journey. You will see that at various stages in your studies, different aspects of this textbook will be more useful. Whether you are learning new key terms and concepts for the first time, reviewing what you have learned or preparing for tests and exams, spending a little time now getting to know your textbook and what it offers will help you reach your learning potential for VCE Psychology.

1. Focus on the Study Design

Each chapter starts with a chapter opening page that will guide you through the **key knowledge** and **key science skills** that are covered within the chapter.

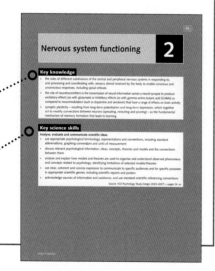

2. Overview of the VCE Psychology Study Design

You will find a book map on page xxii. The book map provides a bird's eye view of the content and how it is interconnected. From this map, the content covered by each chapter is further expanded upon in the chapter maps that sit at the start of each chapter. Each chapter map is an easy-to-use visual navigational tool to gently guide you through the story and connections within each chapter. Each chapter map:

- locates the chapter within the course
- enables you to see how the information in the chapter fits together
- is an easy-to-use navigational tool to guide you through each chapter
- offers a gentle entry into the more complex information.

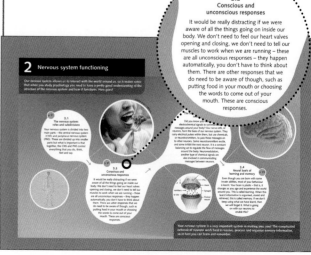

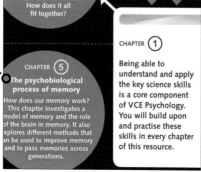

9780170465076

TO THE STUDENT   xv

③ Remember, rehearse and retrieve key terms

We have frontloaded all the new key terms you will meet throughout the chapter at the beginning of each chapter in **Know your key terms**. Use the **flashcards** study tool to encode key terms along with their definitions. Then when you come across a key term in your reading you can retrieve the definition from your memory.

The definitions of these key terms are also found in the glossary at the back of the book.

④ Test your retrieval

At the beginning of each chapter, use the **chapter pre-test** to assist you to retrieve from your memory previously learned concepts. You will assimilate these previously learned concepts into the new concepts as you progress through the chapter. Stronger foundations of knowledge make learning easier.

⑤ Develop your skills

Key science skills are an integral part of Psychology and they are examinable in the external assessment. **Chapter 1** focuses on all the key science skills so this is a good place to start.

Investigations within the textbook focus on developing and practising your key science skills.

Analysing research within chapters enables you to read a piece of psychological research and apply your key science skills to understand, analyse and evaluate it.

To further develop and refine the key science skills set out in the course, complete the activities in the accompanying *VICscience Psychology Skills Workbook*. Signposts to workbook activities are found throughout the textbook.

 WB 2.1.3 EVALUATION OF RESEARCH

⑥ Understand the concepts

It is not only pictures that tell a thousand words, but tables and graphs as well. They are key to strengthening your understanding. Ensure you look carefully at each **figure** and **table** and read the labels and captions so that you can understand what they are telling you.

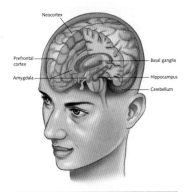

Figure 5.17 Areas of the brain that support long-term implicit and explicit memory

Important ideas, concepts and theories are summarised in **Key concept boxes**. **Concept questions** follow each Key concept box.

These questions will help you to determine whether you have fully understood the content before you progress further in the chapter.

KEY CONCEPTS 8.1b

» Social and emotional wellbeing (SEWB) is a holistic and culturally responsive framework of the factors that influence health outcomes for Aboriginal and Torres Strait Islander peoples and communities, including mental health.

The framework is holistic and multidimensional because a person's SEWB is influenced by the seven domains of connection and their interactions with broader social, cultural, historical and political determinants.

Concept questions 8.1b

Understanding
1. Describe the benefits for the wellbeing of Aboriginal and Torres Strait Islander peoples in using the social and emotional wellbeing (SEWB) framework. **e**

HOT Challenge
2. Create a table or diagram that identifies the similarities and differences between Western definitions of mental wellbeing and the Aboriginal and Torres Strait Islander framework of social and emotional wellbeing. **d**

Concept questions have been carefully structured using cognitive learning strategies to use the power of your brain to maximise your learning efficiency. Five key learning strategies have been signposted:
- Retrieval practice **r**
- Elaboration **e**
- Concrete examples **c**
- Dual coding **d**
- Interleaving **i**

You can find out more about these strategies at https://www.learningscientists.org/posters

If you are feeling confident with the concepts you can give the **HOT Challenges** a go! You can find these at the end of the Concept questions. These questions require higher cognitive application and may need further research. Try to give them a go as they will extend your understanding to a higher level.

⑦ Explore and learn

You will collaborate, explore and discover the psychological world through **investigations** and also come to appreciate the collegial nature of VCE Psychology.

Explore key knowledge and develop, use and demonstrate the key science skills through the Investigations. Investigations provide an opportunity to:
- explore the different methodologies
- use the logbook template provided to build up your logbook
- analyse secondary data
- collect and display your own primary data
- analyse and discuss data
- draw evidence-based conclusions
- discover Psychology for yourself.

INVESTIGATION 10.1 THE INFLUENCE OF MINDFULNESS MEDITATION ON WELLBEING

Scientific investigation methodology

Controlled experiment

Aim

To conduct an investigation into the influence of mindfulness meditation on heart rate and perceived level of stress.

Introduction

This investigation gives you the opportunity to design and conduct your own investigation relating to mental processes and psychological functioning. Remember, that your design must include the generation of primary quantitative data.

Make sure your investigation has an aim, methodology and method, results, discussion and conclusion while complying with safety and ethical guidelines. Remember to use your logbook throughout this investigation. Present

Method

Work out what you need to do in order to answer your research question and test your hypothesis. This will include:
- Population and sample – how big does your sample need to be?
- Variables – dependent, independent, controlled. Are there any extraneous variables you have not accounted for?
- What quantitative data are you going to record and how are you going to record it?
- Are there any ethical or safety issues you have to take into account?

Conclusion

Write a conclusion to respond to the research question.

TO THE STUDENT

8) Complete your outcome

A **dedicated chapter** within your textbook will guide you through the research and investigation outcome. **Chapter 11** provides you with important advice and steps you through Unit 4 Outcome 3. It has been written to help you, so make full use of it.

10) Consolidate your learning

At the end of every chapter, you can consolidate your knowledge using:

- **chapter summary of key concepts** that you have met throughout the chapter. You can download a copy of the concepts by accessing the MindTap icon. Use this to assist you in revising and studying for internal and external assessments.
- **end-of-chapter exam** will help you to retrieve, revise, understand and apply the concepts from the chapter.

A **glossary** of all the key terms plus their definitions can be found at the end of the book.

9) Prepare for tests and exams

Activities appear within each chapter. Each activity is divided into three sections. The first section, *Try it*, challenges you to do something at a low cognitive level. The second section, *Apply it*, asks you about what you have just done at a higher cognitive level, and the third section, *Exam ready*, gets you to use what you have learned so far in the activity to answer an exam-style question.

You can then take the skills that you learn from completing the activities into the **Area of Study reviews** at the end of each Area of Study. These allow you to check your knowledge by completing exam-style questions that are graded for difficulty.

You will find the answers to the activities and Area of Study reviews at the back of the book.

> **EXAM TIP**
> When answering questions regarding EOG, be careful to refer to the device measuring electrical activity of the muscles surrounding

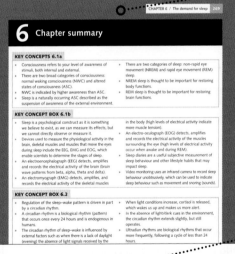

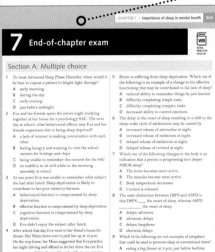

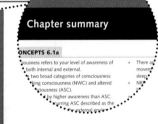

To the teacher

The VCE Psychology course comprises both key knowledge and key science skills. The *VICscience Psychology* suite of products provides you with the perfect resource to teach all the key knowledge and key science skills in an integrated and engaging way and to prepare your students thoroughly for the school-based and external assessments.

1 Follows the Study Design

This textbook has been written so that all of its content closely aligns with the *VCAA VCE Psychology Study Design* (2023–2027). It has been authored and reviewed by experienced Psychology teachers, academics, researchers and First Nations consultants to ensure up-to-date scientific and accurate content for students.

2 Integrates key science skills into the course

Chapter 1 *Scientific research methods* provides an overview of the **key science skills** in the VCE Psychology Study Design. This is built upon in Chapter 11, which steps students through Unit 4 Outcome 3.

In addition, key science skills are integrated into each chapter.

Students get to explore relevant key knowledge and to develop, use and demonstrate the key science skills through the **Investigations**. Investigations provide students an opportunity to:
- explore the different methodologies
- use the logbook template provided to build up their logbook
- analyse secondary data
- collect and display their own primary data
- analyse and discuss data
- draw evidence-based conclusions
- communicate their findings.

Analysing research sections provide students with the opportunity to read authentic psychological research and then apply their understanding of the key science skills by answering the questions that follow.

The *VICscience Psychology Skills Workbook* provides a great source of skill-based activities. Each activity is aligned to the key knowledge and enables students to develop and refine their key science skills. Signposts to the workbook activities are found throughout the textbook. See the chapter teaching plans to see how the textbook and workbook work together.

3 Gives you access to differentiated material

Differentiation is built into each chapter to assist you in helping those students who may struggle with content or skill development and extending those students who want to achieve at a higher level.
- **Chapter maps** provide students with a gentle and visual introduction to each chapter, enabling students to engage with the chapter content prior to entering the chapter.
- **Know your key terms** at the beginning of each chapter presents all the highlighted key terms throughout the chapter in one place. Students can use the flashcards study tool to learn and review key terms with their definitions.
- **Pre-tests** provide students the opportunity to retrieve concepts previously learned that will be revisited and built upon during the chapter.
- **Concept questions** are pitched to be lower-order questions to assist with learning consolidation, but each question set ends with HOT Challenge questions for those

students who would benefit from answering higher-order thinking questions.
- **Worksheets** with associated figures and tables allow students to analyse information in more depth.
- **Activities** are designed to shift students from lower-order thinking in *Try it*, to higher-order thinking in *Apply it*. Exam-style questions are provided in *Exam ready* to illustrate how this concept may be examined.
- **Investigations** provide students with the opportunity to experience psychology research.
- **Weblinks** to external, vetted websites provide extra information.
- **End-of-chapter exams** and **Area of Study reviews** provide students with experience at answering exam-style questions. VCAA-adapted questions are ranked as either Easy, Medium or Hard.

4 Prepare for the exam

Students of VCE Psychology are working towards external assessment at the end of Units 3 & 4. To fully prepare for this exam, students require access to a large number of quality exam-style questions with answers. *VICscience Psychology* gives you the full complement, including the following.
- **End-of-chapter exams** consist of 20 difficulty-graded multiple-choice and three difficulty-graded short-answer questions written in VCAA style. Answers are provided at the back of the book.
- **Area of Study reviews** at the end of each Area of Study provide students with 30 difficulty-graded multiple-choice and five short-answer questions that have been adapted from VCAA exam questions, with answers provided at the back of the book.
- *VICscience Psychology VCE Units 3 & 4 Skills Workbook* develops the skills required by students to confidently answer scientific research questions.

- **examplus (Units 3 & 4 only)** simulates real exam practice and comprises thousands of unseen exam-style and past VCAA exam questions with answers to use in your teaching. Simply select your questions for a quiz, topic test or practice exam and examplus generates a practice test or exam.
- Consider bundling **A+ Study Notes** and **Practice Exams** with your *VICscience Psychology* booklist for the most economical solution for students' exam preparation and readiness.

5 Support for the teacher

There is a wealth of teacher support materials on the MindTap Schools site that accompanies this product. These include:
- **answers** to all textbook questions, activities, case studies, analysing research, investigations (where relevant), end-of-chapter exams and Area of Study reviews.
- **sample SACs** with suggested marking schemes (Units 3 & 4 only)
- **slideshow summaries** to use in a flipped classroom
- **teaching plans** for every chapter showing how all the components of the *VICscience Psychology* suite are integrated to provide your students with a thorough and complete learning experience designed to prepare them for internal and external assessment.

BOOK MAP

UNIT 3 — How does experience affect behaviour and mental processes?

AREA OF STUDY 1

CHAPTER 3
Stress as an example of a psychobiological process

The nervous system is central to the psychological and physiological responses to stress. Two models are presented to explain stress, and strategies are provided to assist in coping with stress.

CHAPTER 2
Nervous system functioning

The nervous system is made up of the brain, spinal cord, neurons and the neurotransmitters that make them all work. How does it all fit together?

AREA OF STUDY 2

CHAPTER 4
Approaches to understand learning

How do we acquire skills and knowledge? This chapter presents different models to answer this question.

CHAPTER 5
The psychobiological process of memory

How does our memory work? This chapter investigates a model of memory and the role of the brain in memory. It also explores different methods that can be used to improve memory and to pass memories across generations.

CHAPTER 1
Scientific research methods

Being able to understand and apply the key science skills is a core component of VCE Psychology. You will build upon and practise these skills in every chapter of this resource.

AREA OF STUDY 3

BOOK MAP xxi

UNIT 4 — How is mental wellbeing supported and maintained?

AREA OF STUDY 1

CHAPTER 6
The demand for sleep
Our bodies and brain replenish through sleep. But what exactly is sleep and what makes a good night's sleep? Is this the same for everyone?

CHAPTER 7
Importance of sleep to mental wellbeing
A good night's sleep is important for mental wellbeing. It can sometimes be difficult, but there are ways to increase your chances of getting a good night's sleep.

AREA OF STUDY 2

CHAPTER 9
Application of a biopsychosocial approach to explain specific phobia
Do you have a really big fear of something? This could be a phobia and your nervous system, learning, memory and environment may have all contributed to the development of this phobia. How can you manage it?

CHAPTER 8
Defining mental wellbeing
What exactly is meant by mental wellbeing? To define it fully you need to take a number of factors into account.

CHAPTER 10
Maintenance of mental wellbeing
How can you maintain your mental wellbeing? The answer will depend on a mixture of biological, psychological, social and cultural factors.

AREA OF STUDY 3

CHAPTER 11
Using scientific inquiry
You now get to apply all the key science skills that you have learnt in your study of Psychology in the completion of Unit 4, Outcome 3. You will carry out a scientific investigation and present your findings as a poster. This chapter provides guidance on how to do this.

Nelson MindTap

An online learning space that provides students with tailored learning experiences.

- Access tools and content that make learning simpler yet smarter to help you achieve mastery.
- Includes an eText with integrated interactives and online assessment.
- Margin links in the student book signpost multimedia student resources found on MindTap.

Weblink

For students:

Nelson MindTap provides you with material that will help you understand, explore, engage and organise the key knowledge and key science skills you have learned about in your textbook. On MindTap, you will find chapter resources such as:

- Interactive ebook
- Key term flashcards
- Chapter pre-test
- Interactive learning activities
- Worksheets
- Slide show of key concepts for each chapter
- Weblinks to online videos and further information
- Templates to assist you in completing statistical analysis of your data
- Downloadable chapter maps and Key concept checklists

For teachers*:

Nelson MindTap allows you to teach in a way that caters to the needs of your students. Monitor student progress and customize your course in a way that makes sense for your classroom. In addition to the resources found on students' MindTap, you will also find teacher support materials such as:

- Teaching plans
- Logbook templates
- A curated list of resources that you can use to develop your own student assessments
- Sample SACs with model answers

* Complimentary access to these resources is only available to teachers who use this book as part of a class set, book hire or booklist. Contact your Cengage Education Consultant for information about access and conditions

9780170465076

Content warning This resource includes discussion of topics that may cause distress to Aboriginal and Torres Strait Islander people, including the discussion of racist policies and trauma associated with events such as massacres and forced removal of children. The content may also include reference to the names, images and voices of people who have died.

1 Scientific research methods

Key science skills

Develop aims and questions, formulate hypotheses and make predictions:
» identify, research and construct aims and questions for investigation
» identify independent, dependent and controlled variables in controlled experiments
» formulate hypotheses to focus investigation
» predict possible outcomes of investigations

Plan and conduct investigations:
» determine appropriate investigation methodology: case study; classification and identification; controlled experiment (within subjects, between subjects, mixed design); correlational study; fieldwork; literature review; modelling; product, process or system development; simulation
» design and conduct investigations; select and use methods appropriate to the investigation, including consideration of sampling technique (random and stratified) and size to achieve representativeness, equipment and procedures, taking into account potential sources of error and uncertainty; determine the type and amount of qualitative and/or quantitative data to be generated or collated
» work independently and collaboratively as appropriate and within identified research constraints, adapting or extending processes as required and recording such modifications

Comply with safety and ethical guidelines:
» demonstrate ethical conduct and apply ethical guidelines when undertaking and reporting investigations
» demonstrate safe laboratory practices when planning and conducting investigations by using risk assessments that are informed by safety data sheets (SDS) and accounting for risks
» apply relevant occupational health and safety guidelines while undertaking practical investigations

Generate, collate and record data:
» systematically generate and record primary data and collate secondary data appropriate to the investigation
» record and summarise both qualitative and quantitative data, including use of a logbook as an authentication of generated or collated data
» organise and present data in useful and meaningful ways, including tables, bar charts and line graphs

Analyse and evaluate data and investigation methods:
- process quantitative data using appropriate mathematical relationships and units, including calculations of percentages, percentage change and measures of central tendencies (mean, median, mode) and demonstrate an understanding of standard deviation as a measure of variability
- identify and analyse experimental data qualitatively, applying where appropriate concepts of: accuracy, precision, repeatability, reproducibility and validity; errors; and certainty in data, including effects of sample size on the quality of data obtained
- identify outliers and contradictory or incomplete data
- repeat experiments to ensure findings are robust
- evaluate investigation methods and possible sources of error or uncertainty and suggest improvements to increase validity and to reduce uncertainty

Construct evidence-based arguments and draw conclusions:
- distinguish between opinion, anecdote and evidence and scientific and non-scientific ideas
- evaluate data to determine the degree to which the evidence supports the aim of the investigation and make recommendations, as appropriate, for modifying or extending the investigation
- evaluate data to determine the degree to which the evidence supports or refutes the initial prediction or hypothesis
- use reasoning to construct scientific arguments and to draw and justify conclusions consistent with evidence base and relevant to the question under investigation
- identify, describe and explain the limitations of conclusions, including identification of further evidence required
- discuss the implications of research findings and proposals, including appropriateness and application of data to different cultural groups and cultural biases in data and conclusions

Analyse, evaluate and communicate scientific ideas:
- use appropriate psychological terminology, representations and conventions, including standard abbreviations, graphing conventions and units of measurement
- discuss relevant psychological information, ideas, concepts, theories and models and the connections between them
- analyse and explain how models and theories are used to organise and understand observed phenomena and concepts related to psychology, identifying limitations of selected models/theories
- critically evaluate and interpret a range of scientific and media texts (including journal articles, mass media communications, opinions, policy documents and reports in the public domain), processes, claims and conclusions related to psychology by considering the quality of available evidence
- analyse and evaluate psychological issues using relevant ethical concepts and principles, including the influence of social, economic, legal and political factors relevant to the selected issue
- use clear, coherent and concise expression to communicate to specific audiences and for specific purposes in appropriate scientific genres, including scientific reports and posters
- acknowledge sources of information and assistance and use standard scientific referencing conventions

Source: VCE Psychology Study Design (2023–2027) p. 12–13

1 Scientific research methods

You have no doubt been learning about the principles of scientific research during your science classes. These principles underpin all science investigations, including all psychological investigations. Scientists always follow the same systematic steps: proposing and investigating hypotheses, collecting and analysing data and drawing evidence-based conclusions from this data.

1.1 The process of psychological research investigations
p. 8

All scientific research starts by asking a question: What? Who? Why? When? Where? Once you have defined your research question, there is a particular process that you can use to design a study to answer the question. Just make sure you record everything you do in your logbook!

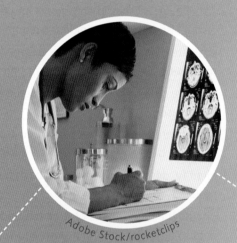
Adobe Stock/rocketclips

1.2 Science investigation methodologies
p. 14

You're probably most familiar with the controlled experiment, however there are many other types of scientific investigation methodologies out there. Each method has its strengths and weaknesses, and is more suited to certain types of research questions than others.

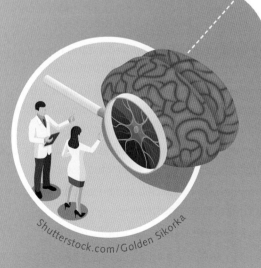
Shutterstock.com/Golden Sikorka

p. 21

1.3
The controlled experiment in detail

In a controlled experiment, it's important that you properly identify the different variables (five in total) within the experiment. Systematically manipulating a particular variable(s) can help us find cause and effect relationships between variables (if they exist). You also need to be careful with the type of experimental design you pick for the experiment as it affects how you choose the participants in your study.

iStock.com/SolStock

p. 29

1.4
Analysing and evaluating research

Now it's time to work out what the data is telling you. Researchers use descriptive statistics to analyse data, and often organise the data into tables and graphs to make it easier to see trends (if there are any). Not only do we have to analyse the data, but we also need to check the quality of the data too to make sure that any conclusions drawn are valid.

Shutterstock.com/Thanakorn.P

Scientific investigation permeates all aspects of science. Only through investigation do we find out new information, but for this information to be valued by the scientific community it must be carried out in a systematic and accepted form. As you continue with your studies of Psychology, always ask questions. You never know, you might just come up with a new area of research!

Test
Chapter 1 pre-test

Flashcards
Chapter 1 flashcards

Slideshow
Chapter 1 Slideshow

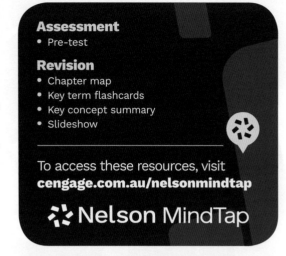

Know your key terms

Accuracy
Aim
Anecdote
Beneficence
Between subjects design
Case study
Conclusion
Confidentiality
Confounding variable
Control group
Controlled experiment
Controlled variable
Convenience sampling
Correlational study
Data
Debriefing
Deception
Dependent variable (DV)
Descriptive statistics
Ethical concepts
Ethical guidelines
Experimental group
External validity
Extraneous variable
Fieldwork
Hypothesis
Independent variable (IV)
Informed consent
Integrity
Internal validity
Justice
Literature review
Logbook
Mean
Measure of central tendency
Median
Mixed design
Mode
Non-maleficence
Non scientific ideas
Opinion
Order effect
Outlier
Population
Precision
Primary data
Psychological construct
Psychological model
Psychological theory
Qualitative data
Quantitative data
Random allocation
Random error
Random sampling
Repeatability
Representative sample
Reproducibility
Respect
Risk assessment
Sample
Sample size
Sampling
Secondary data
Standard deviation (SD)
Stratified sampling
Systematic error
True value
Uncertainty
Validity
Variability
Variable
Voluntary participation
Withdrawal rights
Within-subjects design

Psychological research aims to understand aspects of the mind, brain and behaviour. This is no easy task, because human thoughts, feelings and behaviours are complicated and cannot always be observed directly. Researchers use the scientific method to discover new information about psychological processes. The scientific method is a well-defined, step-by-step process that allows us to generate new knowledge. It is used in all subfields of psychology, including clinical psychology, biological psychology, developmental psychology, cognitive psychology, personality psychology and social psychology. The scientific method is also the process used in other sciences, such as physics, chemistry, biology and environmental science.

Research psychologists develop models and theories to organise and explain the psychological concepts or processes they are interested in. In psychological science, psychological concepts are expressed and defined as psychological constructs. **Psychological constructs** are terms used in psychology that define specific psychological structures, mechanisms and processes that are thought to be the basis of behaviour and mental experiences. For example, the word 'attention' has an everyday sense that people use and understand. However, in psychology the term attention is a psychological construct that has a much more precise definition than its everyday meaning. Unit 4 of the VCE Psychology Study Design asks you to consider 'sleep as a psychological construct'. This means that you need to study sleep as a concept that is precisely defined by a set of psychological, physiological and neurological properties and processes. Psychological science aims to reduce uncertainty about the true nature of the mind, brain and behaviour by developing our understanding of psychological constructs and how they relate in ever more precise ways.

A **psychological theory** is an organised set of interrelated psychological constructs, that describes and/or explains a psychological system, process or experience. For example, a psychological theory of the memory system is made up of a set of constructs, mechanisms and processes that are involved in remembering. Similarly, a psychological theory about racism comprises a set of ideas about the beliefs, attitudes and unconscious biases that drive racist behaviours.

Psychological models are used to make the ideas within a theory more concrete. Models come in a variety of forms. Some examples of models include Ivan Pavlov's model of classical conditioning (see Chapter 4), Richard Lazarus and Susan Folkman's Transactional Model of Stress and Coping (see Chapter 3), and Richard Atkinson and Richard Shiffrin's multi-store model of memory (see Chapter 5). Models and theories are built from our current understanding, so they can have limitations, including incorrect assumptions or oversimplifications. When there are inconsistencies between the predictions of a model or a theory and the results gathered from investigations, this indicates that the model or theory may be incomplete or wrong. With new evidence, researchers update theories and models to describe processes more accurately.

Psychology researchers conduct research investigations to gather evidence to better understand psychological processes. When designing a study, the researcher must choose a **population** of interest that they wish to investigate. This population of interest will relate to the construct they are investigating. For example, the population of interest could be:
» students
» teenagers
» women
» people diagnosed with anxiety.

Often in psychology, the phenomenon being investigated applies to all humans (for example, memory processes), so the population of interest can also be as broad as the entire human population. It isn't practical to include every person from the population of interest in a research investigation, so researchers will recruit a smaller sample of participants. A **sample** is a group of people who are recruited from a larger population of interest for the research. As we will explore later, the validity of research findings for the population of interest depends on how well the sample represents that population.

To generate new knowledge from a study, researchers break down the processes, systems or constructs they are investigating into variables that can be measured. A **variable** is any factor in a study that can vary in its score, amount or type and that can be measured, recorded or manipulated. In psychological science, variables can be factors such as:
» individual characteristics (for example, a personality trait such as extroversion)
» properties of a stimulus (for example, brightness of light)
» behaviours (for example, responses on a memory test)
» processes (for example, excitation of neurons).

Some other examples of variables include height, gender, ethnicity, employment status, reaction time on a hazard perception test and happiness scores on a survey. You will notice that all of these variables are phenomena (that is, things or events) that can:
» be measured
» take a range of values
» change over time
» be different between people or between groups.

Defining and measuring variables allows researchers to investigate how factors or psychological constructs are related to each other. Discovering relationships between variables allows researchers to draw conclusions about psychological processes.

When we interpret the conclusions of scientific research, we must consider whether any claims made are valid. Validity takes two forms: internal validity and external validity.

Internal validity relates to how effective the design and measures used in a study are for understanding the research question **External validity** indicates how well the results of a study can be applied meaningfully to real-world contexts, situations and behaviours. Researchers must make careful decisions when designing a research investigation because their choices affect the validity of the study's conclusions.

This chapter introduces you to the scientific research methods used in psychology. This information will allow you to understand different methods of scientific investigation, develop key science skills and integrate the links between knowledge, theory and practice. Most importantly, it will deepen your understanding of the tools that psychology researchers use to generate new knowledge about the mind, brain and behaviour. We hope that it will spark your interest in this marvellous discipline. In the words of Ivan Pavlov:

"Do not become a mere recorder of facts, but try to penetrate the mystery of their origin."

1.1 The process of psychological research investigations

How do I use the research process to investigate a psychological phenomenon that interests me? In this section, we explore the step-by-step process that researchers use to design a study to answer a question about psychology. These steps are shown in Figure 1.1.

The steps in Figure 1.1 are:
1. decide on a research question that you will attempt to answer
2. choose an aim to focus your research
3. create a broad hypothesis that you wish to test
4. design a study to test the hypothesis
5. generate a specific hypothesis that predicts the results/outcomes of the study
6. collect data
7. analyse data and report results
8. draw evidence-based conclusions about whether the results support the hypothesis
9. identify limitations and make recommendations for modifying or extending the investigation.

Each of these steps is described in detail shortly. However, first it is important to understand the role of your logbook in this process.

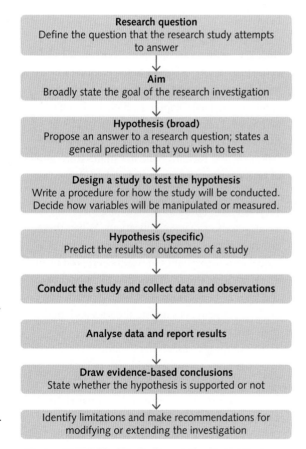

Figure 1.1 The research process, step-by-step

The logbook

Whenever you conduct a research investigation it is crucial that you record each of the steps you take in a **logbook**. Your logbook is a document that you will use throughout the year to record all of your practical work so that it can be assessed. It is particularly important to use your logbook to describe and record the steps involved in developing your student-designed scientific

investigation (Outcome 3 of Unit 4) will use your logbook to:
» describe the process you used to develop your research question, aim and hypothesis
» outline the design of your study and the methods you used
» record the data you collect
» describe your data analysis method and results
» explain the conclusions you have drawn.

Developing your research question, aim and hypothesis

Imagine you are a researcher interested in how meditation training affects wellbeing. Before you begin to think about how you can test your ideas about meditation, you need to find out what is already known about this topic. Be careful about the types of resources you use during your research as some may present **non-scientific ideas** that may resemble real science but aren't supported by scientific evidence. A more common way to research a topic is by conducting a **literature review**. This would involve reading about meditation and its relationship to wellbeing in scientific journals, learning about any existing theories or models that are relevant, and determining what questions need to be addressed. During your literature review, you may discover something you are interested in that is not fully understood (that is, where there is some uncertainty around the issue or around a part of how the process works). For example, you may find some research articles suggesting that meditation training may improve mental wellbeing.

Research question

Once you have surveyed the literature to see what questions may be worthwhile to investigate, you can define your research question more precisely. A research question should express the exact question you are trying to answer with your research, including the population of interest (which could be a particular subpopulation such as 'people with anxiety' or 'teenagers' or could be broad and relate to people in general). An example of a research question is 'Does meditation training improve people's mental wellbeing?'. When choosing a research question, you should consider the unanswered questions that arose during your focused reading or literature review. You will try to answer your research question with your research study, or more specifically you will try to reduce the uncertainty around this question.

Aim

Next, you should create your research **aim**. Your aim is a broad statement about what you intend to investigate. You can think about the process of developing your aim as turning your research question into a statement of your research goal. For example, your aim could be 'To determine whether meditation training improves people's mental wellbeing'.

Hypothesis

While reviewing the literature to develop your research question and aim, you may have come across theories or models that relate to your research question. You may also have developed your own ideas about possible answers to your research question. These ideas about possible answers to your research question are called hypotheses. A **hypothesis** is a statement that expresses a possible (that is, hypothetical) answer to a research question. A hypothesis describes the expected relationship between the variables of interest and should specify the predicted direction of the relationship between them.

A hypothesis can be expressed at two levels:
» a broad hypothesis that makes a general prediction about a relationship between variables in the *population* of interest (sometimes called a research hypothesis)
» a specific hypothesis that makes a prediction about the expected results of a study that relates to a *sample* of participants.

Researchers often begin with a broad hypothesis before they design a study. As they refine their study design, they create a specific hypothesis that can be tested by the study (sometimes called a prediction or operationalised hypothesis). Figure 1.1 (page 8) shows the statement of your broad hypothesis as Step 3 of the research process and the statement of your specific hypothesis as Step 5.

A broad hypothesis can be thought of as a clearly expressed statement that proposes a possible answer to your research question. For example, for our research question 'Does meditation training improve people's mental wellbeing?', the broad hypothesis could be 'Meditation training improves people's mental wellbeing'.

> "Hypothesis" is the singular noun, and "hypotheses" is the plural.

1.1.1 DEVELOPING YOUR RESEARCH QUESTION, AIM AND HYPOTHESIS

Notice how this broad hypothesis refers generally to meditation training (not to a specific implementation of meditation) and to mental wellbeing (not a specific measure of mental wellbeing). It also makes a broad prediction that meditation will improve mental wellbeing and refers to the population of interest (people).

1.1.2 PREDICTING OUTCOMES

The specific hypothesis is a precise statement that makes a prediction about the expected results of your study. For example, in our meditation and mental wellbeing study, imagine you have chosen to:
» define meditation as a daily 5-minute mindfulness session using an online app over one week
» expose one group of participants to the meditation training and to have another group that does not do the training
» measure people's mental wellbeing using a common wellbeing questionnaire called the Positive and Negative Affect Schedule (PANAS; Watson et al., 1988). Higher scores on the PANAS indicate higher levels of mental wellbeing.

In this scenario, your specific hypothesis will state a prediction for how daily use of the meditation app over one week will affect people's scores on the PANAS. For example, your specific hypothesis might predict that 'Daily use of the meditation app for one week will result in higher scores on the PANAS compared to people who do not use the app'. In this example of a specific hypothesis, notice that we have referred to the variables of interest (meditation and mental wellbeing) in terms of how they were implemented or measured in the study. The specific hypothesis also provides a prediction about how mental wellbeing scores will differ between the groups and it makes reference to the sample (people in the two groups).

Notice that although the broad hypothesis and the specific hypothesis have different levels of detail, they both include three core elements; when you are asked to write a hypothesis you must include these three elements, as follows:
» You must refer to the *variables of interest*. As we discuss later in this chapter (p. 22), if your study is a controlled experiment, then these will be the independent variable/s (IV) and the dependent variable/s (DV). However, if your study is a correlational study or some other kind of non-experimental study, then the variables of interest are the two (or more) factors that are thought to be related to each other. Only controlled experiments have variables called IVs and DVs. The different kinds of research investigations and variables are described in detail in section 1.2, Scientific investigation methodologies, page 14.
» You must state the *direction* of the predicted relationship between your variables. For example, the hypothesis could predict that scores for an outcome variable will increase for one group more than another. By including the predicted direction of the outcome, your hypothesis indicates not only that the scores on a variable will change, but also how they will change.
» You must refer to either the *population of interest* or to the *sample*, depending on whether you are stating a broad or specific hypothesis, respectively.

The VCE Psychology Study Design does not distinguish between a broad hypothesis and a specific hypothesis. We have defined these two kinds of hypotheses to show how a hypothesis can be expressed with different levels of precision at different stages of the research design. A broadly expressed hypothesis is a proposed answer to the research question expressed at the population level. A specific hypothesis expresses a prediction for the outcome of a particular study with reference to the sample used in your investigation. If an exam question asks you to write a hypothesis for a particular study, you could express your hypothesis at either of these levels and be marked correct.

Designing a research investigation

Once you have developed a research question, an aim and a broad hypothesis, you are ready to design a research investigation (that is, a study) that you can use to test this hypothesis. The results of your investigation will either provide evidence to support your hypothesis or provide evidence against it. Of course, we can never prove that a hypothesis is true, but a carefully designed research investigation can reduce uncertainty about it.

There are many kinds of research designs that you could choose from to test the hypothesis 'Meditation training improves people's mental wellbeing'. The purpose of a research design is to create a method that allows you to observe and measure the psychological constructs (in this case 'meditation' and 'mental wellbeing'), in order to investigate how they are related.

For example, when designing your investigation of the effect of meditation on mental wellbeing, you need to make concrete decisions about things such as what kind of meditation training will be used, how it will be taught (in-person or online?), and how long people will use it (just 1 hour or several sessions?). You also need to decide exactly how you will define and measure mental wellbeing (which scale or questionnaire will you use to measure this construct?). Will you compare two different groups to each other (one who practises meditation and one who does not), or will you measure the *amount* of meditation people engage in and see if this is related to a measure of mental wellbeing? The kind of design and the kinds of measures you choose will determine the kind of data that you generate. We consider the different research designs in detail later in this chapter (page 26).

Collecting data

Once you have designed your study, you are ready to collect and record your data. To collect data, you need to understand the different kinds of data you could collect and how to select a sample of people to generate the data. You also need to think about how you will record your data so that you can analyse it later.

Kinds of data

Data is the term we give to any information that is collected or used in a scientific investigation. **Primary data** are data that we collect ourselves from a study that we have designed. **Secondary data** are data that have been collected by someone else that we use when conducting a literature review of the existing knowledge on a research topic. In VCE Psychology design your own scientific investigation and Units 3 and 4, you will use primary data for Area of Study 3, Outcome 3 in Unit 4. The investigation can be undertaken in Unit 3 or 4 (or across both) but is assessed as part of Unit 4 Outcome 3.

Primary and secondary data can take different forms, depending on whether or not the information is numerical. **Quantitative data** are numerical. They are recorded in the form of numbers (for example, a response-time measure or a score on a test). They must be collected through systematic and controlled procedures to ensure that the measurements are accurate and precise across people or trials. **Qualitative data** are non-numerical. They are verbal descriptions of states or qualities that are often organised into themes. Examples of qualitative data include descriptions or personal accounts of feelings, attitudes, experiences, behaviours or descriptions of changes to the quality of a variable, such as changes in the experienced intensity of an emotion or vividness of a mental image. In psychological research, qualitative data are often collected through questionnaires or interviews.

Sampling

When a psychology researcher wants to collect data, they need to recruit a sample of participants to take part in their study. The sample will be determined by the population of interest. For example, if a researcher is interested in how meditation affects the mental wellbeing of VCE students, then their population of interest is VCE students. It is not feasible to collect data from all VCE students, so the researcher must select a sample from this population. A sample is a group of people who are recruited from a larger population of interest (Figure 1.2a).

A researcher will conduct a study using a sample of participants to collect data and determine whether they can find an effect or a relationship in this data. However, researchers are not just interested in whether these effects exist in their sample; they are interested in whether an effect or relationship exists in the population from which the sample was drawn. That is, researchers want to be able to *generalise* the results they get from their sample to the entire population of interest.

1.1.3 SAMPLING TECHNIQUES

> The term 'generalise' is not mentioned in the VCE Psychology Study Design but we use it to help you understand the relationship between a sample and a population. It will also be helpful for understanding external validity.

> "Data" is a plural noun, so we write "data are". Datum is the singular noun.

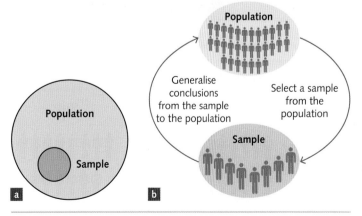

Figure 1.2 a A sample is a subset of a population. **b** A researcher can generalise from the sample to the population if the sample is representative of the population.

Generalisability is the extent to which research findings (found using a sample) can be applied to the population of interest (Figure 1.2b).

For the findings of a study to be generalisable to the population, the sample must be **representative** of the population. **Sampling** is the process of selecting participants from a population of interest to participate in a research investigation. Researchers must consider sampling techniques to recruit an appropriate sample of participants that is representative of the population (a **representative sample**) of interest. Three important sampling methods are convenience sampling, random sampling and stratified sampling.

Convenience sampling is a method of sampling in which the researcher recruits a sample of participants that is convenient to recruit (for example, friends and family members of the researcher, or a class of first-year psychology university students; see Figure 1.3). While it is relatively easy to recruit a convenience sample, it is unlikely that this sample will be an accurate representation of the general population.

For example, imagine a researcher recruits a sample of first-year psychology university students for a study where the population of interest is all people. A sample made up of these participants may have large differences (for example, in age, gender, education, etc.) from the average person in the population of interest.

Random sampling is a sampling technique that uses a chance process to ensure that every member of the population of interest has an equal chance of being selected for the sample. For example, the researcher could assign each member of the population a number and select which participants will participate using a random number generator (a computer program) to select people to be approached for the study. A simple way to think about the random selection process is that it is equivalent to drawing names from a hat.

Stratified sampling is a sampling technique used to ensure that a sample contains the same proportions of participants from each social group (that is, strata or subgroup) present in the population of interest. First, strata are identified within the population of interest. Then the demographics of the population of interest are assessed to determine the number (or the proportion) of people in each stratum. Finally, participants are recruited to the sample from each of the stratum in the same proportions as their numbers in the population (Figure 1.4).

As an example of stratified sampling, consider a researcher studying wellbeing in the general population of Australia who plans to use a stratified sampling method to ensure the age of the sample is representative of the Australian population. The researcher would decide on the strata (how they want to divide up the age groups) and would look up the percentage breakdown of each age group in Australia.

Figure 1.3 Convenience samples consist of participants who are readily available and where no mechanism has been used to ensure that participants are representative of the population.

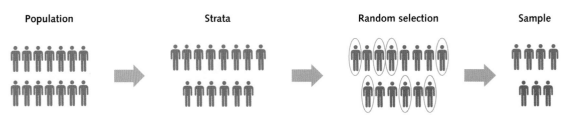

Figure 1.4 Stratified sampling is used to ensure that a sample contains the same proportions of participants from each stratum.

The number of participants recruited from each age group would then be based on that age group's proportion in the larger population.

Sample size

Another important aspect of sampling is obtaining a large enough **sample size**. This means recruiting enough participants for the study. Larger sample sizes will help make the sample more representative of the population of interest, which improves generalisability. With a larger sample, it is more likely a researcher will find any effect or relationship that exists. Samples that are too small also affect the accuracy and precision of any data collected.

Historically, most psychological research has been conducted using samples selected from what the psychology research community terms 'WEIRD' societies. WEIRD stands for Western, educated, industrialised, rich and democratic. Of all published behavioural science research, 96 percent rely on participants recruited from populations that meet the WEIRD criteria (Henrich et al., 2010). A large proportion of psychology research takes place in the US, Europe and Australia, where WEIRD samples are more readily accessible. These samples under-represent a large proportion of the global population. Studies that recruit samples from WEIRD populations may not produce findings that can be generalised to people from other cultures.

Recording your data

The quality of the data you collect is determined by the accuracy and precision of your measurements and how carefully you record your measurements and observations. Be careful to ensure that data are labelled accurately so that they can be analysed later. Your logbook and tools such as spreadsheets are crucial for making accurate and organised records of your observations and measurements, and using them helps you avoid errors.

Analysing data

Data analysis means taking the raw data that you have collected and processing it so that you can report your findings. This step requires high attention to detail to ensure that your data are analysed accurately. For example, imagine in your study of the effects of meditation on mental wellbeing that you have collected:
- » information about whether (or to what extent) participants engaged in meditation
- » sets of responses from participants about their experience of mental wellbeing.

Analysing these data involves matching participants' mental wellbeing scores with the information about their meditation practice so you can describe the relationships you have observed between these variables. This might involve things like calculating the average level of mental wellbeing reported by the group of participants who participated in meditation and comparing this with the average level of mental wellbeing reported by the group of participants who did not participate in meditation. We further discuss specific methods for analysing data later in this chapter (page 38).

Drawing conclusions

After collecting and analysing your data, it is time to interpret your results so that you can determine what conclusions can be drawn. Drawing **conclusions** involves describing and explaining the results of your study and discussing how your findings relate to the aim and the hypothesis of your study. Your conclusion will involve making and justifying claims about whether or not the results support your hypothesis. You must also carefully consider to what extent your results can be generalised to the population of interest. Drawing conclusions involves a careful process of evaluating your claims against the evidence provided by your study. We explore the process of evaluating evidence from research investigations in greater detail in *Evaluating data and investigation methods* on page 38.

Limitations and recommendations

It is important to identify any limitations of your study. This means considering issues and unexpected problems that may have compromised the internal and external validity of the investigation. You should also make recommendations for modifying or extending the investigation in future studies and discuss what further evidence may be required to make conclusions. This may help to enhance the validity of future studies and help other researchers to avoid the pitfalls that you encountered.

We have now covered the nine steps of the research investigation process (Figure 1.1).

The final step (of identifying limitations and making recommendations for modifying or extending the investigation) sets off the whole research process again. Future studies will build on the recommendations, and try to correct and improve on the limitations of existing studies. As the research community continues to generate new knowledge, the cycle continues and the scientific process repeats, with the next study starting again at Step 1 of the process.

One study at a time, researchers improve the collective scientific understanding. We should note that all stages of the research process must be guided by ethical considerations (see page 39).

KEY CONCEPTS 1.1

» A research question defines the question that a research investigation tries to answer.
» An aim is a broad statement about the goal of a research investigation.
» A hypothesis is a proposed answer to a research question (made before the investigation is conducted). It is usually a directional statement about the relationship between the variables and states the expected results of a research investigation.
» A research investigation is used to test a hypothesis.
» Data are any information collected in scientific investigations.
» Primary data are data we collect ourselves from a study.
» Secondary data are data collected by someone else that we use when conducting a literature review.
» Quantitative data are numerical (recorded in the form of numbers).
» Qualitative data are non-numerical (descriptive).
» Sampling is the selection of participants from a population to participate in a research investigation.
» Convenience sampling means recruiting participants that are readily available.
» WEIRD samples (Western, educated, industrialised, rich and democratic) under-represent a large proportion of the population.
» Random sampling gives every member of the population an equal chance of being recruited.
» Stratified sampling ensures that the sample contains the same proportions of participants from each stratum as the proportions in the population of interest.
» Using larger samples can improve representativeness.

1.2 Scientific investigation methodologies

There are many different types of scientific investigation methodologies that researchers can choose from. The ideal scientific investigation is the controlled experiment, but not all research questions can be studied this way. Other kinds of investigations include:
» correlational studies
» case studies
» classification and identification
» fieldwork
» modelling
» simulation
» literature reviews
» product, process or system development.

In this section we begin with a brief outline of controlled experiments and then explore the other kinds of research investigations to understand what makes them different from experiments, and what kinds of research questions they can address. We then explore the controlled experiment in detail in section 1.3 (page 22).

Controlled experiments in brief

A **controlled experiment** is a methodology used to test a hypothesis in which the researcher systematically manipulates (changes) one or more variables to investigate what effect these manipulations have on another variable. Unlike other methodologies, controlled experiments are designed to enable the researcher to draw conclusions about the *causes* of the phenomena or process they are investigating. In psychological science, controlled experiments allow us to determine the causes of behaviours, mental states and psychological processes.

Correlational studies

A **correlational study** is a non-experimental study where the researcher investigates relationships between variables. Unlike in a controlled experiment, the researcher does not try to control or change any of the variables. Instead, they observe and measure the variables as they naturally occur. To conduct a correlational study, the researcher chooses two or more variables they wish to investigate. They then recruit a sample of participants and measure these variables in their sample. The analysis involves describing how (or whether) the two variables are related to each other.

The correlational study is one of the most commonly used research designs in psychology. This is because it is not always possible, desirable or appropriate to investigate some questions with an experiment. Because no manipulation is required in a correlational study, it can be less invasive than a controlled experiment that investigates the same research question. A correlational study can also be useful to inform future work, because it can be used to identify which variables may be more important to study further.

An example of a correlational study in psychology could be an investigation of the relationship between levels of wellbeing and amount of sleep. Researchers could recruit a sample of research participants to report their average sleep duration and complete the PANAS wellbeing survey. Each participant's score on the PANAS (variable 1) and their average sleep duration (variable 2) could then be graphed to observe the relationship between sleep duration and wellbeing. A correlation analysis could also be performed to assess the strength and direction of this relationship mathematically.

What is a correlation?

A correlation is a relationship or association between two variables in a data set. It is a measure of the strength and direction of the relationship between the two variables. Some examples of correlations are:
» people with high values on one variable (for example, longer time spent studying) often having high values on another variable (for example, exam score)
» people with high values on one variable (for example, alcohol consumption) often having low values on another variable (for example, driving accuracy).

Correlations can be visually represented by plotting the values of two variables on a type of graph called a scatterplot. Each point on a scatterplot represents one participant, and the position of the data point on the plot provides information about that person's scores on both variables. Specifically, scores for variable 1 are represented on the x-axis (horizontal) and scores for variable 2 are represented on the y-axis (vertical). The data point that represents the relationship between the two variables for each participant is shown at the point on the scatterplot for which these values intersect.

Figure 1.5 shows a single data point on a scatterplot that represents one participant's scores on two variables. For example, variable 1 could be their score on a sleep quality scale, represented on the horizontal axis with a score of 8, and variable 2 could be their score on the wellbeing scale, represented on the vertical axis with a score of 4.

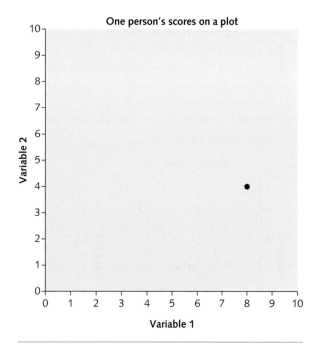

Figure 1.5 A scatterplot showing one person's scores on two variables. For example, variable 1 could be wellbeing and variable 2 could be sleep quality.

In a correlational study, we can record the scores of two variables for a whole sample of participants and display the results on a scatterplot. We can then look for patterns in the plot to see if the sample shows a relationship between the two variables.

Direction of correlation

First, let's explore the direction of a correlation: positive or negative. Figure 1.6 shows three possible types of patterns that could emerge in a data set:
» positive correlation
» no correlation
» negative correlation.

A positive correlation is where the pattern of data slopes up diagonally from left to right on a graph (Figure 1.6a). Positive correlation means that:
» as values on variable 1 *increase*, the values on variable 2 also *increase*
» people who score *high* on one variable are likely to score *high* on the other variable
» people who score *low* on one variable are likely to score *low* on the other variable.

As discussed earlier, a good example of a positive correlation is time spent studying and final exam grade. There is a positive relationship between these variables because students who study more generally score higher in exams than students who study less.

A negative correlation is where the pattern of data slopes downward from left to right on a graph (Figure 1.6b). Negative correlation means that:
» as values on variable 1 *increase*, the values on variable 2 *decrease*
» people who score *high* on variable 1 are likely to score *low* on variable 2

» people who score *low* on variable 1 are likely to score *high* on variable 2.

As discussed earlier, a good example of a negative correlation is alcohol consumption and accuracy on a driving simulation test: as consumption increases, accuracy decreases.

If there is no correlation (Figure 1.6c):
» scores on variable 1 are not related to scores on variable 2
» data points appear on a scatterplot as a random cloud with no particular pattern.

Strength of correlation

The strength of a correlation is an indication of how strongly the variables are related. Figure 1.7 displays scatterplots of positive correlations of different strengths, ranging from strong to weak. Stronger correlations between variables produce data points that are more tightly clustered along the diagonal of the scatterplot.

Correlation vs causation

It is important to know the difference between correlation and causation. If two variables are correlated, there is a relationship between them, but this does not mean that changes in the value of one variable causes changes in the value of the other variable. For example, you could conduct a study that finds a positive correlation between sleep duration and mood. This finding does not necessarily mean that longer sleep causes better wellbeing. Instead, it could be that people who get more sleep also have more structured routines in general and this could be the factor

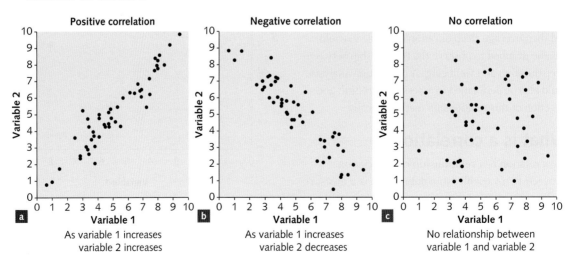

Figure 1.6 Types of correlation

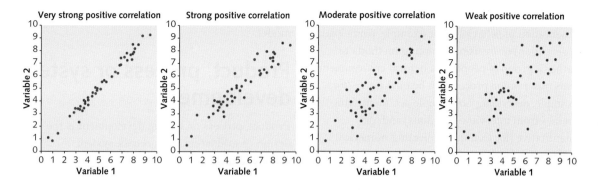

Figure 1.7 Positive correlations of various strengths: very strong, strong, moderate and weak.

responsible for the correlation, rather than the sleep variable measured.

A common phrase in science communication that summarises this point is: 'Correlation does not equal causation'. This is a valuable phrase to remember, both for scientific research and for life in general. Many logical errors arise when evidence of a correlation is confused with causation.

To identify a causal relationship, a controlled experiment is needed.

Case studies

A **case study** is a detailed investigation of one instance of a broader phenomenon. A case study focuses on a particular person, activity, behaviour, event or problem in a real or hypothetical situation.

Case studies can take various forms. For example, a case study may involve:
» direct observation of a situation
» analysis of historical information, which could include a discussion of knowledge learned from the situation or exploration of the causes and consequences of an event
» a real situation or a role-play of an imagined situation (in particular, if recommendations are to be made)
» problem-solving (in particular, if there is a need to develop a new design or methodology).

Case studies are very common in clinical psychology because the investigation of mental disorders often occurs at the level of the individual person. Such case studies may include detailed descriptions of the person, their symptoms and the factors that make up their experience. In social psychology, case studies can be used to examine how groups behave under certain conditions.

An advantage of the case study methodology is that the data are rich and highly detailed. A case study can include the complexities that are encountered in the real world (outside of the laboratory). A disadvantage is that the information is specific to one particular case, so many of the details may be specific to that single case and may not apply to the wider population or to other situations of interest.

Classification and identification

Classification is a scientific activity that seeks to systematically organise phenomena, objects or events into manageable sets. Identification is the process of recognising phenomena as belonging to particular sets or possibly being part of a new or unique set. The *Diagnostic and Statistical Manual of Mental Disorders* (*DSM-5-TR*™) produced by the American Psychological Association is the most common diagnostic classification system for mental disorders. This manual provides a system of classification of mental health disorders into categories based on the presentation of symptoms. When a clinical psychologist uses the *DSM-5-TR*™ criteria to diagnose a patient, they are engaging in the process of identification.

Fieldwork

Most studies are conducted in controlled and carefully manufactured environments, such as the laboratory or classroom, but it is often more appropriate to study psychology in the real world. **Fieldwork** involves observing and interacting with a selected environment beyond the classroom. It is used when researchers want to capture human thoughts, feelings and behaviours in a natural setting.

Fieldwork can be conducted through direct observation of behaviour (for example, participant observation) or by using sampling methods to gather a group of people in a natural environment. This observation process can be qualitative (for example, describing how people act, identifying themes from interview questions) and/or quantitative (for example, collecting numerical information, such as counting how many times a person engages in a particular action). Fieldwork could also be asking people about their opinions or behaviours through qualitative interviews, questionnaires, focus groups or yarning circles. Fieldwork usually aims to determine correlation rather than a causal relationship.

1.2.1 METHODOLOGIES

Modelling and simulation

In psychological research methods, modelling means creating a conceptual, mathematical or physical representation (that is, a model) of a system of concepts, events or processes. Figure 1.8 shows the multi-store model of memory (adapted from Atkinson & Shiffrin, 1968) as an example of a model 4.

Models can be used to test hypotheses or to determine underlying mechanisms and processes in ways that may be unrealistic or even impossible to test with real people.

There is a whole field of modelling where psychological researchers create computer programs to simulate mental processes. This is called computational cognitive modelling. The data generated from these computational models can be compared to data collected in experiments and if the patterns of data match, this is evidence supporting the computational model.

Product, process or system development

Product, process or system development is the design or evaluation of a process, system or artifact to meet a human need. This may involve technological applications as well as scientific knowledge and procedures.

Psychology-focused smartphone applications (apps) are one area where this methodology is flourishing. Smartphone apps are now providing people with information, structures and techniques that can help to improve mental wellbeing or modify behaviour. Some examples are:

» meditation apps such as the headspace app, which aims to help people reduce anxiety and improve their mental health
» habit tracker apps such as the Habitica app, which helps people to track and intentionally modify their habitual behaviours.

Smartphone apps are also being used in psychological research. For example, apps have been developed that deliver short surveys to people on smartphones at regular intervals. This enables experiences or emotions to be measured in real time while the participant is going about their day, and allows surveys to be delivered more frequently. Some apps collect information for researchers, so they facilitate access to more **representative samples**. Case study 1.1 explores an example of one of these applications.

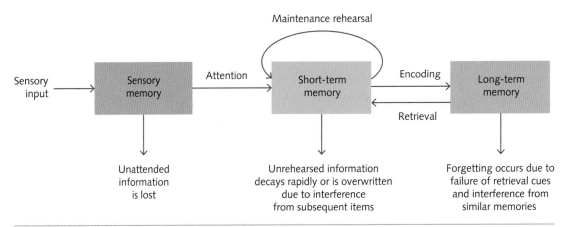

Figure 1.8 The multi-store model of memory (adapted from Atkinson & Shiffrin, 1968) is an example of a model. Models allow psychologists to represent theories or concepts more concretely

CASE STUDY 1.1

SEMA3: an example of a product, process or system development

SEMA3 (Smartphone Ecological Momentary Assessment Version 3) is a smartphone app developed by psychology researchers in Melbourne (Figure 1.9). It is an ideal case study to explore product, process or system development in Psychology while also highlighting other relevant concepts that are important for VCE Psychology. SEMA3 allows researchers to set up short, automated surveys that are sent to participants' smartphones at regular intervals (for example, multiple times per day). Researchers collect information from people using the app, and the repetition of surveys allows them to explore the relationships between responses (for example, relationships between emotions) and changes over time.

An advantage of this methodology is that thoughts, feelings and behaviours can be captured in real time in the real world, instead of in a laboratory. This can improve both the internal validity and external validity of a study.

Being able to get survey responses from participants while they are going about their regular life allows researchers to gain a more realistic assessment of a person's experience. It also avoids some drawbacks of laboratory studies, such as participants being unable to recall information when they are asked questions about the past. These features can help surveys to measure what they intend to measure, thus improving the internal validity of the study.

Many people own smartphones throughout the world, in both advanced and emerging economies (Figure 1.10). By surveying people via their smartphone, researchers are able to reach a more diverse sample of participants for their research studies, including people who they may not have previously had access to. This can improve external validity by allowing researchers to use a sample that more closely resembles the overall human population. Technologies of this kind could help researchers address the common problem that most research is conducted with Western, educated, industrialised, rich and democratic (WEIRD) samples that under-represent a large proportion of the overall population.

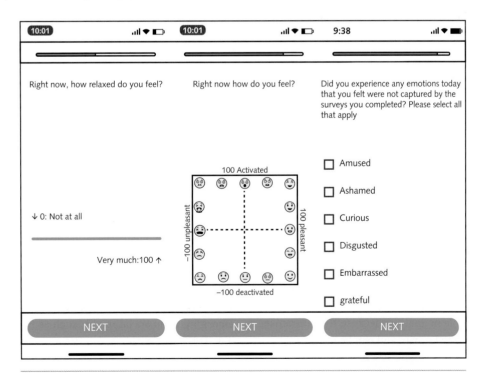

Figure 1.9 Three screens from the SEMA3 app, the Smartphone Ecological Momentary Assessment Application (Version 3)

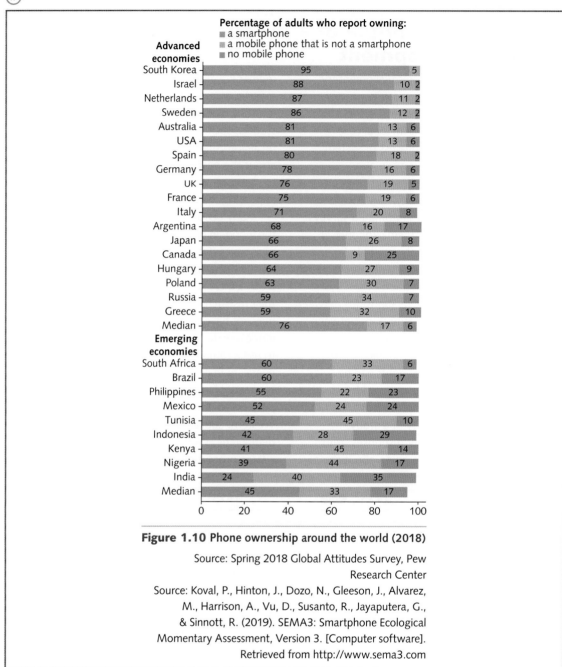

Figure 1.10 Phone ownership around the world (2018)

Source: Spring 2018 Global Attitudes Survey, Pew Research Center
Source: Koval, P., Hinton, J., Dozo, N., Gleeson, J., Alvarez, M., Harrison, A., Vu, D., Susanto, R., Jayaputera, G., & Sinnott, R. (2019). SEMA3: Smartphone Ecological Momentary Assessment, Version 3. [Computer software]. Retrieved from http://www.sema3.com

Literature reviews

A literature review is a report produced by reading scientific research on a particular area and writing a summary. Literature reviews play an important role in analysis because they organise what is already known, and can be used to synthesise new ideas based on the current level of understanding. The main sources of information for conducting research are research articles, review articles, journals and news websites. However, other sources of information are emerging from social media, such as Facebook, Tik Tok and Twitter.

Although these sources are an easy way to obtain information, you should take care to carefully evaluate their suitability and credibility. When reviewing these sources, it's important to be able to distinguish between opinions, anecdotes and evidence. **Opinions**, an individual's thoughts or beliefs, and **anecdotes**, which are personal stories, are not always supported by scientific evidence. They also guide research by identifying knowledge gaps in the literature. This allows researchers to refine current research questions and develop new ones.

KEY CONCEPTS 1.2

- » Each type of research design has specific benefits and limitations.
- » A controlled experiment involves experimental manipulation of a variable to determine the effect on an outcome(s) of interest.
- » A correlational study is a non-experimental investigation of the relationship between variables.
- » Correlations can be visually represented by plotting two variables on a graph.
- » Correlation does not equal causation.
- » A positive correlation is where high scores for one variable occur with high scores for another variable.
- » A negative correlation is where high scores for one variable occur with low scores for another variable.
- » A case study is an investigation of one particular example of an activity, behaviour, event or problem, to acquire knowledge about the process as a whole.
- » Classification and identification are processes used to organise phenomena into categories and identify examples of that categorisation.
- » Fieldwork involves observing and interacting with an environment in the real world.
- » Modelling means creating a representation of an event, process or system of concepts.
- » Product, process or system development is the design or evaluation of a process, system or artifact to meet a human need.
- » A literature review is a report produced by reading scientific research on a particular area and summarising it.

1.3 The controlled experiment in detail

Controlled experiments use a research methodology in which the researcher systematically manipulates one or more variables to investigate how this affects an outcome of interest. The systematic manipulation of variables gives the controlled experiment the unique ability to find cause-and-effect relationships between variables.

The simplest form of controlled experiment consists of two conditions: the *experimental condition* (sometimes called the treatment condition) and the *control condition*.

A randomised controlled trial (RCT) is an example of a controlled experiment design in which one variable is manipulated to find its effect on another variable. RCTs are used by psychologists when they want to determine whether a treatment is effective.

For example, to determine whether practising meditation causes improvements in people's mental wellbeing, we could 'manipulate' people's exposure to meditation to see if this affects their reported level of mental wellbeing. The simplest experimental manipulation would be to create one group of participants who experience meditation and another group of participants who do not. The group that experiences meditation is called the experimental group (or treatment group) and the group that does not is called the control group. The experimental group experiences the experimental (or treatment) condition and the control group experiences the control condition. This design allows us to compare self-reported levels of mental wellbeing between the experimental and control conditions to determine whether meditation improves mental wellbeing.

Of course, drawing a conclusion about the effectiveness of meditation based on this study will depend on whether the two groups differed from each other *only* in the variable of interest (exposure to meditation) and not in any other variable(s) that might also affect mental wellbeing. This brings us to consider the kinds of variables that influence experiment outcomes.

Variables in controlled experiments

A variable is any condition (for example, stimulus, event, quality, trait or characteristic) that can take a range of values and that can be measured or manipulated in a scientific investigation. You need to understand the different types of variables and be able to distinguish between them to understand how controlled experiments work.

Independent and dependent variables

The independent variable and the dependent variable are the two fundamental variables used in controlled experiments. The **independent variable (IV)** is the variable that the researcher manipulates. The **dependent variable (DV)** is the outcome variable that the researcher measures to determine whether manipulating the independent variable had an effect. In our example of an RCT study to determine the effect of meditation on levels of mental wellbeing, the meditation condition (meditation or no meditation) is the IV, and people's scores on a mental wellbeing scale is the DV.

Knowing where these terms come from may help you to remember them. If these variables are causally related (which is what the controlled experiment seeks to determine), then the value of the DV *depends on* the value of the independent variable. In contrast, the IV is not dependent on the DV (it is *independent* of the DV).

In a controlled experiment, participants will be exposed to different conditions and each condition is called a 'level' of the IV. For example, in an experiment where the IV is participation in a mindfulness meditation training program, the two levels of the IV could be:
1 meditation training (experimental condition)
2 no meditation training (control condition).

In an experiment where the IV is caffeine consumption, the two levels of the IV could be
1 caffeinated coffee (experimental condition)
2 decaffeinated coffee (control condition).

In an experiment where the IV is sleep duration, the two levels of the IV could be:
1 restricted sleep (experimental condition)
2 normal sleep (control condition).

Each of these examples has just one independent variable that is manipulated so that it has two levels (conditions). More complex experiments can have the IV at more than two levels. For example, you might want to compare two different kinds of meditation to a no-meditation control condition. The IV then has three levels, one for each kind of meditation and one for the control group. Some experiments have more than one IV, each with two or more levels (for example, the mixed design we consider below).

How do we know which variable in our experiment should be the IV and which variable should be the DV? To answer this, we need to think about which variable *affects* which. Imagine we hypothesise that warm temperatures make people happier. To test this hypothesis in a controlled experiment, temperature would be the IV and score on a happiness survey would be the DV. We are testing whether happiness *depends* on the temperature.

More examples of IVs and DVs

Example 1

Hypothesis (broad): Mindfulness meditation improves people's mental wellbeing
Hypothesis (specific): Teenagers who participate in mindfulness meditation training will score higher on the PANAS wellbeing survey than participants who do not participate in the training.
IV: Engagement in mindfulness meditation training (with two levels: training or no training)
DV: Score on the PANAS wellbeing survey

Example 2

Hypothesis (broad): Caffeine improves students' short-term memory
Hypothesis (specific): Students who consume a caffeinated coffee score better on the Digit Span Memory Test than participants who consume a decaffeinated coffee.

IV: Caffeine consumption (with two levels: Caffeinated coffee or decaffeinated coffee.)
DV: Score on the Digit Span Memory Test

Further points about IVs and DVs

The terms 'independent variable' and 'dependent variable' are only relevant to controlled experiments. You would not use these terms to describe the variables in a correlational study, because correlational studies investigate the relationship between variables, but not the *causal relationship* between variables. In a correlational study we do not assume that one variable, depends on another variable, and there is no manipulation of an IV.

You may be asking what is meant when we say the researcher 'manipulates' the IV. This depends on the type of controlled experiment, but it can mean changing, selecting or controlling the variable.

A common error is to say 'we gave the experimental group the IV'. In our caffeine example, this would incorrectly imply that the coffee itself is the IV. The IV is whether or not people consumed caffeine and the two levels of the IV would be the caffeinated coffee condition and decaffeinated coffee condition.

On a graph, the IV is always placed on the horizontal axis (that is, the x-axis) and the DV is placed on the vertical axis (that is, the y-axis).

Extraneous variables

An **extraneous variable** (EV) is any variable other than the IV that *may* affect the DV. The presence of EVs in an experiment is a problem because they make it difficult to be sure that the IV was responsible for any observed change in the DV.

For example, in a study investigating the effect of meditation training on wellbeing, different participants may:
» experience varying amounts of daily stress in their lives outside of the experiment
» have varying amounts of past experience with different forms of meditation.

These are both examples of EVs. They are factors other than the meditation training intervention that *may* affect people's level of mental wellbeing (the DV).

Note that EVs become a major problem if their effects are distributed differently between the two conditions. When this occurs, an EV becomes a confounding variable.

Confounding variables

A **confounding variable** is a variable other than the IV that *has* systematically affected the DV because its influence is not evenly distributed across the levels of the IV. Confounding variables are a major problem for the internal validity of an experiment because they provide an alternative explanation for the results. (See page 35–36 for more detail on internal validity.)

For example, in our meditation study, the EV of daily stress level would become a confounding variable if stress levels were systematically higher in one of the groups than in the other. This could happen if most people allocated to the meditation group happened to experience high levels of daily stress, but most people in the control group did not. If this were the case and our study finds that there is no beneficial effect of meditation on mental wellbeing, then our results are said to be 'confounded'. This is because we cannot determine whether our failure to find an effect of meditation on mental wellbeing was because meditation is ineffective or because the meditation group also happened to experience much higher stress throughout the experiment than the control group.

Let us consider another example of an EV that becomes a confounding variable. In a study designed to investigate the effect of caffeine on the rate of learning, participants could be divided into two groups:

1. an **experimental group** that receives caffeinated coffee
2. a control group that receives decaffeinated coffee.

The time at which participants are tested could be a confounding variable if the experimenter tested the learning rate of the experimental group in the morning and the control group in the late afternoon. In this example, the effect of caffeine on learning rate is confounded with participants' level of alertness due to the time of day, because both caffeine and time of day affect levels of alertness and the two groups were systematically tested at different times of the day. If the results were that the experimental group learned faster than the control group, the researcher cannot be sure whether participants in the experimental group learned faster due to caffeine, or because they were more alert from being tested earlier in the day (or whether both caffeine and the time of testing influenced the results).

Time of day would *not* be a confounding variable if both groups were tested at a similar time of day. In this case, time of day is still an EV that affects alertness, but it is not confounded with one of the experimental conditions.

Controlled variables

Controlled variables are variables that the researcher holds constant (controls) in an investigation. If an EV is identified before the study is conducted and the experimenter holds its effect constant throughout the experiment, it becomes a controlled variable. It is controlled because its influence has been managed so that it cannot bias the results of the experiment one way or the other. Controlled variables are kept constant to ensure that changes in the DV are caused by the manipulation of the IV, rather than by variation in other variables that are not of interest to the study. Researchers control variables to eliminate or neutralise their potential effects on the results.

A researcher can control variables by trying to keep all aspects of an experiment (except the IV) identical for each condition of the experiment. For example, in the example of analysing the effect of caffeine on the rate of learning, prior caffeine consumption could be a controlled variable; the researcher may instruct participants to abstain from drinking coffee or energy drinks on the day of the experiment. The testing location could be a controlled variable; the researcher may test each participant in the same place to control for variation in the environment, such as noise and distractions. The time of day the participants are tested could also be a controlled variable, to control for level of alertness.

Controlled variables are not variables that are part of an investigation itself. The researcher wants to control variables (keep them constant) so that they do not affect the investigation. Controlled variables should not be confused with the variables of interest in the experiment (that is, the IV and the DV), because *controlling* a variable is not the same as *manipulating* a variable: the aim of controlling a variable is to neutralise its effect, whereas the aim of manipulating a variable is to purposefully make changes to create different conditions (that is, different levels of the IV).

> A controlled variable is not the same as a control group. Be careful not to confuse these terms.

1.3.1 VARIABLES IN CONTROLLED EXPERIMENTS

Group allocation and variables

An important way that researchers control the effects of EVs is by managing how participants are allocated to conditions. Random allocation of participants to groups is used to increase the likelihood that the broad range of potential EVs (such as age, gender, and amount of sleep) will be equally distributed between groups. Alternatively, suppose there is a particular EV that the researcher wants to control. In that case, the researcher may seek to match each of the groups on this EV by ensuring that equal numbers of people with the specific feature (e.g., age group) are in each group. However, even when participants are randomly allocated to groups, there may still be underlying differences between the groups due to chance.

A researcher can match groups on variables such as age or gender to control (that is, neutralise) the effect of the variable on the outcome of the study. For example, to match groups by gender would be to allocate an equal number of males and females to each group. We explore the importance of allocation to groups in experimental designs in detail later in this chapter.

Variables summary

In summary, there are five kinds of variables that are relevant to controlled experiments (Figure 1.11). The independent variable (IV) and dependent variable (DV) are the variables of interest in a controlled experiment. The IV is the variable that the researcher manipulates. The DV is the outcome variable that the researcher measures to determine the effect of the IV. An extraneous variable (EV) is a variable that *may* have an unwanted effect on the DV. An EV becomes a confounding variable when its effect on the DV differs across the levels of the IV. An EV becomes a controlled variable when its effect is held constant by the researcher to prevent it from affecting the DV.

Understanding the variables defined in this section gives us a more comprehensive definition of the controlled experiment: a controlled experiment is a scientific investigation where the researcher systematically manipulates one or more IVs to determine the effect on one or more DVs, while attempting to control (eliminate or neutralise) the influence of EVs. At a minimum, a controlled experiment involves a comparison of outcomes

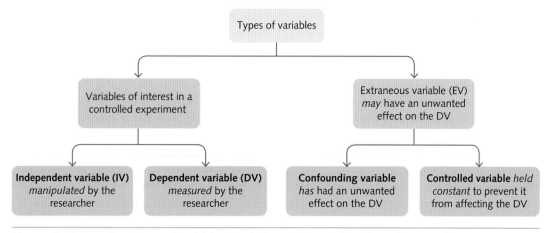

Figure 1.11 Important types of variables: the independent variable, dependent variable, extraneous variable, controlled variable and confounding variable

between at least one experimental condition and one control condition. This design allows the researcher to determine whether a hypothesised causal relationship exists between the IV and the DV.

Kinds of designs for controlled experiments

There are three different designs for controlled experiments: the between subjects design, the within subjects design, and the mixed design. Before we explore each of these designs in detail, it may help to know where these terms come from. In the past, researchers would refer to a person who takes part in a research study as a 'subject' (we now call them 'participants'). In a **between subjects design**, scores are compared *between* different participants. In a **within subjects design**, scores are compared *within* the same participants (that is, each person's score is compared with their own score at a different time). In a **mixed design**, a mix of both types of design is used, with both between subjects and within subjects comparisons.

Between subjects design

The between subjects design is the kind of experimental design we have been using as examples so far in this chapter. In a between subjects controlled experiment, participants are allocated to different groups, each exposed to a different condition. For example, consider a study of the effect of a treatment intervention. The two groups are:

1 the experimental group, which is exposed to the intervention

2 the control group, which is not exposed to the intervention.

In other words, the researcher manipulates the IV by varying the conditions experienced by participants in each group. The researcher measures the DV for each participant in each group and then compares the scores obtained for each group to the other group/s. If the DV scores are significantly different between the groups, this supports the hypothesis that the IV affects the DV. Figure 1.12 shows a flowchart of the between-subjects design.

The randomised controlled trial used as an example at the start of this section (page 21) is an example of a between-subjects controlled experiment.

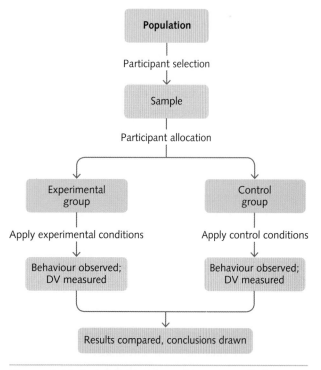

Figure 1.12 A simple between subjects controlled experiment

Let us consider an example of a between-subjects research design to test the effect of a learning program on reading ability. In this experiment, the IV is participation in the learning program. Students are split into an experimental group and a control group. The DV in this study could be the difference in the accuracy of reading a passage of text. The scores on the DV from the experimental group are then compared to the scores on the DV for the control group.

Between subjects design has the limitation that it assumes the groups are relatively similar on the range of extraneous variables that may affect the DV. If the groups are substantially different this creates a confounding variable that can affect the results. For the learning program example, groups could differ substantially if, for example, a Year 6 class was used as the experimental group and a Year 5 class was used as the control group. This potential limitation of differences between groups can be addressed by appropriately assigning participants to groups.

An advantage of the between-subjects design is that there are no **order effects** (see within subjects design, page 27).

Allocation to groups

As noted above, the allocation of participants to groups is important in a between-subjects controlled experiment. One method is random allocation. **Random allocation** to groups uses chance to determine how participants are assigned to groups. For example, the researcher may use a random number generator to generate a number for each participant and then put participants with even numbers in one group and participants with odd numbers in the other. This gives every participant an equal chance of being assigned to each group. Random allocation minimises the likelihood of extraneous participant variables (for example, age or level of stress) becoming confounding variables. This is because when group allocation is random, it is likely that any EVs that may affect the DV will be evenly distributed between the experimental and control groups. Random allocation also protects against unconscious experimenter biases in which they may unintentionally influence group allocation in a way that is advantageous to their hypothesis.

If group assignment is not randomised, but instead relies on existing groups such as two existing classrooms at a school, there could be inherent differences between groups that confound the results. For instance, if the experimental group was a Year 6 class and the control group was a Year 5 class, age could be a confounding variable because it is different between the groups. In the example of a study assessing the effect of a learning program on students' reading ability, age confounds the effect of the intervention because we cannot be sure that the reason the intervention group scores higher on the reading test is because of the intervention; the intervention group could have done better simply due to their age. This issue may be avoided by randomly allocating participants to groups, because randomisation makes it more likely that an approximately equal number of Year 5 and Year 6 students will be in each group. Although randomisation does not guarantee that age distribution will be the same in both groups, it reduces the probability that differences are present.

As an alternative to random allocation, a researcher may match groups on variables such as age or gender to control (that is, neutralise) the effect of this variable on the outcome of the study. This is called a **matched groups** or **matched participants design**. For example, to match groups by age would be to intentionally allocate an equal number of Year 5 and Year 6 children to each group.

Random allocation to groups is not possible when the experimenter is interested in the effect of a characteristic that defines different groups of people. For example, if the researcher is studying the effect of meditation for people who experience anxiety compared to non-anxious people, then group allocation is defined by a person's level of anxiety and cannot be random. We call these kinds of experiments **quasi-experimental designs** because they don't quite meet the standards of a controlled experiment. In quasi-experimental designs, the researcher may choose a matched-participant design to try and control the EVs that may differ between anxious and non-anxious people. In our example, the experimenter may match participants in the non-anxious group to people in the anxious group on factors such as age, gender and prior exposure to meditation.

Exam tip: Random allocation to groups is not the same as random sampling.

✓ **Random allocation** *is a method used to minimise bias in assigning participants to groups for between-subjects experiments.*

✓ **Random sampling** *is a method used to minimise bias in selecting a sample from the population.*

Within subjects design

In a within-subjects controlled experiment, each participant is exposed to both the experimental condition and the control condition. The within subjects design is also known as **repeated measures design** because each participant has to repeat the experiment in order to collect data for both conditions. Each participant can then act as their own control. That is, each participant has one score from their experimental condition to test the intervention or manipulation and another score from their control condition.

For example, a within subjects research design to test the effect of a learning program on reading ability could test the participants' reading ability before and after they participate in a learning program. If test scores are higher after the reading program than before the reading program, this can be seen as evidence that the reading program improves students' reading ability.

A benefit of the within subjects design is that individual differences between people do not influence the results because each participant is compared to themselves (instead of being compared to other participants). People can be very different from each other and the within subjects design overcomes the variation between participants. This variation could include attributes such as gender, ethnicity, personality, ability, education, socioeconomic status, memory, motivation and mood.

A limitation of the within-subjects design is that it is susceptible to the order effect (also known as the practice effect). As each participant has to participate in both conditions, they engage in the testing procedure twice. In the example of a reading test, there could be a systematic improvement to their reading score simply because they have already taken a similar test; they have practised the test once and this makes them better the second time they do it. Another example of the order effect is if somebody had to do a test of reaction time twice in a row and their score in the second test was lower because of fatigue.

The order effect can sometimes be overcome by counterbalancing. Counterbalancing is where the order of the conditions is split, so not everybody completes the same conditions in the same order. For example, half of the participants could undertake the control condition first, followed by the experimental condition, and the other half undertake the experimental condition first, followed by the control condition (Figure 1.13). Counterbalancing averages out any potential order effects across both conditions.

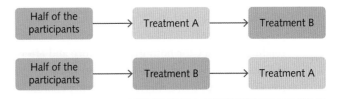

Figure 1.13 In a within-subjects controlled experiment, counterbalancing can be used to minimise order effects.

However, counterbalancing is not always possible due to the nature of particular studies. For instance, it is impossible to counterbalance the study examining the effect of the learning program on reading ability because the same students need to be tested before and after the learning program, and the researcher can not reverse the order of these events.

Some examples where counterbalancing is possible in within subjects designs are the investigation of restricted sleep on wellbeing and the investigation of caffeine consumption on memory. In these studies, the experimenter has control of the order in which participants experience each condition: in the sleep and wellbeing study, half of the participants could experience the restricted sleep condition before the long sleep condition and the other half could experience each condition in the opposite order, and in the caffeine and memory study, half of the participants could experience the caffeinated coffee condition before the decaffeinated coffee condition and the other half could experience these conditions in the opposite order. Of course, the researcher must allow enough time between each condition so that one condition's effects have worn off before testing the following condition.

Mixed design

A mixed design has the elements of both a between-subjects design and a within-subjects design. In its most simple form, a mixed design has two IVs, and each IV has two levels. One of the IVs is a between subjects variable, and the other IV is a within subjects variable.

1.3.2 EXPERIMENTAL DESIGNS FOR CONTROLLED EXPERIMENTS

For example, when studying the effectiveness of the reading program, the ideal design is to compare two groups of students: one that experiences the reading program and one that experiences a control condition in which they continue learning as usual. This is a between subjects variable with two levels (reading program vs. control). The researcher can measure reading accuracy for both groups before and after the reading program. This design allows the researcher to be more confident that any change they see between groups is due to the treatment itself and not simply to improvements that would have occurred during normal learning during that time. It combines the advantage of having a within subjects measure (change in reading performance over time) with the advantage of a between subjects design (e.g., utilising a control group).

It has the disadvantages of being more difficult to carry out and producing results that can be more difficult to analyse.

Table 1.1 A mixed design controlled experiment to study a learning program intervention. The within-subjects factor (time) are the vertical columns of the table and the between-subjects factor (learning group) are the horizontal rows of the table.

		Within subjects Factor (Time)	
		Reading Accuracy Before the Learning Program	Reading Accuracy After the Learning Program
Between-subjects Factor (Learning Group)	Reading Program	Experimental group accuracy before the program	Experimental group accuracy after the program
	Learning as Usual	Control group accuracy before the program	Control group accuracy after the program

KEY CONCEPTS 1.3

- Manipulation of a variable gives the controlled experiment the unique ability to find cause-and-effect relationships between variables.
- The independent variable (IV) is the variable that the researcher manipulates.
- The dependent variable (DV) is the variable that the researcher records to see if it has been affected by a change in the IV.
- IVs and DVs are only relevant to controlled experiments.
- When plotted on a graph, the IV goes on the horizontal axis and the DV goes on the vertical axis.
- An extraneous variable (EV) may affect the DV and influence the results in an unwanted way.
- A confounding variable does affect the DV and influences the results in an unwanted way.
- A controlled variable is held constant in an investigation.
- A controlled variable is not the same as a control group.
- A between subjects design is a controlled experiment where participants are allocated to different groups (for example, the experimental group or the control group).
- Between subjects designs have no order effects.
- Random allocation of participants to groups helps to reduce confounding variable bias in between subjects designs.
- A within subjects design is a controlled experiment where each participant is exposed to both the experimental and the control conditions.
- In within subjects designs, individual differences between people do not influence the results.
- Within subjects designs can suffer from the order effect.
- Counterbalancing can overcome the order effect.
- A mixed design combines the between subjects and within subjects design, with multiple groups and data recorded at multiple times.

Interim summary

So far, this chapter has explored how to precisely create a hypothesis and test it with a study, and the research designs used for research investigations.. We focused much of our attention on the controlled experiment, but we also explored correlational studies and other methodologies and showed why it is not always possible, desirable or appropriate to investigate some questions with an experiment.

Next we will explore some key science skills relevant to analysing and interpreting data collected in research investigations. We will also consider some theoretical concepts relevant to understanding research investigations and data analysis and ethical concepts and guidelines. These are all important foundations for understanding the research process and for ensuring that research investigations validly and ethically assess what they intend to investigate.

1.4 Analysing and evaluating research

Evaluating research includes closely considering different elements of a study and how any data it produces has been processed and analysed. In this section we will first describe how quantitative data can be processed and presented. We will then introduce the important concepts that underlie data analysis, including accuracy, precision, repeatability, reproducibility and validity. Next, we will consider how to evaluate whether findings support a hypothesis, how to identify limitations of a study, and how to understand and evaluate possible sources of error and uncertainty. Finally, we look at both the ethical concepts and guidelines and the health and safety considerations that must apply to all research.

Processing quantitative data

Whether you are deciding how to present your own data or analysing a published research investigation, it's important to understand how data can be processed and presented to best communicate the results.

Displaying data in tables, bar charts and line graphs

Data can be displayed in tables, charts or graphs. These can be used to organise data, compare variables and visualise relationships between them. Figure 1.14 provides an example of a well-constructed table, bar chart and line graph. A table is a grid with horizontal rows and vertical columns that can be used to record and organise data. Statistics that summarise data can also be included in tables.

The terms 'chart' and 'graph' are often used interchangeably, but technically speaking, a chart is a visual representation of data that can take many forms and a graph is a specific type of chart that has two axes representing two variables. A bar chart is a graph that shows data using separated rectangular columns or 'bars' to represent the total number (or other measures such as the mean) for distinct categories of data. It shows how frequently a particular group or score occurs in a data set. A line graph represents the relationship between two variables by a line that connects each data point so that the reader can see the change from point to point and the overall trend. When showing the variables of a controlled experiment on a graph, the IV is plotted on the horizontal axis and the DV is plotted on the vertical axis.

Distribution of data

The distribution of data can be thought of as the shape and symmetry of data when it is plotted on a histogram. A **histogram** is a bar chart that graphs just one variable, where the horizontal axis represents the variable (the score or thing being measured) and the vertical axis lists the frequency at which that score is found in the data set (for example, the number of participants that have that score on that variable).

A data set is described as *normally distributed* when its curve has the shape of a bell; it is highest around the centre (indicating that there is a typical score for most participants) and gradually decreases in a relatively symmetrical pattern as scores move away from the centre (indicating that there are fewer participants who score towards the extremes). An example of normal distribution is the height of a randomly selected sample of people from the population. Figure 1.15 shows a frequency distribution of normally distributed data.

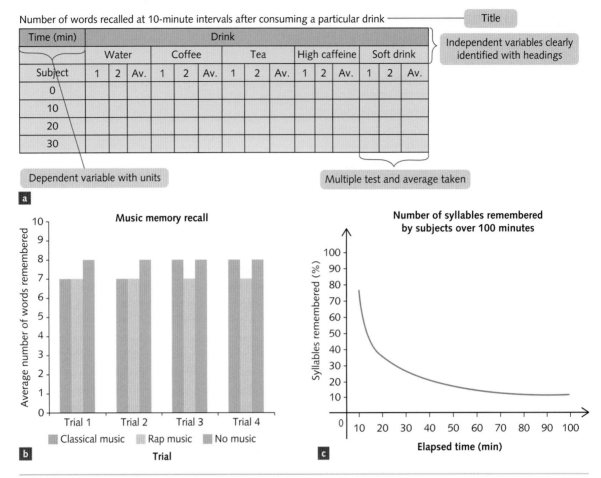

Figure 1.14 Examples of **a** a well-constructed table, **b** a bar chart and **c** a line graph

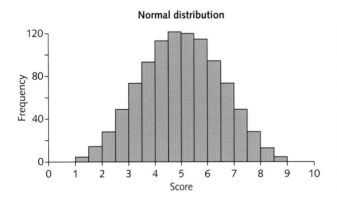

Figure 1.15 A histogram of normally distributed data

1.4.1 DISPLAYING DATA IN TABLES, BAR CHARTS AND LINE GRAPHS

A data set is skewed when the curve is not symmetrical (that is, it is uneven on one side of the curve). An example of something that follows a skewed distribution is income. Most people lie relatively close to the mean, with some variation on either side, but some very wealthy people earn far more than the average person. Figure 1.16 shows two examples of histograms showing data that has a skewed distribution.

The terms 'normally distributed' and 'skewed' can describe one specific data set or they can describe the distribution of the population from which the sample was recruited. For example, imagine we have a data set with one value that is very different from the rest. This unusual value is called an outlier. We may say that the outlier in our data set (obtained from measurements of our sample) could reflect the underlying distribution (the distribution of the data of the whole population) being a skewed distribution, or from a different underlying distribution.

Descriptive statistics

Descriptive statistics summarise the main features of an overall data set. **Descriptive statistics** *describe* the data set by condensing a set of values down to a single numerical value. They can also be visualised in charts, tables and graphs. Descriptive statistics include measures of central tendency such as the mean, mode, and median values and measures of the range and spread of the data such as the standard deviation.

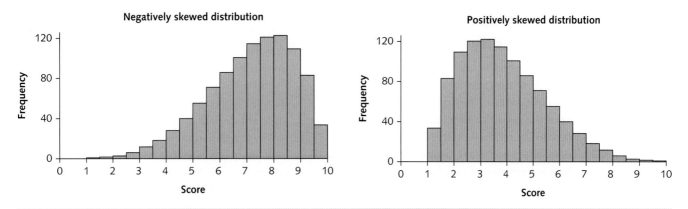

Figure 1.16 Two histograms showing examples of data following a skewed distribution

Measures of central tendency provide a number that describes a 'typical' score around which other scores lie. Central tendency is commonly measured by using three descriptive statistics: the mean, median and mode.

» The **mean** is a measure of central tendency that gives the numerical average of a set of scores, calculated by adding all the scores in a data set and then dividing the total by the number of scores in the set.
» The **median** is a measure of the middle score in a data set. It is calculated by arranging scores in a data set from the highest to the lowest and selecting the middle score.
» The **mode** is a measure of central tendency found by selecting the most frequently occurring score in a set of scores.

Figure 1.17 shows the three measures of central tendency (the mean, median and mode) and demonstrates how to calculate these, using an example data set.

This example shows that the different measures of central tendency can be the same or different values for a data set: in our example, the median and the mode happened to be the same value, but the mean had a slightly different value.

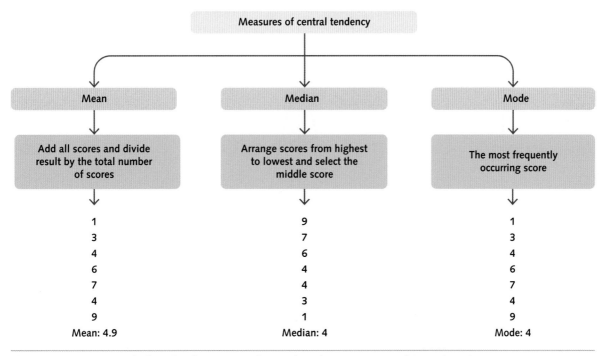

Figure 1.17 How to calculate the three types of central tendency: mean, median and mode

Standard deviation: a measure of variability

While measures of central tendency provide valuable information about the central or typical value in a data set, they provide no information about the *variability* of scores in the data set. **Variability** means how spread out or clustered together the scores are. **Standard deviation (SD)** is a measure of variability that describes the average deviation (or distance) of a set of scores from the mean (that is, the average distance that a set of data points are from the mean). Standard deviation uses a mathematical formula to provide a standardised measure of the average deviation of scores from the mean.

A standard deviation that is a *large* number demonstrates *high variability*, as it indicates that scores have a *large* average distance from the mean. A standard deviation that is a *small* number demonstrates *low variability*, as this indicates that scores have a *small* average distance from the mean.

Figure 1.18 shows the distribution of three data sets plotted on the same graph. The horizontal axis represents the score and the vertical axis can be thought of as the number of participants who have that score. All three data sets have a mean of 50, but the spread of scores (that is, variability) is different for each. The data set represented by the purple line has a low standard deviation (SD = 5), which means most scores are clustered close to the mean (you can see that the distribution is high at the centre and there are not many scores at the extreme values because the distribution does not cover a wide area). The data set represented by the green line has a high standard deviation (SD = 20) and so this data set has scores that are much more spread out. Notice how the green line does not rise very much at the centre, because there are fewer data points around the central value, but the distribution covers a much wider area, meaning there are more scores at the extreme values. The data set represented by the green line has a standard deviation in between the other two data sets (SD = 10).

Standard deviation example

Imagine a study has investigated two forms of psychotherapy to treat anxiety. The first form of psychotherapy is a traditional and well-established form of therapy that is found to decrease anxiety by a relatively consistent amount for all research participants. The second form is a new form of psychotherapy that was found to reduce anxiety by a large amount in some participants and a small amount in other participants, but it increases anxiety in some participants. Both forms of psychotherapy could have the same mean, but the variability measured as standard deviation would be much larger for the new form of psychotherapy.

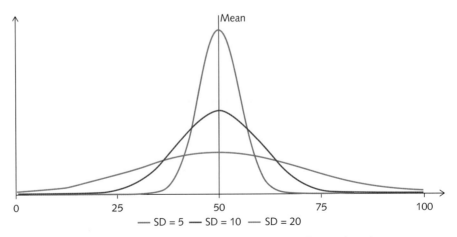

large standard deviation **means** high variability **and** large average distance from the mean
small standard deviation **means** low variability **and** small average distance from the mean

Figure 1.18 Three data sets with different standard deviations. A low standard deviation means that data are clustered close to the mean. As standard deviation increases, so does the spread of data. On this graph, the horizontal x-axis represents a variable (for example, score) and the vertical y-axis represents the frequency with which each score was found.

Percentage and percentage change

To calculate a percentage, you divide the number of the subset by the total number and then multiply the result by 100. Percentage change is the difference between two percentages.

Let us consider the earlier example of an experiment testing whether a new learning program improves children's reading ability. In this example, we will only consider the experimental group (that is, those who participated in the learning program). To pass the reading test, the student must read the passage aloud with no errors. Before the program, we asked 40 students to read a passage aloud and 12 students passed the test. To calculate the percentage who passed the test, we divide the number of students who passed the test (12) by the total number of students (40) and then multiply it by 100.

$$\frac{12}{40} \times 100 = 30\%$$

After the reading program, 28 students passed the reading test. Once again, we divide the number of students who passed the test (28) by the total number of students (40) and then multiply it by 100.

$$\frac{28}{40} \times 100 = 70\%$$

The percentage change is the difference between these two percentages, so we need to subtract the initial percentage value from the final percentage value.

$$70\% - 30\% = 40\%$$

Here the percentage change is 40%, which means that there was a 40% change in the number of students who passed the test after the intervention of the learning program.

Data analysis concepts

Analysing data requires us to look behind the actual data itself. It requires us to also make a judgement on the *quality* of the data collected.

True value

The concept of a true value is something many other concepts build upon. The **true value** can be defined as the value or range of values that would be found if a quantity could be measured perfectly. However, this idea is somewhat more complex to consider in psychology than in other sciences. In psychology, we investigate concepts that are difficult to measure. Often concepts of interest cannot be directly measured, but instead we have to find a way to score the concept by measuring things that reflect the concept. Nevertheless, we can make use of the idea of true value when we analyse numerical data. To make use of the true value idea, we treat our constructs as if they would have a true value if it were possible to measure them perfectly.

Accuracy and precision

In science, the **accuracy** of a measurement means how close it is to the true value of the quantity being measured. While accuracy is not quantifiable, it can be used to describe measurement values as more accurate or less accurate. That means accuracy is a *relative* description, used to compare the accuracy of different values with each other. Researchers aim to make improvements in their research design to make their measures more accurate (that is, closer approximations of the true value).

Precision refers to how close a set of measurement values are to one another. Precision is determined by the repeatability and/or the reproducibility of the measurements obtained using a particular measurement instrument and procedure (for example, equipment for measuring participant response times, a psychological questionnaire or a brain imaging technique). Precision gives no indication of how close the measurements are to the true value and is therefore a separate consideration from accuracy. It is possible for a measurement instrument to produce precise (repeatable and reproducible) measurements, but those measurements may not be accurate. It is also possible to get an accurate measurement of the true value by averaging a set of highly variable (imprecise) measurements.

Figure 1.19 shows a representation of precision and accuracy on a dartboard, where the bullseye in the centre represents the true value and the darts (the black dots) represent individual measurements. This shows measurements can be:
» precise and accurate
» accurate but not precise
» precise but not accurate
» not accurate and not precise.

1.4.2 DESCRIPTIVE STATISTICS

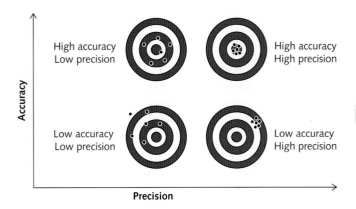

Figure 1.19 A representation of the precision and accuracy of measurements, where the bullseye of the dartboard represents the true value

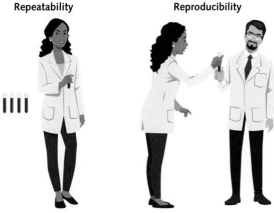

Figure 1.20 Repeatability is the agreement of results or measurements under the same conditions of measurement, and reproducibility is the agreement of results or measurements under changed conditions of measurement.

Repeatability and reproducibility

Figure 1.20 shows examples of repeatability and reproducibility. The VCE Psychology Study Design defines them as follows.

» **Repeatability** is the closeness of the agreement between the results of successive measurements of the same quantity being measured, carried out under *the same* conditions of measurement. These conditions include the same measurement procedure, the same observer, the same measuring instrument used under the same conditions, the same location, and repetition over a short period of time.

» **Reproducibility** is the closeness of the agreement between the results of measurements of the same quantity being measured, carried out under *changed* conditions of measurement. These different conditions include a different method of measurement, different observer, different measuring instrument, different location, different conditions of use, different time and/or different culture(s).

Repeatability and reproducibility both refer to how close successive measurements of the same quantity are. The closer the results are to each other, the higher the repeatability or reproducibility. However, the terms differ in that:

» repeatability is the agreement of results carried out under *the same* conditions of measurement
» reproducibility is the agreement of results carried out under *changed* (different) conditions of measurement.

The conditions of measurement include the measurement procedure, instruments, observers (that is, the people recording the measurement), culture of observers, location of study and timespan between measurements (that is, repetition over a short or long period of time).

Repeating experiments can help to determine whether substantial errors exist, which is important because there will always be a level of error with any tools. Research findings are said to be robust if they are not easily affected by the presence of some errors in the data. Therefore, researchers should repeat their experiments to ensure the findings are robust.

Scientific findings can be considered as reproducible or not reproducible. If results are reproducible, other researchers find the same results when they repeat the experiment. If results are not reproducible, other researchers get different results when they repeat the experiment. Scientific findings that are not reproducible (replicable) may lack credibility, because this suggests that the initial results may not reflect a true, measurable relationship, but may rather have been a result of error or chance.

Validity

Validity means how well the design of a scientific investigation and its measurements provide meaningful and generalisable information about the psychological constructs of interest. The validity of a psychological investigation tells us how well the results from the study participants represent true findings among the population of interest

Another word for 'reproducible' is 'replicable'. The VCE Psychology Study Design uses 'reproducibility' as the key term, but interchangeably uses the term 'replicable' (and its opposite, 'irreplicable', which means not replicable/reproducible).

outside of the study. There are two types of validity: internal validity and external validity.

A psychological investigation has internal validity if it investigates what it sets out to investigate. The internal validity of an investigation depends on:
- how appropriate the investigation design is
- the sampling and participant allocation techniques used
- whether there are extraneous and confounding variables that affect the results.

A lack of internal validity means that the study results deviate from the truth and therefore no conclusions can be drawn.

A psychological investigation has external validity if the results of the research can be applied to similar individuals in a different setting (outside the research context/in the real world). A lack of external validity means that the research results may not apply to individuals or situations outside the study population. External validity can be increased by using sampling techniques that ensure the study population more closely resembles the general population. External validity is also increased by using measures that reflect the way psychological processes operate in natural contexts. For example, measuring memory using lists of words in a lab may not have very high external validity with memory for events in real-world contexts.

If a study is not internally valid, the concept of external validity is irrelevant. If a study has not measured what it intended to measure, it is not possible to generalise its results to the population or make inferences about whether similar results would occur in a real-world context.

Errors

Measurement error is the difference between the measured value and the true value of what is being measured. Two types of measurement errors should be considered when you evaluate the quality of data: random errors and systematic errors. Figure 1.21 shows how errors are categorised.

Random errors

Random errors are unpredictable variations that can happen during measurement. When you take multiple readings of the same thing, random measurement errors cause small variations so that you end up recording a spread of readings. Such errors affect the precision of a measurement (that is, how close a set of measurement values are to one another). Random errors can be caused by limitations of instruments, environmental factors (such as sudden noises or interruptions) and slight variations in procedures.

The effect of random errors can be reduced by making more or repeated measurements and calculating a mean. The average value of repeated measurements can be used as the final value. The resulting reduction in random errors should mean that the final value is a closer approximation of the true value. The effect of random errors can also be reduced by increasing the sample size and/or by refining the measurement method or technique.

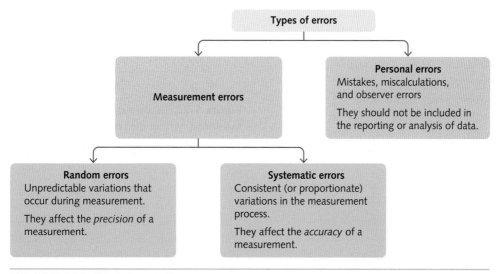

Figure 1.21 Different types of errors

Systematic errors

Systematic errors affect the accuracy of a measurement by causing all of the readings to differ from the true value. They affect the reading by a consistent amount (or by the same proportion) each time a measurement is made. This causes all of the readings to be shifted away from the true value in the same direction.

Systematic errors may be caused by measuring instruments not being correctly calibrated or by environmental interference. Systematic errors can also be caused by observational error if there is a consistent distortion in the way we view things that causes errors that are the same every time (for example, a tall person reading a thermometer from a higher viewpoint and recording a lower measure than the true value every time). Most systematic errors can be reduced by knowing how to use the measurement instrument correctly and being familiar with its limitations. You cannot improve the accuracy of measurements that have systematic errors by repeating those measurements, because systematic errors will be present in all the measurements you make.

Personal errors

Personal errors are not measurement errors. Personal errors are mistakes, miscalculations and observer errors when conducting research. Personal errors should not be included in reporting and analysis of data. Rather, if a researcher makes personal errors, they should repeat the experiment correctly.

Remember: Random errors affect the precision of a measurement. Systematic errors affect the accuracy of a measurement.

Don't confuse measurement errors with personal errors.

Don't confuse the terms 'error' and 'uncertainty'. They are not synonyms.

Uncertainty

The scientific method is designed to reduce the degree of uncertainty about observations, relationships and causes. Researchers use it to reduce uncertainty:
» in the observations they make
» about the relationships between variables of interest
» about the causes of the relationship they are interested in.

The **uncertainty** of a measurement reflects the lack of exact knowledge of the true value of the measurement. All measurements are subject to uncertainty because all measurements have many possible sources of variation. Because all inferences and conclusions in research depend on uncertain measurements, this uncertainty extends to all inferences and conclusions.

Psychological measurement tools (for example, surveys) are used to measure psychological constructs, which are, by nature, not directly measurable. Therefore, there is *always* a degree of uncertainty associated with measurements of psychological constructs.

The concept of uncertainty is especially relevant when evaluating data. You should always be vigilant for things that increase uncertainty, such as possible sources of bias, contradictory data and incomplete data. **Incomplete data** are data that are missing, such as questions without answers or variables without observations. Contradictory data are *incorrect* data: for example, a 5-point scale intended to be scored from 1 'strongly disagree' to 5 'strongly agree', that may have been mis-scored so that 1 represents 'strongly agree' and 5 represents 'strongly disagree', thus reversing the interpretation of the findings. Make sure you carefully consider sources of uncertainty both in data you have collected yourself and in data provided by others.

Outliers

Outliers are data points that differ substantially from the rest of the collected data. Figure 1.22 shows a plot of results where an outlier is present.

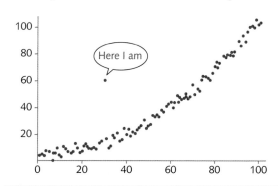

Figure 1.22 Outliers are data points that differ substantially from the rest of the collected data.

An outlier can be caused by a measurement or recording error or it can simply be an uncommon value. Uncommon values can appear in a data set by chance and so are more likely to be found in a larger sample. Therefore, you would expect to find a small number of outliers in a data set of a large sample. An outlier from a measurement or recording error can have a significant impact on the results and can affect the validity of the research.

An outlier can also be a valid measurement that stands out from the rest (that is, a true reading of a rare data point). Such outliers can occur in a data set due to natural variation. An outlier score can be a valid data point from an underlying skewed distribution or other non-normal distribution. In a skewed distribution, the data set may include very high or very low data points that are valid. For example, in a study of income (which follows a skewed distribution in the population), having a very wealthy person in a sample would give a valid outlier score.

If you obtain outliers in research results, you must distinguish between those that are valid measurements and those that result from measurement error. One way to investigate whether an outlier is an error or an uncommon value is by repeating measurements. If the outlier is a valid value, it should remain in the data set on the second repetition of data collection, whereas if the outlier is due to a measurement error, it would be unlikely that the same measurement error would appear again.

When you analyse a study, make sure you reflect on how different types of outliers would affect the testing efforts and validity of the research. If an outlier is just an uncommon value, it can still be considered a valid data point. However, if an outlier is an error, it can compromise the validity of the research.

Sample size and the quality of data

Sample size has a significant impact on the quality of data. Using a larger sample size for a study makes it more likely you will find an effect or a relationship, if one exists. Larger sample sizes improve data quality for several reasons.

Outliers and errors have less impact on the results you get from a larger sample. In small samples, natural variation can significantly influence the results, whereas natural variation is usually averaged out in larger samples. If an error or an outlier is present in a data set from a sample size of 100, it will have less effect than an error or outlier present in a data set from a sample size of 10. The conclusions from the data collected from the larger sample will therefore be less affected by this error, which means that increasing the sample size can improve the quality of the data.

A larger sample size also means that the sample is more likely to be representative of the population because there is more chance that it will include any natural variation and less chance that it will be a unique sample that does not reflect the general population. For these reasons, a larger sample size improves the study's validity and enhances certainty in the data. It follows that a large sample size makes a study more repeatable and reproducible, because it makes it more likely that findings reflect effects present in the real world.

The issue of small sample sizes is a serious problem for psychological research and for scientific research in general. Many historically important scientific studies were conducted with small samples. When other researchers attempted to reproduce these studies to confirm the findings, they got different results. This problem has contributed to what is called 'the replication crisis' (or the reproducibility crisis) and has received significant attention in the past decade. There is a movement in the psychological research community to increase the sample sizes used in studies to address this issue and improve the quality of research findings.

Evaluating data and investigation methods

When you evaluate an investigation, you need to consider both the data and the investigation method itself. To do so, you need to apply the concepts that you have learned in the previous sections.

Evaluating a research design

Students are often asked to *evaluate* a research design for the extended response question on the VCE exam. This type of question often provides an example of a study that you will need to read and evaluate. The question may tell you which concepts to write about or you may need to determine which concepts are relevant. The kinds of things the examiner may be looking for can be found in the key science skills section of the VCE Psychology Study Design; in particular, in the dot points that use the word 'evaluate'. The concepts that you may need to draw on to answer this question include:

WB
1.4.3
QUALITY
OF DATA

Don't confuse errors and outliers. Remember that outliers must be further analysed and accounted for, rather than being automatically dismissed.

1.4.4 KNOW WHAT YOU ARE READING

- » you may be asked to evaluate data to determine the degree to which the evidence supports or refutes the initial prediction or hypothesis, or supports the aim of the investigation
- » you may need to evaluate investigation methods and possible sources of error or uncertainty
- » you may need to suggest improvements to increase validity and to reduce uncertainty
- » when evaluating data, you may need to identify outliers and contradictory or incomplete data; or comment on certainty in data, which includes the effects of sample size on the quality of data
- » you may need to apply the science skills and concepts of accuracy, precision, repeatability, reproducibility, validity and errors
- » you may also need to consider cultural biases that could affect the results (for example, WEIRD samples) or consider ethical concepts and guidelines.

Does the evidence support the aims?

When you interpret the results of a research investigation, it is important to reflect on and critique whether the evidence supports the aims. Ask yourself, 'Has the study answered the question it set out to answer?' Limitations of the study can be difficult to anticipate and are sometimes only identified while or after the research has taken place. For example:
- » after recruitment, the researcher may identify that the sample is not representative of the population; this would affect external validity because the findings are unlikely to generalise to the broader population
- » during the investigation, the researcher may identify unexpected factors that compromise the internal or external validity of the study
- » during data analysis, other problems, such as measurement errors, may be identified.

Identify limitations of conclusions

1.4.5 WHAT IS THE DATA TELLING YOU?

All studies have limitations and all conclusions drawn from studies are inherently limited. The evidence may support the conclusions being drawn but there is always a level of uncertainty and room for improvements in the study design. Limitations are often unavoidable and may be caused by design constraints such as difficulties recruiting participants, small sample size, time constraints or inability to control EVs.

When considering the potential limitations of findings, researchers should consider whether their interpretations of the results would apply in different cultural contexts. Researchers should also consider whether cultural biases are present in their testing methods. For example, people may perform poorly on a test that has been developed and standardised for a different cultural group. The test might assume participants have background knowledge or particular ways of interpreting material; however, some cultural groups may not have been exposed to the testing methods or may interpret them differently through their own cultural context. Also, what is considered normal behavioural, cognitive and socio-emotional functioning can be influenced by cultural contexts, such as traditional customs and beliefs of a particular culture or society.

Ethics

Ethics is a guiding framework that all research investigations must follow. Because psychological research involves human (and sometimes animal) participants, research investigations should only proceed if they can be carried out ethically. Ethical concepts and guidelines provide moral guidance for making decisions about the design and implementation of a research investigation. You must apply your ethical understanding across all units of VCE Psychology, particularly when you design and conduct your own research investigation.

VCE Psychology requires students to know:
- » five **ethical concepts**, which you can use to explore the conduct of psychological investigations and to determine whether a research investigation is ethically acceptable
- » six **ethical guidelines**, which you should consider when you are conducting and evaluating psychological investigations.

The five **ethical concepts** are: beneficence, integrity, justice, non-maleficence and respect (Table 1.2).

The six **ethical guidelines** that underpin psychological research are confidentiality, debriefing, informed consent procedures, use of deception in research, voluntary participation and withdrawal rights (Table 1.3).

Table 1.2 A summary of the five ethical concepts for psychology research

Concept	Description
Beneficence	Having a commitment to do good (and minimise risks and harms)
Integrity	Acting with honesty and transparency
Justice	Ensuring fair distribution of benefits, risks, costs and resources
Non-maleficence	Avoiding harm or ensuring potential harm is outweighed by benefits
Respect	Giving due regard to individual difference and ensuring the right to autonomy and choice

Table 1.3 A summary of the six ethical guidelines for psychology research

Guideline	Description
Confidentiality	Ensuring the privacy of participants' personal information
Debriefing	After the experiment, disclosing the aim, results and conclusions, answering questions and providing support
Informed consent procedures	Ensuring participants understand the nature, purpose and risks of the study before agreeing to participate
Use of deception in research	Concealing aspects of the study (only used when absolutely necessary, and must be accompanied by debriefing)
Voluntary participation	Ensuring there is no coercion or pressure to participate
Withdrawal rights	Allowing participants to discontinue involvement in an experiment, without penalty

Ethical concepts are general in nature and are separate to any specific ethical principles, guidelines, codes or legislation.

1.4.6 ETHICS

The full definitions of the ethical concepts and guidelines as listed in the VCE Psychology Study Design are provided below.

The ethical concepts are:
» **Beneficence**: The commitment to maximising benefits and minimising the risks and harms involved in taking a particular position or course of action.
» **Integrity**: The commitment to searching for knowledge and understanding and the honest reporting of all sources of information and results, whether favourable or unfavourable, in ways that permit scrutiny and contribute to public knowledge and understanding.
» **Justice**: The moral obligation to ensure that there is fair consideration of competing claims; that there is no unfair burden on a particular group from an action; and that there is fair distribution and access to the benefits of an action.
» **Non-maleficence**: Involves avoiding the causations of harm; however, as a position or course of action may involve some degree of harm, the concept of non-maleficence implies that the harm resulting from any position or course of action should not be disproportionate to the benefits from any position or course of action.
» **Respect**: Involves consideration of the extent to which living things have an intrinsic value and/or instrumental value; giving due regard to the welfare, liberty and autonomy, beliefs, perceptions, customs and cultural heritage of both the individual and the collective; consideration of the capacity of living things to make their own decisions; and when living things have diminished capacity to make their own decisions, ensuring that they are empowered where possible and protected as necessary.

Source: VCE Psychology Study Design (2023–2027) p. 20

The ethical guidelines, as listed in the VCE Psychology Study Design, are:
» **Confidentiality**: The privacy, protection and security of a participant's personal information in terms of personal details and the anonymity of individual results, including the removal of identifying elements.
» **Debriefing**: Ensures that, at the end of the experiment, the participant leaves understanding the experimental aim, results and conclusions. Any participant questions are addressed and support is also provided to ensure there is no lasting harm from their involvement in the study. Debriefing is essential for all studies that involve deception.

- **Informed consent** procedures: Ensure participants understand the nature and purpose of the experiment, including potential risks (both physical and psychological), before agreeing to participate in the study. Voluntary written consent should be obtained by the experimenter and if participants are unable to give this consent, then a parent or legal guardian should provide this.
- Use of **deception** in research: Is only permissible when participants knowing the true purpose of the experiment may affect their behaviour while participating in the study and the subsequent validity of the experiment. The use of deception is discouraged in psychological research and used only when necessary.
- **Voluntary participation**: Ensures that there is no coercion of or pressure put on the participant to partake in an experiment and they freely choose to be involved.
- **Withdrawal rights**: Involves a participant being able to discontinue their involvement in an experiment at any time during or after the conclusion of an experiment, without penalty. This may include the removal of the participant's results from the study after the study has been completed.

Source: VCE Psychology Study Design (2023–2027) p. 20–21

Read the exam question carefully to see whether it asks for 'ethical concepts' or 'ethical guidelines'. Where the question asks for something vaguer like 'ethical considerations', you should specify in your response which of your responses are concepts and which are guidelines.

WB 1.4.7 KNOW YOUR KEY TERMS

Safety

Health and safety are important considerations for practical exercises in all sciences. If you are undertaking your own practical research investigations, you must consider any relevant occupational health and safety guidelines. In Victoria, workplaces are covered by the *Occupational Health and Safety (OHS) Act 2004*. As some psychology experiments are conducted in laboratories, researchers must ensure safe laboratory practices when planning and conducting investigations by using **risk assessments**, supported by safety data sheets (SDS), and accounting for risks. SDS are usually only relevant if you are using chemicals as part of your investigation (more relevant to the sciences other than psychology). If your research does not use chemicals, but requires participants to take some actions that may cause harm, you will need to complete a risk assessment form (Figure 1.23). Your school is likely to have one of these documents for you to complete when you conduct your experiment. If you are unsure of either the ethical or the health and safety aspects of your experiment, check with your teacher.

Knowledge

I/we have read and understood the potential hazards and standard handling procedures of all the equipment, chemicals and living organisms.
I/we have read and understood the (Material) Safety Data Sheets for all chemicals used and produced.
I/we have copies of the (Material) Safety Data Sheets of all the chemicals available in or near the laboratory.

Agreement by student(s)

I/we, Bill Wilkins, Mary Newt, Christina Lee, agree to conduct this experiment safely in accordance with school rules and teacher instructions.

Risk assessment

I/we have considered the risks of:

fire	breakage of equipment	electrical shock	radiation
explosion	cuts from equipment	escape of pathogens	waste disposal
chemicals in eyes	sharp objects	heavy lifting	inappropriate behaviour
inhalation of gas/dust	rotating equipment	slipping, tripping, falling	allergies
chemicals on skin	vibration and noise	falling objects	special needs
runaway reaction	pressure	heat and cold	other risks

Assessment by student(s)

I/we have assessed the risks associated with performing this experiment in the classroom on the basis of likelihood and consequences using the School's risk matrix, according to International Organization for Standardization Standard ISO 31000:2009 and the Risk Management Guidelines, HB 436:2013.

I/we consider the inherent level of risk (risk level without control measures) to be:

Low risk **Medium risk** High risk Extreme risk

Control measures:
Always point test tube away from any person.
Add hydrochloric acid slowly and carefully to avoid vigorous reaction and projection of material from test tube.
Dip matches and tapers in water to ensure extinguished before disposal.
Additional measures: safety glasses, gloves

With the specified control measures in place, I/we have found that all the risks are "low risk". Risks will therefore be managed by routine procedures in the classroom, in combination with the specified control measures.

Certification by teacher

I have assessed the risks associated with performing this experiment in the classroom on the basis of likelihood and consequences using the School's risk matrix, according to International Organization for Standardization Standard ISO 31000:2009 and the Risk Management Guidelines, HB 436:2013. I confirm that the risk level and control measures entered by student(s) above are correct and appropriate.

Name: ______ **Signature:** ______ **Date:** ______

Certification by Laboratory Technician

I have assessed the risks associated with preparing the equipment, chemicals and living organisms for this experiment and subsequently cleaning up after the experiment and disposing of wastes, on the basis of likelihood and consequences using the School's risk matrix, according to International Organization for Standardization Standard ISO 31000:2009 and the Risk Management Guidelines, HB 436:2013.

I consider the inherent level of risk (risk level without control measures) to be:

Generated with Student RiskAssess (https://www.riskassess.com.au)

Figure 1.23 An example of a risk assessment form

KEY CONCEPTS 1.4

- » A table records and organises data using horizontal rows and vertical columns.
- » A bar chart uses bars to represent total numbers (or summary statistics) of distinct categories.
- » A line graph displays a line that connects data points to show changes, trends and relationships in the data.
- » A normal distribution of data is relatively symmetrical and takes the shape of a bell.
- » A skewed distribution is a data distribution that is not symmetrical.
- » Descriptive statistics are single numerical values that summarise a data set. They include measures of central tendency (mean, mode, median) and variability (standard deviation).
- » Variability tells us how close or how far apart scores are.
- » A large standard deviation means there is high variability (scores are more spread out from each other) whereas a small standard deviation shows low variability (scores are closer together).
- » The true value is the value or range of values that would be found if the quantity could be measured perfectly.
- » Accuracy is the closeness of a measurement to the true value.
- » Precision is how close a set of measurement values are to one another (not necessarily to the true value).
- » Repeatability is the agreement of results carried out under the same conditions of measurement.
- » Reproducibility is the agreement of results carried out under different conditions of measurement.
- » Researchers should repeat experiments to ensure findings are robust.
- » Findings are generalisable if they can be applied to other people and/or contexts.
- » An investigation is internally valid if it investigates what it set out to investigate.
- » Lack of internal validity implies that the results of the study deviate from the truth and therefore no conclusions can be drawn.
- » An investigation is externally valid if the results of the research can be applied to similar individuals in a different setting.
- » Lack of external validity implies that the results of the research may not apply to individuals who are different from the study population.
- » Personal errors are mistakes, miscalculations and observer errors.
- » Random errors affect the precision of a measurement.
- » Systematic errors affect the accuracy of measurements.
- » The scientific method attempts to reduce uncertainty.
- » Outliers are data points that differ substantially from the rest of the collected data.
- » Contradictory data are incorrect data.
- » Incomplete data are missing data, such as questions without answers or variables without observations.
- » Larger sample sizes improve data quality.
- » The five ethical concepts in VCE Psychology are: beneficence, integrity, justice, non-maleficence and respect.
- » The six ethical guidelines that underpin psychological research are: confidentiality, debriefing, informed consent procedures, use of deception in research, voluntary participation and withdrawal rights.
- » When undertaking any research investigation, you must consider all relevant health and safety guidelines.
- » Use risk assessments and safety data sheets to identify and account for risks in experiments conducted in a laboratory.

1 Chapter summary

KEY CONCEPTS 1.1

- A research question defines the question that a research investigation tries to answer.
- An aim is a broad statement about the goal of a research investigation.
- A hypothesis is a proposed answer to a research question (made before the investigation is conducted). It is usually a directional statement about the relationship between the variables and states the expected results of a research investigation.
- A research investigation is used to test a hypothesis.
- Data are any information collected in scientific investigations.
- Primary data are data we collect ourselves from a study.
- Secondary data are data collected by someone else that we use when conducting a literature review.
- Quantitative data are numerical (recorded in the form of numbers).
- Qualitative data are non-numerical (descriptive).
- Sampling is the selection of participants from a population to participate in a research investigation.
- Convenience sampling means recruiting participants that are readily available.
- WEIRD samples (Western, educated, industrialised, rich and democratic) under-represent a large proportion of the population.
- Random sampling gives every member of the population an equal chance of being recruited.
- Stratified sampling ensures that the sample contains the same proportions of participants from each stratum as the proportions in the population of interest.
- Using larger samples can improve representativeness.

KEY CONCEPTS 1.2

- Each type of research design has specific benefits and limitations.
- A controlled experiment involves experimental manipulation of a variable to determine the effect on an outcome(s) of interest.
- A correlational study is a non-experimental investigation of the relationship between variables.
- Correlations can be visually represented by plotting two variables on a graph.
- Correlation does not equal causation.
- A positive correlation is where high scores for one variable occur with high scores for another variable.
- A negative correlation is where high scores for one variable occur with low scores for another variable.
- A case study is an investigation of one particular example of an activity, behaviour, event or problem, to acquire knowledge about the process as a whole.
- Classification and identification are processes used to organise phenomena into categories and identify examples of that categorisation.
- Fieldwork involves observing and interacting with an environment in the real world.
- Modelling means creating a representation of an event, process or system of concepts.
- Product, process or system development is the design or evaluation of a process, system or artifact to meet a human need.
- A literature review is a report produced by reading scientific research on a particular area and summarising it.

KEY CONCEPTS 1.3

- » Manipulation of a variable gives the controlled experiment the unique ability to find cause-and-effect relationships between variables.
- » The independent variable (IV) is the variable that the researcher manipulates.
- » The dependent variable (DV) is the variable that the researcher records to see if it has been affected by a change in the IV.
- » IVs and DVs are only relevant to controlled experiments.
- » When plotted on a graph, the IV goes on the horizontal axis and the DV goes on the vertical axis.
- » An extraneous variable (EV) may affect the DV and influence the results in an unwanted way.
- » A confounding variable does affect the DV and influences the results in an unwanted way.
- » A controlled variable is held constant in an investigation.
- » A controlled variable is not the same as a control group.
- » A between subjects design is a controlled experiment where participants are allocated to different groups (for example, the experimental group or the control group).
- » Between subjects designs have no order effects.
- » Random allocation of participants to groups helps to reduce confounding variable bias in between subjects designs.
- » A within subjects design is a controlled experiment where each participant is exposed to both the experimental and the control conditions.
- » In within subjects designs, individual differences between people do not influence the results.
- » Within subjects designs can suffer from the order effect.
- » Counterbalancing can overcome the order effect.
- » A mixed design combines the between subjects and within subjects design, with multiple groups and data recorded at multiple times.

KEY CONCEPTS 1.4

- » A table records and organises data using horizontal rows and vertical columns.
- » A bar chart uses bars to represent total numbers (or summary statistics) of distinct categories.
- » A line graph displays a line that connects data points to show changes, trends and relationships in the data.
- » A normal distribution of data is relatively symmetrical and takes the shape of a bell.
- » A skewed distribution is a data distribution that is not symmetrical.
- » Descriptive statistics are single numerical values that summarise a data set. They include measures of central tendency (mean, mode, median) and variability (standard deviation).
- » Variability tells us how close or far apart scores are.
- » A large standard deviation means there is high variability (scores are more spread out from each other) whereas a small standard deviation shows low variability (scores are closer together).
- » The true value is the value or range of values that would be found if the quantity could be measured perfectly.
- » Accuracy is the closeness of a measurement to the true value.
- » Precision is how close a set of measurement values are to one another (not necessarily to the true value).
- » Repeatability is the agreement of results carried out under the same conditions of measurement.
- » Reproducibility is the agreement of results carried out under different conditions of measurement.
- » Researchers should repeat experiments to ensure findings are robust.
- » Findings are generalisable if they can be applied to other people and/or contexts.
- » An investigation is internally valid if it investigates what it set out to investigate.
- » Lack of internal validity implies that the results of the study deviate from the truth and therefore no conclusions can be drawn.
- » An investigation is externally valid if the results of the research can be applied to similar individuals in a different setting.
- » Lack of external validity implies that the results of the research may not apply to individuals who are different from the study population.
- » Personal errors are mistakes, miscalculations and observer errors.
- » Random errors affect the precision of a measurement.
- » Systematic errors affect the accuracy of measurements.
- » The scientific method attempts to reduce uncertainty.

- » Outliers are data points that differ substantially from the rest of the collected data.
- » Contradictory data are incorrect data.
- » Incomplete data are missing data, such as questions without answers or variables without observations.
- » Larger sample sizes improve data quality.
- » The five ethical concepts in VCE Psychology are: beneficence, integrity, justice, non-maleficence and respect.
- » The six ethical guidelines that underpin psychological research are: confidentiality, debriefing, informed consent procedures, use of deception in research, voluntary participation and withdrawal rights.
- » When undertaking any research investigation, you must consider all relevant health and safety guidelines.
- » Use risk assessments and safety data sheets to identify and account for risks in experiments conducted in a laboratory.

Unit 3

How does experience affect behaviour and mental processes?

Area of study 1: How does the nervous system enable psychological functioning?

Area of study 2: How do people learn and remember?

Nervous system functioning 2

Key knowledge

- the roles of different subdivisions of the central and peripheral nervous systems in responding to, and processing and coordinating with, sensory stimuli received by the body to enable conscious and unconscious responses, including spinal reflexes
- the role of neurotransmitters in the transmission of neural information across a neural synapse to produce excitatory effects (as with glutamate) or inhibitory effects (as with gamma-amino butyric acid [GABA]) as compared to neuromodulators (such as dopamine and serotonin) that have a range of effects on brain activity
- synaptic plasticity – resulting from long-term potentiation and long-term depression, which together act to modify connections between neurons (sprouting, rerouting and pruning) – as the fundamental mechanism of memory formation that leads to learning

Key science skills

Analyse, evaluate and communicate scientific ideas

- use appropriate psychological terminology, representations and conventions, including standard abbreviations, graphing conventions and units of measurement
- discuss relevant psychological information, ideas, concepts, theories and models and the connections between them
- analyse and explain how models and theories are used to organise and understand observed phenomena and concepts related to psychology, identifying limitations of selected models/theories
- use clear, coherent and concise expression to communicate to specific audiences and for specific purposes in appropriate scientific genres, including scientific reports and posters
- acknowledge sources of information and assistance, and use standard scientific referencing conventions

Source: VCE Psychology Study Design (2023–2027) + pages 34, 13

2 Nervous system functioning

Our nervous system allows us to interact with the world around us, so it makes sense that when you study psychology you need to have a pretty good understanding of the structure of the nervous system and how it functions. Here goes!

p. 51

2.1
The nervous system: roles and subdivisions

Your nervous system is divided into two main parts – the central nervous system (CNS) and peripheral nervous system (PNS). These are divided up into smaller parts but what is important is that together, the CNS and PNS control everything that you do, think, feel and say.

p. 61

2.2
Conscious and unconscious responses

It would be really distracting if we were aware of all the things going on inside our body. We don't need to feel our heart valves opening and closing, we don't need to tell our muscles to work when we are running – these are all unconscious responses – they happen automatically, you don't have to think about them. There are other responses that we do need to be aware of though, such as putting food in your mouth or choosing the words to come out of your mouth. These are conscious responses.

p. 65

2.3
The transmission of neural information

Did you know your body uses electrochemical signals to communicate messages around your body? Our nerve cells, or neurons, form the basis of our nervous system. They carry electrical pulses within them, but use chemicals, or neurotransmitters, to pass these messages on to other neurons. Some neurotransmitters excite, and some inhibit the next neuron. It is a constant balancing act to regulate the flow of messages around the body. Neuromodulators, another type of chemical signals are also involved in communicating messages between neurons.

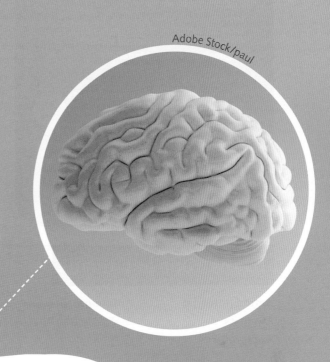
Adobe Stock/paul

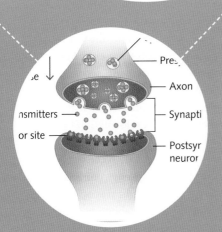

2.4
Neural basis of learning and memory
p. 73

Even though you are born with some innate abilities, most of your behaviour is learnt. Your brain is plastic – that is, it changes as you age and experience the world around you. This is called learning. When this learnt information is organised, stored and retrieved, this is called memory. If we don't keep using what we have learnt, then we will forget it. What is going on with our neurons to enable this?

Your nervous system is a very important system in making you, you! The complicated network of neurons work hard to receive, process and organise sensory information, so in turn you can learn and remember.

Slideshow
Chapter 2 slideshow

Flashcards
Chapter 2 flashcards

Test
Chapter 2 pre-test

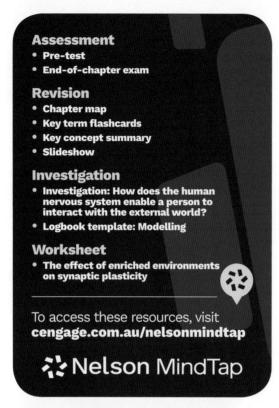

Assessment
- Pre-test
- End-of-chapter exam

Revision
- Chapter map
- Key term flashcards
- Key concept summary
- Slideshow

Investigation
- Investigation: How does the human nervous system enable a person to interact with the external world?
- Logbook template: Modelling

Worksheet
- The effect of enriched environments on synaptic plasticity

To access these resources, visit
cengage.com.au/nelsonmindtap

Nelson MindTap

Know your key terms

Action potential
Agonists
Antagonists
Autonomic nervous system (ANS)
Axon
Axon terminals
Brain
Central nervous system (CNS)
Conscious
Dendrites
Dopamine
Enteric nervous system (ENS)
Excitatory neurotransmitter
Gamma-amino butyric acid (GABA)
Glutamate
Inhibitory neurotransmitter
Long-term depression (LTD)
Long-term potentiation (LTP)
Motor neurons
Nervous system
Neuromodulator
Neurotransmitter
Parasympathetic nervous system
Parkinson's Disease (PD)
Peripheral nervous system (PNS)
Postsynaptic neuron
Presynaptic neuron
Receptor cells
Receptor sites
Rerouting
Reward system
Sensory neurons
Serotonin
Soma
Somatic nervous system (SNS)
Spinal cord
Spinal reflex
Sprouting
Sympathetic nervous system
Synapse
Synaptic cleft
Synaptic plasticity
Synaptic pruning
Synaptic vesicles
Unconscious

The human **nervous system** is an integrated network of neurons, nerves, and associated organs and tissues, including the **brain**, that together coordinate a person's functioning, as they interact with and adapt to the external environment. The neurons are the basic building blocks of the nervous system because they form an interconnected network so that messages can be conveyed rapidly and continuously throughout the nervous system. On their own, neurons are inadequate, but when they form vast interconnected networks, they enable our mental processes and behaviours.

For individuals to function at full capacity and in a coordinated way, the nervous system must fulfil three major functions: first, it must receive sensory information from either our external or internal environment; second, it must process and coordinate this information so that we understand it; third, it must then organise a response to this information by our muscles, organs and glands (Figure 2.1).

The nervous system initially receives information from the external environment in the form of raw energy, such as light waves (enabling vision) and/or soundwaves (enabling hearing).

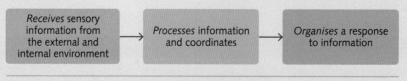

Figure 2.1 Functions of the human nervous system

This sensory energy is converted into electrochemical energy, a single energy form that the nervous system is equipped to process. Messages are then sent via the nervous system, from neuron to neuron in the form of electrochemical messages.

For example, if your phone rings, your nervous system receives this sensory auditory (sound) information from the external environment. It processes and coordinates this information – my phone is ringing and I would like to answer it. On recognising the phone is ringing and making the decision to answer the call, a response is made to reach out and pick up the phone (Figure 2.2).

Figure 2.2 The ability for you to answer your phone when it rings is due to the three main functions of our nervous system: the ability to receive, to process and to respond to sensory stimuli.

2.1 The nervous system: roles and subdivisions

There are two major divisions of the nervous system: the central nervous system (CNS) and the peripheral nervous system (PNS). These divisions contain different physiological structures and components; however, they are interconnected and function together to determine our psychological and physical responses. The CNS and PNS are further divided into structural components and subdivisions (Figures 2.3 and 2.4).

The central nervous system (CNS)

The **central nervous system (CNS)** is composed of the brain and the spinal cord (Figure 2.5). The brain is the 'engine room' of the nervous system. It receives, processes and integrates information from the rest of the body and generates responses to it. The **spinal cord** is part of the CNS consisting of a cable of sensory and motor nerve fibres that extend from the brainstem through a canal in the centre of the spine to the lumbar region of the spine; it receives and transmits sensory information from the **peripheral nervous system (PNS)** to the brain and motor messages from the brain to the PNS.

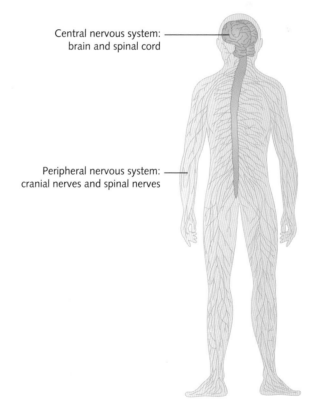

Figure 2.3 The two major divisions of the nervous system and their associated structural components

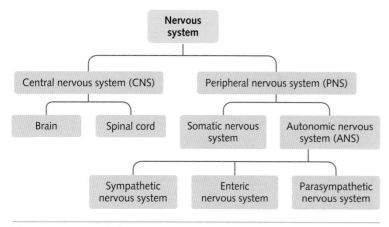

Figure 2.4 The subdivisions of the nervous system

and how we construct our sense of who we are. It enables us to make decisions, solve problems, fall in love or read a book. Thus, everything you do, think or feel can be traced back to these neurons.

Since the brain is the body's master information-processing and decision-making organ, it has an important role in communicating information to the rest of the body. The brain does this by receiving, processing and interpreting information from the body's sensory systems, which it integrates and forms a response to, then sends motor messages out to all parts of the body so that coordinated and appropriate responses can be made. As a result, the brain is responsible for coordinating all the activities of the nervous system. Your brain communicates with the rest of your body through a large cable-like structure called the spinal cord.

The spinal cord: the body's information highway

The spinal cord is made up of *tracts* or long bundles of nerves. There are two types of tracts involved in the transmission of information up and down the spinal cord: *ascending tracts* and *descending tracts*. On the outside of the spinal cord, spinal nerves branch off between the gaps in the vertebrae into the arms, legs and torso (Figure 2.6). These spinal nerves are part of the peripheral nervous system (PNS) and carry messages to and from the spinal cord, connecting the brain with the PNS.

This complex structure allows the spinal cord to fulfil two key roles in communicating neural information.

WB
2.1.1 NERVOUS SYSTEM 'WHAT AM I?'

EXAM TIP
The spine and the spinal cord are different structures and the terms cannot be used interchangeably.

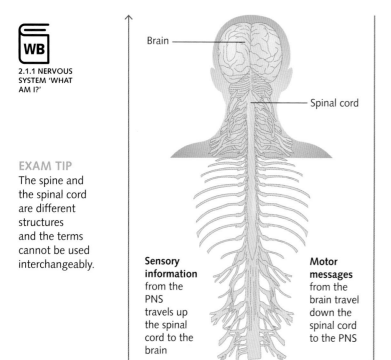

Figure 2.5 The CNS – sensory information travels up the spinal cord to the brain; motor messages travel down to the muscles

The brain: master and commander

The brain is about the size of a large grapefruit and weighs approximately 1.5 kilograms. This complex structure consists of some 100 billion neurons and an almost infinite number of synaptic connections, and contains more than 90 per cent of the CNS's neurons. The mass of neurons we call the human brain is responsible for how we think and feel, how we perceive and react to the world around us,

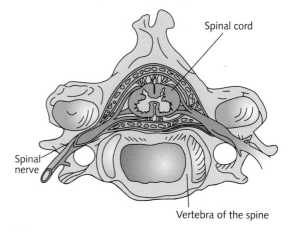

Figure 2.6 A spinal cord section (top view). The spinal column (or backbone) encloses the spinal cord to protect it and consists of 33 vertebrae, each of which has a cavity through which the spinal cord passes. From each vertebra, spinal nerves emerge.

Firstly, the spinal cord receives sensory information from the PNS and transmits it in a chain along the **ascending tracts** of the spinal cord up to the brain for further processing and interpretation. Secondly, it carries motor information from the brain back to the PNS in a chain along the **descending tracts** of the spinal cord where it is relayed to our muscles, organs and glands so they can react in specific ways. This happens so rapidly and continuously that we are barely aware of it. For example, you would not be aware that your shoelaces were tied too tight unless the sensory information about your foot hurting was transmitted to your brain along the ascending tracts of the spinal cord. If you make the decision to bend down to loosen your shoelaces, then the motor instructions would be transmitted from your brain to the PNS along the descending tracts of the spinal cord (Figure 2.7).

The spinal cord plays a crucial role in behaviour. If we did not have a spinal cord to transport sensory information to and motor messages from our brain, we would not experience any bodily sensations and we would not be able to make voluntary movements. We wouldn't feel a feather tickling our toes. We wouldn't be able to pick up a fork to feed ourselves. If neurons in the spinal cord are damaged by infection or injury, the flow of information between the brain and the rest of the body is interrupted. This may result in a loss of sensation in a specific body part or an inability to move a specific body part.

Figure 2.7 The response of loosening your shoelaces is due to the spinal cord transmitting sensory information received from the PNS up ascending tracts to the brain and transmitting motor messages from the brain down descending tracts to the PNS.

🔑 KEY CONCEPTS 2.1a

- » The central nervous system (CNS) is composed of the brain and the spinal cord.
- » The brain is the 'engine room' of the nervous system. It receives and processes information from the rest of the body and generates responses to it.
- » The spinal cord, an intricate and delicate cable of nerve fibres stretching from the base of the brain to the lower back, connects the brain to the rest of the body via its connection to the peripheral nervous system (PNS).
- » The spinal cord receives sensory information from the PNS and transmits it along the **ascending tracts** up to the brain for further processing and interpretation. Motor information from the brain is carried along **descending tracts** where it is relayed to our muscles, organs and glands so they can react.

Concept questions 2.1a

Remembering
1 What are the two main divisions of the nervous system? **r**

Understanding
2 With reference to an example, identify and describe the three main functions of the nervous system. **e**
3 Explain why the brain is referred to as 'the engine room' of the nervous system. **e**
4 Explain the difference between the ascending tracts and the descending tracts of the spinal cord. **e**

Applying
5 How does the function of the spinal cord differ from that of the brain? **e**

HOT Challenge
6 During a diving accident, Sami severed his spinal cord at the neck. With reference to the roles of the central nervous system, describe the likely outcomes of this type of injury. **c**

The peripheral nervous system (PNS)

Many of our body parts are located outside the CNS. Therefore, the CNS needs a support network of information-carrying nerves to connect the various body parts to it. This network is provided by the peripheral nervous system (PNS). The PNS consists of all the nerves (apart from those within the CNS) that extend from the brain and spinal cord throughout the body. These nerves connect with muscles, organs and glands that are located throughout the body (spinal nerves), as well as nerves connected mainly to organs located in the head and neck (cranial nerves). The main role of the PNS is to convey sensory information from the body's internal and external environments to the CNS and transmit motor commands from the CNS to the rest of the body via its connection to the spinal cord.

An example of the way the PNS and CNS work together is when a musician is tuning their guitar and must listen carefully to the pitch of each string. Auditory information about the sound is transmitted from sensory receptors in the ear via the PNS to the brain, which interprets the pitch. A response (to tighten or loosen the string) is then sent from the brain to the spinal cord and on to the PNS, so that the muscles of the hands and arm can make the action (Figure 2.8). Therefore, the CNS and PNS work in an integrated way to enable us to produce the variety of mental processes and behaviours that we execute daily.

The PNS is composed of two subdivisions: the somatic nervous system and the autonomic nervous system (ANS).

> **EXAM TIP**
> Sensory and motor neurons do not 'travel', they are static/stationary: the neural message passes along them to the next neuron.

The somatic nervous system (SNS)

The **somatic nervous system (SNS)** is the subdivision of the PNS that is responsible for sensing stimuli and controlling voluntary responses. Therefore, like all parts of the nervous system, it has both a sensory role and a motor role. Its sensory role consists of **sensory neurons** receiving sensory information from **receptor cells** located throughout the body such as in the skin, joints, eyes and tongue, and transmitting this information inwards to the CNS. Experiencing sensations such as touch, pain, heat, cold, balance, sight, taste, smell and sound are all part of the sensory function of the somatic nervous system. The somatic nervous system also controls the skeletal muscles attached to bones, enabling voluntary movements. To fulfil this motor role, the **motor neurons** receive commands from the CNS and then the information is transferred to the skeletal muscles located throughout the body to enable appropriate and coordinated responses. The somatic nervous system is responsible for all our voluntary actions such as walking, talking, eating and playing sports.

An example of the way the sensory and motor function of your somatic nervous system operates is if your pet cat rubs up against your leg (Figure 2.9). Sensory receptors in your skin register the touch sensation and that information is transmitted by sensory neurons in the somatic nervous system up the spinal cord and on to the brain. The brain then identifies your cat as the source of the sensation and you decide to pick up the cat. The brain then sends messages via motor neurons down the spinal cord to muscles in your legs, arms and hands to enable the coordinated movements required to lean down and pick up your cat.

Figure 2.8 When tuning their guitar, this musician's CNS and PNS are working together to interpret the stimulus and direct the body's muscles in response.

Figure 2.9 The ability to detect the sensation of your cat rubbing up against your leg and making the decision to pick it up is due to the sensory and motor functions of your somatic nervous system.

The autonomic nervous system (ANS)

Imagine that you are strolling along the street and you are about to cross the road. Your somatic nervous system is controlling the many muscles needed to execute this activity. You are in conscious control of these actions because they are voluntary. Your heart muscle, however, is continually receiving messages to 'beat', even though you are not consciously aware of it. It will fluctuate to meet demand. When walking along the street your heart may need to work harder than when resting. Suddenly, a large truck screeches around the corner, catching you by surprise and only just missing you. As a result your heart rate instantly increases. You do not consciously decide to react like this; it is not a voluntary decision. This happens automatically and quickly to aid survival. After this surprising event your heart rate will return to normal. This behaviour is due to your **autonomic nervous system (ANS)**.

The ANS is the division of the PNS that controls the activity level of our internal organs and glands, such as our heart, pupils, salivary glands, adrenal glands, stomach and bladder, all of which are essential for survival. It contains nerves that transmit motor messages from the brain to the body's involuntary (or smooth) muscles that control the activity level of these internal organs and glands. It also transmits messages back to the brain about the activity level of these organs and glands.

The word 'autonomic' means 'self-regulating'; consistent with its name, the ANS functions automatically. The activities controlled by the ANS are involuntary and operate continuously. Imagine if you had to consciously think about controlling the activities of your ANS such as keeping your heart beating, controlling your breathing or even ensuring that your internal organs are functioning. By automatically regulating the activity level of survival-related internal systems, the ANS leaves our mind free to focus on other external or internal stimuli. For example, can you imagine what might happen if you were sitting your final Psychology exam and, at the same time, you also had to concentrate on controlling your heartbeat and remembering to breathe? Because our ANS functions independently and constantly during the exam, we do not need to focus on controlling our vital organs and body systems, leaving us free to consciously focus on other important external and internal stimuli required to complete the exam effectively (Figure 2.10).

2.1.2 NERVOUS SYSTEM CONCEPT MAP

Figure 2.10 This student does not have to control the functions of her ANS, leaving her free to consciously focus on her exam.

The activity of the ANS ensures that our changing energy requirements are met and our body is maintained in proper working order according to the demands placed on it. The messages relayed by the autonomic nervous system automatically increase, decrease or otherwise regulate the activity of our internal organs and glands. The ANS energises us so we are aroused when we need to be, and so that we can physically deal with a variety of situations whether they be vigorous, demanding, threatening or stressful. For example, if you are woken by someone shouting 'There's an intruder in the house!', sensory neurons in your ears would send this message to your brain. Your brain would interpret this as a threat and would send an emergency command to the ANS. The ANS can increase arousal levels by activating the involuntary muscles that control our internal organs and glands. As a result our body's internal resources can be used to provide the extra energy that is required to deal with the 'threatening' situation. The ANS also has the essential role of lowering our arousal level and calming us when less energetic behaviour is required. The three divisions of the ANS that perform these functions are the **sympathetic nervous system**, the **parasympathetic nervous system** and the **enteric nervous system**.

KEY CONCEPTS 2.1b

» The PNS consists of all the nerves outside the CNS; that is, all the nerves that extend throughout the rest of the body.
» The main role of the PNS is to convey sensory information from the body's internal and external environments to the CNS and transmit motor commands from the CNS to the rest of the body.
» The PNS comprises the somatic nervous system (SNS), which controls voluntary responses, and the autonomic nervous system (ANS), which controls the activity level of our internal organs and glands.

Concept questions 2.1b

Remembering

1. What are the two branches of the Peripheral Nervous system?

Understanding

2. Explain the function of sensory and motor neurons.
3. Explain the difference between the autonomic nervous system and the somatic nervous system.

Applying

4. With reference to your own example, explain how the somatic nervous system operates in terms of its sensory and motor functions.

HOT Challenge

5. Identify which division of the PNS (somatic or autonomic) is at work during each of the following actions.
   a. It starts raining and you put up your umbrella.
   b. The sun comes out and your pupils constrict.
   c. Your heart rate increases when you go for a jog.
   d. You roll up your sleeves when it gets hot.
   e. Your tummy grumbles at the smell of food.

2.1.3 EVALUATION OF RESEARCH

The sympathetic nervous system: arousing effect

The **sympathetic nervous system** is the branch of the ANS that dominates when we experience heightened emotions, during times of vigorous physical activity, or when we perceive a threat. When the sympathetic nervous system is activated, it enables the use of the body's internal resources to provide the extra energy that is required for increased physical activity or to deal with intense emotions (such as fear) or stressful and threatening stimuli and situations. Just as a car's accelerator pumps more petrol through the engine to provide energy for the car to move faster, the sympathetic nervous system changes the activity level of our internal systems, so we have a sudden increase in our energy levels when needed.

When we engage in physical activity such as playing sport or if we experience intense fear when watching a scary movie, our sympathetic nervous system is dominant. It prepares our body for action by increasing the activity level of some bodily systems while slowing down others (Figure 2.11 and Table 2.1).

For example, imagine if you were walking in the bush and suddenly you saw a snake moving fast towards you (Figure 2.12). This perceived threat would automatically activate our sympathetic nervous system preparing our body for emergency action. Initially a message would be sent to the adrenal glands to stimulate the release of the stress hormones adrenaline and noradrenaline into our bloodstream. Our heart rate and blood pressure would then increase,

Figure 2.11 For these people, their sympathetic nervous system would be dominant when watching a scary movie.

Figure 2.12 If you perceived this snake coming towards you as a threat, then your sympathetic nervous system would automatically be activated.

causing the blood carrying the adrenal hormones to move rapidly around our body. As a result, a range of other bodily systems are activated in order to deal with the threat of the snake. Those bodily systems not essentials for immediate survival like salivation and digestion are inhibited. Extra oxygen would be available because of the increase in our heart rate and breathing rate, and this extra oxygen would also help to convert the fats and sugars that have been released into energy. Our skeletal muscles would benefit from this increase in energy-charged blood, enabling them to move at a fast rate to respond to the snake. Our pupils would dilate to allow extra light to enter the eye so that we can see our surrounding environment and any other perceived obstacles. Our sweat glands would also produce more perspiration and our mouth would feel quite dry because the activity level of our salivary glands has slowed. We would also be unaware that our bladder has relaxed and that our digestion has slowed as our stomach decreases its contraction rate.

Therefore, the sympathetic nervous system energises us and physically prepares us for emergency action.

Sometimes the sympathetic nervous system can stay 'switched on' even once a perceived stressor or threat has passed. When this happens, stress is experienced, a physiological and psychological response (Chapter 3). If a person remains stressed for a prolonged period, there is danger of developing a range of physiological and psychological disorders that could negatively affect their health and wellbeing.

The parasympathetic nervous system: calming effect

The **parasympathetic nervous system** has the opposite effect of the sympathetic nervous system: it maintains a level of homeostasis. It keeps our internal systems in a balanced and healthy state by maintaining vital unconscious functions such as heart rate, breathing rate, blood pressure and digestion at their regular level of operation.

Once the need for high arousal is over and it is no longer necessary for the sympathetic nervous system to dominate, the parasympathetic nervous system is activated and a normal level of arousal is experienced. For example, heart and breathing rates would decrease and return to their resting level of functioning, our pupils would constrict, our sweat glands would inhibit the amount of perspiration that is produced and the activity level of our salivary glands would return to normal so our mouth no longer feels dry. Likewise, the functioning of other internal organs including the liver, stomach and bladder would all return to their normal levels of functioning (Figure 2.13 and Table 2.1). In this way, the parasympathetic nervous system acts like our car's brakes. We apply the brakes when we want our car to slow down, and the parasympathetic nervous system usually dominates when we need to be less energised or active.

The sympathetic nervous system and parasympathetic nervous system not only differ in their effects on arousal levels and internal bodily systems, but also in their response rate.

Increased levels of adrenaline and other hormones (that arouse and prepare the body for sudden increases in activity) instantly enter our bloodstream once the sympathetic nervous system has been activated (Figure 2.14). However, these hormones take some time to disappear, which is why the effects of the parasympathetic nervous system take longer to occur. The effects of the sympathetic nervous system and high levels of arousal usually do not fade for 20 or 30 minutes after the need for extra energy has passed.

Both the sympathetic and parasympathetic nervous systems directly influence the activity level of our internal organs, glands and non-skeletal muscles.

The enteric nervous system: gut control

Described as a mesh of sensory and motor neurons lining the wall of the digestive organs (Breedlove, 2014), the **enteric nervous system (ENS)** is the third and largest part of the autonomic nervous system. It receives and sends messages to the sympathetic and parasympathetic nervous system and is responsible for controlling the many functions of the digestive system. Due to its connection to the peripheral nervous system, it can communicate directly with the CNS. It can also function independently of the CNS and so is sometimes called the 'second brain'.

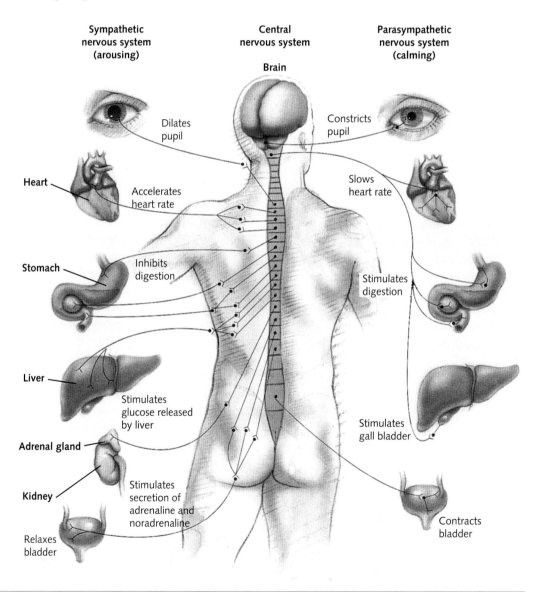

Figure 2.13 The sympathetic and parasympathetic divisions of the ANS have different effects on organs to arouse and increase energy or calm and conserve energy.

Table 2.1 Physiological changes of internal organs and glands due to the activation of the sympathetic and parasympathetic nervous systems

Internal organs and glands	Sympathetic response	Parasympathetic response
Adrenal glands attached to kidneys	Releases hormones (adrenaline and noradrenaline) to stimulate or repress activity of internal organs	Decreases hormone activity
Heart	Increases heart rate and pumps blood faster	Decreases heart rate to normal functioning
Lungs	Relaxes the airways	Constricts airways
Liver	Produces and releases increased level of sugar	Returns to normal functioning
Stomach	Decreases contractions and slows digestion	Stimulates digestion
Salivary glands	Decreases production of saliva	Normal production of saliva
Sweat glands	Increases perspiration	Inhibits perspiration
Gall bladder	Inhibits production of bile	Stimulates gall bladder to produce bile
Eyes	Dilates pupils to take in more light	Constricts pupils; gives near vision
Bladder	Slows contraction rate and relaxes	Contracts, returning to normal functioning
Sex organs	Stimulated	Unstimulated

Figure 2.14 For these athletes, their sympathetic nervous system is dominant while they are running in the race, and their parasympathetic nervous system dominates once the race has finished.

There are extensive two-way neural connections between the ENS and central nervous system (CNS), and this forms the gut–brain axis (discussed further in Chapter 3).

Some functions of the ENS are to control the movement of food through the digestive system as well as the mixing of food with digestive enzymes, and the regulation of the secretions of digestive enzymes and other chemicals associated with digestion

The enteric nervous system controls the actions of the gastrointestinal tract without conscious control of the CNS. For example, the ENS receives sensory information about the acidity of the stomach and is able to control acidity levels. When food arrives at the small intestine, digestive enzymes are secreted from the pancreas directly into the small intestine; this is done without any conscious control.

In some instances, information about the gastrointestinal tract does reach our conscious awareness, such as feeling hungry or full after a meal or experiencing gut pain. Our brain is then able to send messages to our somatic nervous system and either start looking for food or eat if hungry and stop eating if full (Furness, 2007).

INVESTIGATION 2.1 HOW DOES THE HUMAN NERVOUS SYSTEM ENABLE A PERSON TO INTERACT WITH THE EXTERNAL WORLD?

Scientific investigation methodology
Modelling

Aim
To create a model that demonstrates how the functioning of the human nervous system enables a person to interact with the external world.

Logbook template
Modelling

Introduction

Investigate and create a physical or conceptual model (with labels and annotations) that demonstrates how the functioning of the human nervous system enables a person to interact with the external world. Students could have a choice of (or may be allocated) one of the following topics from the Chapter 2 Key Knowledge on which to create their model:

- **subdivisions of the central and peripheral nervous systems**
- **the sympathetic versus the parasympathetic nervous systems**
- **the enteric nervous system**
- **spinal reflexes**
- **the structure of a neuron**
- **the transmission of neural information**
- **excitatory and inhibitory effects of neurotransmitters**
- **synaptic plasticity**
- **long-term potentiation and long-term depression.**

Results

Present your model to the class, explaining how it represents the part of the system that you are modelling.

KEY CONCEPTS 2.1c

» The ANS is divided into the sympathetic nervous system, the parasympathetic nervous system and the enteric nervous system.
» The sympathetic nervous system stimulates arousal and activity levels.
» The parasympathetic nervous system calms and lowers arousal levels.
» The enteric nervous system controls the activity of the digestive system.

Concept questions 2.1c

Remembering

1 Name the three divisions of the autonomic nervous system and describe the role of each. **r**

Understanding

2 Describe the effects that can occur when our sympathetic nervous system is activated for a prolonged period. **e**

3 Explain why the effects of the sympathetic nervous system last longer than those of the parasympathetic nervous system. **e**

4 What functions are under the control of the ENS? **r**

Applying

5 While playing in her football grand final, Isabel is likely to be experiencing a high level of arousal. Explain three physiological responses that she might experience and how they will assist her in playing in the grand final. **c**

6 Once the football grand final has finished, Isabel's physiological responses take some time to return to normal. Explain why this occurs and identify the division of the nervous system responsible for this. **c**

HOT Challenge

7 Identify the division of the autonomic nervous system that is most at work during the following situations. **c**

a Your breathing rate slows after doing 20 star jumps.
b Your pupils dilate when you become scared.
c Your pupils return to normal size when you are no longer scared.
d You sweat when it gets hot.
e Your heart rate elevates as you enter your Psychology exam.
f Your stomach digests your lunch.

8 Imagine you walk past a take-away food shop and can smell the fried chips. Outline the interaction between the ENS and the CNS in this scenario. **e**

2.2 Conscious and unconscious responses

There are two types of responses to stimuli – these include conscious and unconscious responses.

Conscious responses

Conscious responses are those that occur in response to sensory stimuli that involves awareness. As human beings we are moving all the time, whether we are consciously aware of this or not. For example, we blink, talk and walk, our muscles contract and twitch, our balance and body position shifts and changes. Most of our responses to sensory stimuli are voluntary and occur as a result of communication between the brain and the somatic nervous system. We are aware of stimuli and make a conscious decision about what our response will be. For example, if we are cooking soup and we decide to do a 'taste test', our brain makes this decision and sends messages via the spinal cord to the motor neurons in our somatic nervous system to muscles in our arm and hand, which enables us to pick up a spoon and taste the soup. Sensory neurons in our somatic nervous system send messages about the taste of the soup to the brain. We might decide that the soup requires more salt, so the brain then sends a message back via the spinal cord to the motor neurons to enable a response of picking up the salt and adding more to the soup (Figure 2.15).

Figure 2.15 The activity of 'taste testing' while cooking and adding more ingredients as needed is a conscious response that requires the brain to direct the movements of the somatic nervous system.

This voluntary response requires conscious thought and therefore the interaction of the somatic nervous system with the brain. In this instance, the brain guides the actions of the somatic nervous system.

Sensory information is required for us to plan and execute movements. We are constantly exposed to all types of sensory information such as light, soundwaves and pressure. Sensory receptors in our sense organs are the first point of contact for receiving this information. Once the sensory information is transmitted via the spinal cord to the brain, specialised areas of the brain are then responsible for processing the information. Once the brain has planned the execution of a response, the information is sent back via the spinal cord to the somatic nervous system where motor neurons controlling the relevant muscles can implement the response. As you can see, making a conscious and planned response to sensory stimuli is a complex process that involves a range of nervous system divisions (Carter et al., 2014).

Unconscious responses

As we saw earlier in the chapter, the ANS is responsible for controlling involuntary muscles that regulate the activity level of our internal organs and glands, such as heart rate and perspiration. Sensory stimuli is received by the brain and **unconscious** responses are controlled by the ANS. In this case we are not consciously aware of the sensory stimulus nor the responses that are made. The activities controlled by the ANS are mostly self-regulating and generally operate constantly, and not normally under our conscious control, whether we are awake or asleep. An independent and constantly functioning ANS means that the vital organs and body systems we depend on for physical survival are kept functioning more or less automatically and without any conscious effort.

2.2.1 CONSCIOUS AND UNCONSCIOUS RESPONSES OF THE NERVOUS SYSTEM

Weblink
2-minute neuroscience: withdrawal reflex

2-minute neuroscience: knee-jerk reflex

EXAM TIP
The spinal reflex is controlled by the central nervous system and the somatic nervous system. Don't make the mistake of thinking that the spinal reflex is automatic so is under the control of the autonomic nervous system.

Unconscious responses: the spinal reflex

As discussed earlier in the chapter, the somatic nervous system (a division of the PNS), via its communication with the brain, is responsible for all voluntary responses. However, not all responses performed by the somatic nervous system are conscious voluntary responses involving the brain. Some simple responses that are essential for survival occur automatically by the somatic nervous system and the spinal cord, and therefore operate independently from the brain.

For example, if while walking barefoot, your foot lands on a nail, you lift your up foot instantly. Such a rapid response occurs independently of the brain. A response like this is deemed essential for survival and involves communication only between neurons in the spinal cord and somatic nervous system; as a result, the response can occur quickly in order to minimise harm. These simple involuntary and unconscious responses to stimuli are known as **spinal reflexes**. Spinal reflexes are 'hardwired' into our nervous system and aid our survival because they allow us to make rapid responses to stimuli that may be potentially dangerous, without us having to think about it (that is, they are unconscious responses).

The simplest stimulus–response behaviour pattern, the spinal reflex occurs in the spinal cord, independent of the brain, and generally involves three types of specialised neurons: sensory neurons, interneurons and motor neurons (discussed in more detail later in the chapter).

An example of the spinal reflex is the withdrawal reflex that occurs when we experience an intense sensation, such as the example of standing on a nail (Figure 2.16). Initially, an intense sensation is detected in your foot by a sensory receptor. Instantly, the sensory receptor fires off a message via the sensory neurons to your spinal cord and activates an interneuron, which in turn transmits a message via the motor neurons, causing your leg muscles to contract and allowing you to withdraw your foot from the sharp nail.

Sometimes a spinal reflex involves only two neurons, for example the patella 'knee-jerk' reflex or stretch reflex (Figure 2.17). This reflex is often used by doctors to test spinal nerve function. It occurs when the patella tendon, located under the kneecap, is struck. The patella nerves are directly connected to the spinal cord, so when pressure is applied to the patella tendon, its sensory nerve receptors fire, carrying information along the sensory neuron to the motor neurons in the spinal cord, which in turn activates the thigh muscle. This causes the lower half of the leg to automatically kick out. Only two neurons are activated and as a result, the response is rapid (Barnes, 2013; Carter et al., 2014).

In both examples, your body initiates a rapid and instant spinal reflex, reacting automatically to protect itself, independently of the brain (Barnes, 2013).

By bypassing the brain, spinal reflexes help us automatically adapt to changing conditions, which in turn assists our survival. This automatic response system allows quicker reaction times when confronted with sudden hazards. However, even a simple spinal reflex usually triggers more complex activity. The spinal cord will normally send secondary signals to the brain during the reflex action so that the brain becomes aware of the actions it has taken. As your foot pulls away from the nail you have stepped on, you will feel the pain and think, 'Ouch! What was that?' Not all reflexes involve the activation of a muscle.

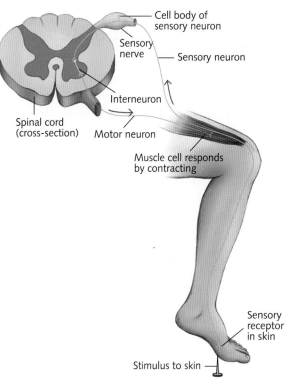

Figure 2.16 An example of a spinal reflex is the withdrawal reflex experienced when we step on something sharp.

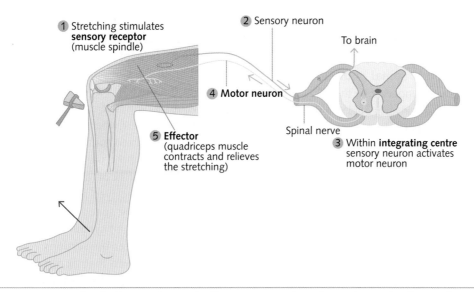

Figure 2.17 An example of a knee-jerk spinal reflex occurring when the tendon under the patella is struck

ACTIVITY 2.1 CONSCIOUS AND UNCONSCIOUS RESPONSES

Try it

1. Close your eyes and reach into your pencil case or pocket (wherever you keep pens!). Without looking, find a highlighter or a special pen.

   Even with your eyes closed you could probably find the item you were aiming for. Here is what was going on.

Sensory receptors in your fingers detected a stimulus. The sensory messages were sent via the somatic division of the peripheral nervous system to the spinal cord. The message, now in the central nervous system, was sent to the brain for processing. If the brain recognised the information as belonging to the item you were aiming for then a motor message would be sent from the brain (central nervous system) via the spinal cord (central nervous system) to the muscles in the fingers to contract and pick up the item. This is the motor role of the somatic division of the peripheral nervous system. If the brain did not recognise the sensory information as belonging to the item you were aiming for, then the message sent from the central nervous system to the somatic nervous system would be to release the muscles to let go of the item. This is an example of a conscious response.

Apply it

2. If, when you reached into your pencil case with your eyes closed, your finger accidentally touched the sharp end of your compass, your hand would automatically pull away from the compass. Construct a flow chart that describes the sequence of nervous system responses for this unconscious response known as a spinal reflex.

Exam ready

3. Sasha is going for her nightly run when she feels that she has done the laces on her sneaker up too tightly. She bends down to adjust her laces. Which of the following is correct?

   A. Sasha's autonomic nervous system is responsible for her sensing that her shoe laces were too tight.

   B. Sasha's central nervous system initiated a spinal reflex for her to adjust her shoelaces.

   C. Sasha's somatic nervous system was responsible for the voluntary control of the muscles in her hands to adjust her shoelaces.

   D. Sasha's somatic nervous system was responsible for the involuntary control of the muscles in her hands to adjust her shoelaces.

ANALYSING RESEARCH 2.1

Reflex control could improve walking after incomplete spinal injuries

A research study supported by the US National Institutes of Health investigated whether people with incomplete spinal injuries can control hyperactive reflexes they experience by using a training technique that suppresses these reflexes. While some injuries of the spinal cord involve complete paralysis below the point of damage, other common injuries of the spinal cord are considered 'incomplete injuries' as they involve less disability, with some feeling or movement still evident below the point of damage. Incomplete injury of the spinal cord can result in classic symptoms that include the exaggeration of some spinal reflexes leading to muscle stiffness, an unusual pattern of movements and a reduced ability to walk.

The study by Aiko Thompson and Jonathan Wolpaw is the first to investigate if people are able to consciously control their 'knee-jerk' reflex that can become exaggerated during incomplete spinal injuries. They investigated this by using an H-reflex, a response caused by electrical stimulation of the nerves behind the knee, which is similar to the automatic tendon stretch that occurs during a 'knee-jerk' reflex. The study involved 13 participants with the classic symptoms of an incomplete spinal injury. The participant injuries had occurred somewhere in the 8 months to 50 years before the study.

Initially, baseline data of the reflex response was taken over a 2-week period, while the participant received an electrical stimulation to the calf muscle of their weaker leg. Over the next 10 weeks, nine participants underwent training that encouraged them to suppress their reflex when it was stimulated. The training occurred for three sessions per week and involved the participants viewing the size of their reflex on a monitor. The other four participants received no training. The participants' walking speed over a 10-metre distance and their gait symmetry was measured both before and after the sessions.

The result showed that six out of the nine participants who received the training were able to suppress their reflexes. Their walking speed increased by 59 per cent on average and their gait became more symmetrical. Three of the participants who received training and the four participants who received no training were unable to suppress their reflexes and showed no improvements in walking speed or symmetry.

While this was a small study, the results are encouraging and offer hope that patients with incomplete spinal cord injuries may be able to improve their walking speed and symmetry by controlling reflexes that become overexaggerated.

Source: adapted from NIH/National Institute of Neurological Disorders and Stroke. (2013, February 5). Reflex control could improve walking after incomplete spinal injuries. *ScienceDaily*.

Questions

1. Explain the difference between a complete and an incomplete spinal injury.
2. What was the aim of this research?
3. Identify the independent variable(s) (IV) and the dependent variable(s) (DV) of the research.
4. What is the purpose of the baseline data?
5. Was there a control group in this experiment? Explain your answer.
6. Explain the results that were obtained.

HOT Challenge

7. Identify one limitation associated with this study.

KEY CONCEPTS 2.2

» The somatic nervous system is responsible for all voluntary responses through communication with the brain.
» Sensory information is sent via the spinal cord to the brain which determines a response and then sends this message via the spinal cord back to the somatic nervous system and the motor neurons to enable a response.
» Not all responses performed by the somatic nervous system are conscious, voluntary responses involving the brain.
» Spinal reflexes occur automatically and only involve the spinal cord and the somatic nervous system. They are quick responses that are essential for survival.
» Spinal reflexes usually involve an interneuron making a connection between a sensory neuron and a motor neuron.
» The simplest spinal reflex involves the connection between a sensory neuron and a motor neuron.

Concept questions 2.2

Remembering
1 Which divisions of the nervous system are responsible when you withdraw your hand from a hot pan? **r**

Understanding
2 Explain the purpose of a spinal reflex. **e**
3 Explain why the activities of the autonomic nervous system are mostly self-regulating. **e**

Applying
4 Name another reflex that does not involve the activation of a muscle by a motor neuron. **r**
5 When a baby is born, a number of tests are carried out immediately. One such test is the grasping reflex, whereby the nurse will place their finger in the newborn's hand and the reflex behaviour is for the newborn to grasp the finger. What does this test for? **c**

HOT Challenge
6 Ivan was baking some biscuits for his afternoon snack. When he pulled the biscuit tray from the hot oven his arm accidentally brushed against the oven door. Ivan immediately moved his arm away. Use a diagram or flowchart to identify and explain the processes involved in Ivan's spinal reflex. In your answer, mention the role of sensory neurons, motor neurons and interneurons. **d**
7 The somatic nervous system is sometimes described as controlling voluntary responses. Provide an example of when the somatic nervous system controls an involuntary response. **c**

2.3 The transmission of neural information

2.3.1 THE STRUCTURE AND FUNCTION OF NEURONS

For the complex communication system that is the nervous system to function effectively, the neurons must be able to work together to transmit information around the body. Neurons are the basic building blocks of the nervous system and are specialised to receive, process and transmit information. The nervous system is made from billions of interconnected neurons that communicate with each other in an electrical and chemical form, transmitting information all around the brain and body.

The structure of neurons

No two neurons are exactly alike in size or shape, but most are composed of common basic structural parts that play a specific role in communicating information throughout the body (Figure 2.18). Each neuron has a **soma** (cell body) that has threadlike branches extending out called **dendrites**.

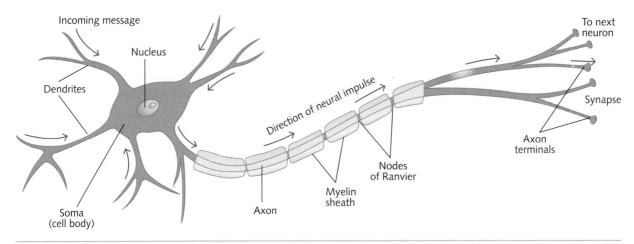

Figure 2.18 The basic structural parts of a typical neuron

Dendrites are responsible for receiving the message in the form of a chemical called a **neurotransmitter**, from other neurons and converting it to an electrical neural impulse (**action potential**) and transmitting the message inwards to the soma. The more dendrites a neuron has, the more information it receives. Extending from the other end of the soma is a single protrusion called an **axon**. Axons are thin fibres that carry messages, in the form of electrical neural impulses, away from the soma. Some axons are also coated with a myelin sheath, an insulating fatty layer that protects the axon and speeds neural transmission of information down the axon.

The end of the axon branches into a number of **axon terminals**. These axon terminals contain many sacs called **synaptic vesicles** that contain neurotransmitters. When the nerve impulse reaches the axon terminals, these vesicles release the neurotransmitters into the space found between neurons, called the **synaptic cleft**, to be carried to the dendrites on the next neuron.

Neurotransmitters: chemical messengers at the synapse

Neurons do not physically connect with each other, so how do they transmit messages from one neuron to another? The answer is via chemical messengers called neurotransmitters.

The message that travels within a neuron is primarily electrical. In contrast, the communication between neurons is via neurotransmitters and therefore is chemical. This communication between neurons takes place at the **synapse**. The synapse is made up of the axon terminals of the **presynaptic neuron** (sending neuron), the dendrites of the **postsynaptic neuron** (receiving neuron) and the microscopic gap between the two neurons (synaptic cleft) (Figure 2.19).

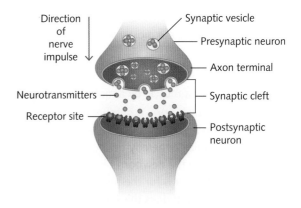

Figure 2.19 When a nerve impulse arrives at an axon terminal, the synaptic vesicles move and fuse with the cell membrane and release neurotransmitters. These neurotransmitters cross the synaptic cleft to influence the next neuron.

When an action potential reaches the tips of the axon terminals of the presynaptic neuron, neurotransmitters are released from synaptic vesicles into the synaptic cleft. The neurotransmitter will then carry the chemical message across the synaptic cleft to the receptor site on the dendrite of the postsynaptic neuron. The **receptor sites** in the postsynaptic membrane can be thought of as a locks with a specific shape that will match a specific neurotransmitter – the key. If the neurotransmitter has a complementary shape to match the receptor it

Weblink
2-minute neuroscience: the neuron

2.3.2
NEUROTRANSMITTERS

will bind (attach) to the receptor on the postsynaptic neuron. The neurotransmitters do not enter the postsynaptic neuron, but the attachment of the neurotransmitter to the receptors alters the activity of the postsynaptic neuron (Figure 2.20). This process may excite the postsynaptic neuron and make it more likely to fire an action potential or inhibit it and make it less likely to fire an action potential.

Any excess neurotransmitter that has not attached to a receptor can be reabsorbed by the presynaptic neurons (Breedlove & Watson, 2013) in a process called re-uptake and stored for later use, or broken down by enzymes in the synaptic cleft. Neurotransmitters therefore enable communication between neurons around the nervous system.

Most people will be aware of the effects of taking a painkiller on the transmission of pain messages through the nervous system. Like neurotransmitters, drugs have their own molecular shape and can be classified as either agonists or antagonists. Drugs that are classified as **agonists** are known to either increase the

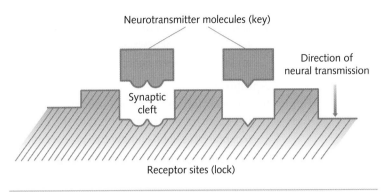

Figure 2.20 The 'lock-and-key' process at the synapse. The neurotransmitter fits its specific receptor site in a similar way to a key fitting a lock.

release of neurotransmitters or imitate certain neurotransmitters by binding to their receptor sites and causing the postsynaptic neuron to respond in the same way as it would to the neurotransmitter. By contrast, drugs that are classified as **antagonists** are known to either inhibit the release of neurotransmitters or block the receptor sites for these neurotransmitters (Gazzaniga et al., 2013).

KEY CONCEPTS 2.3a

- » Neurons are individual nerve cells that are specialised to receive, process and transmit electrochemical information.
- » The synapse is made up of the axon terminals of the presynaptic neuron, the dendrites of the postsynaptic neuron and the space between the presynaptic and postsynaptic neurons, called the synaptic cleft.
- » The synapse is the site of chemical communication between neurons, with neurotransmitters carrying the message across the synaptic cleft to the receptor sites on the dendrite of the postsynaptic neuron.
- » A neurotransmitter must bind (attach) to its own specific shaped receptor site on the receiving neuron to affect the next neuron, in a similar way to a key fitting a lock.
- » Agonists are drugs that either increase the release of neurotransmitters or imitate certain neurotransmitters by binding to their receptor sites and causing the postsynaptic neuron to respond in the same way as it would to the neurotransmitter.
- » Antagonists are drugs that inhibit the release of neurotransmitters or block the receptor sites for these neurotransmitters.

Concept questions 2.3a

Remembering
1. State the name given to the neuron receiving a message from another neuron.
2. In general, information is first received by a neuron at what structure?
3. Outline the role of neurotransmitters on a postsynaptic neuron.

Understanding
4. Explain the role of the synaptic vesicles.
5. What happens to any neurotransmitters remaining in the synaptic cleft?
6. Explain the difference between agonist and antagonist drugs.

> 7 Identify the difference in energy used to transmit a neural message within a neuron and then between neurons. e

Applying

8 The lock and key process is used to describe the action of neurotransmitters at their receptor sites. Explain how this process applies to the action of a neurotransmitter. e

HOT Challenge

9 A certain drug is known to bind to the receptor sites of a specific neurotransmitter. Compare the possible actions of this drug. Provide a scenario when a drug that can bind to a receptor site might be used. e

The excitatory and inhibitory effects of neurotransmitters

There are more than 100 different neurotransmitters (and their receptors) located throughout the brain. These neurotransmitters are responsible for the vital physical and psychological functions we are capable of, such as behaviours, movements, emotions and thoughts.

Neurotransmitters may act in one of two ways when they arrive at the post-synaptic neuron. Some neurotransmitters have an excitatory effect. This means that the postsynaptic neuron is stimulated to pass on the impulse. Other neurotransmitters have an inhibitory effect. This means that the postsynaptic neuron is blocked from passing on the impulse. At any instant, a single neuron may receive hundreds or thousands of messages from adjacent neurons. However, whether the result of synaptic transmission will be excitatory or inhibitory depends on the type of neurotransmitter used, the receptor they interact with and its location in the brain (Gazzaniga et al., 2013).

Glutamate: the primary excitatory neurotransmitter

The primary **excitatory neurotransmitter** in the CNS is **glutamate**. It is extensively found throughout neurons and synapses in the brain, in particular in the hippocampus, the main learning and memory centre of the brain, and the outer layers of the cerebral cortex. Given that glutamate is an excitatory neurotransmitter it is involved in high-speed neural transmission. When glutamate is released at the synapse it binds to glutamate receptor sites located on the postsynaptic neuron. As a result it activates these receptor sites and makes it more likely for the postsynaptic neuron to fire an action potential.

Because of its widespread location throughout the CNS, it is no surprise that glutamate has a key role in the neurological functioning characteristic of a normal brain, including cognition, memory, learning, behaviour, movement and sensations (Sundaram, et al., 2012).

Gamma-amino butyric acid (GABA): the primary inhibitory neurotransmitter

The primary **inhibitory neurotransmitter** is **gamma-amino butyric acid (GABA)**, which is extensively found throughout the nervous system. Given that GABA is an inhibitory neurotransmitter, its overall effect is to calm and slow neural transmission and the body's response, therefore it counteracts the excitability of neurons. When GABA activates its receptor sites, the cells that have those receptors are inhibited and therefore less likely to fire an action potential (Rosenzweig et al., 1999).

As a result of its effects, GABA has been identified as one of the neurotransmitters influential in the development of anxiety disorders. A lack of the neurotransmitter GABA can lead to overstimulation of our neurons, resulting in increased heart rate, respiration or blood pressure, which are features of the stress response associated with anxiety. Treatments for anxiety often target the enhancement of GABA neurotransmission, to inhibit the hyper- and over-excited bodily responses seen in anxiety.

A balancing act: glutamate versus GABA

As we have seen, glutamate and GABA have opposing effects. Glutamate has an excitatory effect, causing neurons to fire an action potential and to speed up neural transmission. GABA has the opposite effect by causing neurons to inhibit the firing of an action potential and slow down neural transmission. A reasonable assumption can be made that if there is too much glutamate and not enough GABA, neural transmission may be overstimulated. Neural transmission requires a balance between these neurotransmitters so that they can work together to regulate the flow of messages around the brain, just as traffic lights regulate the flow of traffic (Figure 2.21) (University of Utah, 2016).

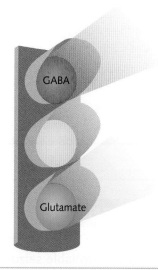

Figure 2.21 GABA and glutamate regulate the flow of messages like traffic lights regulate the flow of traffic. GABA acts like a red traffic light by inhibiting action potentials from firing, while glutamate acts like a green traffic light by allowing action potentials to fire.

ACTIVITY 2.2 GLUTAMATE AND GABA BALANCE

Try it

Glutamate and GABA need to be in balance. It is like running a bath: glutamate is like the hot water and GABA is like the cold water. You need both hot water and cold water but they need to be in balance to have your bath at the optimal temperature.

1. Use red and blue pens or pencils to copy and colour this diagram.

a. Colour one tap blue and label it glutamate. Blue represents the excitatory nature of glutamate.

b. Colour the other tap red and label it GABA. Red represents the inhibitory nature of GABA.

c. Then colour the bath in both colours to show the balance in temperature to demonstrate the balance of glutamate and GABA in the body.

Apply it

2. GABA is our body's natural anti-anxiety treatment. In GABAergic inhibitory neurons, glutamate is converted into GABA, so if you are low in glutamate, you are likely to also be low in GABA. Outline one possible consequence of low levels of glutamate.

Exam ready

3 Which one of the following is correct about glutamate and GABA?

A Glutamate is an inhibitory neurotransmitter while GABA is an excitatory neurotransmitter.

B Glutamate makes the postsynaptic neuron less likely to function, whereas GABA makes the postsynaptic neuron more likely to function.

C Glutamate makes the postsynaptic neuron more likely to fire as it has an excitatory effect, whereas GABA makes the postsynaptic neuron less likely to fire as it has an inhibitory effect.

D Glutamate makes the postsynaptic neuron more likely to fire as it has an inhibitory effect, whereas GABA makes the postsynaptic neuron less likely to fire as it has an excitatory effect.

ANALYSING RESEARCH 2.2

New medication offers hope to patients with frequent, uncontrollable seizures

A study carried out at the Johns Hopkins University School of Medicine has investigated a new type of anti-epilepsy medication to reduce seizures. Epilepsy is a group of brain disorders featuring disturbances in the electrical activity of the brain, causing recurrent seizures that may include loss of consciousness. The study, led by professor of neurology Gregory Krauss, has investigated the effects of perampanel, a new class of drugs that appears to inhibit the glutamate receptor called AMPA, therefore decreasing the excitatory responses in the brain and reducing seizures. In the past, research has targeted a variety of drugs to modify glutamate receptors with the aim of controlling a range of diseases; however, results have not offered much hope, with major side effects such as sleepiness and even comas occurring.

This study included 700 people with uncontrolled partial-onset seizures (the most common form in epilepsy). These seizures can involve anything from twitching to confusion to convulsions. At the beginning of the study, the participants were experiencing about 10 seizures a day and were all taking one to three anti-epileptic drugs as part of their treatment plan.

Each participant was assigned by Krauss and his colleagues to add to their treatment plan either a placebo or 2 milligrams, 4 milligrams or 8 milligrams per day of the drug perampanel.

The researchers found that the higher the dose, the better the results. Roughly one-third of participants who were given 8 milligrams of the saw the frequency of their seizures fall by more than 50 per cent. Four milligrams of perampanel was the lowest effective dose.

While the most common side effect was dizziness, these results do offer hope to people with epilepsy as well as people suffering from some drug addictions and ALS, a form of motor neuron disease in which glutamate is believed to play a role.

Source: Adapted from Johns Hopkins Medicine. (2012). New medication offers hope to patients with frequent, uncontrollable seizures, Science Daily.

Questions

1 Does the drug perampanel act as an agonist or an antagonist? Explain your answer.
2 What was the aim of this research?
3 Identify the independent variable(s) and dependent variable(s) of the research.
4 What is a placebo? Explain the purpose of using a placebo in this study.
5 Explain the results that were obtained.

HOT Challenge

6 Outline one ethical concept that would need to be considered when carrying out this experiment using a placebo.

Neuromodulators: chemical messengers beyond the synapse

The neurotransmitters GABA and glutamate, discussed in the preceding sections, can be described as fast acting, localised – that is, acting at a single synapse – and short lived. By short lived, we mean the neurotransmitter is released into the synaptic cleft, attaches to postsynaptic receptors and any excess neurotransmitter is reabsorbed or broken down by enzymes in the synaptic cleft. This contrasts with the action of another group of neurotransmitters called **neuromodulators**. Neuromodulators are capable of affecting a large number of neurons at the same time. They can also influence the effects of other chemical messengers. Neuromodulators are released by a neuron and will diffuse (spread) through large areas of the brain; they are slow acting but will bring about long-lasting change to the neurons and synapses affected. Neuromodulators will bind to receptors on neurons; however, those neurons may be some distance from the neuron releasing the neuromodulator (Figure 2.22). Neuromodulators are responsible for a range of human behaviour related to sleep, pain, motivation and voluntary movements. Examples of neuromodulators include dopamine and serotonin.

Dopamine and serotonin

Dopamine is a neuromodulator that is responsible for signalling that a reward is available. Dopamine has long been considered as an important factor in the brain's interpretation and processing of rewarding experiences. Scientists have hypothesised that dopamine is not responsible for the pleasure we experience when receiving a reward, but is more likely to be important in motivating behaviours. These are things such as searching for food and water, and learning new behaviours that will bring about a reward, as well as attending to and remembering the environmental stimuli that are linked to the reward (Neuroscientifically Challenged, 2015).

When it is released in areas of the brain, called the **reward system**, dopamine can produce states of motivation (Gray, 2011). Studies have shown that it is released before the reward is obtained and that if the expected reward is larger than expected, then more dopamine is released (Gray, 2011). However, if a behaviour is so well learned and the reward is accurately predicted, then dopamine is no longer released (Neuroscientifically Challenged, 2017).

The feeling we experience when dopamine is released is desirable and this motivates us to repeat the behaviour that caused the dopamine to be released.

Dopamine is also responsible for initiating movement towards the reward, called approach behaviour. Dopamine's role in initiating movement can help explain some of the impaired movement symptoms (slowness of movement and difficulty in initiating voluntary and spontaneous movements) associated with **Parkinson's Disease (PD)**, which is thought to be due to a lack of dopamine (Whitten, 2022).

Treatment of PD aims to restore the insufficient production of dopamine. The standard and most widely used treatment is a substance called levodopa (L-dopa).

2.3.3 NERVOUS SYSTEM 'MATCH THE PAIRS'

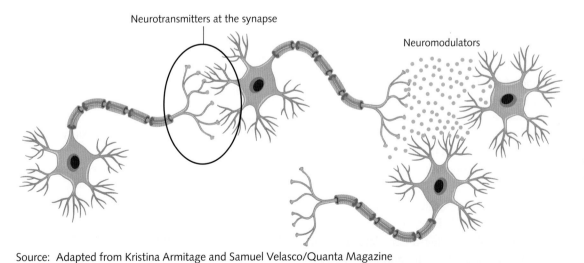

Source: Adapted from Kristina Armitage and Samuel Velasco/Quanta Magazine

Figure 2.22 Neuromodulators can spread over a large area and bind to receptors on neurons that are a large distance from the releasing neuron.

EXAM TIP
When asked to outline two differences between two objects, for each difference you must write a sentence that contrasts the two objects.

This substance is a type of amino acid (these make up proteins) that is converted to dopamine by neurons in the brain. When dopamine levels are raised, symptoms of PD are reduced.

Serotonin, a chemical produced in the CNS as well as the intestines, is another major neuromodulator. It is involved in many processes including pain, sleep and mood regulation. Lower levels of serotonin have been associated with mood disorders, anxiety and sleep disorders.

A study on serotonin levels and the sleep–wake cycle in mice, found that mice with low levels of serotonin experienced interruptions to their sleep. These mice had frequent waking during what would normally be a sleep stage. This suggests that serotonin is involved in maintaining the sleep–wake cycle, by promoting continuous episodes of sleep and wakefulness (Nakamaru-Ogiso et al., 2012).

Some drugs have been developed to 'boost' the effects of serotonin and improve mood. One well-known example is the antidepressant Prozac (a selective serotonin re-uptake inhibitor), which acts by preventing the re-uptake of serotonin after it has been released, resulting in greater than usual amounts of serotonin being available. The effects are instant, accounting for the immediate improvement in mood.

ANALYSING RESEARCH 2.3

Levodopa better than newer drugs for long-term treatment of Parkinson's, largest ever trial shows

The largest ever trial of Parkinson's Disease (PD) treatment has been published in *The Lancet*, revealing that the most widely used treatment of levodopa is still the best initial treatment for most patients when compared with a newer class of drugs including dopamine agonists (DA) and monoamine oxidase type B inhibitors (MAOBI).

The trial was interested in comparing the drugs because the standard use of levodopa over the long term can lead to movement problems and dyskinesia, which can be less evident with the use of the new alternatives of DA and MAOBI. However, the use of these alternative drugs can lead to a range of other side effects that have an impact on quality of life, such as nausea, hallucinations and sleep disturbances.

The trial included 1620 people with early PD, who were randomly assigned to either receive levodopa or an alternative drug (DA or MAOBI). The trial involved obtaining self-report data over a period of seven years. While the difference found between the two drugs in terms of mobility and quality of life was small, it did indicate that the use of levodopa was more beneficial, with patients receiving the levodopa treatment reporting significantly better scores on a variety of scales that measured factors such as body discomfort and experiences in daily living. The trial indicates that when comparing levodopa to DA and MAOBI, on the benefits, side effects, overall quality of life as well as cost, then levodopa is still considered the best initial treatment.

Recently, treatment for PD has focused on these newer drugs for initial treatment; however, treatment for PD worldwide is likely to change as a result of this trial. When compared to past studies, not only is this trial the largest ever performed on PD treatment, other studies have not included long-term follow-up and they have not included self-report data; rather, they have focused on clinicians' assessments of motor symptoms.

Source: Based on *The Lancet* (2014), Long-term effectiveness of dopamine agonists and monoamine oxidase B inhibitors compared with levodopa as initial treatment for Parkinson's disease (PD MED): a large, open-label, pragmatic randomised trial. Vol.384 (9949) https://www.thelancet.com/journals/lancet/article/PIIS0140-6736(14)60683-8/fulltext

Questions

1. Write a hypothesis for the study.
2. Identify the independent variable(s) and dependent variable(s).
3. Define the term 'random allocation'. Was it used for this trial? Explain.
4. The trial obtained self-report data. Explain what self-report data are and describe one strength and one limitation of obtaining this type of data.
5. Is this a longitudinal study? Explain your answer.
6. Based on the results, what conclusion(s) can be made about this trial?

HOT Challenge

7. Explain two benefits of this trial when compared to previous studies of this type.

KEY CONCEPTS 2.3b

» Neurotransmitters may act in one of two ways when they reach the postsynaptic neuron.
» Some neurotransmitters have an excitatory effect (e.g. glutamate) that speeds up neural transmission by stimulating the postsynaptic neuron, making it more likely for it to fire and pass on a neural impulse.
» Some neurotransmitters have an inhibitory effect (e.g. GABA) by slowing down neural transmission and making it less likely for the postsynaptic neuron to fire and pass on the neural impulse.
» Neurotransmitters are localised, fast acting, and short lived.
» Neuromodulators are slow acting and have far-reaching, long lasting effects on areas of the brain.
» Dopamine and serotonin can act as neuromodulators.
» Dopamine is important in motivating behaviour in search of rewards, in learning how to predict when rewards are likely and in initiating movements.
» Serotonin influences pain, mood and sleep.

Concept questions 2.3b

Remembering
1 Name the primary excitatory neurotransmitter. **r**
2 Name the primary inhibitory neurotransmitter. **r**

Understanding
3 Explain how GABA and glutamate are essential in regulating the flow of messages around our nervous system. **e**
4 How is a neuromodulator different from a neurotransmitter? **e**

Applying
5 What happens to dopamine levels when we receive an unexpected reward? How does this influence our future behaviours? **e**
6 We know that alcohol activates GABA receptors. Using this information, describe the likely effects of drinking alcohol. **e**

HOT Challenge
7 Explain the role that GABA and glutamate are believed to play in epileptic seizures. **c**

2.4 Neural basis of learning and memory

Although we are born with some innate abilities that enhance our chances of survival (for example, sucking and grasping), most human behaviour is learned. Most of our behaviours (and changes in behaviour) are the result of experience, which is stored in our memory system. Learning and memory are closely related concepts. If information is not learned, it cannot be remembered and, if information cannot be remembered, learning does not occur.

Learning begins at birth and continues throughout our lifetime. It enables us to function daily and adapt to changes that are constantly occurring in the world around us. Learning is dependent on memory, which is an active information processing system that receives, organises, stores and recovers information when needed. Memory begins as you take in information from both the internal and external environment via our senses. Our nervous system converts this sensory information into impulses of electrochemical energy and transmits it to our brain where it is processed, interpreted and stored for future use (Figure 2.23). In Chapter 5 you will investigate how our memory system operates to encode, store and retrieve information and the different memory systems thought to be involved in different aspects of memory.

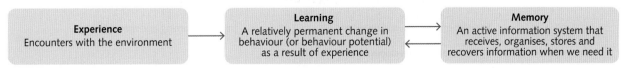

Figure 2.23 Learning is dependent on memory.

Researchers have been investigating the changes that occur in our brain as a result of learning, and how memories are formed. However, given the complex nature of the nervous system, particularly the brain, there are many questions still surrounding the mechanisms involved in learning and memory, and more research is needed to fully understand the biological basis of these processes.

Effects of experience on neural development

Some earlier studies have demonstrated the influence of experience on neural development including those on visual deprivation. For example, animals reared in the dark were found to have fewer synapses (Cragg, 1975) and fewer dendrites (Valverde, 1971) in their primary visual cortices (the area of the brain that receives and initially processes visual stimuli) than animals reared in their natural light-filled environment. The animals reared in the dark were also unable to judge depth correctly (Walk & Walters, 1973) and had deficits in their pattern vision as adults (Tees, 1968).

Other studies into the influence of early exposure to enriched environments (environments offering a variety of sensory and intellectual stimulation) on animal neural development support the belief that the external environment plays a crucial role in neural development. These studies consistently suggest that enriched experience results in greater synapse formation (Figure 2.24). In one study, rats were raised with other rats in enriched environments (complex cages) that stimulated their senses. They were found to have thicker cortices with more dendritic development (Greenough & Volkmar, 1973) and more synapses per neuron (Turner & Greenough, 1983) than rats raised by themselves in simple cages (Figure 2.25) where sensory stimulation was lacking. In fact, there was 23 per cent more branching of the dendrites in the cells of the cerebellum (a brain structure involved in the memory of learned skills and actions) of the enriched rats when compared to the rats in simple cages. Similarly, rats raised in an enriched environment showed increased branching of dendrites in the primary visual area of the cerebral cortex (Greenough & Juraska, 1979; Volkmar & Greenough, 1972). These studies suggest that learning through experience can play a big part in how neuron structure changes and synapses form.

The brain appears to change when we learn and in turn form memories, with exposure to stimuli encouraging this process. Learning any musical instrument, for example the guitar, requires effort, particularly when starting out. The learning experience for most instruments requires you to engage various parts of the brain, including cognitive, sensory and motor areas. When you begin guitar lessons, you may find it difficult to operate both hands at once, because one hand is strumming the strings and the other is fingering the notes. If you practise and keep at it, soon the neural pathways used to produce the behaviour

Worksheet
The effect of enriched environments on synaptic plasticity

Enriched environment

↓
Sensory stimulation
↓
Greater synapse function
↓
Strong neural circuits
↓
Strengthens learning and memory
↓
Positive brain development

Deprived environment

↓
Limited sensory stimulation
↓
Limited synapse function
↓
Weak neural circuits
↓
Impedes learning and memory

Figure 2.24 Our brain changes in response to experience.

will begin to fire repeatedly. The more often the neurons in the circuitry fire, the stronger the connection between the neurons will become.

If you practise the guitar regularly, eventually you may become more proficient. If you have learned to read music, for example, the neural circuits involved in this behaviour will have developed 'bushier' dendrites and the neurons will have extended their axons, creating connections with more neurons and a better organised neural network. Over time, if your brain was scanned using neuroimaging technology, most likely some areas of your brain will have changed in size or structure; for example, the areas of the brain involved in the motor coordination used in the strumming and fingering of the notes.

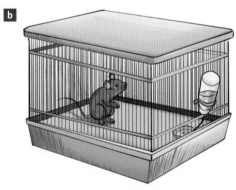

Figure 2.25 Rats raised in **a** an enriched environment show greater synapse formation and thicker cortices (with more dendrite development) than **b** rats raised in simple cages.

Synaptic changes as a mechanism for learning and memory

For many years neuroscientists have understood that the number of neurons in the adult brain does not increase significantly with age or with use; that is, we do not acquire an abundance of new neurons when we learn and form memories. So, if there is no large-scale change to the number of neurons in the brain, something else must happen when we learn and form memories.

In 1949 Donald Hebb, a Canadian psychologist, attempted to explain what might be happening at the level of the neuron when learning was occurring. Hebb proposed that increased activation of groups of neurons by sensory stimulation caused these groups of neurons to be connected to each other (Breedlove & Watson, 2014). He theorised that when one neuron activates another or 'fires together' they become more closely linked, or 'wired together'. This is commonly referred to as Hebb's rules – 'cells that fire together, wire together'. Further to this, Hebb proposed that when the one neuron activates another neuron, a change occurs at the synapse between the two neurons. This change strengthens the synapse, and therefore any further activation of the first neuron will easily activate the second neuron (Gazzaniga et al., 2013).

Hebb went on to propose that the strengthened synapse formed the physical basis of memory (Kolb & Whishaw, 2003).

Hebb's theory was debated for many years and, although not all details are supported, most researchers now agree that when learning takes place a physical change occurs in the synapse between neurons (Hawkins & Bower, 1989). These changes result in the laying down of new neural circuits, or neural pathways, through which information can travel around the brain.

Synaptic plasticity

The changes to synapses described by Hebb in the 1940s are now referred to as **synaptic plasticity**. Synaptic plasticity is a type of neural plasticity, and this term is used to describe the changes that occur to the synapse, which can lead to either an increase or a decrease in activity between neurons.

Synaptic plasticity can involve structural changes, for example:
» the growth of new synaptic connections (sprouting) or the pruning away of existing connections (pruning)
» a change in the number of receptors in a postsynaptic neuron.

Or they can be physiological, for example:
» changes in the ability of the postsynaptic neuron to be excited by neurotransmitters
» changes to the amount of neurotransmitters released by the presynaptic neurons.

Synaptic plasticity is now thought to be the basis of learning and memory because it enables a flexible, functioning nervous system.

The two processes involved in synaptic plasticity or changes to the connections between neurons are long-term potentiation (*strengthening*) (LTP) and long-term depression (*weakening*) (LTD). These processes are particularly important for learning and memory when they occur in synapses in the hippocampus of the brain.

The process of long-term potentiation (LTP)

Long-term potentiation (LTP) is a long-lasting strengthening of neural connections at the synapse as a result of repeated stimulations from a presynaptic to postsynaptic neuron during learning. This produces the neural changes that underlie the formation of memory. LTP leads to a long-lasting increase in neural excitability at the synapse along specific neural pathways because of changes in the efficacy (effectiveness) of synaptic transmission. LTP is a form of synaptic plasticity that is dependent on the activity between two neurons. This means that it is experience dependent; it will not occur if there is insufficient stimulation.

During LTP, changes occur in the presynaptic neuron. These changes include more glutamate (the main excitatory transmitter) being produced and released, further exciting the neurons. To simplify the concept of LTP somewhat: when glutamate is released by the presynaptic neuron, this causes excitation of the postsynaptic neuron. It does this through binding with receptors. New receptors are synthesised, allowing more receptors to emerge on the postsynaptic neuron. As a result, both the presynaptic neuron and the postsynaptic neuron become more efficient at transmitting neural messages (Figure 2.26).

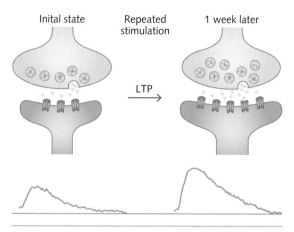

Figure 2.26 The number of receptor sites on the postsynaptic site increases, the number of neurotransmitters being released by the presynaptic neuron increases, and the resulting action potential (shown in red) is stronger when LTP takes place.

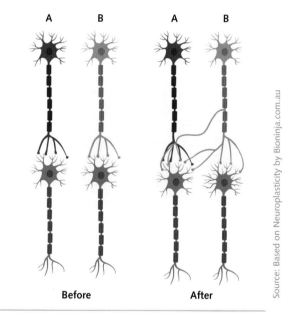

Figure 2.27 Structural changes to the synapse as a result of LTP, when pathway B is stimulated'

As well as the physiological changes that occur at the synapse as a result of long-term potentiation, structural changes also occur. These structural changes are the development of more dendritic spines on the postsynaptic neuron as well as new axon terminals in the presynaptic neuron. This structural change is called **sprouting** (Figure 2.27).

So how could this intricate process act to form new memories? Long-term potentiation, like learning, is not just dependent on increased stimulation from one particular neuron, but on a repeated stimulus from several sources. It is thought that when a particular stimulus is repeatedly presented, so is a particular circuit of neurons stimulated and therefore the circuitry becomes more efficient at transmitting messages. With repetition, the activation of that circuit results in learning and these efficient circuits of neurons form the basis of your memories or enable the storage of your memories. The brain is an enormously intricate and complicated organ. Rather than a one-to-one line of stimulating neurons, it involves a complex web of interacting neurons. The way in which the neurons interact appears to have global effects, meaning that behaviour is the result of a great many neurons; not single circuits.

Brain injury can cause some neural pathways and synapses to be damaged. When this happens, an undamaged neuron that has a synapse with the damaged neuron, will form a new synapse with other undamaged neurons. This results in a **rerouting** of the neural pathway. Through LTP these new pathways will be strengthened and the previously lost function is taken over by an undamaged area (Figure 2.28).

The process of long-term depression (LTD)

Long-term depression (LTD) is a change to the connection between neurons that results in a long-lasting reduction in the strength of a neural response due to persistent weak stimulation. LTD occurs when the efficacy of the synaptic transmission between two neurons is reduced. LTP is a form of synaptic plasticity and will occur if the message being sent from the presynaptic neuron is weak (because less glutamate is released). The result will be a reduced action potential, as the postsynaptic neuron becomes less responsive to glutamate.

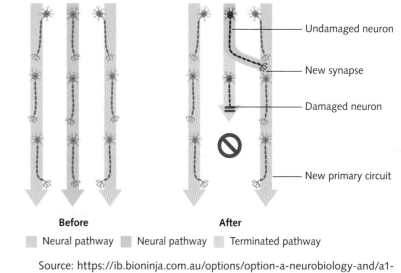

Figure 2.28 Structural changes as a result of rerouting of synapses

The process of LTD may be involved in forgetting, particularly when it occurs in the hippocampus. This is because, rather than improving the ability of neurons to communicate, LTD reduces the transmission of information between neurons. While it may seem to go against logic, LTD may be just as important in learning and memory as LTP, because it may help to weaken and prune back unused synapses (**synaptic pruning**) – or those that are not stimulated. By removing unused or unnecessary synapses, the brain becomes more efficient and therefore so do the learning and memory processes (Figure 2.29). The resources to keep the cells healthy are distributed among fewer neurons, and energy is also more effectively used in the neural pathways that are left. Table 2.2 compares the key characteristics of LTP and LTD.

2.4.1 NEUROLOGICAL BASIS OF MEMORY AND LEARNING

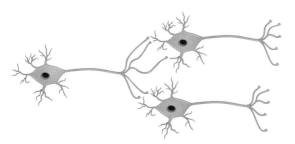
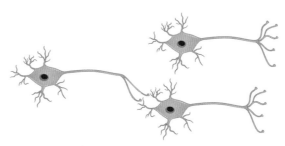

Figure 2.29 Synaptic pruning due to reduced transmission as a result of LTD

Table 2.2 Comparing long-term potentiation and long-term depression

Long-term potentiation (LTP)	Long-term depression (LTD)
1. Increased glutamate is produced and released by the presynaptic neuron.	1. Decreased glutamate is produced and released by the presynaptic neuron.
2. Synapses are strengthened. Receptors in the postsynaptic neuron are increased (number and efficiency), allowing more glutamate to enter the cell, increasing the neuron's excitatory response.	2. Synapses are weakened. Receptors in the postsynaptic neuron are decreased, reducing the amount of glutamate entering the cell, weakening the neuron's response.
3. Involved in memory and learning as neurons become more efficient at transmitting information and sprouting or rerouting may take place.	3. Involved in memory and learning as neurons become less efficient at transmitting messages and synaptic pruning may take place.

ACTIVITY 2.3 SYNAPTIC PLASTICITY

Try it

1. To make synaptic plasticity happen, follow these steps.

    Step 1: Time yourself writing your first name with your dominant hand. Write it as neatly as possible.

    Step 2: Time yourself writing your first name with your non-dominant hand. Write it as neatly as possible.

    Repeat Step 2 10 times and record your time for each attempt. Make sure you try to form the letters as neatly as possible each time.

    Keep doing this every day and the times (and neatness) should improve as the neural pathways strengthen and long-term potentiation takes place.

    Now wait a week and time yourself writing your first name with your non-dominant hand again.

Apply it

2. Jorje is studying VCE Drama. One of the requirements for VCE Drama is to deliver a solo performance in front of a panel appointed by the VCAA. Explain how long-term potentiation would be involved when Jorje practised his monologue for his solo performance.

Exam ready

3. As years went by after VCE, Jorje struggled to remember the words of the monologue he had once memorised for his solo performance. He made mistakes with certain lines and forgot entire sections of his monologue.

    In terms of synaptic plasticity, Jorje's decreased recall was likely a result of:

    A long-term potentiation.
    B long-term depression.
    C a lack of rehearsal.
    D a lack of cues to aid in recall.

CASE STUDY 2.1

How video games can affect your brain

You can't say you weren't warned! In the 1990s researchers warned us that playing video games could affect our brain development. They said that video games would only stimulate the brain regions that controlled vision and movement. Other parts of the brain responsible for behavior, emotion and learning would not be stimulated and may not develop as much as they would under other conditions.

In 1998 the scientific journal *Nature* reported that the playing of video games causes the release of dopamine, the "feel-good" neurotransmitter that promotes pleasure. It went on to say that while playing video games the amount of dopamine released was similar to the amount released by an intravenous injection of amphetamine or methylphenidate – drugs that are used to stimulate the central nervous system and to treat ADHD (Koepp et al., 1998). Video games are created to have unpredictable outcomes and rewards. It is this unpredictability that contributes to players spending hours in a game, believing that next time they will be successful.

Dr Tom A. Hummer, from the department of psychiatry at Indiana University School of Medicine in Indianapolis reports that "The prefrontal cortex—the locus of judgment, decision-making and impulse control—undergoes major reorganization during adolescence." He is referring to the executive control centre, the part of the brain that weighs risks and rewards. This part of the brain does not reach maturity until between the ages of 25 and 30. It is this part of the brain that weighs up the pros and cons of immediate rewards (such as those from online gaming) against longer-term rewards (such as putting in hours of study to achieve an A+ on an exam). When faced with such a choice it is easy to see why some young people spend hours playing online games instead of attending to longer term goals, and their basic needs such as eating, sleeping and keeping clean.

This is a Catch-22 situation. Adolescents are not able to weigh the negative consequences of excessive online gaming due to the immaturity of their frontal lobes. But excessive online gaming impacts the development and maturation of their frontal lobes. The powerful release of dopamine effectively shuts down the prefrontal regions and allows them to play online games for anything up to 36 hours at a time (Greenfield, 2017).

The NPD Group, a global market research firm, has found that nine out of 10 children play video games. That's a lot of kids! Nearly 64 million of them. Some of them are exposed to computers and smart phones before they can even talk or walk. This is a critical period for these children, their brains are being wired with skills essential for survival, and these essential skills do not include online gaming. Many researchers believe that excessive gaming before 22 years of age can physically rewire the brain.

We know that if you keep doing the same thing over and over again, such as online gaming, you change your brain. You also get better at what you are doing. So, by performing the repetitive actions required by online gaming you create and strengthen those particular neural pathways between different parts of your brain. By continuing with these actions (because let's face it you are really good at this) the neural pathways become stronger and stronger. This is what is meant by "use it or lose it". If you cease to use these neural pathways they will be lost (or pruned).

To study the impact of online gaming on young people researchers in China gathered a sample of 18 college students who spent an average of 10 hours a day playing online games like World of Warcraft. The researchers performed MRI (magnetic resonance imaging) on the brains of these students. The results were compared to a group of students (control group) who spent an average of less than two hours a day playing online games. The results were surprising. The students who spent more than 10 hours a day playing online games had significant less grey matter in their brains. The grey matter is the part of the brain that controls thinking.

Diagnostic and Statistical Manual of Mental Disorders, *DSM-V*, is the standard classification of mental disorders used by mental health professionals. *DSM-V* states that more clinical research and experience needs to be done on **internet gaming disorder** before a classification can be included as a formal disorder. This is despite the mounting research evidence about the cognitive, behavioural and neurological impact of online gaming. Some researchers argue that it is its own distinct psychiatric disorder while other researchers are convinced that it is a symptom of another psychiatric disorder.

Online gaming has addictive qualities. There is no disagreement about this. Video game makers are very savvy and have tapped into the human brain's desire for instant gratification, fast pace and unpredictability. As already discussed, the pleasure centre of the brain is flooded with dopamine while playing video games. This provides gamers with a temporary pleasurable rush but also gives the brain the message to produce less of this crucial neurotransmitter. So, paradoxically, gamers actually end up with less dopamine than they need. If gamers are prevented from playing online games they go through withdrawal symptoms manifesting as behavioural problems and aggression (Greenfield, 2017).

It may seem contradictory now to say that not all gaming is detrimental. Video games can enhance visual perception, information processing, tracking multiple objects, mentally rotating objects and storing and manipulating objects in the memory centres of the brain, improving the ability to switch between tasks as well as strategising. Many of these skills are desired by many of today's employers. The model used for video game production could also be used in the development of best-practice educational resources to maximise learning. These resources could provide students with achievable levels of challenge, reward student effort and practice with acknowledgement of incremental progress, all fueled by dopamine (Willis, 2011).

Anthony Rosner was like many 17-year-old boys. He spent up to 18 hours each day building empires and leading raids. Anthony was a hero in the online gaming World of Warcraft community hiding behind his online avatar, Sevrin. For two years he further and further immersed himself in his fantasy world. During this time his real life grew more and more distant. He saw friends as time wasting until they eventually drifted away, he saw school as irrelevant and he saw health and cleanliness as unnecessary. After all, his online companions could not smell him! As a result Anthony put on weight and became more and more lazy. His grades suffered, he missed assignments, and he almost failed to complete his first year of college.

This was the wakeup call that Anthony needed. He realized that he had gotten into university, was pursuing his dream of becoming a film director, but he was knowingly throwing it away. His academic advisor gave him two options: either complete all of his essays for the first year within a span of three weeks or fail and retake the first year. "I didn't want to let myself or my parents down, so I uninstalled World of Warcraft and focused on my work," says Anthony.

Incredibly, Anthony began to live his life again. He joined a gym to regain his health, he started DJing at university and reconnected with friends to restart his social life. "I couldn't believe what I had been missing," he says.

It was World of Warcraft that inspired Anthony to achieve his dream of making and directing films. You can view Anthony's documentary, **IRL – In Real Life**, in the weblink. IRL – In Real Life recounts Anthony's adventures with Sevrin as well as how he learned to break free from gaming. This film has been viewed by more than 1 million people worldwide.

Today, Anthony plays World of Warcraft occasionally. He does not allow it to dominate his life any more. As Anthony says "it is far more rewarding to achieve your potential in real life".

If you think you or someone you know may have a gaming addiction, here are the warning signs?
1. Spends a lot of time on the computer.
2. Becomes defensive when asked about what they are doing on the computer.
3. Loses track of time spent on the computer and in general.
4. Would rather spend time on the computer than with friends or family.
5. Loses interest in hobbies.
6. Is socially isolated and moody.
7. Has more online friends than real life friends.
8. Does not keep up with schoolwork; grades fall.
9. Starts spending large amounts of money on unexplained activities.
10. Tries to hide the amount of time spent online.

References

Gentile, D.A., Bailey, K., Bavelier, D., Brockmyer, J.F., Cash, H., Coyne, S.M., Doan, A., Grant, D.S., Green, C.S., Griffiths, M., Markle, T., Petry, N.M., Prot, S., Rae, C.D., Rehbein, F., Rich, M., Sullivan, D., Woolley, E., Young, K. (2017) Internet Gaming Disorder in Children and Adolescents. *Pediatrics,140*(Suppl 2), S81-S85. doi: 10.1542/peds.2016-1758H. PMID: 29093038

Greenfield, D. N. (2018). Treatment Considerations in Internet and Video Game Addiction. *Child and Adolescent Psychiatric Clinics of North America, 27*(2), 327–344. https://doi.org/10.1016/j.chc.2017.11.007

Koepp, M. J., Gunn, R. N., Lawrence, A. D., Cunningham, V. J., Dagher, A., Jones, T., Brooks, D. J., Bench, C. J., & Grasby, P. M. (1998). Evidence for striatal dopamine release during a video game. *Nature, 393*(6682), 266–268. https://doi.org/10.1038/30498

Willis, J. (2011). *A Neurologist Makes the Case for the Video Game Model as a Learning Tool*. Edutopia; George Lucas Educational Foundation. https://www.edutopia.org/blog/neurologist-makes-case-video-game-model-learning-tool

Source: Paturel, A (June/July 2014). Game theory: the effects of video games on the brain. Mental Health, American Academy of Neurology, with permissions Wolters Kluwer

Weblink
IRL - In Real Life

Questions

1. What were the signs that gaming was a problem for Anthony?
2. Using your knowledge of dopamine and its role in rewarding behaviour, explain Anthony's behaviour.
3. From the information presented in the article, why does gaming become a problem?
4. Does any of this article resonate with you?
5. Although in the article LTP and LTD are not mentioned by name, identify where in the article they are described.

KEY CONCEPTS 2.4

» Learning is a relatively permanent change in behaviour (or behaviour potential) due to experience, adapting to changes that are constantly occurring in the world around you. It begins at birth and continues throughout your lifetime.
» Memory is an active information processing system that receives, organises, stores and recovers information when needed.
» For learning to occur, neurons need the ability to communicate.
» Synaptic plasticity can involve structural changes of the synapse by the growth of new synaptic connections, the pruning away of existing connections or a change in the number of receptors in a postsynaptic neuron.
» Synaptic plasticity can involve physiological changes of the synapse such as changes in the ability of the postsynaptic neuron to be excited by neurotransmitters or changes to the production of neurotransmitters by the presynaptic neuron.
» Long-term potentiation (LTP) and long-term depression (LTD) are forms of synaptic plasticity.
» LTP results in a long-lasting strengthening of neural connections at the synapse as a result of repeated stimulations from a presynaptic to postsynaptic neuron during learning.
» LTP has been demonstrated to occur in cells of the hippocampus producing the neural changes that underlie the formation of memory.
» LTD results in a long-lasting reduction in the strength of a neural response due to persistent weak stimulation or no stimulation.
» LTP can result in sprouting and rerouting. LTD can result in synaptic pruning.

Concept questions 2.4

Remembering

1. How does the presynaptic neuron change as a result of LTP? **r**
2. How does the postsynaptic neuron change as a result of LTP? **r**

Understanding

3. How does the process of LTP result in learning and the formation of memories? **e**

Applying

4. What effect has raising rats in an enriched environment had on their brains? **r**
5. Describe one way the neural development of an animal raised in an enriched environment would differ from that of an animal raised in a deprived environment. **r**
6. Outline two changes which can be described as synaptic plasticity. **r**
7. What is one advantage of the process of LTD to learning and memory? **r**

HOT Challenge

8. Why might the growth of new synapses and the pruning of unused synaptic connections be beneficial to an organism when creating neural pathways for learning? **e**
9. How does Hebb's rule 'cells that fire together wire together' fit with the theory of LTP? **e**

2 Chapter Summary

KEY CONCEPTS 2.1a

- » The central nervous system (CNS) is composed of the brain and the spinal cord.
- » The brain is the 'engine room' of the nervous system. It receives and processes information from the rest of the body and generates responses to it.
- » The spinal cord, an intricate and delicate cable of nerve fibres stretching from the base of the brain to the lower back, connects the brain to the rest of the body via its connection to the peripheral nervous system (PNS).
- » The spinal cord receives sensory information from the PNS and transmits it along the *ascending tracts* up to the brain for further processing and interpretation. Motor information from the brain is carried along *descending tracts* where it is relayed to our muscles, organs and glands so they can react.

KEY CONCEPTS 2.1b

- » The PNS consists of all the nerves outside the CNS; that is, all the nerves that extend throughout the rest of the body.
- » The main role of the PNS is to convey sensory information from the body's internal and external environments to the CNS and transmit motor commands from the CNS to the rest of the body.
- » The PNS comprises the somatic nervous system (SNS), which controls voluntary responses, and the autonomic nervous system, which that controls the activity level of our internal organs and glands.

KEY CONCEPTS 2.1c

- » The ANS is divided into the sympathetic nervous system, the parasympathetic nervous system and the enteric nervous system.
- » The sympathetic nervous system stimulates arousal and activity levels.
- » The parasympathetic nervous system calms and lowers arousal levels.
- » The enteric nervous system controls the activity of the digestive system.

KEY CONCEPTS 2.2

- » The somatic nervous system is responsible for all voluntary responses through communication with the brain.
- » Sensory information is sent via the spinal cord to the brain, which determines a response and then sends this message via the spinal cord back to the somatic nervous system and the motor neurons to enable a response.
- » Not all responses performed by the somatic nervous system are conscious, voluntary responses involving the brain.
- » Spinal reflexes occur automatically and only involve the spinal cord and the somatic nervous system. They are quick responses that are essential for survival.
- » Spinal reflexes usually involve an interneuron making a connection between a sensory neuron and a motor neuron.
- » The simplest spinal reflex involves the connection between a sensory neuron and a motor neuron.

KEY CONCEPTS 2.3a

- » Neurons are individual nerve cells that are specialised to receive, process and transmit electrochemical information.
- » The synapse is made up of the axon terminals of the presynaptic neuron, the dendrites of the postsynaptic neuron and the space between the presynaptic and postsynaptic neurons, called the synaptic cleft.
- » The synapse is the site of chemical communication between neurons, with neurotransmitters carrying the

message across the synaptic cleft to the receptor sites on the dendrite of the postsynaptic neuron.
» A neurotransmitter must bind (attach) to its own specific shaped receptor site on the receiving neuron to affect the next neuron, in a similar way to a key fitting in a lock.
» Agonists are chemicals that either increase the release of neurotransmitters or imitate certain neurotransmitters by binding to their receptor sites and causing the postsynaptic neuron to respond in the same way as it would to the neurotransmitter.
» Antagonists are chemicals that inhibit the release of neurotransmitters or block the receptor sites for those neurotransmitters.

KEY CONCEPTS 2.3b

» Neurotransmitters may act in one of two ways when they reach the postsynaptic neuron.
» Some neurotransmitters have an excitatory effect (e.g. glutamate) that speeds up neural transmission by stimulating the postsynaptic neuron, making it more likely for it to fire and pass on a neural impulse.
» Some neurotransmitters have an inhibitory effect (e.g. GABA) by slowing down neural transmission and making it less likely for the postsynaptic neuron to fire and pass on the neural impulse.
» Neurotransmitters are localised, fast acting and short lived.
» Neuromodulators are slow acting and have far-reaching, long-lasting effects on areas of the brain.
» Dopamine and serotonin can act as neuromodulators.
» Dopamine is important in motivating behaviour in search of rewards, in learning how to predict when rewards are likely and in initiating movements.
» Serotonin influences pain, mood and sleep.

KEY CONCEPTS 2.4

» Learning is a relatively permanent change in behaviour (or behaviour potential) due to experience, adapting to changes that are constantly occurring in the world around you. It begins at birth and continues throughout your lifetime.
» Memory is an active information processing system that receives, organises, stores and recovers information when needed.
» For learning to occur, neurons need the ability to communicate.
» Synaptic plasticity can involve structural changes of the synapse by the growth of new synaptic connections, the pruning away of existing connections or a change in the number of receptors in a postsynaptic neuron.
» Synaptic plasticity can involve physiological changes of the synapse such as changes in the ability of the postsynaptic neuron to be excited by neurotransmitters or changes to the production of neurotransmitters by the presynaptic neuron.
» Long-term potentiation (LTP) and long-term depression (LTD) are forms of synaptic plasticity.
» LTP results in a long-lasting strengthening of neural connections at the synapse as a result of repeated stimulations from a presynaptic to postsynaptic neuron during learning.
» LTP has been demonstrated to occur in cells of the hippocampus producing the neural changes that underlie the formation of memory.
» LTD results in a long-lasting reduction in the strength of a neural response due to persistent weak stimulation or no stimulation.
» LTP can result in sprouting and rerouting. LTD can result in synaptic pruning.

2 End-of-chapter exam

Section A: Multiple choice

1. The main function of the spinal cord is to:
   A transmit neural impulses to and from the brain
   B control the body's voluntary (skeletal) muscles
   C store and release neurotransmitters
   D coordinate movements of body parts

   ©VCAA 2020 Q1 ADAPTED MEDIUM

2. Which statement correctly identifies the role of the postsynaptic neuron?
   A where neurotransmitters are stored
   B location of reuptake of neurotransmitters
   C transmits the chemical message
   D neurotransmitters bind to receptor sites

3. The peripheral nervous system is composed of the:
   A brain and spinal cord
   B sympathetic, parasympathetic and enteric nervous systems
   C somatic and autonomic nervous systems
   D parasympathetic nervous system and spinal cord

4. Crossing the road, Sam gets a fright when he narrowly misses a moving car. Which of the following physiological changes is Sam most likely to experience?
   A an increased heart rate and respiration rate
   B an increased heart rate and decreased perspiration
   C a decreased heart rate and dilated pupils
   D a drop in his blood pressure

5. Which branch of Sam's nervous system is dominant in question 4?
   A the peripheral nervous system
   B the spinal cord
   C the parasympathetic division of the autonomic nervous system
   D the sympathetic division of the autonomic nervous system

6. Corrine was walking along the beach when she accidently stepped on a jagged seashell. Instantly she withdrew her foot from the shell. Which of the following pathway sequences is correct when describing what happened to Corrine?
   A interneuron, sensory neuron, motor neuron
   B sensory neuron, motor neuron, interneuron
   C motor neuron, interneuron, sensory neuron
   D sensory neuron, interneuron, motor neuron

7. Jake was doing the ironing when he accidentally drew the hot iron over his finger. Jake quickly pulled his finger away from the iron in a reflex action. In this instance:
   A a spinal reflex occurred in Jake's cranial nerves, when a sensory neuron carried a neural impulse to a motor neuron, via an interneuron, leading to the withdrawal of Jake's hand from the iron
   B a spinal reflex occurred in Jake's spinal cord, when a sensory neuron carried a neural impulse to a motor neuron, via an interneuron, leading to the withdrawal of Jake's hand from the iron
   C a spinal reflex occurred in Jake's spinal cord when an interneuron caused a motor neuron to automatically activate a sensory neuron, leading to the withdrawal of Jake's hand from the iron
   D the activation of the parasympathetic nervous system caused Jake's hand to automatically withdraw from the iron

8. Which statement is true for neurotransmitters?
   A neurotransmitters are only found in the brain
   B the most abundant excitatory neurotransmitter is GABA
   C neurotransmitters are released into the bloodstream to act on distant sites
   D neurotransmitters cross the synaptic cleft to bind to receptor sites on the postsynaptic neuron

9 Which statement about the neurotransmitter glutamate is correct?
   A glutamate is an excitatory neurotransmitter that calms down neural activity
   B glutamate is an excitatory neurotransmitter that blocks activity in the postsynaptic neuron
   C glutamate is an inhibitory neurotransmitter that causes slow neural transmission
   D glutamate is an excitatory neurotransmitter that makes it more likely for the postsynaptic neuron to fire an action potential

10 The difference between LTP and LTD is that:
   A LTD results in more excitation of the neurons and LTP results in less efficient neural transmissions
   B LTD results in less excitation of the neurons and LTP results in more efficient neural transmission
   C LTD only occurs in the hippocampus, whereas LTP occurs everywhere in the brain
   D LTD is faster acting than LTP

11 Synaptic plasticity can include:
   A modification of the strength or efficacy of synaptic transmission
   B production of new synaptic connections or the pruning of unused connections
   C increasing the number of receptor sites on the membrane of a postsynaptic neuron
   D All of the above

12 Which is the correct statement?
   A a neuromodulator only acts at a synapse where it is released
   B a neuromodulator can travel a distance from the presynaptic neuron and act in other areas of the brain
   C a neurotransmitter is slower acting than a neuromodulator
   D a neuromodulator does not bind to receptor sites on neurons.

13 Which statement about dopamine is the most correct?
   A It only acts as a neurotransmitter.
   B It is a neuromodulator responsible for sleep and mood regulation.
   C It is the neuromodulator responsible for feelings of pleasure
   D It is the neuromodulator involved in the reward circuit of the brain

©VCAA 2013 Q26 ADAPTED EASY

14 A neuroscientist who is studying the effects of a stimulating environments on the structures of a rat's brain is most likely to observe:
   A a decrease in total weight of the brain
   B a decrease in overall number of synapses
   C bushier dendrites
   D pruning of dendritic branches of neurons in the hippocampus

©VCAA 2016 Q58 ADAPTED EASY

15 What change occurs in neurons as a result of long-term depression?
   A a decrease in the release of the main inhibitory neurotransmitter glutamate
   B an increase in the formation of dendritic spines to allow for easier communication between neurons
   C more neurotransmitters being produced and released by presynaptic neurons, which act on the receptor sites of postsynaptic neurons
   D less neurotransmitters being produced and released by presynaptic neurons, which act on the receptor sites of postsynaptic neurons

16 Which substance is linked to the reward system?
   A glutamate
   B GABA
   C dopamine
   D adrenaline

17 During learning, the axon terminals of some neurons will:
   A transmit neural impulses towards the synapses with other neurons
   B integrate and process incoming information from other connecting neurons
   C release neurotransmitters into the synaptic gap
   D receive neurotransmitters across the synaptic gap

18 A researcher was investigating the effects of a GABA agonist in the treatment of anxiety.
   A GABA agonist will:
   A block the GABA receptors on neurons
   B promote the reuptake of GABA at the synapse
   C decrease the release of GABA into the synapse
   D mimic the action of GABA

19 Jonah is meditating. Which nervous system is most likely to be dominant?
- A the somatic nervous system
- B the sympathetic nervous system
- C the peripheral nervous system
- D the parasympathetic nervous system

20 The motor function of the somatic nervous system can be demonstrated by:
- A moving your hand away from a hot stove reflexively.
- B the activation of the flight-or-fight-or-freeze response.
- C experiencing the sensation of heat when holding a cup of coffee.
- D the homeostatic response to an increase in body temperature.

SECTION B: Short answer

1 Julie was standing barefooted in the gardens when she suddenly lifted her foot from the grass. She then felt a pain sensation and realised she had been bitten by an ant.
- **a** What division of the nervous system was involved when she lifted her leg? [1 mark]
- **b** Explain why Julie felt the pain after she had lifted her leg. [2 marks]

[Total = 3 marks]

2 Trudy learned to speak Spanish as a second language when she was a child, but as an adult, she did not hear or speak Spanish for many years. When Trudy retired, she travelled to Spain and started speaking Spanish again.

Identify and explain whether long-term potentiation or long-term depression was likely to be involved in each of the following stages.
- **a** learning to speak Spanish as a child [3 marks]
- **b** not speaking Spanish as an adult [3 marks]
- **c** speaking Spanish in retirement. [3 marks]

[Total = 9 marks]

3 Outline **three** changes to a neuron that occur because of long-term potentiation. [3 marks]

Stress as an example of a psychobiological process

3

Key knowledge

- internal and external stressors causing psychological and physiological stress responses, including the flight-or-fight-or-freeze response in acute stress and the role of cortisol in chronic stress
- the gut–brain axis (GBA) as an area of emerging research, with reference to the interaction of gut microbiota with stress and the nervous system in the control of psychological processes and behaviour
- the explanatory power of Hans Selye's General Adaptation Syndrome as a biological model of stress, including alarm reaction (shock/counter shock), resistance and exhaustion
- the explanatory power of Richard Lazarus and Susan Folkman's Transactional Model of Stress and Coping to explain stress as a psychological process (primary and secondary appraisal only)
- use of strategies (approach and avoidance) for coping with stress and improving mental wellbeing, including context-specific effectiveness and coping flexibility

Key science skills

Develop aims and questions, formulate hypotheses and make predictions
- formulate hypotheses to focus investigation

Comply with safety and ethical guidelines
- demonstrate ethical conduct and apply ethical guidelines when undertaking and reporting investigations

Generate, collate and record data
- systematically generate and record primary data, and collate secondary data, appropriate to the investigation
- record and summarise both qualitative and quantitative data, including use of a logbook as an authentication of generated or collated data
- organise and present data in useful and meaningful ways, including tables, bar charts and line graphs

Analyse and evaluate data and investigation methods
- process quantitative data using appropriate mathematical relationships and units, including calculations of percentages, percentage change and measures of central tendencies (mean, median and mode), and demonstrate an understanding of standard deviation as a measure of variability
- evaluate investigation methods and possible sources of error or uncertainty, and suggest improvements to increase validity and to reduce uncertainty

Construct evidence-based arguments and draw conclusions
- evaluate data to determine the degree to which the evidence supports or refutes the initial prediction or hypothesis

Source: VCE Psychology Study Design (2023–2027) pages 34, 12–13

3 Stress as an example of a psychobiological process

Let's face it, VCE is pretty good at causing you stress, so it can help to understand what exactly is going on in your body and to try and reduce it.

p. 91

3.1
Stress and stressors

Anything that causes you stress is called a stressor. Whether stressors originate from inside or outside your body, the result is the same – your sympathetic nervous system is activated, the hormone cortisol is released, and you get stressed. Stress equips your body with biological and psychological responses to cope with whatever you perceive as a threat.

p. 101

3.2
Selye's General Adaptation Syndrome

GAS describes the biological changes that your body exhibits when you are stressed – increased heart rate, alertness and then exhaustion. But GAS, like any model, has its strengths and limitations.

3.3
Stress as a psychological process
p. 104

Stress can manifest itself as an overwhelming inability to cope. We find it difficult to recall information, cry, get angry, panicky and feel helpless. The Transactional Model of Stress and Coping provides a framework to explain what is happening when we are stressed. Again, this model has strengths and limitations.

3.5
Coping strategies
p.115

But wait, help is at hand! There are coping strategies that can help you manage your stress. You can either deal with the stressor head-on, or avoid it. The former is preferable to the latter, as avoiding it usually ends up adding to your stress.

3.4
The gut–brain axis
p. 109

When you are stressed, one of the places in your body that feels it is your stomach. Getting a nervous tummy is completely normal. This is because your gut and your brain communicate with each other through the enteric nervous system, mainly through the vagus nerve. It also appears that all the bacteria that live in your gut play a huge part in how you feel.

Coping with stress is part of everyday life. When you have your stress under control you can learn and remember at optimal levels. This is especially important in VCE.

Slidehow
Chapter 3 slideshow

Flashcards
Chapter 3 flashcards

Test
Chapter 3 pre-test

Know your key terms

Acute stress
Adrenaline
Alarm-reaction stage
Approach strategy
Avoidance strategy
Chronic stress
Context-specific effectiveness
Coping
Coping flexibility
Coping skills
Coping strategy
Cortisol
Countershock
Distress
Eustress
Exhaustion stage
External stressor
Flight-or-fight-or-freeze response (FFF)
Freeze response
General adaptation syndrome (GAS)
Gut–brain axis (GBA)
Gut microbiota
Internal stressor
Noradrenaline
Physiological stress response
Primary appraisal
Psychological stress response
Resistance stage
Secondary appraisal
Shock
Stress
Stress responses
Stressor
Transactional Model of Stress and Coping
Vagus nerve

Assessment
- Pre-test
- End-of-chapter exam

Revision
- Chapter map
- Key term flashcards
- Key concept summary
- Slideshow

Investigation
- Investigation: The effectiveness of strategies for coping with stress
- Data calculator
- Logbook template: Fieldwork

Worksheet
- Prevalence and effectiveness of coping strategies

To access these resources, visit cengage.com.au/nelsonmindtap

Nelson MindTap

Stress. We've either all said or heard someone say 'I'm so stressed' and over the past few years you may have heard it more than ever before. On 11 March, 2020, the World Health Organization officially declared the COVID-19 outbreak a global pandemic. After this announcement, we saw sales of hand sanitiser sky-rocket, people lining up to panic buy toilet paper, masks became part of our everyday life, we faced restrictions on where we could go, and many of us experienced lockdowns and remote learning for the first time ever. And while all this was happening, it was accompanied by a 24-hour news cycle coming through our TVs, phones and devices. Hours and hours of reports, press conferences and lots of unknowns about the virus, when we would return to school, when we could see friends, have celebrations and lots of questions about the overall future. Many of us felt and experienced short-term stress and long-term stress during this time, as it was without a doubt a very stressful experience (Figure 3.1). The good news is that stress is exactly the right thing that you should have been feeling through all of that. Many of us feel like we should not be stressed, but here is a secret: stress is actually a normal and expected part of life and sometimes, can even be beneficial!

Figure 3.1 Many students experienced stress during the lockdowns of the pandemic.

3.1 Stress and stressors

Stress refers to the physiological and psychological responses that a person experiences when confronted with a situation that is threatening or challenging. *Physiological* refers to all the processes that are involved with the functioning of an organism (in the body). Stress is experienced when a person perceives that the demands of the threatening or challenging situation exceed their ability to cope. For a person to experience stress, they must be exposed to a stressor. A **stressor** is any person, object or event that challenges or threatens them, thus causing a feeling of stress. Stressors are categorised as either internal (coming from the inside) or external stressors (coming from the outside).

Internal and external stressors

Internal stressors are factors that originate within a person and can be psychological or biological. Biological stressors include things related to your physical wellbeing such as illness, pain, nutritional health, and adequate sleep or rest. Psychological internal stressors are stress-inducing thoughts or behaviours that come from our own psychological mindset and expectations. This could include things such as feelings of worry and anxiety, negative self-talk and low self-esteem, comparing ourselves to others, feelings of anger or fear, pessimism and negative attitudes.

External stressors are factors that originate outside the body. These can be social, cultural or physical environmental conditions, and are often forces that you cannot control. Examples of external stressors include:
» life events such as marriage, divorce, death or loss of a loved one, pregnancy and childbirth
» daily pressures including meeting deadlines for school or work, worrying about finances, commuting in stressful traffic or public transport, small conflicts in relationships, busy schedules, and losing personal items such as keys, wallets or phones
» adjusting to a new culture such as moving to a new country or even adjusting to the culture of a new school or workplace
» caregiving of children or people with severe illness or disability
» bullying or harassment
» witnessing or experiencing violence or trauma
» extreme heat, cold or noise
» being involved in a car accident
» experiencing a natural disaster or catastrophe
» other global events such as a pandemic.

Regardless of the origin of the stress or type of stressor, stress activates the sympathetic nervous system, and the physiological changes it triggers in the body are the same (this chapter will explore these changes in detail). However, at a psychological level (our thoughts, feelings and behaviours), even though we may be exposed to the same stressor as others, perception of stimuli is an individual, subjective experience influenced by a range of factors, including past experiences, personality, belief systems, culture, educational background and genetic factors. What represents a stressor for one person may not be perceived as a stressor by another person, therefore the type of stress experienced and the intensity of the stress varies between individuals. For example, you may become quite anxious and feel threatened if you are asked to deliver a speech at a school assembly, but your friend may feel quite excited at the prospect of doing this.

Psychologists generally agree that stress can manifest itself in two forms: eustress and distress. **Distress** is the form of stress that most often comes to people's minds when they think of stress. It is a negative psychological response to a stressor, as indicated by the presence of negative psychological states such as anger, anxiety, fear, or feelings of helplessness and hopelessness (Simmons & Nelson, 2001). Distress is considered to be a negative form of stress because it impedes our ability to perform and cope at an optimal level. We experience distress when we feel we have no control over a situation and we feel overwhelmed and unable to manage its demands. For example, when a person loses their job they often experience emotions such as anger, worry and despair as they wonder about what happened and how they will take care of their finances in the future.

3.1.1 INTRODUCTION TO STRESS

3.1.2 SOURCES OF STRESS

Weblink
How to make stress your friend

On the other hand, **eustress** is a positive psychological response to a stressor as indicated by the presence of positive psychological states (Simmons & Nelson, 2001). Activities such as playing in a tennis match, skydiving, waiting to meet a role model or planning your next birthday party are examples of stressors that tend to produce more positive psychological states such as excitement, enthusiasm and optimism (Figure 3.2). They also produce the same physiological responses in the body as stressors that result in feelings of distress (Figure 3.3). Experiencing eustress can be beneficial because it increases our alertness and energises us so we are ready to meet the demands of challenging situations, without causing physical harm to the body in the same way that distress can. Eustress is normally enjoyable and can motivate us to perform as well as possible or to change undesirable behaviours. For example, you may feel anxious about your end-of-year exams. However, your desire to succeed and the thought of all the enjoyable things you can do when the exams are over may motivate you to study when you would rather be socialising with your friends. Psychologists recommend that this change in thinking about a stressor from a negative to a positive experience can be a very powerful tool – see weblink *How to make stress your friend*.

Figure 3.2 Many enjoyable activities produce eustress.

Acute and chronic stress

Stress can also be categorised as acute or chronic stress. **Acute stress**, which is the most common form of stress, is the body's immediate response to a perceived stressor. Since our environment is constantly changing, we must constantly change our behaviour to suit these new circumstances. Acute stress is typically caused by the daily demands and pressures encountered in everyday life. Regardless of whether we perceive these stressors to be challenging or threatening, our immediate response is to become highly aroused. So, acute stress can be intense but, because it usually appears and disappears over a short period of time, it doesn't have enough time to damage us psychologically or physically. Mild acute stress can even be beneficial because it can motivate and energise us, so it keeps us active and alert. Acute stress can be thrilling and exciting in small doses. However, if we are repeatedly exposed to the same stressor, or repeatedly exposed to many different stressors over an extended period of time, acute stress may develop into chronic stress.

Chronic stress is the body's response to a persistent or long-term stressor. It involves ongoing demands, pressures and worries that we do not feel we have under control and that we feel will not end. Unlike acute stress, chronic stress often does not appear to be intense and it is generally experienced as a continual feeling of unease, despair or hopelessness. Often individuals become so used to living with a stressor (or number of stressors) that they do not notice they are experiencing an ongoing state of higher-than-normal physiological arousal that characterises

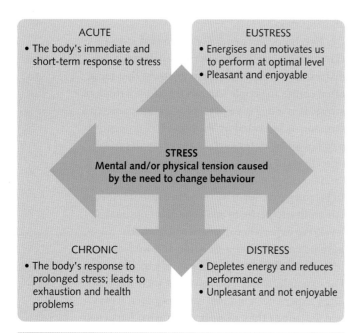

Figure 3.3 Stress can be experienced in the form of eustress or distress. It can be acute or chronic.

chronic stress. Because it lingers for a much longer period of time, chronic stress depletes our body's natural resources and exhausts us, leaving us vulnerable to a host of physical and mental problems. Figure 3.3 shows the relationship between the different forms of stress.

Weblink
Acute vs chronic stress: Signs it's time to seek professional help

KEY CONCEPTS 3.1a

- » Stress is the psychological and physiological responses that a person experiences when confronted with a situation that is challenging or threatening, and is perceived to exceed their ability to cope with the stressor.
- » A stressor is a person, object or event that causes feelings of stress.
- » Stressors can be internal (coming from within a person) or external (coming from the outside environment).
- » The psychological response to stress is subjective and can be negative or positive.
- » The physiological response is the same irrespective of whether the stressful situation is perceived to be positive or negative.
- » Stress can be acute (short term) or chronic (long term).

Concept questions 3.1a

Remembering
1. What is stress? r
2. What is the difference between acute stress and chronic stress? r

Understanding
3. Explain why acute stress could be beneficial. e
4. Identify the difference between internal stressors and external stressors. Provide an example of each from your own experience. e

Applying
5. Tien has a part-time job working in a popular fast-food restaurant while he is also doing VCE in Year 11. Recently, Tien was sick and missed five days of school. Last night, he was up until 2 a.m. trying to catch up on his notes on all his classes. Currently, he is at work in the kitchen on a very busy day. It is extremely hot in the kitchen and his manager, who Tien does not have a great relationship with, has made a few negative comments about how Tien missed shifts last week due to illness. Tien feels scared that he will lose his job because he feels that he cannot be relied on like his workmate, Archie, who has never missed a shift at work since he started.
Draw up a table like the one below to identify internal and external stressors that are contributing to Tien's feelings of stress. c

Internal stressors	External stressors

Stress responses

Stress responses consist of a set of physical and psychological responses that are automatically set in motion when the sympathetic nervous system is activated following the perception of a threat. Stress responses enable us to harness all necessary physiological and psychological resources to help combat the stressor. They help the body and mind to function at their optimal levels when attempting to cope with a threat. Consequently, they help us adapt to our changed circumstance and this aids our chance of survival. This is why stress is referred to as a psychobiological response, because it is the result of both biological and psychological factors and it also has psychological and biological consequences. Figure 3.4 shows the relationship between a stressor, stress and a stress reaction.

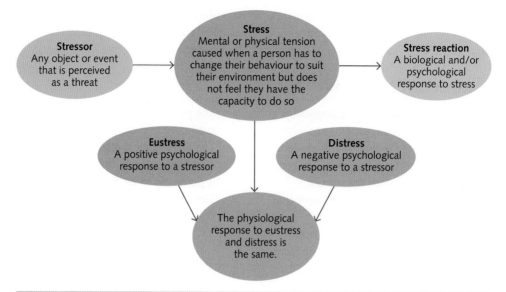

Figure 3.4 The relationship between a stressor, stress and a stress reaction

Stress as a biological process

Psychological responses to stress are influenced by a variety of internal and external factors to determine whether we view a stimulus as stressful. However, the biological responses to stress tend to be stereotypical and follow the same patterns, regardless of the type and severity of the stressor. Two models that describe the biological processes that occur in response to stress are the flight-or-fight-or-freeze response and the General Adaptation Syndrome (GAS).

Weblink
How stress affects your body

Flight-or-fight-or-freeze response

Imagine you are standing on a plane. The entire back of the plane is open with nothing but the sky in front of you. You have a parachute strapped to your back and are about to take that one step that will send you out of the plane and hurtling through the open air towards the ground. Whether this is the best or worst day of your life, your body is likely to show the same symptoms; your heart starts pounding, your breath quickens, your skin goes pale, your pupils widen, your legs or hands might start trembling, your palms start sweating, and those metaphorical butterflies would be fluttering up a storm in your stomach (Figure 3.5). These are just some of the physiological symptoms of stress that you are likely already familiar with. You may get the same feelings before a test, before public speaking, or when you are about to ask someone out for the first time. On the other hand, you may have experienced

Figure 3.5 Skydiving can be a person's favourite thing in the world, or the most terrifying. Regardless, the flight-or-fight-or-freeze response produces the same physiological symptoms whether the skydive brings euphoria or panic.

a different type of response. The one where you find yourself frozen to the spot, unable to move or get the words out of your mouth that you have been practising for weeks. This immobility is also a physiological response to stress.

All these symptoms are part of the **flight-or-fight-or-freeze response (FFF)**. This is a concept that was introduced by Walter Bradford Cannon in the 1920s. He theorised that animals react to threats with a general activation of the sympathetic nervous system, which prepares the body for fighting the threat or fleeing from it. Research since then has added the freeze response, a reaction from the parasympathetic nervous system that appears to occur when animals feel there is no hope of defeating the stressor or escaping to safety (Barlow, 2002). The flight-or-fight-or-freeze response is an adaptive response, meaning it allows the body to automatically adapt and survive in the face of a stressor, without the need for conscious awareness from the brain. You may have noticed this yourself when you have gotten out of the way of immediate danger such as an oncoming car before you have even registered what is happening.

To understand the flight-or-fight-or-freeze response, we must know how the autonomic nervous system (ANS) works, in particular the way in which the two branches of the sympathetic and parasympathetic nervous systems work in harmony to deal with threats and recover from them. As we saw in Chapter 2, the ANS is independent of the rest of the nervous system, meaning that it functions without conscious, voluntary control. Because the ANS supplies nerves to muscles, organs and glands all over the body, it has a significant influence on the activity of most tissues and organ systems in the body. Therefore, it makes a significant contribution to homeostasis and in preparing the body for the flight-or-fight-or-freeze response (McCorry, 2007).

The ANS is made up of two branches, the sympathetic and parasympathetic nervous systems. Both systems are *tonically active*, this means that they are both providing some level of nervous input into the muscles, organs and glands at all times. However, each system may become dominant in different circumstances. The sympathetic nervous system becomes dominant in response to a perceived threat or danger during the flight-or-fight-or-freeze response and during exercise. The parasympathetic nervous system dominates during resting periods to save energy and to maintain basic body functions; often referred to as the 'rest and digest' period. The parasympathetic nervous system also appears to be dominant in the 'freeze' reaction of the flight-or-fight-or-freeze response in the face of a stressor (McCorry, 2007).

Flight-or-fight reactions

The flight-or-fight reactions are incredibly important for survival in that they allow us to respond to threatening situations very quickly. Humans have evolved the flight-or-fight response to enable the body to react quickly when threatened. Our ancestors would have faced different threats than we have today but needed the same physiological response. Back in the day, hunting was dangerous but necessary for food. What if they were out and came across a lion? The choice? Protect the food necessary for survival of their tribe and fight the lion to win, or run very very fast and make an escape. These days we don't have lions walking around the grocery store when we do our shopping but our bodies are still able to activate the flight-or-fight response very quickly in dangerous situations such as having to jump out of the way of an oncoming car, or to stop someone falling from a dangerous height.

Regardless of the situation, the perception of a threat activates your sympathetic nervous system. When activated, the sympathetic nervous system sends signals to the adrenal glands, which are located in the body just above each kidney. The adrenal gland secretes the hormones **adrenaline, noradrenaline** (also referred to as epinephrine and norepinephrine) and cortisol into the bloodstream. These hormones prepare the body for flight or fight to enable the body to react quickly. Some physiological effects that these hormones produce include:

» increased heart rate and blood pressure: this enables more blood and oxygen to get to the necessary muscles and organs to either fight or flee
» increased breathing rate: this also provides more oxygen to the muscles and also to the brain, which helps with quick thinking
» increased glucose secretion: the liver increases the availability of glucose (sugar) to the muscles to provide more energy for use
» dilation of the pupils: the pupils dilate to allow eyes to take in more light and improve eyesight
» redistribution of blood: blood is directed away from extremities (hands and feet) so that it can go to the larger skeletal muscles for necessary movement
» suppression of functions that are not immediately needed: systems such as digestion and sexual functioning are limited so that resources can be redistributed to organs that are more vital for fighting or fleeing.

The flight-or-fight response is considered to be an adaptive response because the body is automatically prepared – both physically and psychologically – for flight or fight. Once you perceive the threat to be gone, the flight-or-fight response is switched off. Your parasympathetic nervous system now becomes dominant and activates the relaxation response, the counterpart of the flight-or-fight response, which calms your body by reversing the effects of the sympathetic nervous system, returning it to its normal level of functioning.

Freeze response

Like the flight-or-fight response, the **freeze response** is also an adaptive response that enhances our survival chances.

3.1.3 FLIGHT-OR-FIGHT-OR-FREEZE RESPONSE

Similar to flight-or-fight, the freeze response happens automatically without conscious thought; it is not a choice made by the person or animal. Think back to the scenario of the hunter and the lion. Had the hunter experienced the freeze response, they would have become immobilised and unable to move. This may have helped the hunter avoid detection so that the lion didn't see them or perceive the hunter as a threat. It also could have bought the hunter some time to enable their body to conserve energy and make a plan for a safe escape. This is actually a common situation in the animal kingdom, where the freeze response is often seen in action. A mouse that has been caught by a cat may freeze and become completely immobile. The cat, perceiving that the animal is dead, may take a break from hunting it or toying with it, and the mouse would be able to make an escape. Situations like these have led scientists to believe that the freeze response involves the activation of the sympathetic and parasympathetic nervous systems at the same time, so that either system can quickly become dominant as the situation requires it (Roelofs, 2017). This is often associated with the analogy of having one foot on the accelerator and the other foot on the brake. The parasympathetic nervous system is the brake, and while it is dominant it allows the body to prepare for action. As the sympathetic nervous system is also activated the body is ready to spring into action once the 'brake' of the parasympathetic nervous system is removed.

During the freeze response, while the parasympathetic nervous system is dominant, the:
» heart rate reduces
» blood pressure drops
» heightened muscle tension leads to reduced body mobility (movement)
» vocalisations are reduced (the ability to make speech or sounds)
» people experience heightened attention and awareness.

While the freeze response can be adaptive as a defence measure, there are situations when it can be maladaptive, as in, not helpful to the organism. There are times when the situation can be so overwhelming that the freeze response prevents the animal or person from taking further action. In 2007, a police officer in the Netherlands became severely injured during a knife attack. When the attack began, the police officer froze for a moment and then decided to defend herself. Later analysis of the attack revealed that if the officer had frozen for any longer, it is likely that more people would have been injured. However, had the police officer responded with the fight response prior to freezing, she may have prevented injury to herself (Roelofs, 2007). It is not hard to imagine the effect of the freeze response if it were to become activated in someone sitting an exam – they would be unable to move, they may feel faint from the sudden drop in blood pressure, and they would feel very tense while valuable time passed by. Usually, the freeze response is short lived, and the person is able to move forward and do their best work.

However, there are times when the trauma of a situation is so profound that it may lead the body to escalate to the next level of defence and enter *tonic immobility*. Tonic immobility differs from the freeze response in that it does not prepare the body for action and often occurs after flight or fight has failed or is not possible. Tonic immobility is most often observed in survivors of physical or sexual assault, plane and car crash victims, during wild animal attacks, and in soldiers who experienced a trauma. People who experience tonic immobility describe experiences of fear, immobility (being unable to move), feeling cold, being unable to feel pain, shaking, eye closure and dissociation from the situation (Kozlowska et al., 2015; Roelofs, 2017). While freezing occurs as an organism encounters a threat to prepare the body for further defensive actions, tonic immobility occurs during the threatening situation and appears to be a defence mechanism to reduce pain during an event (Sakai, 2021).

Role of cortisol

Cortisol is the primary stress hormone produced by the adrenal glands (located on top of each kidney) and it is directly secreted into the bloodstream for quick transportation throughout the body. Cortisol is essential to the maintenance of homeostasis (the body's processes for maintaining a healthy and balanced state). Cortisol is involved in a range of biological functions. These include glucose metabolism, regulation of blood pressure, insulin release for blood sugar maintenance, immune function and anti-inflammatory reactions, and central nervous system activation. Cortisol levels normally fluctuate throughout the day and night in a regular pattern. Although cortisol secretion varies among individuals, normally cortisol levels are higher in the morning, decrease during the day and are low at night. Cortisol is normally released in response to events or circumstances, such as waking up in the morning, exercise and in response to a stressor. Cortisol is often called the 'stress hormone' because both eustress and distress trigger its release into the bloodstream.

Cortisol is released by the hypothalamic–pituitary–adrenal (HPA) axis in your brain. When encountering a stressor, the immediate response of flight-or-fight-or-freeze is activated by the sympathetic nervous system. Shortly after this initial response, the HPA axis is stimulated and the hypothalamus (a region in your brain responsible for hormone production and regulating many important bodily processes) releases a hormone called corticotropin-releasing hormone (CRH) into the bloodstream. This hormone stimulates the pituitary gland (just below your hypothalamus in the brain) to release adrenocorticotropic hormone (ACTH), which travels down to the adrenal glands to stimulate the release of cortisol into the bloodstream (Figure 3.6). The HPA axis also monitors the amount of cortisol in the blood through a feedback loop to inhibit the release of cortisol and shut off the stress response to protect the body against too much activity in the HPA axis (Guy-Evans, 2021).

A short-term increase in cortisol levels affects the body in a number of positive ways. For example, cortisol floods the body with glucose and increases blood sugar levels to provide the body with an immediate burst of energy. It also mobilises energy by selecting the right type and amount of carbohydrate, fat or protein needed by the body to meet the physiological demands placed on it. In terms of our survival, it is vital that the adrenal glands secrete more cortisol during the flight-or-fight-or-freeze response so that the adaptive changes necessary to energise the body and prepare it for action occur. Ideally, the stressful event is resolved and then the parasympathetic nervous system takes over (becomes dominant) and calms the body down, returns cortisol levels to normal, and returns the body to a more stable level.

However, life is not always neat and easy and often we find ourselves still dealing with a stressor, even after the initial flight-or-fight-or-freeze response. The effects of adrenaline and noradrenaline do not last long; however, cortisol is released by a different biological pathway and is released over a longer period. This allows the body to continue to deal with stress for longer, such as with chronic stress. While this can be helpful as an adaptive response to the stress, it can also result in too much cortisol remaining in the bloodstream. This causes a biochemical and hormonal imbalance that can have several negative effects, including the suppression of the immune system, which reduces the body's ability to fight illness and infection (Figure 3.7). Other negative consequences of high levels of cortisol are impaired cognitive performance, suppressed thyroid function, blood sugar imbalances, decreased bone density and muscle tissue, higher blood pressure, lowered inflammatory responses in the body, slowed wound healing and increased weight gain due to accumulation of abdominal fat.

Furthermore, these effects can also introduce a new stressor to an individual, in the form of a health problem. If the adrenal glands become chronically fatigued because of prolonged activation, inadequate levels of cortisol will be secreted and this can also lead to health problems such as fatigue, sleep disturbances, emotional hypersensitivity, anxiety or mild depression. Either way, prolonged abnormal levels of cortisol in the bloodstream cause health problems that exacerbate a person's level of stress.

3.1.4 EFFECTS OF STRESS

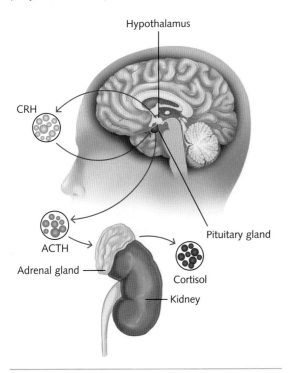

Figure 3.6 When a stressor is perceived, the HPA axis is activated from the hypothalamus, to the pituitary gland, down to the adrenal glands to stimulate the release of cortisol into the bloodstream.

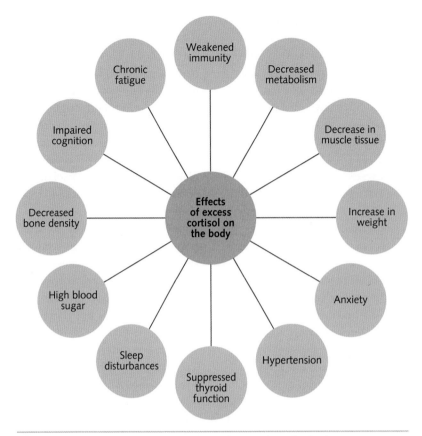

Figure 3.7 The effects of excess levels of cortisol on the body

ACTIVITY 3.1 THE STRESS RESPONSE

Try it

1. Below is a list of activities that lead to acute stress in most people. Find classmates who find the below activities eustressful or distressful and discuss the feelings associated with each activity.
   » Going on a waterslide
   » Riding on a roller-coaster
   » Handling exotic animals (snakes, spiders)
   » Skydiving
   » Bungee jumping
   » Public speaking or performing on stage

Apply it

2. When confronted by acutely stressful situations, we exhibit a 'non-specific stress response'. With reference to eustress and distress, explain why the stress response is considered 'non-specific'.

Exam ready

3. Which one of the following is not a result of long-term cortisol release?
   A. Cortisol assists in the metabolism of glucose to provide us with an immediate burst of energy.
   B. Cortisol suppresses the immune system, leaving us vulnerable to additional stressors.
   C. Cortisol slows wound healing.
   D. Cortisol causes sleep disturbances.

ANALYSING RESEARCH 3.1

Can meditation reduce chronic stress?

The Max Planck Institute for Human Cognitive and Brain Sciences in Germany and the Social Neuroscience Research Group of the Max Planck Society in Berlin conducted a study to determine whether practices such as mindfulness, gratitude and compassion could reduce the effects of chronic stress. Previous research conducted by the groups showed that training people in these techniques had positive effects on stressful days but the data was based on self-reports. For this long-term study, scientists wanted data that was more objective.

Meditation and chronic stress

The study aimed to test long-term effects of meditation on chronic stress levels. To gain objective data, the researchers tested the level of cortisol in individuals' hair. Cortisol is the main stress hormone that is released during stress, and particularly supports the body during times of chronic stress. One way to measure cortisol levels in the body is through testing levels present in hair. On average, a person's hair grows a centimetre per month, so changes in cortisol levels can be tracked through testing changes over time.

This study contained 332 participants who were each trained in different types of meditation for a period of nine months. For each three-month period, they trained in a specific type of meditation either focusing on attention and mindfulness, compassion and gratitude, or taking on the perspective of their own thoughts or the thoughts of others. Their training was for 30 minutes a day, five days a week using a tailor-made app and online platform. The study used counterbalancing so that different participants completed the three types of meditation training in a different order.

To measure the levels of stress, the study analysed the amount of cortisol in the hair at T0 (at the start of the study) and T1, T2 and T3, which is every three months. The level of cortisol was measured in the first three centimetres of hair, starting at the scalp.

The findings

In the first three months, there was a slight reduction in cortisol levels as analysed in the hair samples. However, by the sixth-month testing period the amount of cortisol in the participants' hair had decreased on average, 25 per cent. In the final three months (up to nine months total), the level of cortisol remained at low levels. These results were consistent regardless of order in which participants trained in the different types of meditation.

These findings suggest that while meditation has been shown in the past to reduce stress in the moment (acute stress), long-term meditation practices appear to be effective in reducing the effects of long-term stress (chronic stress).

"There are many diseases worldwide, including depression, that are directly or indirectly related to long-term stress," explains Lara Puhlmann, one of the members of the research team. "We need to work on counteracting the effects of chronic stress in a preventative way. Our study uses physiological measurements to [support the theory] that meditation-based training interventions can alleviate general stress levels even in healthy individuals."

Source: Adapted from Muller, V. (2021). Max Planck Institute. (2021, October 9). Hair samples show meditation training reduces long-term stress. *Neuroscience News*. Original article Pulhman, L. et al (2021).

Original article: Pulhman, L. et al. (2021). Contemplative mental training reduces hair glucocorticoid levels in a randomized clinical trial. *Psychosomatic Medicine, 83*, 894–905.

Questions

1. What was the aim of this study?
2. Identify the dependent and independent variable(s).
3. Using dot points, outline the method.
4. Identify the experimental design used in this study and outline one advantage of using this type of design for this study.
5. Explain why the experimenters used counterbalancing in the study.
6. What were the results of the study?
7. What conclusion(s) can be drawn from the study?

KEY CONCEPTS 3.1b

» The flight-or-fight-or-freeze response is an automatic physiological reaction to a stressor that involves the sympathetic and parasympathetic branches of the autonomic nervous system.
» The response prepares the body to:
  – flee (flight): escape the stressor
  – fight: confront the stressor
  – freeze: immobilise to evade detection and prepare in the face of a stressor.
» The flight-or-fight responses are activated when the sympathetic nervous system is dominant and adrenaline causes physical changes to flood the body with energy in order to flee or confront the stressor.
» The freeze response is activated when the parasympathetic nervous system is dominant, and is characterised by reduced heart rate and hyper-awareness. It aids in avoiding detection and preparing the body for further defensive actions.
» Cortisol is the main stress hormone and continues to be released after the FFF response to allow the body to adapt to stress for prolonged periods of time. It provides the body with energy; however, it can be harmful if it persists in high levels in the bloodstream for prolonged periods of time as it suppresses the immune system and increases the likelihood of illness.

Concept questions 3.1b

Remembering
1. Why is stress referred to as a psychobiological response?
2. Identify the stress hormones involved in the flight-or-fight-or-freeze response.
3. Which branch of the autonomic nervous system is dominant in the:
   a. flight-or-fight response?
   b. freeze response?
4. Identify two physiological responses you may experience in the freeze response. Explain how these responses might aid your survival in a life-threatening situation.
5. Identify two physiological responses you may experience in the flight-or-fight response. Explain how these responses might aid your survival in a life-threatening situation.

Understanding
6. Explain how too much cortisol in your bloodstream can impact your health. Give two examples.
7. What is the role of cortisol during chronic stress compared to acute stress?
8. What is the overall purpose of the physiological changes that occur in the flight-or-fight response?

Applying
9. Carina is getting ready for her first day of university. She has tried on multiple outfits and packed and repacked her bag six times. Carina told her mum that she had butterflies in her stomach because she is so excited to be starting this new chapter in her life. How would you categorise Carina's stress and how would it present in her body?

HOT Challenge
10. Tasha was working as a volunteer in Bangladesh for Oxfam International for a 12-month period. She was excited about the work but was struggling with day-to-day life. She did not speak the language and each day would find herself in a stressful situation due to not being able to communicate. Adding to this, traffic in the city was so bad it would take her over an hour to travel the 10 kilometres from her home to her workplace by car. She also didn't know many people in the city and was missing her friends and family very much. Using psychological terminology, explain why Tasha might decide to leave the position after 9 months instead of 12.

3.2 Selye's General Adaptation Syndrome

The **General Adaptation Syndrome (GAS)** is used to describe the physiological changes that the body automatically goes through when it responds to stress. From the 1930s through to the 1970s, Hans Selye conducted years of research on the short- and long-term effects of stress. He received several nominations for the Nobel Prize for his ground breaking research. Selye conducted much of his research on rats, exposing them to different stressors such as extreme heat and cold, and injecting them with different toxins, enzymes and hormones. Throughout his studies he noted that regardless of the type of stressor, the rats all ultimately experienced similar effects and illnesses in a similar pattern. He later observed similar effects in human patients with diverse health issues such as infections, heart disease and cancer (Rice, 2012). As a result of his studies, Selye concluded stress to be a non-specific physiological response, which can be brought on by internal and external stressors. As the reactions in the body are the same regardless of whether the conditions are pleasant or unpleasant, he developed a model of stress which became the General Adaptation Syndrome (GAS).

The GAS is the body's typical response pattern in terms of resistance to stress over time. Selye proposed that a bodily mechanism called 'adaptation' was required to respond to both eustress and distress. He noticed organisms respond to stress with a similar sequence of reactions to any traumatic event. Based on this, he concluded that humans react to any stress (real, symbolic or imagined) by putting into motion a set of physiological responses that attempt to alleviate the impact of the stressor.

According to Selye, the GAS consists of three stages: an alarm-reaction stage, a stage of resistance and a stage of exhaustion (Selye, 1976).

Alarm-reaction stage

The **alarm-reaction stage** comprises two substages: **shock** and **countershock**. When we first perceive a threat, we go into a state of shock, our resistance level falls below normal and our body acts as though it is injured. Body temperature and blood pressure drop and our muscles temporarily lose tone. These physical effects of shock reduce the individual's ability to deal with the stressor, and they feel momentarily helpless. For example, a VCE student may feel threatened as their critical exam period looms closer and they may enter the shock stage, and feel they cannot cope with the demand. If this happens, they may stop studying, have trouble concentrating and underperform.

After the shock stage, the body rebounds and enters the stage of counatershock. Countershock is characterised by the activation of the sympathetic nervous system and the release of adrenaline and noradrenaline. This leads to physiological effects such as increased heart rate and respiration so that glucose and oxygen can be delivered to the muscles. Cortisol is also released by the HPA axis a little later to increase the body's resistance to the stressor; however, it plays a more significant role in the next stage. In the example of a student in a critical exam period, when countershock kicks in, our student may feel that they have received energy and now have the necessary resources to stay awake, alert and able to function above their normal level.

Resistance stage

Following the alarm-reaction stage, if the stressor remains, the body enters the **resistance stage** as it attempts to stabilise its internal systems and fight the stressor. When we first move into the stage of resistance, the symptoms of the alarm-reaction stage subside. Although the body is better able to cope with the initial stressor, because physiological arousal remains higher than normal (but lower than during the alarm-reaction stage), continued high cortisol levels suppress our immune system and the body is more susceptible to 'wear and tear' and is at an increased risk of illness. A person may experience catching frequent colds or flu or have a cut or wound that may heal slower than normal.

Often in this stage, people are focused on managing the current stressor, and may ignore other social or personal commitments. The body is using a lot of resources to increase the resistance to the most significant stressor, which may reduce the ability to manage any additional stressors. For example, if the current stressor is upcoming exams that the student is preparing for, they will have less resistance to cope with any arising conflicts with friends.

Weblink
The GAS model

EXAM TIP
Many students often get confused between illness in the resistance and exhaustion stages of the GAS model. In a scenario-based question, keep an eye out for words like 'cold' or 'flu' – this usually indicates the resistance stage, particularly if the person is able to maintain their day-to-day functioning in some manner.

During the resistance stage, the student in the critical exam period will continue with the energetic approach to study they achieved in the countershock stage of alarm-reaction; however, after a period of time their bodily resources can become depleted, and this leads to exhaustion.

Exhaustion stage

If, during the stage of resistance, we defeat the stressor, our parasympathetic nervous system takes control, reversing the effects of the sympathetic nervous system and returning our body to its normal level of functioning. If we are unable to defeat the original stressor during the first two stages, we enter the **exhaustion stage**, in which our body's resources are drained and cortisol levels are depleted. A person in the exhaustion stage could experience fatigue and be at an increased risk of mental health disorders such as anxiety and depression. They are also at an increased risk of more serious physical illness such as diabetes, heart disease and high blood pressure.

As the VCE student undertaking exams enters this stage, they may become sick and physically worn out. They may also be experiencing depression or anxiety that can disrupt their day-to-day functioning.

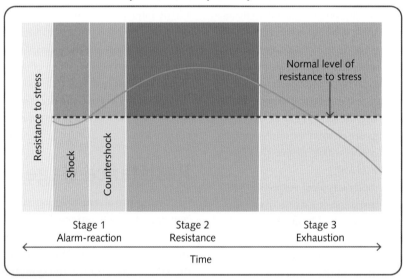

WB
3.2.1 THE GENERAL ADAPTATION SYNDROME

WB
3.2.2 THE GAS: A CASE STUDY

Stage 1: Alarm-reaction — When a threat is perceived, the body enters a state of 'shock'. Our resistance level to stress momentarily falls below normal. Then the body rebounds and enters the stage of 'countershock'. Activation of the sympathetic nervous system results in physiological changes that prepare the body for action (flight-or-fight) by producing a higher-than-normal arousal level that increases the body's resistance to the stressor.

Stage 2: Resistance — If the threat remains, the body enters the stage of resistance. Arousal is higher than normal (but lower than Stage 1) as the body attempts to stabilise its internal systems and cope with the threat. This higher arousal weakens the immune system and signs of illness appear because we are unable to cope with additional stressors.

Stage 3: Exhaustion — If the original threat continues, the body's resources are drained and stress hormones are depleted. Our immune system is depleted and the body moves into the stage of exhaustion and we become ill.

Figure 3.8 Hans Selye (1956) proposed that there are three stages in the body's reaction to stress, which he called the General Adaptation Syndrome.

When Selye examined animals in the later stages of the GAS, he found that their adrenal glands were enlarged and discoloured. He also found that there was intense shrinkage of the spleen and lymph nodes and many animals also suffered from bleeding ulcers deep in their stomach. Stress can have similar effects in humans. Stress can also disrupt our body's immune system, which mobilises defences, such as white blood cells, against invading microbes and other disease agents (Ader & Cohen, 1993). When our immune system does not function normally, we are more susceptible to long-term illness or disease. Because the immune system is regulated in part by the brain, stress and upsetting emotions can affect the immune system in ways that increase this susceptibility (Miller, 1998). For example, during major exam times, divorce, bereavement or job loss, our immune system can be compromised by higher than normal levels of cortisol and we fall ill. Figure 3.8 summarises Selye's proposal of the three stages of GAS.

ACTIVITY 3.2 EXPLAINING SELYE'S GAS

Try it

1. Draw Selye's General Adaptation Syndrome, similar to Figure 3.8. Annotate each stage of the GAS to explain how the autonomic nervous system brings about the changes in our resistance to stress at each stage.

Apply it

Over the course of the term, Oscar has been very busy with school commitments and extracurricular activities. On top of his normal VCE subjects, Oscar completes a language on the weekends. Additionally, Oscar has been selected for his football competition's representative side, so he now trains three nights a week and plays a game every Sunday.

During the first half of the term, even though he felt stressed and tired some of the time, Oscar managed his commitments quite well. However, by the last few weeks of term, Oscar started suffering severe migraines and had to take some days off school.

2. a Which stage of the GAS was Oscar most likely in during the first half of the term? Justify your response.
   b Which stage of the GAS was Oscar most likely in during the last few weeks of term? Justify your response.

Exam ready

3. Respectively, the vertical axis of the GAS and horizontal axis of the GAS are

A the time taken to combat the stressor; our level of resistance to stress.

B our level of resistance to stress; the amount of cortisol released at each stage.

C the amount of cortisol released at each stage; the time taken to combat the stressor.

D our level of resistance to stress; the time taken to combat the stressor.

Strengths and limitations of the GAS

Selye was considered a pioneer of stress research and his studies were extensive and spanned decades. The GAS model was developed as a result of this research and is supported by empirical evidence. The GAS model identifies the biological processes associated with the stress response, and vitally identifies the different biological processes dependent on the stage of stress and whether the stress is acute or chronic. Moreover, the GAS model was the first to highlight that stress has a major impact on the immune system and to make the important connection between stress and increased risk of illness.

However, there is criticism that the GAS model places too much emphasis on the biology of the stress response and does not take into account the important psychological factors that explain the subjective stress experience of individuals. Furthermore, the GAS model fails to acknowledge the unique environmental and biological factors of an individual that could impact their response to stress. In fact, more recent research indicates that different stressors activate different biomarkers and regions of the brain, therefore stress may not be as non-specific as proposed by Selye. And finally, while Selye's research was extensive, it was predominantly based on animal research (though included some human observation) and the results cannot be generalised to human populations.

KEY CONCEPTS 3.2

» Selye's GAS model outlines the non-specific physiological response to stress. It has three stages: the alarm-reaction stage (including shock and countershock), the resistance stage and the exhaustion stage.
» The GAS model was the first to link the stress response to a compromised immune system and identifies the biological processes of the stress response; however, it fails to acknowledge the subjective psychological experience of the stress response and is predominantly based on animal research.

Concept questions 3.2

Remembering
1. Outline the three stages of the GAS model. **r**
2. Explain the difference between shock and countershock. **e**
3. Describe the role of cortisol in each stage of the General Adaptation Syndrome. **r**

Understanding
4. According to the General Adaptation Syndrome, why does prolonged stress cause physical illness? **e**
5. Draw a Venn diagram to explain the similarities and differences between the flight-or-fight-or-freeze model and the GAS model. **d**

Applying
6. Why do you think that Selye's research in the GAS was largely based on animals instead of humans?
7. Why is it beneficial that the relationship between stress and the immune system has been established?

HOT Challenge
8. Iman is a Year 12 student who has always been very ambitious and is determined to get a very high ATAR score to pursue her dream of becoming a neurosurgeon. At the start of Year 12 she decided she would study for 4 hours every night and 10 hours on weekends. She also had a part-time job two days a week. During the year, Iman experienced some colds but nothing too serious. Iman continued to put a lot of pressure on herself and took very few breaks in the study period before and during exams. In the last week of the exam period, Iman felt extremely fatigued and overcome with a sickness that she could not seem to shake. She also struggled to concentrate when she was trying to study. According to the model, assess which stages of the GAS Iman was in at different times throughout the year. Justify your response. **c**

3.3 Stress as a psychological process

Psychologically, stress has a range of cognitive, emotional and behavioural effects. When we are stressed, especially if the stress is intense and prolonged, cognitive functions such as perception, thinking, attention, memory and learning are negatively affected. Instead of thinking logically and rationally, we may become confused and indecisive. Our sense of being is under threat by the stressor and our lack of confidence in our ability to manage the stressor may be magnified to the point where we become dysfunctional and unable to complete our normal daily activities. We may find it difficult to focus our thoughts and solve problems. Our emotions may become unbalanced and difficult to control. We may experience an increase in negative moods such as frustration, grief, anxiety, apathy, anger and aggression. Some people may also experience panic attacks and feelings of hopelessness and helplessness. Stress may also bring about a range of negative behavioural changes. These can include disrupted sleeping patterns, avoiding people or situations by withdrawing from social activities, and an increase in maladaptive behaviours such as drug and alcohol abuse, or eating too much or too little.

Weblink
Stress pulls us apart

ANALYSING RESEARCH 3.2

How stress can lead to inequality

A study carried out by scientists at Switzerland's Ecole Polytechnique Fédérale de Lausanne shows how stress could actually be both a consequence and a cause of social and economic inequality, affecting our confidence to compete with others and make financial decisions. The research also shows how cortisol levels in the bloodstream may contribute to this inequality.

Stress and confidence

To test the effect of stress on confidence, more than 200 university students with a mean age of 19 years were recruited to take two online tests: one to assess their IQ, and one to measure their trait anxiety (how prone a person is to see the world as threatening and worrisome). A week later, about half of the study's participants underwent a standard psychological procedure designed to cause acute social stress, such as going through a mock job interview and performing mental arithmetic tasks before an expressionless audience. The other half of the participants did not undergo the stress-inducing procedure. All participants were then given two options in a game where they could win money: they could either take their chances in a lottery, or they could use their IQ score to compete with that of another, unknown participant's; the one with the higher IQ score would be the winner.

In the non-stressed group, approximately 60 per cent of participants chose the IQ score competition over the lottery, showing overall high confidence in the participants, regardless of their trait anxiety scores. But in the group that experienced stress before the money game, the competitive confidence of participants varied depending on their trait anxiety scores. In people with very low anxiety, stress actually increased their competitive confidence compared to their unstressed counterparts; in highly anxious individuals, it dropped.

The findings suggest that stress affects a person's competitive confidence by raising or suppressing an individual's confidence depending on their predisposition to anxiety. When stressed, people with low trait anxiety experience a boost in competing confidence. People with high trait anxiety experience less competing confidence when stressed.

Stress and cortisol

The researchers also found that the effects of stress on confidence were influenced by the hormone cortisol, which is normally released from the adrenal glands in response to stress. The team examined saliva samples from stressed participants for the presence of cortisol. In people with low anxiety, those that showed higher confidence also showed a higher cortisol response to stress. But in highly anxious people, high cortisol levels were associated with lower confidence, which suggests that the behavioural effects of stress are linked to a biological mechanism.

The researchers argued that the findings can be seen as a simulation of confidence in social competition and the way it relates to socioeconomic inequality. Studies have shown that, in areas with wide socioeconomic inequality (for example, a wide rich–poor gap), people on the low end of the social ladder often experience high levels of stress as a consequence. "People often interpret self-confidence as competence," says Carmen Sandi, leader of the research team. "So if the stress of, say, a job interview, makes a person over-confident, they will be more likely to be hired – even though they might not be more competent than other candidates. This would be the case for people with low anxiety."

The researchers claimed that their results suggest that far from being only a product of competitive inequality, stress must now also be regarded as a direct cause of it. In other words, stress can become a major obstacle in overcoming socioeconomic inequality by trapping highly anxious individuals in a self-perpetuating loop of low competitive confidence.

Source: adapted from Ecole Polytechnique Fédérale de Lausanne. (2015, February 18). How stress can lead to inequality ScienceDaily.

Questions

1. What was the aim of this study?
2. Identify the dependent and independent variable(s).
3. Using dot points, outline the method.
4. What were the results of the study?
5. What conclusion(s) can be drawn from the study?

Lazarus and Folkman's Transactional Model of Stress and Coping

In 1984, Richard Lazarus and Susan Folkman introduced a model that explained the mental process, or cognitive appraisal, that influences our response to stressors. This model, the **Transactional Model of Stress and Coping**, proposes that stress is a subjective experience that varies between individuals depending on how they interpret the stressor and perceive their own ability to cope with it. The transactional model focuses on how a person interacts with their external environment, and stress is viewed as the result of how a person appraises (evaluates) a situation and their ability to cope with it. According to this model, stress is experienced when there is an imbalance between the demands of a situation and the person's evaluation of their ability to cope with these demands. Stress is experienced if the demands on an individual exceed their perception of their coping resources, even if that stressor is not life-threatening or it poses only a perceived (not an actual) threat. Unlike Selye's biological explanation of the stress response, Lazarus and Folkman suggest that stress responses are directed by our cognitive appraisal of the stressor as either a challenge or a threat and also by our assessment of the personal and social resources we have to cope with the stressor. In other words, our stress response depends on emotions and psychological factors unique to the individual.

In this model, Lazarus and Folkman suggest that the **psychological stress response** is a two-way process that comprises a transaction between an individual and their environment. It involves the production of stressors by the environment and the subjective response, or cognitive appraisal, an individual makes about their ability to cope with these stressors. The process of cognitive appraisal consists of two sequentially linked stages: primary and secondary appraisal.

Primary appraisal: how significant or threatening is the event?

The first stage of the transactional model is known as **primary appraisal** and is the stage in which the person evaluates whether the stimulus (stressor) is stressful and if so, in what way. Thus, primary appraisal involves two steps. First, the individual appraises the severity of the stressor. In this stage, the person may decide that the stimulus is:

» irrelevant: the stimulus has no significance for the person and will not cause stress
» benign-positive: the stimulus is judged to be neutral or positive for the person
» stressful: the stimulus is judged to be significant and relevant to the person and a source of stress.

If the person appraises the stimulus as stressful, then the next step of primary appraisal is to determine the way in which the stressor is stressful. Specifically, this involves evaluating whether the stressor is a:

» harm/loss: the stressor has already caused harm or damage to the individual (for example, the person has had an accident that has caused pain and resulted in a broken leg)
» threat: the stressor could cause harm or damage to the individual in the future (for example, the person cannot work due to their broken leg and has lost income)
» challenge: the stressor could be perceived as a potential for growth or a good opportunity for the individual (for example, the person was thinking about changing to a less physically demanding job and this provides an opportunity to update their resume and start applying for new and exciting roles).

Note that an appraisal of harm/loss or threat would both cause distress to the individual, whereas an appraisal of the stressor as a challenge would cause the individual to experience eustress.

Secondary appraisal: how can I cope with the event?

The second stage of the transactional model is referred to as **secondary appraisal** and is when an individual assesses whether they can cope with the stressor. People can assess their own internal and external resources available to them to evaluate whether they can meet the demands of the stressor. If someone believes they have the resources to cope, then they will make a positive secondary appraisal resulting in no stress or eustress. However, if a person feels that the demands of the stressor are greater than their resources to cope, they will make a negative secondary appraisal and experience distress. For example, being asked to speak at a whole school assembly could potentially be a very stressful situation. One person may initially appraise the situation as stressful, but then realise that they have a lot of time to prepare and teachers who they can ask for help with their presentation. They might see the experience as an opportunity to grow. On the other hand, someone who experiences 'stage fright' may have the same time and resources, but their internal coping strategies could be very different to the other person and make it very difficult for them to meet the demands of the task, causing them to feel distressed. Figure 3.9 shows the sequence of events that follow as a result of the presence of stressors, the involvement of the two stages of cognitive appraisal and finally the stress response.

3.3.1 LAZARUS AND FOLKMAN'S TRANSACTIONAL MODEL OF STRESS AND COPING

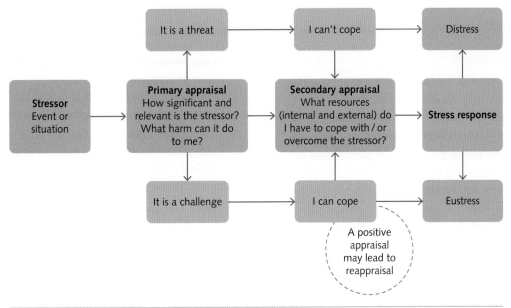

Figure 3.9 How we appraise a situation or event influences the type of stress we experience, how much stress we experience and how effectively we respond to a stressor. If, during primary appraisal, we appraise the stressor as a threat, we experience distress. If we appraise it as a challenge, we experience eustress. If we appraise it as neither, we experience no stress at all.

ACTIVITY 3.3 LAZARUS AND FOLKMAN'S TRANSACTIONAL MODEL

Try it

1 Match the component of Lazarus and Folkman's transactional model to the explanation of the component.

Component of the transactional model		Explanation of the component	
1	Challenge	A	An appraisal we make when the stressor has no significance to us
2	Secondary appraisal	B	The situation is appraised as stressful and provides us with a chance for personal growth
3	Primary appraisal	C	An evaluation of our coping resources
4	Threat	D	The situation has been appraised as stressful and the stressor has already caused damage
5	Harm/loss	E	An appraisal we make when the stressor is perceived as harmless or beneficial
6	Benign/positive	F	An evaluation of the nature of the stressor
7	Irrelevant	G	The situation is appraised as stressful and has the potential to cause damage in the future

EXAM TIP
When breaking down a person's response to a stressor, be sure to use the language of appraisal to ensure you indicate that you are aware that both primary and secondary appraisals involve an individual making judgements related to their individual circumstances.

Apply it

Jake is a Year 12 student who has ambitions of playing football professionally. However, during the first game of the season at his local club, Jake suffered a serious knee injury. This was a major setback to Jake's aspirations to play AFL.

2 Assuming that Jake has appraised the situation as stressful, identify and explain two primary appraisals that Jake may form.

Exam ready

3 Which of the following is a key distinction between Lazarus and Folkman's Transactional Model of Stress and Coping and Selye's General Adaptation Syndrome?

A The transactional model provides us with a biological model for understanding how we respond to stress, while the GAS provides us with a psychological model for understanding how we respond to stress.

B The transactional model was studied using rats, with generalisations made to humans, while the GAS was studied using human participants.

C The transactional model provides us with a psychological model for understanding how we respond to stress, while the GAS provides us with a biological model for understanding how we respond to stress.

D The transactional model explains how adrenaline influences our response to stress, while the GAS explains the role of cortisol in our response to stress.

Strengths and limitations of the Transactional Model

Lazarus and Folkman's model is useful because it distinguishes between eustress and distress, and it acknowledges that stress is a subjective experience. By recognising that different individuals have different responses to stress, it helps us understand the role personal interpretation and appraisal play in the stress response. It also highlights that people can change their appraisal of a stressor and their response to it, therefore the individual plays an active role in their stress response. Unlike Selye's biological model of stress, the transactional model considers the role a range of psychological factors such as personality, motivation, confidence and sense of self play in the stress response. However, because it does not consider the automatic physical responses the body has to a stressor, it has been criticised as being too simplistic. Also, because of its subjective nature and the variability of individual responses, Lazarus and Folkman's model is difficult to test through experimental research. Some critics also point out that it is difficult to separate primary and secondary appraisal, and they suggest that secondary appraisal actually happens simultaneously with primary appraisal.

KEY CONCEPTS 3.3

» Lazarus and Folkman's Transactional Model of Stress and Coping views stress from a psychological perspective.
» The model considers the unique characteristics of an individual that make up their subjective stress response in a transaction with their external environment.
» The model consists of two major stages: primary appraisal and secondary appraisal.
» During primary appraisal, a person evaluates whether a stimulus is stressful and whether the stressor is harmful, a threat or a challenge.
» During secondary appraisal, a person evaluates the internal and external resources available to cope with the stressor.
» Strengths: focuses on the subjective experiences of the individual and acknowledges that people have different responses to the same stressors; focuses on the cognitive, affective and behavioural (psychological) responses to stress; considers a range of different psychological factors in the stress response.

» Limitations: does not refer to or consider the biological responses to stress; primary and secondary appraisal may not occur as two separate processes; model may be too simplistic as most people are not aware of the appraisal process as it occurs.

Concept questions 3.3

Remembering
1. Describe the two processes that individuals undergo during a stress response, according to the Transactional Model of Stress and Coping. **r**
2. What is the difference between evaluating the stress as harm/loss and a threat? At what stage of the model does this evaluation occur? **r**

Understanding
3. Imagine you were Lazarus or Folkman. Justify why you have named the model the Transactional Model of Stress and Coping.
4. Identify two strengths of the Transactional Model of Stress and Coping. **r**

Applying
5. As a model to explain the stress response, how does Lazarus and Folkman's Transactional Model of Stress and Coping differ from Selye's GAS model? **e**

HOT Challenge
6. Kent is feeling stressed about an upcoming assessment for Psychology. Provide some examples of potential internal and external resources available to Kent that he could consider during his secondary appraisal process. **c**
7. With reference to Figure 3.9, construct two similar flowcharts to represent Kent's stressful situation: one where he appraises that he cannot cope with the stressor, and a separate one where he appraises that he can cope with the stressor. **d**

3.4 The gut–brain axis

Butterflies in your stomach. Comfort food. Tummy doing backflips. Nervous poos. When you think about it, it's pretty clear that there is a link between your brain and your gut. What you may not know is that scientists have established that this link is so strong and complex that the gut is now often referred to as the second brain! Early studies investigating the connection between the gut and the brain focused on the physical responses of digestion such as automatic salivation in response to food, or the way in which the stomach lets the brain know if we are hungry or full. But more recently, scientists have been exploring the way in which the gut and brain communicate through the **gut–brain axis (GBA)**, and the role this connection plays in influencing our mental wellbeing and behaviour.

The GBA is the network of bidirectional (two-way) neural pathways that enable communication between bacteria in the gastrointestinal (GI) tract and the brain.

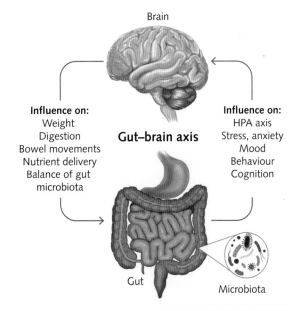

Figure 3.10 The gut–brain axis provides two-way communication between the gut and the brain. Research into the GBA is a new field in psychology in exploring how the two influence each other, and opens up new and interesting ways to manage stress and maintain mental wellbeing.

The GBA includes the central nervous system, the sympathetic and parasympathetic branches of the autonomic nervous system, the enteric nervous system, the vagus nerve, and the gut microbiota. It is also linked to the immune and endocrine (hormonal) systems (Figure 3.10).

The enteric nervous system (ENS) is a branch of the autonomic nervous system which functions independently of the central nervous system to manage the functions of the digestive system. The ENS is highly complex and is estimated to have 100–500 million neurons (Breit et al., 2018). At around nine metres long, it is an extensive mesh-like network of neurons embedded throughout two layers of gut tissue within the gastrointestinal tract including the oesophagus, stomach, anus, and both the small and large intestines (Figure 3.11). The ENS has many functions and regulates the major processes of the gut including:

» controlling motility: the stretching and contractions of the muscles of the GI tract that enable food to move through the digestive tract
» detecting nutrients for the body to use
» regulating fluids and blood flow within the enteric nervous system
» immune and defence responses (against toxic foods and bacteria)
» maintaining the chemistry of the gut (includes hormones, digestive acids and neurotransmitters)
» communicating with the central nervous system.

Communication between the gut and the brain happens primarily through the **vagus nerve**. The vagus nerve is the longest nerve in the human body running from the brain stem all the way down to the intestines and is the main contributor to the parasympathetic nervous system. It connects and innervates (provides neural energy to) many organs such as the gut, heart, lungs and liver. The vagus nerve connects with most of the gastrointestinal (GI) system and is the major communication route between the gut and the brain. Communication between the gut and the vagus nerve is bidirectional, meaning that neural signals go both ways between the two major organs; however, this communication is not divided evenly. Ninety per cent of the vagus nerve fibres are afferent (sensory) connections that send signals 'up' from the gut to the brain. Sensory information from the gut includes information such as pain, movement in the muscles that enable food to move for digestion, and tension in muscles and cells that indicate levels of fullness. These afferent nerves also communicate information about hormones, neurotransmitters and other chemical signals produced in the gut to the brain. The remaining 10 per cent of nerve fibres send efferent (motor) signals from the brain to the gut such as the release of saliva and stomach acids, information about incoming food and required changes in movements to aid digestion. They also send messages about flight-or-fight-or-freeze and the return to homeostasis (Breit et al., 2018). The gut communicates with the brain about many of its processes related to digestion, but psychologists are most interested in the microbiota that lives in our gut and how it interacts and influences the brain.

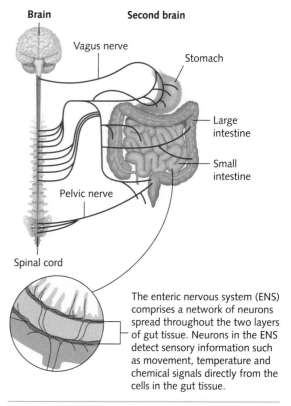

The enteric nervous system (ENS) comprises a network of neurons spread throughout the two layers of gut tissue. Neurons in the ENS detect sensory information such as movement, temperature and chemical signals directly from the cells in the gut tissue.

Figure 3.11 The interaction of the gut microbiota and the central nervous system. Neurons in the enteric nervous system detect sensory information directly from cell walls in the gut, convert the information into action potentials, and then transmit those signals to the vagus nerve. The vagus nerve communicates the signals to the brain.

The interaction of the microbiota and the nervous system

The **gut microbiota** (also referred to as microbiome) is the highly diverse and dynamic system of almost 100 trillion bacteria and other microorganisms that live in the human gastrointestinal (GI) tract. There are approximately 1000 different species of bacteria in the microbiota that have co-evolved with humans over thousands of years to form a mutually beneficial relationship (Thursby & Juge, 2017). This relationship is so beneficial that some scientists argue that the GBA should in fact be called the microbiome–gut–brain axis.

A healthy gut microbiota is extremely diverse, has a balance between 'good' and 'bad' bacteria, and is seen as a strong indicator of good physical and mental health. The microbiota has evolved to respond to changes in the internal and external environment to maintain health and wellbeing. However, the microbiota is influenced by several factors including genetics, diet, metabolism, age, illness, geography, changes in diet, antibiotic treatment and stress (Foster & McVey Neufeld, 2013). Dysbiosis (when the gut bacteria become less diverse or there is no longer a healthy balance of bacteria) can cause a range of digestive illnesses and reduce the effectiveness of the immune system overall. Thus, any disruption in the health of the microbiome can increase the body's susceptibility to disease. Scientists are now discovering, it can also impact a range of psychological processes such as cognitive responses, feelings of stress and anxiety, feelings of depression and some social behaviours.

So how do messages from these bacteria in your gut get all the way to the brain to be able to influence the nervous system? As stated earlier, the enteric nervous system is a very large network of neurons that are embedded throughout the two layers of the gut tissues in the GI system. Because these neurons exist in the walls of the intestines (and all throughout the GI system), they are able to detect sensory information directly from the gut in relation to movement, temperature and chemical stimuli such as nutrients, bacteria and bacterial by-products (chemicals produced by the actions of bacteria). Once detected, this information is converted into action potentials and transmitted to the vagus nerve. The vagus nerve then transmits that information to the brain, thus linking the signals in the intestines to the brain (Breit et al., 2018; Duke University, 2018).

3.4.1
THE GUT–BRAIN AXIS (GBA)

Weblink
The gut–brain connection

ACTIVITY 3.4 GBA FLOWCHART

Try it

1. Create a flowchart to explain the gut–brain axis. In your flowchart, ensure you include the following terms: gut microbiota; CNS; stress; vagus nerve; psychological effects; cortisol; gastrointestinal system; enteric nervous system; brain.

Apply it

A growing area of research is the use of faecal transplants (receiving faeces from a 'healthy' donor) to treat various mental health conditions. An analysis of 21 studies showed that the use of faecal transplants reduced people's experiences of anxiety and depression.

2. Using your knowledge of the gut–brain axis, explain how a faecal transplant could improve a person's mental health.

Exam ready

3. Which one of the following statements about the gut–brain axis is incorrect?

A. Communication between the gut and the brain occurs primarily through the vagus nerve.

B. The gut–brain axis is an example of a bidirectional relationship, with the gut and brain influencing each other.

C. Acute stress has a more pronounced influence on the gut–brain axis than chronic stress.

D. Disruptions to the microbiota appear to affect psychological functioning.

The impact on psychological processes and behaviour

The gut is called the second brain because psychologists have found evidence that shows that the health of the microbiota can influence our psychological processes such as the way we think, feel and behave.

The study of the GBA and its influence on the CNS is still a relatively new field in psychology and most of the research has been conducted using animal research, though there have been some promising studies involving humans. The most common animal used in research on the GBA is mice. Scientists study the influence of the microbiome by breeding germ-free mice (mice that have no gut microbiota at all) and by manipulating and making changes to the gut microbiota in mice and humans. The microbiota can be manipulated through positive intervention with probiotics – living organisms found naturally in some foods. These probiotics can be provided through supplements, which are also known as 'good' or 'friendly' bacteria because they fight against harmful bacteria and prevent them from settling in the gut. Conversely, antibiotic treatment is also used to manipulate the microbiota in a negative manner, as they can harm healthy gut bacteria, thus influencing the whole microbiome.

Through this research, scientists have found that the health of the microbiota can influence our social behaviour, cognition and feelings of depression, anxiety and stress.

Social behaviour

Studies with mice found that those born without a microbiome (referred to as germ-free mice) were more likely to display antisocial behaviours, such as avoiding other mice. However, they also found that transferring faecal matter (poo) from other mice with healthy microbes into the germ-free mice changed the activity of certain neurons that are influenced by the gut and improved the social behaviours of the mice (Wu et al., 2021). In humans, researchers have found that those who have more diverse gut microbiota tend also to have larger social networks. Though researchers do recognise that gastrointestinal illnesses and some mental disorders are likely to limit someone's social interactions (Nguyen et al., 2021).

Cognition

There is increasing evidence supporting the idea that changes in the gut microbiota can influence cognitive function. Animal studies have demonstrated that mice with no microbiota and those who had long-term antibiotic treatment performed worse on memory-related tasks, which suggests a connection between memory and the microbiome. While there is limited research in humans, probiotic treatments appear to have beneficial effects on memory and cognition. One study in healthy elderly patients found significant improvement in cognitive test performance compared with a placebo group (Cryan et al., 2019).

Depression

Depression is associated with instability in the HPA axis, and evidence has determined that the gut microbiota plays a role in the regulation and dysregulation of the HPA axis (Bastiaanssen et al., 2020). In humans, some studies have found that people diagnosed with depressive disorders are more likely to also have a disrupted microbiome (Horne & Foster, 2018) (Figure 3.12). Adding to this, participants in studies involving probiotic treatment to manipulate the microbiota in a positive manner, reported improved mood and reduced distress levels after 30 days (Foster & Neufeld, 2013).

Anxiety

Studies have shown that the presence of specific unhealthy bacteria and infection in the microbiome can increase anxiety-like behaviours in mice. There have been few human studies specifically related to anxiety; however, so far, intervention with probiotics appears to show some improvement in symptoms of anxiety (Cryan et al., 2019).

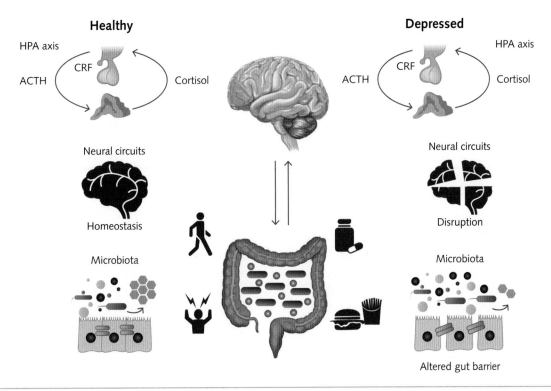

Figure 3.12 Scientists have established that the gut microbiota plays a role in the regulation of the HPA axis. Thus, when there is dysregulation in the microbiota, this is communicated to the central nervous system via the gut–brain axis and affects the gut–brain axis. This, is turn, can affect mental health through increasing the risk of depression, anxiety and stress.

Stress and the microbiota

Acute stress appears to have a limited effect on the microbiome; however, chronic stress has been well-established to increase the risk of illness and disease. As you learned in Section 3.2. Repeated or prolonged activation of the HPA axis has been linked to an increased risk of GI disorders such as inflammatory bowel disease (including Crohn's disease and ulcerative colitis), irritable bowel syndrome and coeliac disease. These diseases have been associated with the reduction of 'good' bacteria and reduced diversity in the microbiome for many years. Thus, the psychological and **physiological stress responses** are clearly linked to dysbiosis in the gut microbiota (De Palma et al., 2014).

However, the relationship between stress and the microbiota is complex. Research indicates that not only does stress appear to disrupt the health of the gut and microbiota, but a disrupted microbiota appears to influence and increase stress activity as well. Due to the connection between the gut microbiota and the CNS via the gut–brain axis, the microbiota is able to influence the regulation and sensitivity of the HPA axis. Research has found that mice with disrupted microbiota had increased stress reactivity (via the HPA axis) in response to an environmental stressor (Foster & Neufeld, 2013). Furthermore, studies suggest that stress in early life can have long-term consequences on a person's sensitivity to stress. Researchers found that mice who experienced stress in their early life were more likely to have a hyperactive HPA axis – meaning that the HPA axis is more sensitive to what may be perceived as a stressor and therefore may be activated more frequently and in situations where it is not necessary. They also had increased sensitivity to movements in the gut and bacterial imbalance. However, in both mice and humans, studies have repeatedly demonstrated that the positive use of probiotics and a healthy gut-friendly diet, appear to improve the negative effects of stress as well as other health issues (De Palma et al., 2014).

3.4.2 THE HUMAN MICROBIOME PROJECT

When examining the link between the GBA and thoughts, feelings and behaviours, it is a bit like the chicken and the egg scenario – as in, which came first? While studies show that changes in the microbiota can influence the brain, it has also been shown that stress, chronic depression and anxiety can influence the microbiota. Thus, it appears that much like the vagus nerve that connects them, the relationship between the microbiota and the brain is a dynamic and bidirectional relationship that continues to provide feedback and impact both components. The GBA is an emerging field in psychology and as such, we don't yet fully understand the relationship between the GBA and the CNS. Furthermore, the type of research used so far to study the connection between our two brains has many limitations in the ability to establish clear cause and effect relationships, especially in humans. However, it is clear that this is a new and very exciting field of study which has great potential in improving mental health and wellbeing.

KEY CONCEPTS 3.4

- The gut–brain axis is the network of bidirectional communication pathways that enable communication between the microbiota and the central nervous system (brain).
- Part of the GBA, the enteric nervous system, involves millions of neurons within the digestive tract that regulate major functions of the gut and communicate with the central nervous system to regulate these functions.
- Communication between the gut and the brain occurs primarily through the vagus nerve.
- Disruptions to the microbiota appear to affect psychological functioning and have been linked to changes in social behaviour, stress, anxiety and depression.
- Communication is bidirectional between the microbiota and brain, and they appear to influence each other. For example, disruptions in the microbiota can contribute to feelings of stress, depression and anxiety. However, disorders such as chronic stress, depression and anxiety appear to negatively influence the health of the gut microbiota.
- The use of probiotics appears to be a promising field in the treatment of disruptions to the microbiota and psychological disorders.
- The GBA is an emerging field of psychological research and most research in this field has been conducted using animal research and therefore has limitations.

Concept questions 3.4

Remembering
1. The enteric nervous system is part of the _____ nervous system.
2. List the components of the gut–brain axis.
3. What is the vagus nerve?
4. Identify three factors that can influence the gut microbiota.

Understanding
5. Draw a flowchart to depict the way in which the gut microbiota communicate with the brain.
6. Why is the gut called the 'second brain'?

Applying
7. Buckley had a serious infection and was required to take a course of antibiotics for a prolonged period of time to treat it. However, after Buckley's initial infection healed, he found that he was experiencing some issues with his digestion and gut. What are some psychological symptoms that Buckley might also be experiencing?
8. As part of his treatment, Buckley's doctor has recommended to him that he add probiotics and probiotic food into his diet. Why would his doctor recommend this?

HOT Challenge
9. As an emerging area of study, research on the GBA relies heavily on the use of animal research and correlational research. Using your knowledge of research methods, evaluate the benefits and limitations of each type of research.

3.5 Coping strategies

No one is immune to stress; it is an integral part of our lives. Obviously, the simplest way of coping with stress is to modify or remove its source. This is often impossible, which is why learning to manage stress is so important. This is where coping comes in. **Coping** is a process involving constantly changing thoughts and behaviours so we can manage the demands (internal and/or external) of stressors we appraise as taxing or exceeding our resources (Lazarus & Folkman, 1984). While our body has an innate physiological stress response in terms of the flight-or-fight-or-freeze response, our coping response must be learned. If we are going to effectively cope with stress and maintain good health and wellbeing, we must develop a set of coping skills. **Coping skills** consist of behaviours or techniques that help us solve problems or meet demands. Generally, coping does not consist of a single response, but a series of responses that are repeated in order to handle the remaining, continuing or changed nature of a stressor (Snyder, 2001). Coping is a complex process that varies according to the demands of the stressor, the individual's appraisal of the situation, and the personal and social resources available.

Context-specific effectiveness

By now you probably know that when you are sick your body needs different things depending on what type of illness you have and how severe it is. Sometimes it is healing to go for a gentle walk, get some sun and eat some breakfast, whereas other times your body needs you to stay in bed all day with a hot water bottle, can only handle soup, and needs to sleep a lot. Just like your body needs different strategies for different illnesses, different stressors and the circumstances surrounding them need coping strategies that are specific to the context in order to help you manage the stress.

Context-specific effectiveness refers to a coping strategy being more effective when the strategy is well matched to the stressful situation. As well as situational factors, a context-specific coping strategy also takes into account the unique characteristics of the individual's personality, interests, past experiences, knowledge, skills and access to external resources and support.

For example, Harper and Karim are in the same psychology class and have a SAC on the nervous system next week. Harper attended all classes and has no other SACs that week. Karim was absent for the lesson on spinal reflexes and has a Literature SAC two days before. While both students are nervous and want to do well, each will need to adapt their coping strategies to suit their circumstances. Harper plans to complete two practice SAC questions and have her sister test her with flashcards for 20 minutes each night. Although Karim needs to access his available resources to learn the spinal reflex for the first time, he will initially focus on preparing for the Literature SAC as he is more stressed about it. He will then have time to concentrate on Psychology as he has a lot of study periods and no work shifts. In this case, while both Harper and Karim seem to have the same stressor, their individual circumstances and knowledge differ and they therefore have to use different strategies in order to 'match' the demands of the stressor.

Coping flexibility

Moreover, to cope effectively under different circumstances, a person must demonstrate that they have **coping flexibility**. Coping flexibility refers to an individual's ability to stop an ineffective coping strategy (or evaluate their coping process) and implement an alternative effective coping strategy (or adapt their coping process). Coping flexibility is a strong predictor of psychological health and is a good measure of a person's adaptive ability. It determines the extent to which a person can cope with changes in their circumstances and how creative they can be in using available resources to manage their stress.

For example, in the previous scenario, if Karim were to discover that he had an excursion for Japanese the day before his Psychology SAC, this would change his situation. He would no longer have the study periods he was relying on. To cope effectively, he must change his original plan and begin studying for Psychology much earlier to meet the demands of the stressor and achieve success.

Stress research suggests that people who have learned a variety of coping skills generally handle demands and solve problems more easily and effectively than those who have not. They are less likely to experience distress and more likely to be confident in their coping ability. Coping flexibility has been shown to have several health benefits. These include higher levels of positive adjustment and lower levels of symptoms of burnout, decreases in the severity of anxiety symptoms and increases in quality of life (Cheng, 2003; Cheng & Cheung, 2005; Gan et al., 2007). Coping flexibility is also associated with a decreased likelihood of experiencing increases in depressive symptoms following the occurrence of stressful life events. On the other hand, lower levels of coping flexibility have been found to predict increases in depressive symptoms over time (Fresco et al., 2006).

Approach strategies

People use a range of **coping strategies** when they encounter a stressor. Some strategies are adaptive and helpful in meeting the demands of a stressor. Other strategies may provide a distraction, which could either be a healthy or unhealthy approach, depending on the circumstances. One way that psychologists categorise these different types of coping responses is as either approach strategies or avoidance strategies.

Approach strategies target the stressor or the response to the stressor in practical ways. They consist of behavioural or psychological responses designed to change (remove or diminish) the nature of the stressor or how one thinks about it. Approach strategies help us adapt to the changes and demands a stressor introduces so that living with it becomes tolerable. Approach strategies are considered to be active strategies because they involve an awareness of the stressor, followed by attempts to reduce the negative outcome of the stressful situation. They require a person to directly confront the stressor, make a realistic appraisal of it, evaluate alternative courses of action, recognise and change any unhealthy emotional responses they may have to it, and do something to try to prevent any adverse effects it may have on them physically or psychologically. Examples include seeking support from others, making plans to deal with the stressor, seeking information about the stressor and goal setting (Littleton et al., 2007) (Figure 3.13).

Approach strategies are considered to be the most effective way of coping with stress. By attempting to cope with a stressor, the individual gains a sense of control over their situation, which leads to a feeling that they can manage it. Approach strategies also allow people to practise problem-solving, which may be of use when confronted with future stressors (Min et al., 2007).

Figure 3.13 Creating to-do lists, making a schedule or creating a plan to deal with a stressor are some examples of approach strategies.

INVESTIGATION 3.1 THE EFFECTIVENESS OF STRATEGIES FOR COPING WITH STRESS

Scientific investigation methodology
Fieldwork

Aim
To compare the Perceived Stress Scale results before and after adopting strategies for coping with stress.

Logbook template
Fieldwork

Introduction
The Perceived Stress Scale (PSS) was developed to measure the degree to which situations in a person's life are perceived as stressful. Psychological stress is defined as the extent to which persons perceive that their demands exceed their ability to cope. The PSS is a set of questions designed to understand how unpredictable, uncontrollable and overloaded respondents find their lives. It is a measure of the degree to which situations in one's life are appraised as stressful. The questions in the PSS ask about a person's feelings and thoughts during the last month. In each case, the respondent will be asked to indicate how often they felt or thought a certain way.

Data calculator

Stress research suggests that people who have learned a variety of coping skills generally handle demands and solve problems more easily and effectively than those who have not. They are less likely to experience distress and more likely to be confident in their coping ability.

Method
This fieldwork investigation requires you to access a number of participants to complete the PSS questionnaire before and after completing one month of selected strategies for coping with stress. Strategies could include exercise, meditation, good sleeping and/or eating habits, or any other self-care approaches to manage stress. The PSS questionnaire may be accessed at the weblink.

Weblink
Perceived Stress Scale

When undertaking this investigation, you are required to develop a research question, state an aim, formulate a hypothesis, and plan an appropriate method to answer the research question, while complying with safety and ethical guidelines. Your investigation must generate primary quantitative data. You will maintain a logbook for this investigation.

Results
Collate the results for all participants and process the quantitative data obtained, using appropriate mathematical calculations and units (for example, descriptive statistics such as a mean). Organise, present and interpret data using an appropriate table and/or graph.

Discussion
1. State whether the hypothesis was supported or refuted.
2. Describe the influence of the strategy utilised to cope with stress on the PSS results.
3. Discuss any implications of the results.
4. State if it is possible to generalise any of the results to the wider population.
5. Discuss any potential extraneous variables and how they may have affected the data.
6. Suggest future improvements to address the extraneous variables if the investigation was to be repeated.

Conclusion
Write a conclusion to respond to the research question.

Avoidance strategies

3.5.1 STRATEGIES FOR COPING WITH STRESS

Worksheet
Prevalence and effectiveness of coping strategies

Avoidance strategies involve choosing your response to a stressor based on trying to either cognitively or behaviourally avoid or escape painful or threatening thoughts, feelings, memories or sensations associated with the stressor. They are used when we feel we have little or no control over our situation and we feel we cannot manage it. Typical examples of avoidance strategies include withdrawing from others, denying the stressor exists, disengaging from one's thoughts and feelings about the stressor and self-destructive behaviour (Littleton et al., 2007). These strategies can involve doing something (for example, someone who excessively washes their hands to try to avoid possible contamination from germs) or not doing something (for example, when someone avoids a person because they don't want to have an awkward conversation), but they do not involve the person directly confronting the stressor.

Avoidance coping strategies are usually ineffective because they don't encourage people to seek social support or engage in problem-solving activities (Min et al., 2007). Research consistently shows that less adaptive avoidant coping responses contribute to poorer health outcomes (Futa et al., 2003) and suggests that avoidance coping is the strongest predictor of adverse future wellbeing (Gibbons et al., 2011). The manner in which individuals react to or cope with stressful situations influences the long-term impact of those stressors, and differences in coping are important contributors to psychological adjustment (Min et al., 2007).

Avoidance coping is also considered to be a maladaptive coping mechanism because it does not attempt to defeat the stressor. It can be counterproductive because it may lead to activities (for example, excessive alcohol consumption or binge eating) or mental states (for example, withdrawal) that keep the person from confronting their stressor or successfully adapting to future stressful events.

It is important to note that avoidance strategies are not always a negative approach and may decrease psychological stress in the short term. In moments of high arousal, some avoidance strategies, such as exercising, meditation, watching a movie or taking a nap, may provide space for a person to relax and provide them with the energy and clarity to take on a more approach-based response as needed.

However, avoidance strategies generally add to long-term stress because the underlying problem remains unsolved. Instead of alleviating a person's stress, avoidance strategies generally maintain or increase the stress. They do not provide a person with any sense of control over their stressful situation and research suggests that they often lead to a further lack of self-confidence and self-esteem when dealing with future stressors. For example, if coping strategies adopted during childhood, such as withdrawal and a tendency to dissociate, are used in adulthood, they can impair a person's ability to develop more adaptive social, cognitive and emotional coping mechanisms needed for the successful handling of stressful situations in adult life (Briere, 2002).

Whether we choose one type of strategy over another will be determined partly by our personality and also by the type of stressful event and our appraisal of it. However, using approach strategies, while reducing the use of avoidance coping, may lead to a decline in stress (Kao & Craigie, 2013).

ACTIVITY 3.5 SAC PREPARATION AUDIT

Try it

Imagine it is two weeks until your next Psychology SAC. Spend a few minutes jotting down the techniques that you use and behaviours that you demonstrate when preparing for SACs; these can be useful (completing practice questions) or not so useful (checking your phone or watching TV instead).

1. Once you've got your list of techniques and behaviours, state whether they are approach strategies or avoidance strategies.

Apply it

It is the night before the Psychology exam and Mitch knows that he should be studying. Mitch makes a deal with himself: he'll watch one episode of his favourite TV show and then he'll start reading his notes. Not wanting to deal with his studies, Mitch watches another 2 hours of TV.

Mitch eventually musters up the will to start studying. He gets his phone, picks a playlist to listen to but then gets distracted by his social media feed. After half an hour or so, Mitch tries to read over his notes, but he can't stay focused. Mitch messages his best friend, Toby, for tips on how to study for the exam.

Toby sends Mitch a list of ways he could study for the exam, which include creating mind maps, completing past exam questions and writing dot-point summaries of his key terms. Mitch downloads a few past exams and works his way through the multiple-choice questions and finds this very helpful.

2   Identify one approach strategy and one avoidance strategy used by Mitch when studying for the Psychology exam.

Exam ready

3   An example of coping flexibility used by Mitch to support him in studying for his exam was

A   picking a playlist on his phone and scrolling through social media.

B   realising that he needed additional support, sending Toby a message and asking for strategies to help him study.

C   making a deal with himself that he will watch one episode of his favourite show and then read over his notes.

D   trying to read over his notes.

3.5.2 EVALUATION OF RESEARCH

Seven practical ways to cope with stress

By now you know that stress is a very normal part of everyday life and something that affects everybody in different ways. However, we also know that too much stress can be detrimental and it is important to find ways to cope with stress. Here are some practical strategies that you can use to cope with stress.

1   Exercise: Exercise is one of the best tools we have for coping with stress. This is because exercise can help you 'use up' excess stress hormones such as cortisol, which helps reduce the build-up in the body. Exercise also results in the release of beta-endorphins. These are 'feel good' neurotransmitters that lift your mood. Exercise also has the added bonus of providing a good distraction from your stress and can provide some much-needed social connection when exercising with others.

2   Meditate: It might have its roots in ancient wisdoms, but science has well and truly caught up to determine that meditation can reduce psychological stress and anxiety – even just five minutes a day! Try to set aside a time to sit in a quiet space and breathe. Focus on the present moment and let go of any thoughts that arise. If you're not sure where to start, there are thousands of free guided meditations on YouTube, and the Smiling Mind app is free and has lots of different types of meditations, including ones designed for students.

3   Connect with others: Strong social connections can improve our resilience to stress. It might be tempting to hide away when you're feeling stressed but research has demonstrated that spending time with others, particularly loved ones, can help reduce feelings of stress and often provide a helpful perspective. Added bonus! Some studies have also found that hugs and physical affection can protect people from the physiological effects of stress. So get out there and start hugging! With consent, of course.

4   Eat well: As you have seen, there is a very strong connection between your gut and your brain. Having a diet filled with a diverse range of fruits, vegetables and probiotic foods (fermented foods like yoghurt, kimchi and sauerkraut) and also limited in processed sugars and fast food can help you support your gut microbiome and reduce your reactivity to stress.

Vitamins and nutrients can also support your overall wellbeing in times when your immune system may be suppressed from high levels of cortisol. A good diet can protect you from the harmful effects of stress.

5 Prioritise your sleep: You will learn all about the importance of sleep and the psychological impacts of poor sleep in Unit 4 but in the meantime, just know that the quality and amount of sleep you get can affect your mood, energy, concentration and overall functioning. Stress can impact how well you sleep so take steps to try to protect your sleep by having good sleep routines to help you get the 9 hours recommended for teenagers.

6 Live, laugh, love: Ha ha! No, really. Often when people feel stressed they spend so much time focused on the stressor that they forget to do the activities that bring them joy. Listen to music, dance in your bedroom, go outside and kick a ball with friends, paint, sew, cook, play guitar, rollerskate, go for a swim, play a game, hug your dog or watch a really funny show. Not only do these activities provide a good break from the stressor, research has shown that creative activities and laughter can help reduce feelings of stress.

7 Seek help: It has been a difficult few years. If new stressors are challenging your ability to cope or you find that your strategies aren't helping you reduce your stress, take your self-care to the next level by seeking some support from a psychologist or a counsellor. They can help you identify factors that may contribute to your stress and help you develop a plan to manage your stress in other ways. Everyone needs some extra support now and then – take care of yourself by bringing in reinforcements to support you.

KEY CONCEPTS 3.5

» Coping is the process of managing the demands of stressors and involves a range of different behaviours and techniques.
» A coping strategy is more likely to be effective if the strategy is well matched to the specific context of the situation (context-specific effectiveness).
» Coping flexibility refers to the ability to be able to adapt and change coping strategies as needed. It is seen as a strong predictor of mental health.
» Approach strategies are coping strategies that manage stress by confronting and directly engaging with the stressor.
» Avoidance strategies are stress management strategies that involve actively avoiding thinking about or engaging with the stressor and diverting attention elsewhere.
» Avoidance strategies are often maladaptive but can be adaptive in certain circumstances.

Concept questions 3.5

Remembering
1 How do psychologists define coping?
2 What does context-specific coping effectiveness refer to?

Understanding
3 Describe a time when you have demonstrated (or seen someone else demonstrate) coping flexibility.
4 Explain the difference between an approach strategy and an avoidance strategy as a means of coping with stress. Provide an example for each.
5 Provide two reasons why avoidance strategies are considered to be less effective for coping with stress than approach strategies.

Applying

6 As a student you will have experienced many stressors throughout your school years. Provide at least five examples of approach strategies and avoidance strategies that students may use in response to stress. **e**

7 Riley needs to write a report for their boss over the weekend but is so stressed that they are struggling to concentrate to get the work completed. Develop a plan for Riley to meet the demands of the stressor that uses a range of coping strategies. **c**

HOT Challenge

8 Write a paragraph that evaluates the effectiveness of using avoidance strategies as a means of dealing with a stressor. Use examples and psychological terminology throughout your response. **r**

3 Chapter summary

KEY CONCEPTS 3.1a

» Stress is the psychological and physiological responses that a person experiences when confronted with a situation that is challenging or threatening, and is perceived to exceed their ability to cope with the stressor.
» A stressor is a person, object or event that causes feelings of stress.
» Stressors can be internal (coming from within a person) or external (coming from the outside environment).
» The psychological response to stress is subjective and can be negative or positive.

» The physiological response is the same irrespective of whether the stressful situation is perceived to be positive or negative.
» Stress can be acute (short term) or chronic (long term).
» The flight-or-fight-or-freeze response is an automatic physiological reaction to a stressor that involves the sympathetic and parasympathetic branches of the autonomic nervous system.
» The response prepares the body to:
  – flee (flight): escape the stressor
  – fight: confront the stressor
  – freeze: immobilise to evade detection and prepare in the face of a stressor.

KEY CONCEPTS 3.1b

» The flight-or-fight responses are activated when the sympathetic nervous system is dominant and adrenaline causes physical changes to flood the body with energy in order to flee or confront the stressor.
» The freeze response is activated when the parasympathetic nervous system is dominant and is characterised by reduced heart rate and hyper-awareness. It aids in avoiding detection and preparing the body for further defensive actions.

» Cortisol is the main stress hormone and continues to be released after the FFF response to allow the body to adapt to stress for prolonged periods of time. It provides the body with energy; however, it can be harmful if it persists in high levels in the bloodstream for prolonged periods of time as it suppresses the immune system and increases the likelihood of illness.

KEY CONCEPTS 3.2

» Selye's GAS model outlines the non-specific physiological response to stress. It has three stages: the alarm-reaction stage (including shock and countershock), the resistance stage and the exhaustion stage.

» The GAS model was the first to link the stress response to a compromised immune system and identifies the biological processes of the stress response; however, it fails to acknowledge the subjective psychological experience of the stress response and is predominantly based on animal research.

KEY CONCEPTS 3.3

» Lazarus and Folkman's Transactional Model of Stress and Coping views stress from a psychological perspective.
» The model considers the unique characteristics of an individual that make up their subjective stress response in a transaction with their external environment.
» The model consists of two major stages: primary appraisal and secondary appraisal.
» During primary appraisal, a person evaluates whether a stimulus is stressful and whether the stressor is harmful, a threat or a challenge.
» During secondary appraisal, a person evaluates the internal and external resources available to cope with the stressor.
» Strengths: focuses on the subjective experiences of the individual and acknowledges that people have different responses to the same stressors; focuses on the cognitive, affective and behavioural (psychological) responses to stress; considers a range of different psychological factors in the stress response.
» Limitations: does not refer to or consider the biological responses to stress; primary and secondary appraisal may not occur as two separate processes; model may be too simplistic as most people are not aware of the appraisal process as it occurs.

KEY CONCEPTS 3.4

» The gut–brain axis is the network of bidirectional communication pathways that enable communication between the microbiota and the central nervous system (brain).
» Part of the GBA, the enteric nervous system, involves millions of neurons within the digestive tract that regulate major functions of the gut and communicate with the central nervous system to regulate these functions.
» Communication between the gut and the brain occurs primarily through the vagus nerve.
» Disruptions to the microbiota appear to affect psychological functioning and have been linked to changes in social behaviour, stress, anxiety and depression.
» Communication is bidirectional between the microbiota and brain and they appear to influence each other; for example, disruptions in the microbiota can contribute to feelings of stress, depression and anxiety. However, disorders such as chronic stress, depression and anxiety appear to negatively influence the health of the gut microbiota.
» The use of probiotics appears to be a promising field in the treatment of disruptions to the microbiota and psychological disorders.
» The GBA is an emerging field of psychological research and most research in this field has been conducted using animal research and therefore has limitations.

KEY CONCEPTS 3.5

» Coping is the process of managing the demands of stressors and involves a range of different behaviours and techniques.
» A coping strategy is more likely to be effective if the strategy is well matched to the specific context of the situation (context-specific effectiveness).
» Coping flexibility refers to the ability to be able to adapt and change coping strategies as needed. It is seen as a strong predictor of mental health.
» Approach strategies are coping strategies that manage stress by confronting and directly engaging with the stressor.
» Avoidance strategies are stress management strategies that involve actively avoiding thinking about or engaging with the stressor and diverting attention elsewhere.
» Avoidance strategies are often maladaptive but can be adaptive in certain circumstances.

3 End-of-chapter exam

Section A: Multiple choice

1. Which of the following situations can cause acute stress?
   A inadvertently sending your car into a skid
   B skiing downhill on a difficult slope
   C making a presentation before a roomful of people
   D any of the above.

   ©VCAA 2002 E1Q2 ADAPTED EASY

2. The General Adaptation Syndrome has three stages. Which of the following statements gives these stages in the correct order?
   A shock, alarm, resistance
   B resistance, alarm, exhaustion
   C exhaustion, resistance, alarm
   D alarm, resistance, exhaustion

3. Which of the following is not an example of an external stressor?
   A stomach pain
   B extreme heat
   C conflict with a friend
   D a rollercoaster ride

   ©VCAA 2018 Q4 ADAPTED MEDIUM

4. A role of cortisol during prolonged stress is to
   A activate the HPA axis.
   B stimulate the parasympathetic nervous system.
   C maintain homeostasis.
   D suppress the immune system.

5. According to the Transactional Model of Stress and Coping, the two processes that individuals undergo when attempting to cope with stress are
   A approach and avoidance.
   B alarm and resistance.
   C primary and secondary appraisal.
   D stress and stressors.

   ©VCAA 2017 Q18 ADAPTED MEDIUM

6. Which of the following physiological responses may be experienced by a person when a scary clown jumps out at them?
   A decreased muscle tension and increased perspiration
   B relaxed bladder and dilated pupils
   C increased adrenaline and stimulated digestion
   D stimulated digestion and relaxed bladder.

   ©VCAA 2021 Q3 ADAPTED HARD

7. Shanti wakes up to the sound of something scratching at her bedroom door and becomes so terrified that she cannot move. Shanti is likely experiencing
   A the flight-or-fight-or-freeze response, which is controlled by the somatic nervous system.
   B an inability to move due to parasympathetic dominance.
   C activation of the autonomic nervous system in preparation to fight.
   D a heightened heart rate with sympathetic nervous system activation.

8. Psychologists describe coping as
   A a non-specific response to a stressor.
   B the behavioural and psychological responses a person makes to a stressor that are intended to manage the stressor and reduce the physical and psychological stress related to it.
   C a process that requires a person to constantly change their thoughts and behaviour so they can manage stressors they feel they don't have the resources to deal with.
   D a response to a stressor based on trying to either cognitively or behaviourally avoid or escape painful or threatening thoughts, feelings, memories or sensations associated with the stressor.

9 One evening, Russell, who was very tired, was curled up on his beanbag watching late-night television. He suddenly noticed a large spider crawling across the television screen. Immediately, Russell jumped to his feet and ran out of the room. What response was activated when Russell saw the spider?
   A the sympathetic nervous system response
   B the parasympathetic nervous system response
   C the autonomic nervous system response
   D the flight-or-fight-or-freeze response.

10 Which of the following influence the gut microbiota?
   A diet
   B stress
   C illness
   D all of the above

11 Which of the following is an example of primary appraisal according to Lazarus and Folkman's Transactional Model of Stress and Coping?
   A Determining the extent to which additional resources are needed to cope.
   B Evaluating the potential impact of the stressor.
   C Judging the usefulness of the coping resources that are available.
   D Any exchange between the individual and their environment.

©VCAA 2012 E2Q24 ADAPTED MEDIUM

12 The night before a Psychology SAC, Tarquin realised that there was an entire topic that he had not studied. He began crying and was unable to sleep that night. According to Lazarus and Folkman's Transactional Model of Stress and Coping, it is likely that Tarquin's appraisal of this situation was
   A a challenge.
   B irrelevant.
   C a threat.
   D eustress.

13 Communication between the gut and the brain is
   A afferent.
   B efferent.
   C from the brain down to the gut.
   D bidirectional.

14 Identify the correct order of communication between the gut microbiota and the central nervous system.
   A Neurons in the ENS detect sensory information, sensory information is communicated to the brain via the HPA axis.
   B The gut microbiota produces chemical signals in the gut, neurons in the ENS detect sensory information, the information is converted into action potentials, the information is communicated to the brain via the vagus nerve.
   C Sensory information is detected in the vagus nerve, the vagus nerve sends the signal to the ENS, the ENS transmits sensory information to the brain.
   D The gut microbiota produces motor information, which is detected by sensory neurons in the ENS, the motor signals are transmitted to the brain via the vagus nerve.

15 According to Hans Selye, stress is
   A a non-specific physiological response.
   B the result of internal stressors.
   C the result of external stressors.
   D negative and unpleasant.

©VCAA 2005 E1Q16 ADAPTED EASY

16 The first stage of General Adaptation Syndrome (GAS) is divided into two parts: _____ and _____.
   A resistance, exhaustion
   B shock, flight-or-fight
   C alarm, resistance
   D shock, countershock

17 Dysbiosis is
   A the result of stress.
   B when the gut bacteria become less diverse or there is no longer a healthy balance of bacteria.
   C the term used to refer to a healthy microbiota.
   D the absence of a microbiota.

18 Rafael's unemployment was making him feel stressed. He applied for jobs advertised on a job website but this did not result in him getting a job. He went to see a careers counsellor who was able to suggest some other strategies for finding a new job. Which of the following statements applies to the coping strategies used by Rafael?
   A They have context-specific effectiveness and will help him avoid stressful situations.
   B They have context-specific effectiveness and demonstrate coping flexibility.
   C They are approach strategies, that demonstrate secondary appraisal.
   D They are approach strategies that do not demonstrate coping flexibility.

19 The physiological arousal response in which the body prepares to combat a real or perceived threat is called
   A   the sympathetic nervous system.
   B   the flight-or-fight-or-freeze response.
   C   the alarm-reaction stage of the General Adaptation Syndrome.
   D   stress.

20 The body enters the exhaustion stage because it
   A   has been sustaining high levels of physiological arousal for some time.
   B   experiences fatigue and an increased susceptibility to serious illness.
   C   experiences a drop in temperature and blood pressure.
   D   has become susceptible to wear and tear.

Section B: Short answer

1 According to Selye's General Adaptation Syndrome, why might a VCE student fall ill just before their final exams? [3 marks]

2 Why are approach strategies considered to be an adaptive response to stress and avoidance strategies considered to be a maladaptive response to stress? [2 marks]

3 Identify two limitations of Lazarus and Folkman's Transactional Model of Stress and Coping. [2 marks]

Unit 3, Area of Study 1 review

Section A: Multiple choice

Question 1 ©VCAA 2021 Q3 ADAPTED HARD
Sarah wakes up to the sound of something scratching at the bedroom window and becomes so frightened she feels her heart racing faster and her palms begin to sweat. Sarah's inability to move is triggered by the
A parasympathetic nervous system.
B sympathetic nervous system.
C autonomic nervous system.
D somatic nervous system.

Question 2 ©VCAA 2020 Q2 ADAPTED EASY
Which statement about conscious or unconscious responses by the nervous system is incorrect?
A A conscious response by the nervous system involves deciding to pick up a pen and write.
B A conscious response by the nervous system is voluntary and attention is given to the stimulus.
C An unconscious response by the nervous system is involuntary and can be regulated by the somatic or autonomic nervous systems.
D An unconscious response by the nervous system is unintentional and is always regulated by the autonomic nervous system.

Question 3 ©VCAA 2020 Q3 ADAPTED EASY
Douglas is a paramedic attending to a patient who has been unconscious for at least 30 minutes. Douglas records his observations of the patient on a clipboard. He accidentally cuts his finger on the side of the clipboard and drops it.
Which of the following identifies what is happening to Douglas at the neural level as he drops his clipboard?

	Response	Postsynaptic neuron
A	muscles contract	postsynaptic dendrites receive the message
B	spinal reflex initiated	postsynaptic axon terminals send the message
C	muscles relax	postsynaptic axon terminals receive the message
D	spinal reflex initiated	postsynaptic dendrites send the message

Question 4 ©VCAA 2019 Q1 ADAPTED EASY
Which of the following is not true of spinal reflex?
A It is the body's survival response.
B It is an involuntary response to harmful stimuli.
C It is an automatic response that occurs in the spinal cord.
D It is regulated by the autonomic nervous system.

Question 5 ©VCAA 2019 Q2 ADAPTED HARD
When someone pricks their finger and immediately withdraws it, which of the following statements is incorrect about their response?
A It is an example of an innate nervous system response.
B It illustrates how the spinal cord makes decisions about movement.
C It shows the unconscious response involved in the coordination of the reflex.
D It shows the role of the brain in the responses of the autonomic nervous system.

Question 6 ©VCAA 2019 Q3 ADAPTED MEDIUM
What would be the effect on the transmission of neuronal messages if GABA was released?
A It would bind to the postsynaptic dendrite and make it more likely to fire.
B It would bind to the postsynaptic axon terminal and make it more likely to fire.
C It would bind to the postsynaptic dendrite and make it less likely to fire.
D It would bind to the postsynaptic axon terminal and make it more likely to fire.

Question 7 ©VCAA 2021 Q28 ADAPTED

Jas used to play 'Call of Duty' for 2 hours a day with his friends. As he played more, he got better at playing the game. When a new game came out, Jas and his friends stopped playing Call of Duty and began playing the new game. After a year, Jas decided to play Call of Duty again and found he was much slower and performed much worse than before.

This can be explained by:

A long-term depression, which strengthens the synaptic connections that allow him to move his fingers.
B long-term depression, which decreases the synaptic communication as he reduces the frequency of play.
C long-term potentiation, which increases synaptic communication when he presses the buttons on the controller.
D long-term potentiation, which decreases the synaptic transmission speed of the neurons involved in him playing the video game.

Question 8 ©VCAA 2019 Q9 ADAPTED EASY

Megan began tennis lessons when she was eight years old. Each time she practised, she played more accurately and made fewer mistakes. However, after two years, she lost interest and stopped her lessons. As an adult she decided to play tennis with her friends, telling them she was good and had played as a child, but she was quite inaccurate with her performance.

In terms of neural plasticity, Megan's decreased accuracy was likely a result of

A rehearsal.
B relearning.
C long-term depression.
D long-term potentiation.

Question 9 ©VCAA 2021 Q2 ADAPTED MEDIUM

Which of the following paired options does not describe a role of a dendrite and an axon terminal?

	Role of dendrite	Role of axon terminal
A	grows during neuroplasticity	conducts electrical signals from dendrites
B	collects chemical signals from other neurons	transmits information to other neurons
C	transmits information to adjoining neurons	stores neurotransmitters in pockets ready for release
D	receives information from presynaptic neurons	releases neurotransmitters into synapses

Question 10 ©VCAA 2020 Q5 ADAPTED HARD

Which one of the following accurately describes the characteristics of neural plasticity of a neuron?

A Dendrites 'grow' whereas axon terminals are 'pruned'.
B Myelin coating the axon terminals improves signal transmission.
C Long-term potentiation is the long-lasting decrease in neural transmissions.
D Long-term depression occurs when neural pathways are not sufficiently stimulated.

Question 11 ©VCAA 2020 Q11 ADAPTED HARD

Which of the following identifies a similarity between long-term potentiation and long-term depression?

A Both result from increased stimulation of neural pathways.
B Both are experience-dependent and require environmental factors to induce.
C Both result from decreased stimulation.
D They are short-term effects of memory and learning.

Question 12 MEDIUM

Which of the following is true regarding modulatory neurotransmitters?

A They act at singular synapses to regulate excitation and inhibition of postsynaptic neurons.
B They are released into the bloodstream to create fast-acting, long-lasting change.
C They can affect multiple neurons at the same time and have long-lasting effects.
D Dopamine is an example of how they can make localised effects on subsequent neurons.

Question 13 ©VCAA 2020 Q8 ADAPTED HARD

Mohamed is experiencing stress in response to news he has received. The stress Mohamed is experiencing is more likely to be distress if Mohamed has

A increased arousal momentarily, decreased motivation and an elevated heart rate.
B increased alertness momentarily, increased motivation and an elevated heart rate.
C heightened arousal for several hours and decreased motivation, and he feels overwhelmed.
D elevated alertness for several hours and increased cortisol levels, and he feels confident he can manage the news.

Question 14

A psychologist wanted to investigate people's responses to being told to give a presentation to 100 people in a week's time.

Before the presentation, Zoe was quite stressed and caught the flu. According to Selye's general adaptation syndrome, Zoe was most likely experiencing

A shock.
B resistance.
C exhaustion.
D countershock.

Question 15 ©VCAA 2020 Q10 ADAPTED HARD

Which one of the following is not a role of cortisol in the stress response, according to Selye's general adaptation syndrome?

A It stops the immune system from functioning through prolonged presence in the bloodstream.
B It increases glucose in the bloodstream and reduces inflammation.
C It reactivates functions that are non-essential in a flight-or-fight-or-freeze response.
D It provides the initial alert about a perceived threat, through the release of adrenaline.

Question 16 ©VCAA 2019 Q7 ADAPTED HARD

One limitation of the general adaptation syndrome is that the model

A fails to explain the outcome if coping resources are inadequate.
B does not account for the physiological aspects of stress.
C does not recognise that the individual and the environment both play a role in the stress response.
D is unable to be researched experimentally because of the subjective nature of it.

Question 17 ©VCAA 2019 Q9 ADAPTED MEDIUM

Malin is experiencing a constant state of stress and has become fatigued. Which of the following most accurately identifies the stage of Selye's general adaptation syndrome that Malin is in and the reason that supports this stage?

	Stage	Reason
A	shock	Malin's ability to cope with the stressor is immobilised.
B	resistance	Continued cortisol release weakens Malin's coping ability, resulting in his body being fatigued.
C	exhaustion	Malin's body's resources are depleted, resulting in vulnerability to a range of serious physical disorders.
D	resistance	Increased adrenaline in Malin's bloodstream results in his body becoming susceptible to illnesses.

Use the following information to answer questions 18–19.

Shanna decided to purchase a new set of ballet pointe shoes before her upcoming performance, as her old ones are good but are close to being worn out. When they arrived, she noticed that they were damaged, and she was informed that a replacement would not be available until after the performance.

Question 18 ©VCAA 2021 Q5 ADAPTED EASY

In terms of Lazarus and Folkman's Transactional Model of Stress and Coping, a possible secondary appraisal for Shanna might be

A feeling disappointed because her new shoes cannot be worn.
B that she could cope with the delay by using her existing pointe shoes.
C feelings of loss at not being able to perform because her new shoes are damaged.
D that this is an opportunity to seek another supplier from whom she can purchase alternative pointe shoes.

Question 19 ©VCAA 2021 Q5 ADAPTED MEDIUM

Which one of the following would not be an effective approach coping strategy that Shanna could use while preparing for the ballet performance?

A Research other shops to find new pointe shoes.
B Go for a long run to take her mind off things instead of attending dance practice.
C Discuss with her dance school the concerns she has about performing without the new shoes.
D Avoid overuse of her old pointe shoes to ensure they are safe for use at the performance.

Question 20 ©VCAA 2021 Q19 ADAPTED

Dushani received the news that her audition for the lead role in the school play was unsuccessful. Dushani was upset by this news as she recently quit her part-time job to allow her time to work on the play.

According to Lazarus and Folkman's Transactional Model of Stress and Coping, which one of the following may best describe Dushani's primary appraisal of the news?

A She views it as a challenge to improve for next time.
B She sees the situation as a threat as she has not gotten the role and now has no job.
C She views it as irrelevant as she did not care about getting the role.
D She decides that she does not have the resources to cope due to the stress of the news.

Question 21 ©VCAA 2018 Q28 ADAPTED MEDIUM

Anh's husband of 40 years died following a long illness. Anh felt stressed, had difficulty sleeping, struggled to go shopping on her own and sometimes forgot to pay her bills, which affected her levels of anxiety and independence.

In terms of Lazarus and Folkman's transactional model, Anh's secondary appraisal is likely to be

A insufficient resources to cope with the stressor.
B problem-focused coping.
C emotion-focused coping.
D distress.

Use the following information to answer Questions 22–24.

Selin was anxious about competing in the last netball game before the finals. If her team won, it would progress to the finals. Selin was new to the sport and doubted her abilities. She was worried that she would play poorly and let her team down. When it came time for the game to start, she was so nervous that she looked out at the crowd, which made her miss a pass from her teammate. However, despite her worries, Selin's team won the game and made it to finals.

Question 22 ©VCAA 2019 Q4 ADAPTED EASY

Selin was most likely experiencing distress because

A she was worried about hurting herself during the game and missing finals.
B she felt excited about progressing to the finals if her team won.
C the stress of doing a good job interfered with her concentration.
D she felt nervous about not playing well but knew it was only a game.

Question 23 ©VCAA 2019 Q6 ADAPTED MEDIUM

According to Lazarus and Folkman's Transactional Model of Stress and Coping, an example of Selin undertaking primary appraisal before the game would be if she thought

A of the crowd cheering her on.
B of the tips given to her by her coach.
C of the situation as good practice for the finals.
D that she was not as experienced as the other players on her team.

Question 24 ©VCAA 2019 Q5 ADAPTED MEDIUM

Which of the following identifies the functioning of Selin's autonomic nervous system and a resulting physiological response when she realised her team made it to the finals?

	Autonomic nervous system functioning		Physiological response
	Parasympathetic nervous system	**Sympathetic nervous system**	
A	dominant	non-dominant	decreased salivation
B	non-dominant	dominant	increased blood pressure
C	inactive	dominant	movement of skeletal muscles
D	inactive	active	constricted pupils

Question 25

Robyn has recently experienced a divorce from her partner Matt after 5 years of marriage. She has been experiencing anxiety because of this and has started to get constant colds and flus and digestive issues. She feels she is always unwell. Which of the following is a likely explanation for Robyn's illnesses?

A Robyn is experiencing exhaustion leading to chronic illness.
B Robyn's chronic stress is affecting her gut - microbiota and as a result is weakening her immune system.
C Chronic stress increases adrenaline in the body, which suppresses the immune system to divert energy to deal with the stressor.
D Robyn has too great a range of gut bacteria as a result of the stressor, which is leading to ongoing illness.

Question 26 ©VCAA 2021 Q6 ADAPTED

Jacinta is a 16-year-old student whose family travels around Australia for work. Each time Jacinta moves, she is forced to change schools. Jacinta has been finding each move progressively more difficult. At her latest school, Jacinta has made little effort to adjust to the new environment. Jacinta is not

A using coping flexibility as she is unable to fit in.
B having difficulty adapting to a new environment.
C feeling distress caused by the constant shifting of environment.
D using avoidance-based strategies to deal with the stressor of constantly changing schools.

Unit 3, Area of study 1 review

Question 27 ©VCAA 2020 Q45 ADAPTED EASY

Prithvi lost his job two years before he intended to retire and this had a negative impact on his mood and ability to cope. He did not pay two electricity bills despite having sufficient funds. He became withdrawn while at his golf club and soon stopped playing. Now all he does is sit on the couch and watch cartoons from his childhood.

Which one of the following statements applies to the coping strategy used by Prithvi?

A   It is an approach strategy that does not demonstrate coping flexibility.
B   It has context-specific effectiveness and demonstrates coping flexibility.
C   It is an avoidance strategy that does not demonstrate coping flexibility.
D   Given the difficulty older people have finding a job, it demonstrates coping inflexibility.

Use the following information to answer Questions 28–30.

Ali conducted research to find out whether tailored training in different approaches to dealing with stressors would lead to greater coping flexibility in adolescents. Twenty participants were exposed to simulations of a stressful scenario. In group A, the participants were pre-exposed to a 1-hour training session in coping strategies 1 week prior to the simulation; group B had no prior training. Participants were asked to describe how they would deal with the stressor, each time being prompted to explain how they would approach the situation if the initial coping mechanism did not work. At the end of the experiment, the number of effective coping mechanisms was tallied, with the higher value corresponding to higher coping flexibility.

Question 28 ©VCAA 2019 Q10 ADAPTED HARD

The dependent variable and its measurement technique was

A   coping flexibility as measured through the total number of coping strategies.
B   stress level score.
C   levels of arousal during the simulations.
D   change score calculated as the difference between coping mechanism frequency.

Question 29 ©VCAA 2019 Q11 ADAPTED HARD

Which one of the following was a confounding variable in Ali's research?

A   using different participants in each group, leading to experimenter effects.
B   using only 20 participants, as this does not allow for generalisation of the results.
C   telling participants to use a particular coping strategy, as this may bias participants.
D   not debriefing participants on the nature of the study.

Question 30 ©VCAA 2019 Q12 ADAPTED MEDIUM

Ali decided to do a further study to see if approach or avoidance-based coping strategies would be more effective in individuals dealing with stressors. Predict what the likely results will be.

A   Avoidance-based strategies would be more effective because they allow the individual to deal directly with the stressor.
B   Approach strategies would be more effective as the individual can overcome the direct cause of the stressor.
C   Avoidance-based strategies would be more effective because they allow the individual to ignore their feelings of stress resulting from the stimulus.
D   Both are equally effective because both allow the individual to address the stressor.

Section B: Short answer

Question 1 ©VCAA 2020 SECTION B Q1 ADAPTED

Outline two similarities between the sympathetic nervous system response and the spinal reflex. [4 marks]

Question 2 ©VCAA 2019 SECTION B Q1 ADAPTED

Luca was standing near a campfire with his friends when he accidentally stepped back and bumped the fire with his legs. Luca automatically pulled his legs away from the fire.

The human nervous system has two major divisions. Identify the subdivision of one of these major divisions that activated Luca's responses and outline how the subdivision was involved in Luca's responses. [5 marks]

Question 3 ©VCAA 2020 SECTION B Q2 ADAPTED

For her Psychology investigation Bec decided to test if reporting on student performance affects the stress levels of students. Bec selected 40 students from her year level based on by asking for volunteers and randomly assigned participants to one of 2 groups. Group 1, the experimental group, was told that the test was going towards their report and group 2, the control group, was not told of any bearing of the test results. The heart rate of participants in both groups was measured prior to them being told about the test to provide a baseline, and again 5 minutes into the test. Means for the two measurements for each group were calculated. The results are shown in the graph below.

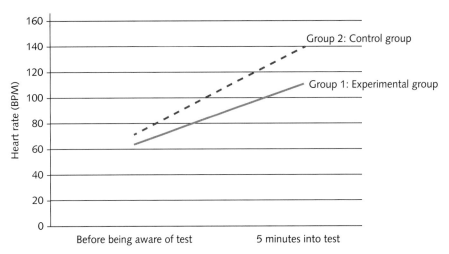

Mean baseline heart rate measured prior to being aware of test and 5 minutes into test

a. Write a possible hypothesis for Bec's experiment. [1 mark]

b. In terms of the nervous system, what is a potential limitation of using heart rate as a measure in this experiment? [2 marks]

c. Using Lazarus and Folkman's Transactional Model of Stress and Coping, identify the most likely primary appraisal made by participants in the experimental group and by participants in the control group.

Justify your response with reference to the graph and the independent variable. [6 marks]

d. At the conclusion of her practical investigation, Bec realises that, entirely by chance, seven participants allocated to the experimental group were at risk of failing the unit whereas the control group contained only one participant who was at risk of failure.

Explain the problem created by this uneven allocation of students between groups. What does this mean in terms of generalisability of results? [4 marks]

[Total = 13 marks]

Question 4 ©VCAA 2018 SECTION B Q8 ADAPTED

Geoff and Lia are 22-year-old students in their final year of university. Geoff and Lia both work part time and the workload is starting to become quite overwhelming on top of the already constant pressures of university: both decide to quit their jobs to focus on university. Geoff comes from a wealthy family and as such, his parents choose to help him by paying for his university accommodation and giving him an allowance of $300 a week; however, he misses his friends and his financial independence. Lia is the first in her family to go to university and her parents are both on the minimum wage and cannot afford to help Lia support herself. Lia is receiving welfare payments, but they are only enough to cover accommodation. She is becoming more and more in debt, which is causing her stress and leading to her grades slipping.

Using relevant theories, compare the stress responses of Geoff and Lia to quitting their jobs to focus on university, referring to both the psychological and physiological stress responses, including strengths and limitations of both.

[10 marks]

Approaches to understand learning

4

Key knowledge

- » behaviourist approaches to learning, as illustrated by classical conditioning as a three-phase process (before conditioning, during conditioning and after conditioning) that results in the involuntary association between a neutral stimulus and unconditioned stimulus to produce a conditioned response, and operant conditioning as a three-phase process (antecedent, behaviour and consequence) involving reinforcement (positive and negative) and punishment (positive and negative)
- » social-cognitive approaches to learning, as illustrated by observational learning as a process involving attention, retention, reproduction, motivation and reinforcement
- » approaches to learning that situate the learner within a system, as illustrated by Aboriginal and Torres Strait Islander ways of knowing where learning is viewed as being embedded in relationships where the learner is part of a multimodal system of knowledge patterned on *Country*

Key science skills

Analyse and evaluate data and investigation methods
- » evaluate investigation methods and possible sources of error or uncertainty, and suggest improvements to increase validity and to reduce uncertainty

Analyse, evaluate and communicate scientific ideas
- » use appropriate psychological terminology, representations and conventions, including standard abbreviations, graphing conventions and units of measurement
- » discuss relevant psychological information, ideas, concepts, theories and models and the connections between them
- » critically evaluate and interpret a range of scientific and media texts (including journal articles, mass media communications, opinions, policy documents and reports in the public domain), processes, claims and conclusions related to psychology by considering the quality of available evidence
- » analyse and evaluate psychological issues using relevant ethical concepts and guidelines, including the influence of social, economic, legal and political factors relevant to the selected issue

Source: VCE Psychology Study Design (2023–2027) + pages 35, 13

4 Approaches to understand learning

Learning is the process that allows all living organisms to survive and thrive, and on which human cultures are built. For humans, learning begins in the womb and encompasses everything from learned reflex responses to learning about theories of learning!

Adobe Stock/Victoria

4.1 Approaches to understand learning
p. 137

Researchers have attempted to identify patterns in the way we learn to try to explain how we are able do it. We explore three models to explain learning: the behaviourist, social-cognitive and Aboriginal and Torres Strait Islander peoples' models.

Adobe Stock/Maria

4.2 Behaviourist approaches to learning
p. 137

Behaviourist approaches include classical and operant conditioning. They explain how we learn associations between stimuli in the environment and behavioural responses. The behaviourist approach is used by advertisers to associate their product with famous people or pleasurable feelings.

4.3
The social-cognitive approach to learning

p. 160

Humans are social creatures, and this comes across in the social-cognitive model of learning, which is based around the idea that we learn by observing and interacting with others. This means that you can learn some aspects of how to play football by watching AFL or AFLW matches.

Source: Matt Webb and the Bawaka Collective

4.4
Aboriginal and Torres Strait Islander approaches to learning

p. 166

Australia's First Nation peoples have passed down their knowledges for at least 65,000 years using methods that embed knowledge in the relationships between people and the multimodal system of *Country*.

Adobe Stock/Jacob Lund

Learning though observation is an important part of both Western and Aboriginal and Torres Strait Islander approaches to learning. However, this is understood differently in each culture. Read on to learn more!

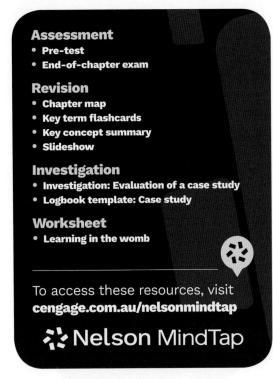

Assessment
- Pre-test
- End-of-chapter exam

Revision
- Chapter map
- Key term flashcards
- Key concept summary
- Slideshow

Investigation
- Investigation: Evaluation of a case study
- Logbook template: Case study

Worksheet
- Learning in the womb

To access these resources, visit cengage.com.au/nelsonmindtap

Nelson MindTap

Slideshow — Chapter 4 slideshow
Flashcards — Chapter 4 flashcards
Test — Chapter 4 pre-test

Know your key terms

- Antecedent
- Attention
- Behaviourism
- Behaviourist approach
- Classical conditioning
- Conditioned response (CR)
- Conditioned stimulus (CS)
- Consequence
- *Country*
- Involuntary association
- Learning
- Motivation and reinforcement
- Multimodal system
- Negative punishment
- Negative reinforcement
- Neutral stimulus (NS)
- Observational learning
- Operant conditioning
- Patterned on Country
- Positive punishment
- Positive reinforcement
- Punishment
- Reinforcement
- Reproduction
- Response
- Retention
- Social-cognitive approach
- Songlines/Songspirals
- Stimulus
- Three-phase process of classical conditioning
- Three-phase process of operant conditioning
- Unconditioned response (UCR)
- Unconditioned stimulus (UCS)
- Voluntary behaviour
- Ways of knowing

We often think of learning as something that we start doing from birth. However, research from the University of Helsinki provides fascinating evidence of language learning within the womb. It has long been known the developing foetus acquires the ability to perceive sounds from the outside world during the last trimester of pregnancy and is especially attuned to the sounds of their mother's voice. To test the hypothesis that the foundations of language learning begin within the womb, the researchers recorded the sounds of 'words' they had made up (for example, 'taataataa') and asked a group of mothers to play two 4-minute recordings of the nonsense words each day during the last trimester of their pregnancy. The researchers used nonsense words so that they would be distinct from real words the foetus would hear spoken naturally. Then, within the first month of life, the mums brought their newborns into the lab where they were wired up with EEG sensors on their heads (see Figure 4.1a). The researchers played the babies recordings of the nonsense words and variations of them. They then compared the EEG signals from the trained babies to those of a matched control group of babies who were never exposed to the nonsense words. Incredibly, the EEG recordings for the trained babies showed the telltale signal of recognition for the nonsense words that was absent from the EEG signals for the babies in the control group (see Figure 4.1b) (Partanen et al., 2013). The research provides evidence that babies' brains can learn the sounds of words before birth, and that these memories provide the foundation for learning their meanings when babies are born. The research brings a whole new meaning to 'lifelong learning'.

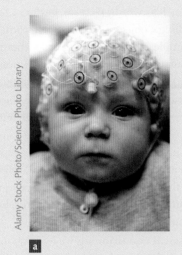
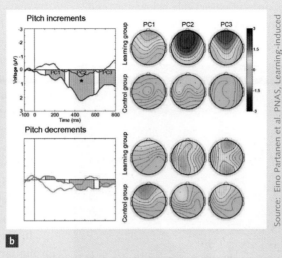

worksheet
Learning in the womb

Figure 4.1 a Infant wearing an EEG net. **b** Infant neural activity measured by EEG in response to variants of trained words. EEG signal was significantly stronger (shown by the areas in red and orange) for babies exposed to the sounds of words in the womb than for the control group.

4.1 Approaches to understand learning

Learning is fundamental to the survival and flourishing of all animals, including humans. In fact, the most basic forms of learning are shared throughout the animal kingdom from the simplest organisms right through to humans.

We can define **learning** as the biological, cognitive and social processes through which an individual makes meaning from their experiences, resulting in long-lasting changes in their nervous system, **behaviour**, abilities and knowledge.

In this chapter, we will explore three different approaches to understand learning and consider how they relate to each other.

» We begin with the **behaviourist approach** to understand how individuals learn associations between stimuli in the environment and between stimuli and their own behaviours.

» We then explore the **social-cognitive approach** to understand how learning occurs through observing and thinking about the behaviours of others.

» Finally, we explore Aboriginal and Torres Strait Islander peoples' **ways of knowing**. We see how knowledge is kept, developed and transferred over generations through stories, songs, dance, art and cultural practices that are embedded in networks of interrelationships between people and *Country*.

4.1.1 DIFFERENT APPROACHES

4.2 Behaviourist approaches to learning

Psychology is a very young science compared to other scientific fields. The first psychology laboratories were established in the late 1800s in Europe and then later in Britain and America. The earliest psychology researchers tried to understand the mind by measuring peoples' subjective reports of their perceptions, and they developed theories about internal mental states to explain human thoughts and behaviours.

Early in the 20th century, a young American psychologist named John B. Watson (Figure 4.2) argued for a different approach. He believed that a true science of psychology must be based on objective measures of observable behaviours and the environmental events that cause behaviour (Watson, 1913).

Figure 4.2 American psychologist John B. Watson, founder of the school of Behaviourism, circa 1920

Watson called his approach behaviourism, and it became the dominant school of thought in psychology for much of the 20th century. Behaviourism produced the two behaviourist approaches to learning that we explore in this chapter – classical conditioning and operant conditioning.

Behaviourists describe learning as a process of learning associations between environmental events and behavioural responses. Associations between two events – for example, hearing a sound and seeing a bird – are called stimulus–stimulus associations, and associations between events and a person's behaviour – for example, hearing a sound and turning towards it – are called stimulus–response associations. The terms 'stimulus' and 'response' are defined as follows.

» **Stimulus:** Any event that occurs in the environment or within a learner that the learner's nervous system can detect. Stimuli (plural) can be visual, auditory, gustatory (taste), tactile (touch) and olfactory (smell) sensations, as well as sensations like pain, sense of balance/body posture, and temperature. Stimuli can be simple sensory events, such as a particular sound or colour, but can also include more complex events such as words, pictures, gestures, faces, living things, objects and situational contexts (for example, a room, building or location). In some situations, a thought or feeling (mood or emotion) can also be considered a stimulus.

» **Response:** A measurable behaviour produced by a learner (human or other animal) as an outcome of perceiving a particular stimulus. Responses include **involuntary behaviours** such as spinal reflexes and autonomic (physiological) reflexes, such as the flight-or-fight-freeze response, as well as **voluntary behaviours** such as approaching or moving away from a stimulus.

In **classical conditioning** we learn about involuntary behavioural responses to stimuli that have been associated with either harmful or beneficial stimuli in the past. In **operant conditioning** we learn to produce voluntary behavioural responses based on their past association with either gaining a pleasant stimulus or avoiding an unpleasant stimulus.

Watson built behaviourism on the foundation of classical conditioning. His inspiration came not from psychology, but from the work of Ivan Pavlov, a Russian physiologist who first described the process of classical conditioning when he was researching the process of digestion.

Ivan Pavlov and the discovery of classical conditioning

In the late 1800s, Pavlov (Figure 4.3) was studying the physiology of digestion using dogs. He was interested in the production of saliva as the first stage of digestion. Salivation is an autonomic reflex response stimulated by the presence of food in the mouth. Recall the definition of a reflex from Chapter 2.

Pavlov began by giving his dogs powdered meat and measuring the salivation response. He collected saliva from tubes inserted into the salivary ducts inside the cheeks of his dogs. You can see Pavlov's experimental equipment in Figure 4.4.

Pavlov noticed that once his dogs had some experience with this procedure, they started to salivate *before* he had given them

Figure 4.3 Russian physiologist Ivan Pavlov, who discovered the principles of classical conditioning, circa 1890

4.2.1 CLASSICAL CONDITIONING PROCESS

the meat powder. In fact, they seemed to start salivating just at the sound of his footsteps approaching the lab. Pavlov called these anticipatory salivation responses 'psychic secretions' because it seemed as if his dogs had learned to predict the future!

Pavlov was intrigued by this and shifted his studies from digestion to learning (Pavlov, 1927; 1928).

Pavlov's research question became: How can a reflex like salivation occur in response to an event that does not naturally cause it? Pavlov hypothesised that his dogs seemed to have learned a predictive relationship between the meat powder and the environmental events that occurred regularly before the meat was presented. For example, it seemed that his dogs had learned a simple conditional rule, something like – '*if* footsteps, *then* food'.

Pavlov developed a broad hypothesis about what was going on. It seemed that a biologically neutral event – such as the sound of footsteps – can become a signal for a biologically significant event – such as the presentation of food – through repeated experience of the neutral event immediately before the biologically significant event.

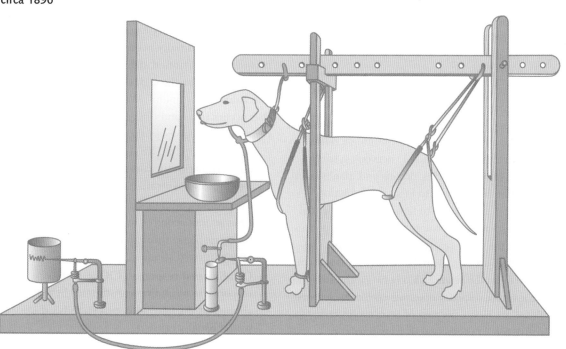

Figure 4.4 Diagram of experimental apparatus used by Ivan Pavlov. A tube carries saliva from the dog's salivary duct to a container that activates a recording device (far left).

In this way, the repeated occurrence of the sound of his footsteps before the presentation of food resulted in the footsteps acquiring the ability to cause the reflex salivation response on their own, before the food was presented.

Pavlov called this kind of learning **conditioning** because of the conditional 'if–then' stimulus–response associations that are learned. We now call it **classical conditioning**, because it was the first kind of conditioned learning described (classic), and it provides a foundation for other kinds of conditioned learning that we will explore later in this chapter.

Classical conditioning terminology

Before we learn how Pavlov tested his hypothesis, it helps to introduce the formal terminology of classical conditioning. We can group the terms into two sets:
1 Terms that refer to the innate stimulus and response components of a reflex
2 Terms that refer to the learned stimulus and response components that become associated with a reflex.

1 The reflex: unconditioned stimulus (UCS) and unconditioned response (UCR)

Classical conditioning always has a reflex behaviour at its core. A reflex is an innate involuntary response to a particular stimulus. Reflexes include **somatic reflexes** and autonomic reflexes. Somatic reflexes are produced by the somatic nervous system and affect skeletal muscles, such as the withdrawal reflex. Autonomic reflexes are produced by the autonomic nervous system and affect smooth muscles, organs and glands, such as salivation in response to food, pupil dilation/constriction in response to light, as well as sexual arousal responses and emotional arousal responses, including the flight-or-fight-freeze response.

Reflexes are hard-wired in the nervous system from birth, so the relationship between the stimulus and the reflex response does not need to be learned. Reflexes are innate because of their evolutionary adaptive role in protecting us from harm or promoting survival.

In classical conditioning, the stimulus and response components of the reflex are referred to as the unconditioned stimulus and unconditioned response respectively. 'Unconditioned' means the stimulus and response are *not learned*.

» **Unconditioned stimulus (UCS)** is a biologically significant stimulus (for example, food) that naturally causes a reflex response (for example, salivation).
» **Unconditioned response (UCR)** is a specific involuntary response (reflex) that occurs to a specific biologically significant stimulus (for example, pupil constriction in response to light).

For example, in the withdrawal reflex the unconditioned stimuli are the afferent pain signals caused by the heat or the sharp object, and the unconditioned response is the efferent withdrawal response. In the flight-or-fight-or-freeze response, the unconditioned stimuli are the afferent signals from a stimulus such as a sudden loud sound, a sudden large movement, or a strong force applied to the body that reaches the amygdala, and the unconditioned response is the cascade of physiological responses that prepare the body to fight, escape or freeze.

2 Learned components: neutral stimulus, conditioned stimulus, conditioned response

In classical conditioning, our bodies learn to produce a reflex response to a stimulus that would not naturally cause it. We call the stimulus that does not naturally cause a reflex a **neutral stimulus**. An initially neutral stimulus can become a **conditioned stimulus** that can cause a reflex response through association with an unconditioned stimulus. We call the reflex response that occurs to a conditioned stimulus a conditioned response. This gives us the following definitions.

» **Neutral stimulus (NS)** is a stimulus that does not naturally cause a reflex response (biologically neutral).
» **Conditioned stimulus (CS)** is a *previously neutral stimulus* that acquires the ability to cause a reflex response through its association with an unconditioned stimulus.
» **Conditioned response (CR)** is a reflex response to a conditioned stimulus in the absence of the unconditioned stimulus that would usually cause it.

Pavlov's hypothesis about how his dogs learned to salivate in response to the sound of his approaching footsteps can now be re-expressed using the terms of classical conditioning, as follows.

Pavlov's footsteps are an initially neutral stimulus (NS) that become associated with the unconditioned stimulus (UCS) of food and the unconditioned response (UCR) of salivation through the footsteps repeatedly occurring just before the food. The learned association between the NS and UCS–UCR (reflex) causes the footsteps to become a conditioned stimulus that produces a conditioned response (CR) of salivation before the food is presented.

Pavlov's three-phases of classical conditioning

Pavlov designed a series of experiments to test his hypothesis by systematically controlling the events that occurred before he presented the dogs with food. He selected the sound of a bell as an initially neutral stimulus to associate with the unconditioned stimulus of food. He tested the hypothesis by sounding the bell before he presented the food. This way he could determine how many trials it would take until the bell became a conditioned stimulus that could produce a conditioned salivation response.

The process of training a conditioned response is called conditioning. Each presentation of the neutral stimulus before the conditioned stimulus is called a conditioning trial (or an acquisition trial). Pavlov described a three-phase process to produce a classically conditioned salivation response. Table 4.1 outlines the **three phase process of classical conditioning**.

It is important to notice that the CR *is the same reflex behaviour* that previously occurred only in the presence of the UCS. We call the reflex response a CR when it occurs to the CS alone, and we call it a UCR when it occurs in the presence of the UCS.

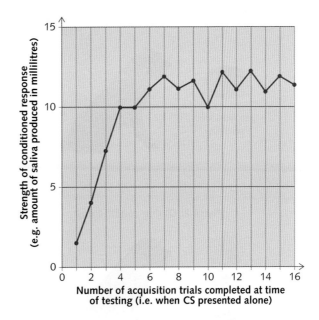

Source: Adapted from Weiten, W. (2007). Psycholog: Themes and Variatons 7th Edition, Thomason Wadsworth

Figure 4.5 Graph showing the acquisition of a conditioned salivation response

Figure 4.5 shows a graph similar to those used by Pavlov to record the acquisition of a conditioned salivation response. Each data point on the graph represents a test conducted during phase 3 of the conditioning process, when the conditioned stimulus is presented alone. The horizontal *x*-axis shows the number of conditioning trials (phase 2) that have occurred before each phase 3 test. The vertical *y*-axis shows the strength of the conditioned salivation response as millilitres of saliva. Notice how the strength of the CR increases rapidly over the first five or so conditioning trials, and then flattens out when it reaches maximum strength.

EXAM TIP
When asked to identify the components of classical conditioning in a scenario, ask yourself the following questions: (1) What is the preexisting stimulus-response reflex association? The answer to this question will give you the UCS and UCR components; (2) What biologically neutral stimulus is paired repeatedly with the UCS? The answer to this will give you the initially neutral stimulus that becomes a conditioned stimulus; (3) What response occurs to the conditioned stimulus alone after conditioning? This gives you the conditioned response.

Table 4.1 The three phases of classical conditioning in Pavlov's experiment to train a classically conditioned salivation response to the sound of a bell

Phase 1: Before conditioning	Phase 2: During conditioning	Phase 3: After conditioning
Phase 1 describes the relationships that exist between stimuli and responses before any conditioning trials occur. At this stage, the bell is the neutral stimulus (NS) because it does not naturally cause salivation. The food is the unconditioned stimulus (UCS) because it naturally causes salivation. Salivation is the unconditioned response (UCR) because it is a reflex that occurs naturally in response to food.	Phase 2 describes the process of learning the association between the NS and the UCS. On each conditioning trial, Pavlov sounded the bell (NS) *immediately before* the presentation of the food (the UCS) and the natural salivation response (UCR).	Phase 3 describes the process used to test whether the bell (NS) had become a conditioned stimulus (CS). Pavlov did this by sounding the bell on its own, without the food (UCS), and measuring the salivation response. If the dogs produced saliva in response to the bell, then the bell has become a conditioned stimulus (CS) that causes a conditioned response (CR) of salivation.

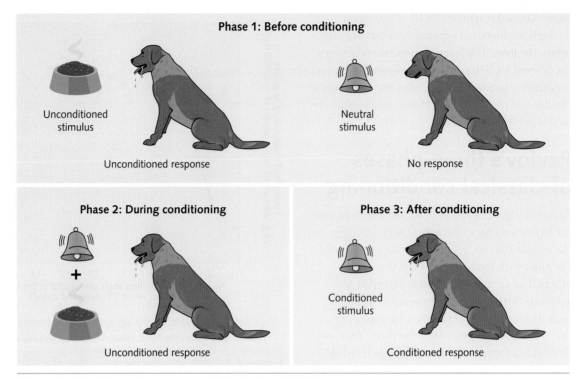

Figure 4.6 The three-phase process of classical conditioning

The three-phase process of classical conditioning is shown in diagrammatic form in Figure 4.6.

What is learned in classical conditioning?

Although Pavlov described classical conditioning as learning a conditional relationship between the CS and the UCS, this does not mean that he believed his dogs represented this knowledge consciously. Instead, classical conditioning produces a learned **involuntary association** between the CS and the UCS that produces an automatic conditioned response to the CS.

The learned association between the CS and the UCS is involuntary because it is learned regardless of whether the learner is aware of the association between the two stimuli, and because the conditioned response is automatic and outside of the learner's conscious control.

From this, we can define classical conditioning as a form of associative learning in which an involuntary association is learned between a neutral stimulus and an unconditioned stimulus so that the neutral stimulus becomes a conditioned stimulus that can cause a conditioned reflex response.

The learning is represented implicitly in the learner's nervous system by the strengthening of neural connections between the neurons that fire in response to the neutral stimulus and the neurons that fire in response to the unconditioned stimulus and unconditioned response. You might recognise this as the process of long-term potentiation (LTP) that was described in Chapter 2. In Chapter 5 you will learn how classical conditioning through LTP is a form of implicit memory.

KEY CONCEPTS 4.2a

» Behaviourist approaches to learning are based on behaviourism, an approach to psychology introduced by John B. Watson in the early 20th century that focuses on measuring observable behavioural responses to stimuli.
» Watson's behaviourism was inspired by the work of Russian physiologist Ivan Pavlov, who described the process of classical conditioning through experiments investigating dogs' learned salivation responses to stimuli other than food.
» The three-phase process of classical conditioning involves three stages in the acquisition of a conditioned response:
1 Before conditioning
2 During conditioning
3 After conditioning.
» Classical conditioning is the process of learning an involuntary association between an initially neutral stimulus (NS) and an unconditioned stimulus (UCS) so that the neutral stimulus becomes a conditioned stimulus that can cause a conditioned reflex response in the absence of the unconditioned stimulus.

Concept questions 4.2a

Remembering
1 Which two classical conditioning terms describe the stimulus and response components of a reflex? **r**
2 Define 'unconditioned'. **r**
3 Complete this sentence: Classical conditioning is a behaviourist approach to learning because … **r**
4 What is the name of the neural mechanism that enables learning of CS–UCS associations? **i**

Understanding
5 Explain the difference between a neutral stimulus and an unconditioned stimulus. **e**
6 Explain the difference between an unconditioned response and a conditioned response. **e**
7 Explain what is learned in classical conditioning. **e**

Applying
8 Walking in the park one day, I see someone has discarded the wrapper for my favourite chocolate bar on the path ahead. This causes me to crave chocolate. Using the terminology of classical conditioning, explain why I respond in this way to an inedible piece of litter. **c**
9 A young child is taken to the doctor to receive an immunisation and they find the needle painful. Ever since this experience the child becomes distressed when they approach the doctor's office. Apply the principles of classical conditioning to explain the child's learned fear of the doctor's office. **c**
10 map the three-phase process of classical conditioning onto the neural process of long-term potentiation described in Chapter 2. **i**

Classical conditioning	Long-term potentiation (LTP)
Before conditioning	
During conditioning	
After conditioning	

HOT Challenge
11 Jai and his friend Kyle are discussing whether their teacher could classically condition their class to salivate to the sound of the school bell. Kyle doesn't think the teacher can use their class and they will need to test students who don't study Psychology and are unaware of Pavlov's study. Jai thinks they can classically condition their class. With reference to the autonomic nervous system and what you know about the process of classical conditioning, who do you think is correct? **c**

From Pavlov to Watson and the 'Little Albert' experiment

Watson greatly admired Pavlov's careful and systematic experiments. He believed that all human learning and development could be explained as the result of learned associations between environmental stimuli and innate emotional responses, such as fear or love. The following quotation captures Watson's belief that human infants are born with only basic reflexes that enable survival and that they learn everything else through conditioning.

> 'Give me a dozen healthy infants, well-formed, and my own specified world to bring them up in and I'll guarantee to take any one at random and train him to become any type of specialist I might select – doctor, lawyer, artist, merchant-chief and, yes, even beggar-man and thief, regardless of his talents, penchants, tendencies, abilities, vocations and the race of his ancestors.'

Watson, J. B. (1924). *Behaviorism*. New York: People's Institute Publishing Company. p. 104

To test his theories, Watson needed to prove that he could produce a classically conditioned emotional response in a human infant. This is exactly what he set out to do with his research assistant Rosalie Rayner in 1920 (Watson & Rayner, 1920).

Analysing research 4.1 explains how Watson and Rayner demonstrated a classically conditioned fear response to a tame white rat in an infant they called 'Little Albert'.

4.2.3 ETHICS IN LEARNING RESEARCH: 'LITTLE ALBERT'

ANALYSING RESEARCH 4.1

Conditioned emotional reactions in a human infant (Watson and Rayner, 1920)

Watson and Rayner (1920) describe how they acquired a male infant from the hospital nursery with permission from his mother. They protected the infant's identity by referring to him as 'Little Albert'. They began their testing when Albert was 8 months old, demonstrating that he was not innately fearful of a wide range of stimuli. Watson and Rayner recorded Albert's responses to being suddenly confronted with a dog, a burning newspaper, a monkey, adults wearing masks and a tame white rat. They reported that Albert showed curiosity to all of these stimuli, not fear. However, Albert did show a strong and immediate fear response to the sound of the metal bar being hit. Watson captured all of Little Albert's responses on film, which you can watch at the weblink *The Little Albert experiment*. Figure 4.7 shows a still image from the original film.

Watson and Rayner selected the tame white rat as an initially neutral stimulus that would become a conditioned stimulus. Their aim was to produce a conditioned fear response to the white rat by repeatedly associating it with the sound of the metal bar being hit. The experimental work to establish a classically conditioned fear response began when Albert was 11 months old.

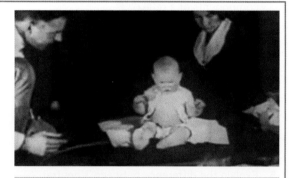

Figure 4.7 John Watson and Rosalie Rayner during their experiment to condition Little Albert to fear a white rat, 1920

They described the procedure as follows:

"(1) White rat suddenly taken from the basket and presented to Albert. He began to reach for rat with left hand. Just as his hand touched the animal the bar was struck immediately behind his head. The infant jumped violently and fell forward, burying his face in the mattress. He did not cry, however. (2) Just as the right hand touched the rat the bar was again struck. Again the infant jumped violently, fell forward and began to whimper. In order not to disturb the child too seriously no further tests were given for one week."

Watson & Rayner, 1920

Weblink
The Little Albert experiment

One week later they presented the rat again, *first without the loud sound*. Albert did not startle but was wary of the rat, cautiously reaching to touch it and then withdrawing his hand quickly. Watson and Rayner then presented the rat before the loud sound three more times, reporting that Albert startled and became distressed each time. The rat was then presented alone and Albert showed the same startled and distressed response. Two more conditioning trials were conducted, then one final test with the rat alone. Again, they reported a strong conditioned fear response to the white rat alone.

Little Albert was brought back again one week later. No further conditioning trials were conducted during this session. Instead, Watson and Rayner first gave Albert some building blocks and he played with them happily. This suggested that Albert had not developed a generalised fear of the testing room itself. Then they presented the rat alone, and Albert immediately showed a strong fear response and attempted to move away from the rat. The conditioned fear response had been maintained without further training from the previous week.

Albert was brought back again after 1 month (12.5 months old) with no further conditioning and presented with the white rat (CS) alone. This time, Albert showed extreme caution and avoidance of the rat, but was not extremely distressed. It seemed that the conditioned response remained but was weaker.

Watson and Rayner wrote that they intended to do further studies to see if they could eliminate Albert's conditioned fear response. They knew that Pavlov had described **extinction** of the conditioned response by presenting the CS (bell) alone repeatedly without reintroducing the UCS (the food). Pavlov had found that it took many extinction trials over weeks to achieve lasting extinction. For example, the conditioned response would often reappear at the beginning of a new day after having been extinguished the previous day. He called this **spontaneous recovery** of the conditioned response. Gradually spontaneous recovery would weaken and eventually disappear.

Watson and Rayner wrote that they intended to try a number of approaches to extinguish the fear response, including teaching Albert to associate the rat with safe experiences such as play, or with something pleasant such as eating sweets. Unfortunately, Watson and Rayner were not able to study extinction with Albert because his mother withdrew him from the experiment unexpectedly.

Source: adapted from Watson, J. B. & Rayner, R. (1920). Conditioned emotional reactions. *Journal of Experimental Psychology,* 3(1), 1–14. https://doi.org/10.1037/h0069608

Questions

1 Use the table below to identify the three phases of classical conditioning in the description of Watson and Rayner's (1920) experiment with Little Albert.

2 Describe two ethical guidelines that are apparent in Watson and Rayner's study.

Before conditioning	
During conditioning	
After conditioning	

INVESTIGATION 4.1 EVALUATION OF A CASE STUDY

Scientific investigation methodology
Case study

Aim
To analyse and evaluate the Little Albert Case study

Logbook template
Case study

Introduction
The Little Albert experiment described in Analysing research 4.1 is an example of a case study. Reread Analysing research 4.1 and rewatch the Little Albert Experiment available through the weblink on page 144. When you have done this, use the questions that follow as a guide to evaluate this case study.

Questions

1. The Little Albert experiment is a case study. What is a case study?
2. Write a research question that Watson and Rayner might have been trying to answer in their experiment.
3. What was the purpose of exposing 'Little Albert' to the dog and white rat prior to the conditioning trials?
4. Evaluate the experimental method in terms of
    a. repeat experiments to ensure the findings are robust
    b. any possible sources of error.
5. Analyse whether the ethical concept of beneficence was followed with reference to the experiment.
6. What could have been some consequences for Little Albert as a result of this experiment as he got older?

The survival value of classical conditioning

Although classical conditioning does not provide the complete explanation for human psychological development that Watson hoped, there is no doubt that it is fundamental to the survival and development of all animals. Classical conditioning has been shown to occur across the animal kingdom, from the simplest single-celled organisms to humans (LeDoux, 2022).

Learning through classical conditioning has been preserved across evolution because the survival of all animals depends on learning which stimuli in the environment predict threatening situations and which stimuli predict situations that benefit survival. As Pavlov noted, when we learn an association between a conditioned stimulus and an unconditioned stimulus, the conditioned response prepares our body for what is likely to happen next.

Learning about stimuli that predict danger is particularly important for survival. Fear is an adaptive emotional response when it leads to behaviours that promote safety, such as avoiding walking over a high cliff or moving away from a potentially poisonous snake or spider. As we will see when we discuss operant conditioning, classically conditioned responses are important in learning adaptive voluntary behaviours. For this reason, psychologists have continued to study the acquisition of conditioned fear responses since the Little Albert study (Fullana et al., 2020). Studies of fear conditioning have helped us understand the brain mechanisms involved in fear learning and why this adaptive process sometimes causes maladaptive fear responses in anxiety disorders, including phobias.

Analysing research 4.2 explains how fear conditioning studies are conducted and their findings.

ANALYSING RESEARCH 4.2

From Little Albert to fMRI: contemporary approaches to fear conditioning in humans

In a typical fear conditioning study, participants are first presented with a series of pictures on a computer screen. The pictures may be of two different coloured squares (red and blue), or they may be of objects or living things that vary in their natural association with threat (e.g. flowers or snakes). During the conditioning phase, half of the pictures in each category are paired with an aversive (unpleasant) unconditioned stimulus, usually an electric shock applied to the hand or arm or a sudden loud burst of noise. The other half of the pictures are never paired with an aversive stimulus. The researchers measure sympathetic nervous system arousal during the experiment using the skin conductance response (SCR).

The SCR measures the potential of the skin to conduct electricity, which increases in response to sweating. As you know from Chapter 3, increased sweating is associated with activation of the sympathetic nervous system in response to a threat. So, increased SCR readings indicate an increased response to perceived threat. The SCR is measured by attaching two small electrodes to the palm of a participant's hand or fingers, as shown in Figure 4.8.

Figure 4.8 Equipment used to measure skin conductance responses in studies of fear conditioning. The electrodes on the fingers detect changes in the electrical conductance of the skin due to sweating.

After the conditioning trials, the researchers use the SCR to detect whether a conditioned fear response has been learned for stimuli that were paired with the aversive stimulus by presenting all the pictures from the conditioning phase again, without the aversive stimulus. The results show that pictures that were paired with the aversive stimulus produce higher SCR readings than pictures that were not paired with the aversive stimulus.

Functional magnetic resonance imaging (fMRI) studies of the brain during fear conditioning studies show that the amygdala, a small almond-shaped structure deep within each temporal lobe, is active during unconditioned fear responses to electric shock and when conditioned stimuli (pictures previously paired with electric shock) are presented without shock. Studies of patients with lesions to the amygdala show that these patients do not acquire conditioned fear responses.

Interestingly, fear conditioning studies show that non-phobic people acquire conditioned fear responses to pictures of snakes and spiders much more quickly than they do to pictures of pleasant or neutral stimuli such as flowers or coffee mugs. The conditioned fear response to pictures of snakes and spiders is also much stronger and harder to extinguish than a conditioned fear response to pictures of pleasant or neutral objects.

Why does this happen? The theory is that our evolutionary history with threat-related stimuli like snakes and spiders has prepared us to learn fear responses to these stimuli quickly and has made them hard to extinguish. In Chapter 9, you will explore how conditioned fear responses can become maladaptive when people experience a phobia to a specific object or situation, and how classical conditioning principles can also be used in treatment.

Source: Adapted from Delgado, M. R., Olsson, A., & Phelps, E. A. (2006). Extending animal models of fear conditioning to humans. *Biological psychology, 73*(1), 39–48 © Elsevier.

Questions

1. Apply the three phases of classical conditioning to a typical fear conditioning study. Identify the UCS and UCR, and the NS, CS and CR components.
2. What kind of research design is used in fear conditioning studies?
3. Explain how the skin conductance response (SCR) and fMRI data are used as dependent measures of fear conditioning.
4. Describe how unconditioned stimuli and unconditioned responses relate to the stress response described in Chapter 3.

HOT Challenge

5. Humans and other animals learn conditioned taste aversions to foods or drinks that they have consumed followed by an experience of nausea and vomiting. It takes only one experience like this for the mere sight or smell of the substance to cause a strong disgust response and feeling of nausea. Apply the three-phase process of classical conditioning to explain how learned taste aversions are acquired.
6. Explain why the association between the food and nausea is learned after a single exposure.

EXAM TIP
When analysing operant conditioning scenarios in an exam, it is helpful to remember that in operant conditioning the term 'positive' means to add and 'negative' means to take away (remove).

Operant conditioning

Classical conditioning is a very powerful model of associative learning, but it can only explain learning of involuntary stimulus–response associations. Behaviourist psychologists realised that behavioural models of learning also needed to explain how environmental stimuli cause voluntary behaviours.

This challenge was taken up by another influential American psychologist Burrhus Frederick Skinner (B. F. Skinner) (Figure 4.9). Skinner devoted his career to explaining how environmental stimuli control our voluntary behaviours through a process he called operant conditioning. Although Skinner undertook much of his work from the 1950s through to the 1980s, his work continues to have great significance today. Let's take a look.

Figure 4.9 American psychologist B. F. Skinner, whose work has led us to understand the principles of operant conditioning

Skinner referred to voluntary behaviours as operant behaviours. This term reflects his belief that our voluntary behaviours *operate* (i.e. act) on the environment *to generate* **consequences** (Skinner, 1953). Operant conditioning is a form of associative learning in which the learner learns associations between their voluntary behaviours and rewarding or punishing consequences.

Reinforcement and punishment

The idea behind operant conditioning is very simple yet very powerful – we are more likely to reproduce behaviours that have been associated with pleasant or *rewarding* consequences, and we are less likely to reproduce behaviours that have been associated with unpleasant consequences.

In the formal language of operant conditioning, rewarding consequences *strengthen* a behaviour through **reinforcement**, whereas unpleasant consequences *weaken* a behaviour through **punishment**.

This sounds quite straightforward, until we consider that reinforcement and punishment can both take positive and negative forms. Working out whether a particular consequence of a behaviour is an example of positive or negative reinforcement or positive or negative punishment depends on whether the consequence involves *adding* or *removing* a rewarding (pleasant) or punishing (unpleasant) stimulus.

Positive reinforcement

In **positive reinforcement**, the learner receives something *rewarding* as a consequence of their behaviour. For example, a teenager might be praised by a parent for tidying their room without having to be asked. In this case, receiving praise is a rewarding consequence that reinforces (strengthens) the behaviour, making it more likely to occur in future.

Negative reinforcement

In **negative reinforcement**, the learner experiences something *unpleasant* being removed as a consequence of their behaviour. For example, a teenager who has tidied their room without being asked might be rewarded by their parents removing the ban they had placed on the teenager's social media use. In this case, the ban on social media is an unpleasant stimulus and removing it is a rewarding consequence that reinforces the behaviour, making it more likely to occur in future.

In positive reinforcement, a rewarding stimulus is *added* (+) as a consequence of behaviour. In negative reinforcement, an unpleasant stimulus is *subtracted* (−), or taken away, as a consequence of behaviour. Both kinds of reinforcement are rewarding outcomes for the learner and act to *strengthen* the behaviour they follow, making it more likely to occur in future (Table 4.2).

Table 4.2 The two kinds of reinforcement

Type of reinforcement	Definition
Positive reinforcement	The *addition* of a rewarding stimulus as a consequence of a behaviour, making the behaviour more likely in future
Negative reinforcement	The *removal* of an aversive stimulus as a consequence of a behaviour, making the behaviour more likely in future

Positive punishment

In **positive punishment**, the learner experiences an unpleasant stimulus as a consequence of their behaviour (Table 4.3). For example, when a teenager gets home from a party after the time set by their parents, the parents may express anger as a consequence. The expression of anger is an unpleasant stimulus that causes feelings of shame or guilt in the teenager. The anger is a punishing consequence that weakens the behaviour, making it less likely that the teenager will come home late from a party in future.

Negative punishment

In **negative punishment**, the learner experiences the loss of a rewarding stimulus as a consequence of their behaviour. For example, when a teenager comes home late from a party, their parents may confiscate the teenager's mobile phone for a day as a consequence. Removal of the mobile phone takes away something that is rewarding, causing unpleasant feelings about the loss of something that is valued. The loss of the mobile phone is a punishing consequence that weakens the behaviour, making it less likely to occur in future.

In positive punishment, an unpleasant stimulus is *added* (+) as a consequence of behaviour. In negative punishment a rewarding stimulus is *subtracted* (−) as a consequence of behaviour. Both kinds of punishment act to *weaken* the behaviour they follow, making it less likely to occur in future (Table 4.3).

Table 4.3 The two kinds of punishment

Type of punishment	Definition
Positive punishment	The *addition* of an unpleasant stimulus as a consequence of a behaviour, making the behaviour less likely in future
Negative punishment	The *removal* of a rewarding stimulus as a consequence of a behaviour, making the behaviour less likely in future

Whenever you are presented with an operant conditioning scenario you can ask yourself the following three questions to work out what kind of operant conditioning is involved:

1. What is the meaning for the learner of the stimulus that is used in the consequence? Is it a pleasant or unpleasant stimulus?
2. Is the stimulus being *given* to the learner or *taken away* from them?
3. Is the combined effect of 1 and 2 a good outcome for the learner (reinforcing), or is it a bad outcome for the learner (punishing)?

Table 4.4 shows how these three questions can be used to identify the kind of operant conditioning in a scenario. The headings for the two columns show the options for answering Question 1.

The headings for the two rows show the options for answering Question 2. The answer to Question 3 is given within the four cells of the table, with each cell representing one of the four methods of operant conditioning.

4.2.4 OPERANT CONDITIONING

Table 4.4 How to identify the kind of operant conditioning in a scenario

		Meaning of stimulus for learner	
		Rewarding (pleasant)	Aversive (unpleasant)
Stimulus action	Given	POSITIVE REINFORCEMENT (Behaviour is strengthened)	POSITIVE PUNISHMENT (Behaviour is weakened)
	Taken away	NEGATIVE PUNISHMENT (Behaviour is weakened)	NEGATIVE REINFORCEMENT (Behaviour is strengthened)

Measuring the effects of reinforcement and punishment

Skinner's goal was to use the principles of operant conditioning to predict and control voluntary behaviours. He conducted his work using laboratory animals, such as rats and pigeons, and designed specialised laboratory equipment to enable precise control of the animals' experiences. Skinner improved the precision of his measurements by pioneering the use of computers so he could present stimuli and record responses automatically.

Skinner's primary tool for studying learning was the Skinner box. A Skinner box is a specialised animal enclosure used to control the stimuli in the animal's environment and for recording their responses. Skinner boxes are still used today to study learning in animals. Figure 4.10 shows the components of a typical Skinner box.

The lever on the wall of the box provided a simple mechanism for measuring voluntary behaviour. For example, Skinner could measure how often a rat pressed the lever.

The food dispenser provided a mechanism to present or withhold a rewarding stimulus (food), and the electric grid provided a mechanism to present (or turn off) an unpleasant stimulus (pain). Table 4.5 shows an example of each kind of operant conditioning used in the Skinner box.

The diagram of the Skinner box in Figure 4.10 also shows lights and speakers for producing visual or auditory stimuli. These could be used to signal an upcoming rewarding or unpleasant stimulus. We explore how these environmental signals are used in the three-phase process of operant conditioning next.

Table 4.5 Operant conditioning of consequences for a rat pressing a lever in a Skinner box

Type of operant conditioning	Example
Positive reinforcement	Pressing the lever causes a food pellet to be released into the rat's food bowl
Negative reinforcement	Pressing the lever causes an electric shock under the animal's feet to be turned off
Positive punishment	Pressing the lever causes an electric shock to the animal's feet
Negative punishment	Pressing the lever causes a heat lamp that provided warmth to be turned off

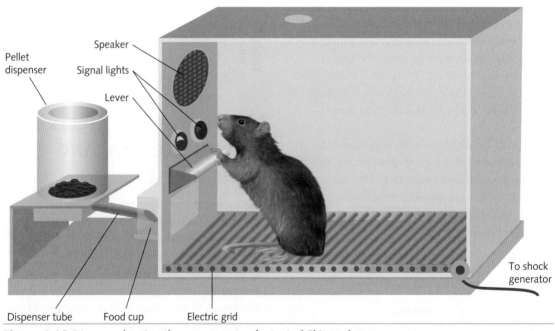

Figure 4.10 Diagram showing the components of a typical Skinner box

ACTIVITY 4.1 HOW TO IMPROVE STUDY HABITS USING OPERANT CONDITIONING

Try it

Future behaviours depend on the consequences of the behaviour. You are more likely to repeat a behaviour if the consequence is desirable. This is called reinforcement. You are less likely to repeat a behaviour if the consequence is undesirable. This is called punishment.

Reinforcement and punishment can be split into positive and negative.

Positive reinforcement:
For every set of key terms you define for your notes give yourself a treat – take a short walk outside, eat your favourite snack, listen to your favourite song. Adding in something desirable makes it more likely you will complete another set of key terms.

Negative reinforcement:
Every set of key terms you define for your notes means you will need less time to prepare for the SAC. Removal of something undesirable, like cramming, makes it more likely you will complete another set of key terms.

Positive punishment:
This is not recommended as an effective strategy to promote learning, as it only promotes negative feelings in the learner.

Negative punishment:
Every time you forget to complete a set of key terms you give your phone to your parents for the night. Removal of something desirable like your phone should make it less likely that you forget to complete your key terms in future.

Apply it

Miley has trouble sitting and concentrating on her work. When she manages to sit for five minutes of focused work, her teacher puts a sticker in Miley's student diary. Miley loves stickers! When Miley fidgets and cannot sit still in class, her teacher takes time from Miley's recess.

Identify and explain the two types of consequences Miley's teacher uses.

Exam ready

At the end of the lesson, Miley's teacher has a headache. She takes Panadol and her headache goes away. The next time she has a headache she is likely to take Panadol again.

This scenario demonstrates

A   Positive reinforcement
B   Negative reinforcement
C   Positive punishment
D   Negative punishment

KEY CONCEPTS 4.2b

» B. F. Skinner was a behaviourist psychologist who described the principles of operant conditioning.
» Operant conditioning is a form of associative learning in which voluntary behaviours are influenced by the consequences associated with them – behaviours are strengthened through reinforcement and weakened through punishment.
» Consequences for voluntary behaviours can be of four different kinds:
    1   Positive reinforcement (addition of a rewarding stimulus)
    2   Negative reinforcement (removal of an unpleasant stimulus)
    3   Positive punishment (addition of an unpleasant stimulus)
    4   Negative punishment (removal of a rewarding stimulus).
» Skinner studied the effects of reinforcement and punishment on the voluntary behaviours of rats and pigeons using a special enclosure called a Skinner box.
» The Skinner box allowed precise recording of behaviours and control of the stimuli and consequences associated with behaviours.

Concept questions 4.2b

Remembering

1   What kind of operant conditioning occurs when a rewarding stimulus is removed as a consequence of behaviour? r
2   What kind of operant conditioning occurs when an unpleasant stimulus is removed as a consequence of behaviour? r

Understanding

3   Give an example of a stimulus that could be used to punish behaviour if it was removed as a consequence of behaviour and that could reinforce behaviour if it were given as a consequence. e

Applying

4 Apply the terms of operant conditioning to the following scenarios.

   a A netball player receives an angry look from her teammate when she drops the ball

   b A student who has been misbehaving in class is sent out of the room for 10 minutes of 'time-out'

   c A prisoner is tortured until they confess

   d A driver receives a fine for speeding

   e A child receives additional playtime for completing their homework

   f A woman trips on the footpath while walking and looking at her mobile phone

   g A fruit picker receives $5 for every kilo of grapes they pick

   h A gamer receives points for completing each level of the game

   i A child cries until she receives attention from her parents

   j A parent nags their child until they take the rubbish bin out.

HOT Challenge

5 In the example in Question **4i** where the child cries to receive attention from her parents, both the parents and the child are influenced by principles of operant conditioning. Use the correct operant conditioning terms to explain what the parent is likely to do next time the child cries.

The three-phase process of operant conditioning – learning our ABCs

4.2.5 THE ABCS OF OPERANT CONDITIONING

Skinner's goal was not only to understand how reinforcement and punishment affect the production of voluntary behaviours, but also to control and predict behaviours through manipulating environmental stimuli.

This sounds like an evil plan for world domination, but Skinner's careful analysis of how environmental stimuli can come to control our behaviours has helped us understand how we form habitual behaviours, and how we can change unwanted habits. To understand how, we need to know the ABCs of operant conditioning.

We already know the Bs and Cs of operant conditioning – these stand for **B**ehaviours and their **C**onsequences. What about the **A**s? The **A** stands for the antecedents. An **antecedent** is another word for cue – it is a stimulus that comes *before* a behaviour and its consequence (*ante* comes from the Latin, meaning 'before' or 'prior to'). Together, the antecedent, behaviour and consequence give us the **three-phase process of operant conditioning** (also known as the ABC model of operant conditioning). Figure 4.11 shows the relationship between the antecedent, behaviour and consequence in the model.

To understand how the ABC model works, let's apply it to a simple Skinner box experiment. Imagine that we want to teach a rat when to press a lever to receive a food pellet. How can we do this? The three components we need to solve this are given below:

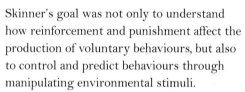

Figure 4.11 The three-phase process of operant conditioning, or **A**ntecedent-**B**ehaviour-**C**onsequence (ABC) model

1. The antecedent is a stimulus that occurs *before* a behaviour and consequence. In our Skinner box example, the antecedent could be the red light inside the Skinner box turning on.
2. The behaviour is the voluntary behaviour produced by the learner. In our Skinner box, the behaviour is pressing the lever.
3. The consequence is the outcome that follows the behaviour. In our Skinner box, the outcome of lever-pressing is that a food pellet is released into the rat's food bowl (positive reinforcement).

If we want our rat to learn to press the lever, then we can provide a food pellet as positive reinforcement each time the rat presses the lever. This relates to the behaviour and consequence parts of the model. But what is the role of the antecedent?

The antecedent is important if we want to control *when* the rat presses the lever. To do this, we can structure the rat's learning experience so that pressing the lever only results in food when the red light is on. This way, the rat will learn through experience that pressing the lever is only reinforced when the red light is on. That is, the red light becomes a signal or cue for the availability of the reward if the behaviour is produced. Skinner found that when he could control the antecedents, he could also predict and control when a behaviour occurred.

The following scenarios illustrate the ABC model in action, and some ways in which human behaviour is sometimes just like the behaviour of a rat in a Skinner box!

Scenario 1: Finley is two years old and is learning about which fruits to eat from the family garden. When he picks the green strawberries and eats them, they taste sour. When he picks the red strawberries, they taste sweet. Finley learns to only pick and eat the red strawberries.

In this scenario, eating red strawberries is associated with a rewarding outcome (yum!) and eating green strawberries is associated with a punishing outcome (yuk!). The colour of the strawberry becomes an antecedent for behaviour – *redness* signals that picking and eating the strawberry will be positively reinforced, whereas *greenness* signals that avoiding the strawberries will be negatively reinforced (through avoiding the 'yuk' experience associated with eating green strawberries).

Scenario 2: Gemma knows that when her history teacher, Mr Jackson, looks at her a certain way he is about to yell at her to stop talking during class. Gemma has learned to stop talking whenever she sees Mr Jackson give her 'that look'.

In this scenario, talking in class is associated with positive punishment (Mr Jackson yelling), and 'that look' from Mr Jackson occurs just before Mr Jackson yells at Gemma for talking in class. In this way, 'that look' serves as an antecedent that signals punishment if Gemma continues to talk or negative reinforcement if Gemma stops talking (avoidance of Mr Jackson yelling).

Scenario 3: Hugo looks forward to playing his favourite video game each afternoon when he gets home from school. However, this afternoon his mum says he can't play until he has done an hour of study for the maths test he has tomorrow. Hugo sits down to study at his desk and sees the gaming console out of the corner of his eye. The sight of the gaming console makes it hard for him to concentrate on studying. He finds his thoughts keep going to which level of the game he will reach today and which superpower he might gain. As soon as an hour is up, he closes his books and starts the game. He becomes engrossed in the game for the next couple of hours and does not want to stop when his mum calls him for dinner.

In this scenario, there are some strong antecedents for gaming that cause Hugo to become distracted from his homework. The time of day and the sight of the gaming console on his desk have both become strongly associated with the rewards of playing the game and provide strong cues to play. As you learned in Chapter 2, dopamine is released in anticipation of rewards. If we map the dopamine response to the three-phase process of operant conditioning, it is the antecedent stimuli that cue the release of dopamine.

From these scenarios it is clear that learning stimulus–response associations between antecedents, behavioural responses and their consequences allows us to respond adaptively to seek rewards and avoid punishments. Antecedents are powerful drivers of habitual voluntary behaviours. Habits can be adaptive. For example, leaving my running shoes beside my bed each night provides an antecedent to prompt me to go for my morning run. However, habits can also be maladaptive. For example, when people find it difficult to control their gambling behaviour or to moderate their smoking, maladaptive behaviours are maintained through the strong associations formed between antecedent stimuli and the strongly rewarding consequences of winning big on the pokies or the stimulant effect of smoking.

In the case of gambling behaviour, passing a sign advertising the poker machines in the local pub whilst out shopping becomes a powerful antecedent stimulus to gamble, rather than maintaining the adaptive goal of doing the shopping. This occurs through the learned association between the sign and the rewarding consequences of gambling. Once inside the gambling venue the antecedents multiply – including the sounds and visual stimuli associated with the machines, the lighting in the venue and familiar faces. All of these stimuli act as antecedents and strengthen the desire to gamble through their past association with the excitement of gambling. Situations like this can prompt us to wonder about just how much of our 'voluntary behaviour' is actually under our control and how much is under the control of environmental cues.

KEY CONCEPTS 4.2c

» The three-phase process of operant conditioning (ABC model) describes how environmental stimuli that are present when we experience consequences for our behaviours can become cues for us to produce voluntary behaviours to gain rewards or avoid punishment.
» An antecedent is an environmental stimulus that has become associated with the consequences of a voluntary behaviour.
» Antecedent stimuli cue behaviours that are associated with rewarding consequences through the release of dopamine in anticipation of a reward.

Concept questions 4.2c

Remembering
1 What are the three components of the three-phase process of operant conditioning? (r)

Understanding
2 Explain the role of the antecedent in the three-phase process of operant conditioning. (e)

Applying
3 Apply the three-phase process of operant conditioning to explain why Scruffy the dog pesters his owner to feed him at the same time each day. (c)

HOT Challenge
4 Give an example of one healthy and one unhealthy behaviour that has become habitual for you and apply the three-phase process of operant conditioning to identify the antecedents, behaviours and consequences. (c)

4.2.6 USING OPERANT CONDITIONING IN ANIMAL TRAINING

The three-phase process of operant conditioning and animal training

The three-phase process of operant conditioning gives us a powerful tool for teaching desirable behaviours and reducing unwanted behaviours by controlling the antecedents. A familiar example of the ABC model in action is when we teach a pet dog to sit on command.

In this example, the antecedent is the word 'sit', the behaviour is the dog *sitting in response to the verbal command*, and the consequence is a reward for the behaviour *if it was performed in response to the command*. The most effective form of reinforcement is positive reinforcement, such as verbal praise and/or a food treat. Rewarding the dog *only* when it sits in response to the command will help the dog learn that 'sit' is a cue for a specific behaviour that is associated with a reward.

Teaching a dog to sit is one thing – training a seal to perform complex behaviours in response to verbal commands, hand gestures or a whistle is another thing entirely! Explore Case study 4.1 and the accompanying video to learn how marine mammal trainers use the three-phase processes of operant and classical conditioning when they are training seals.

CASE STUDY 4.1

Teaching Groucho the Australian fur seal to brush his teeth

Skinner's work with training behaviours in laboratory animals has strongly influenced animal training models outside of the laboratory. One of the earliest applications of the ABC model to animal training was for the care and training of captive marine mammals (Gillaspy et al., 2014).

Marine mammal trainers know that effective training begins with building a positive relationship with the animal (Brando, 2020; Ramirez, 2012). In seal training, this starts with learning a simple classically conditioned association between the sound of a whistle and receiving a piece of fish. Once a seal learns that the whistle means reward, the trainer can use the whistle to pinpoint particular voluntary behaviours they want to reinforce.

For example, blowing the whistle immediately when the seal produces a desirable behaviour causes the seal to stop to receive the reward, reinforcing the behaviour that occurred just before the whistle. The seal also learns that the trainer is an antecedent for rewards that follow certain behaviours – an important part of building a positive training relationship.

The trainer can now begin building associations between specific verbal commands, hand gestures or objects as antecedents (cues) for specific behaviours they want the seal to perform on command. A useful place to begin is to teach the seal a behaviour called 'targeting'. Targeting requires the seal to press their nose to the soft, rounded end of a training tool called a targeting stick and to keep it there while the trainer moves the stick.

Targeting provides a foundation for training many other complex behaviours. For example, once a seal can target successfully, the trainer can lead the seal to a platform that they use for training. Figure 4.12 shows a trainer with a targeting stick. Later, the seal can also be trained to target on the trainer's hand or fingers.

Targeting is not a natural behaviour and is unlikely to happen spontaneously to be rewarded.

Figure 4.12 Seal trainer with a targeting stick

So, the trainer starts by rewarding small steps towards the full behaviour. Training starts with the trainer holding the targeting stick in front of them. Seals are naturally curious, and are likely to approach the targeting stick, if they trust the trainer. The trainer can immediately signal their approval of the approach behaviour by blowing the whistle and giving the seal a treat. The trainer then only rewards the seal again if they approach the targeting stick even more closely. By systematically giving and withholding the reward the trainer can gradually shape the seal's behaviour towards keeping her nose on the target while it is moved.

Applying the ABC model, the targeting stick is the antecedent, the seal holding its nose to the end of the targeting stick while it is moved is the behaviour, and a piece of fish provides the positive reinforcement.

Skilled animal trainers avoid negative reinforcement and positive punishment in training because they produce negative feelings towards training. If punishment is necessary to manage a problem behaviour, the trainers prefer negative rather than positive punishment. Negative punishment is easy – the trainer just needs to turn their back on a misbehaving animal and walk away, taking their supply of fish with them!

Weblink
Teaching Groucho the Australian fur seal to have his teeth brushed

Trainers use video recordings of training sessions to identify the antecedents that were present in the environment when the problem behaviour occurred and the consequences that reinforce the behaviour. They use this information in future training sessions to remove the antecedents and/or the reinforcer.

Watch the short video at the weblink provided and apply the ABC model to explain how the trainer teaches Groucho the Australian fur seal to have his teeth brushed.

Questions

1. In the video, the trainer shows Groucho the electric toothbrush and rewards him for showing interest in it. Apply the ABC model to identify the antecedent, behaviour and consequence in this situation.
2. The trainer demonstrates targeting by training Groucho to press his nose to the trainer's open hand. Apply the ABC model to identify the antecedent, behaviour and consequence.
3. Explain how the trainer shapes Groucho's targeting behaviour so he learns to open his mouth in response to a hand gesture.
4. Explain why a trainer turning their back and walking away from a misbehaving animal is a form of negative punishment.
5. Explain why positive punishment and negative reinforcement are likely to cause negative feelings towards the trainer and training.

Behaviour modification: The ABC of changing our unwanted behaviours

The ABC model of operant conditioning isn't only used to train animals. Psychologists use behaviour modification programs based on the ABC model to help people change unwanted behaviours and to promote alternative positive behaviours. Behaviour modification programs start with analysing the behaviour to identify the antecedent stimuli that trigger the behaviour and the consequences that reinforce and maintain the behaviour.

A behaviour modification program can then be designed to focus on three strategies:
1. Modifying the environment to remove or weaken the antecedents
2. Modifying the environment with new antecedents that cue the desired behaviour
3. Designing rewards that motivate the positive behaviour.

Changing serious problem behaviours that affect our mental and/or physical wellbeing requires the expertise and support of a psychologist. However, all of us can learn how to apply the basic principles of the ABC model to change everyday unwanted behaviours – like procrastinating to avoid studying or overuse of social media. Take a look at Case study 4.2 to learn how.

CASE STUDY 4.2

Using behaviour modification strategies to change social media habits

Ava is 18 years old and has started studying psychology at university. One of her psychology assignments asks her to apply behaviour modification strategies to change an unwanted behaviour. Ava knows she is having trouble managing her time and is not studying enough. She suspects that her social media use might be the problem, so chooses to focus on reducing her social media use during times she should be studying (Figure 4.13).

Figure 4.13 The amount of time we spend looking at our phones is controlled by many antecedents and consequences.

Table 4.6 shows the table Ava is given to complete for her assignment. The first two columns define the steps for behaviour change. Ava fills out the last column to apply the ABC model to reduce her social media use when she should be studying.

Table 4.6 Application of the three-phase process of operant conditioning (ABC model) in a behaviour modification program

Behaviour modification step	Description of strategy to apply at each step	Ava's response
1 Identifying behaviour to change	Describe the behaviour you want to change.	I want to reduce social media use during times when I should be studying.
2 Behaviour analysis	» Keep a diary for a week to record time spent engaging in the unwanted behaviour (baseline to measure progress against)	» Diary shows I spend an average of three hours per day using social media and one hour studying
	» Identify antecedents – when and where does the behaviour occur? What stimuli and thoughts occur before the behaviour?	**Antecedents** » Seeing my phone or hearing a notification are antecedents that cause me to pick it up. » I feel a need to 'quickly check' my social media before I start studying to make sure I have not missed something.
	» Identify rewarding consequences that maintain the behaviour.	**Consequences** » I feel happy when people like my posts and when watching cute cat videos. » I feel relieved that I have not missed anything while checking my social media.
3 Goal setting	Phrase goal in terms of increasing a desired behaviour. Make it specific, achievable and measurable.	To increase the time I spend studying to 3 hours per day and reduce checking social media when I should be studying.
4 Action planning	» Define achievable plan to meet goal.	» I created a timetable with 2 × 2-hour study blocks each day from Monday to Friday, scheduled to fit with classes, part-time job and weekend socialising. I want to complete 3 × 30-minute study sessions in each 2-hour block with 10 minutes, break between each session. » I allocated 2 × 30-minute sessions each day to check social media outside of study times.

4 Action planning cont'd	» Modify the environment to weaken or remove antecedents of unwanted behaviour.	**Remove or weaken antecedents of problem behaviour** » I remove my phone from my room when studying and have deactivated notifications.
	» Create antecedents that promote the positive behaviour.	**Create antecedents for desired behaviour** » I have created a chart on my bedroom wall with my timetable and study sessions that is clearly visible when I enter the room. I have a kitchen timer sitting on my desk that I can set to time each 30-minute study period.
	» Create a reward structure for the positive behaviour.	**Create rewards for positive behaviour** » I place a gold star on my chart for each 30-minute study session I complete. If I achieve all six 30-minute study sessions in a day, then I can have an extra half hour of social media use.
5 Monitoring	Keep track of progress towards goal each day.	I track my progress on my wallchart each day.

Questions

1. Ava notes that she feels a need to 'quickly check' social media before studying to make sure she has not missed anything. Explain how this is an antecedent for behaviour.
2. Ava notes that she 'feels relieved' to know she has not missed anything when she checks her social media. Identify the kind of reinforcement that is maintaining the checking behaviour in this example.
3. Explain how Ava has modified her environment to remove the antecedents of the unwanted behaviour and how she has created antecedents that promote studying.
4. Explain how Ava has created a reward structure to motivate her study sessions.

4.2.7 THE PSYCHOLOGY OF ADVERTISING

Weblink
Google Android's Friends Furever advertisement 2015

Classical and operant conditioning in the psychology of advertising

The psychology of classical and operant conditioning is central to many advertising campaigns. Classical conditioning is used to create involuntary emotional responses to brand names by pairing the brand with stimuli that cause pleasurable feelings. The brand name then becomes a powerful antecedent stimulus that prompts voluntary behaviours associated with purchasing the product or with changing an unhealthy behaviour (for example, in anti-smoking campaigns).

Research has shown that associating a product with feelings of joy and happiness is the most effective way to increase the popularity of a brand (Berg et al., 2015). Google Android used this

Figure 4.14 Still image from the Friends Furever campaign, Google Android, 2015

simple classical conditioning principle in its 2015 'Friends Furever' campaign – one of the most watched advertisements of all time (Figure 4.14). The advertisement works simply by associating the brand with the feelings of happiness and joy people feel when watching the captivating images of unlikely animal friends and listening to the feel-good soundtrack.

Table 4.7 shows how we can apply the three-phase process of classical conditioning to understand why the advertisement works.

Table 4.7 Three-phase process of classical conditioning applied to the Friends Furever advertising campaign

Phase 1	Before conditioning 1 Initially neutral stimulus (NS) 2 Natural reflex (UCS–UCR)	1 Google Android brand 2 Cute animals and happy music (UCSs); feelings of happiness and joy (UCR)
Phase 2	During conditioning	Google Android brand (NS) is paired with the images and music (UCS) that cause feelings of happiness and joy (UCR).
Phase 3	After conditioning 1 Conditioned stimulus (CS) 2 Conditioned response (CR)	1 Google Android becomes a conditioned stimulus (CS) that creates 2 a conditioned response of feelings of happiness and joy (CR)

🔑 KEY CONCEPTS 4.2d

- » The three-phase process of operant conditioning (ABC model) can be used to train animals.
- » Psychologists use behaviour modification programs based on the ABC model to help people change unwanted behaviours and to promote alternative positive behaviours.
- » Behaviour modification steps are: Identify the behaviour to change; Behaviour analysis; Goal setting; Action planning and Monitoring.
- » Behavioural models of classical and operant conditioning can be used in advertising campaigns.

Concept questions 4.2d

Remembering

1 What is the most effective consequence to use to train an animal to produce a behaviour? **r**

Understanding

2 Why is it important to identify and modify the antecedents when designing a behaviour modification program? **e**

Applying

3 Use the three-phase process of operant conditioning to explain how a farmer could train a sheepdog to jump up onto a quad bike. **c**

HOT Challenge

4 Watch the Samsung advertisement in the weblink *Samsung: Be together* and apply the three-phase process of classical conditioning to identify the NS, UCS, UCR, CS and CR by using the following table.

Phase 1	Before conditioning	
Phase 2	During conditioning	
Phase 3	After conditioning	

5 Advertising campaigns that aim to change unhealthy or dangerous behaviours work best when they cause emotional responses of fear and disgust (Tannenbaum et al., 2015). These unpleasant emotions cause us to avoid stimuli that have been associated with them. This has been used very effectively in anti-smoking campaigns that use graphic images of disease and suffering on the packaging of cigarettes (Clayton et al., 2017).

Weblink
Samsung:
Be together

Analyse the examples of anti-smoking cigarette packaging in Figure 4.15 to answer these questions.
a Explain how the three-phase process of *classical conditioning* applies to the use of graphic images on cigarette packaging.
b Apply the three-phase process of *operant conditioning* to explain how the combination of graphic images and health information on the packets is designed to change harmful behaviours and promote healthy behaviours.

Figure 4.15 Cigarette packaging images and messages are designed to promote quitting smoking.

4.3 The social-cognitive approach to learning

Behaviourist approaches to learning have helped us understand the powerful influence that the principles of classical and operant conditioning have on many aspects of our lives. But there is much more to learning than simply learning conditioned associations, especially for human learners. We learn socially, through observing and interacting with other people. We also have advanced cognitive abilities to remember and think about what we have learned.

Albert Bandura and social-cognitive learning

Albert Bandura's groundbreaking work during the 1960s explored how children learn aggressive behaviours through observing the behaviours of adult role models (Figure 4.16). This early work provided the foundation for Bandura's social-cognitive model of learning and continues to influence our understanding of learning in important ways today.

Learning through observing others

The ability to learn socially through observing another person supercharges human learning and increases our opportunities for learning. It means that we don't need to have direct experience of a situation to learn – and that can be really handy when you need to learn about things that are dangerous!

Children learn much of what they know through observing parents, siblings, peers and teachers. This can occur through intentional demonstration or through observing everyday

Figure 4.16 Social-cognitive psychologist Albert Bandura

interactions. All of us continue to learn by observing others throughout our lives as well.

Bandura (1972) defined **observational learning** as *changes in behaviour and knowledge that are acquired through observing the behaviours of a model*. He defined a model as someone who performs actions in a situation that informs observers about how to behave in that situation (Ahn et al., 2019).

The social-cognitive model of learning describes four cognitive processes that occur during learning.

1. **Attention** involves the learner focusing their awareness on the behaviours of the model. Learning cannot occur without attending to the model.
2. **Retention** involves the process of forming memory representations for the behaviours we observe. The memory representations take two forms.
    » The learner encodes visual, auditory and motor imagery of the situation and behaviours.
    » The learner encodes symbolic (language-based) representations as they think about the meaning of the behaviour, the situation and the consequences.
3. **Reproduction** is the process that occurs when the observer rehearses the behaviour they have observed in their mind, and when they practise the behaviour by physically performing the sequence of actions that was observed.
4. **Motivation and reinforcement** relates to how the consequences that were observed for the behaviour affect the learner's decision about whether to reproduce the behaviour. Observed behaviours that were followed by reinforcement are likely to be reproduced, whereas behaviours that were followed by punishing consequences are not likely to be reproduced.

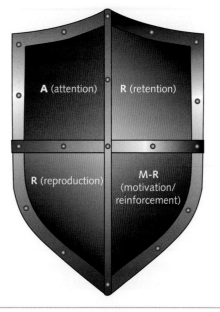

Figure 4.17 An acronym and a visual image may help you remember the four cognitive processes engaged in observational learning: A = attention, R = retention, R = reproduction, M-R = motivation/reinforcement.

Bandura developed his model based on a series of experiments of observational learning in children. One his most influential experiments was a 1965 study of learned aggression in children, which has become known as the 'Bobo doll experiment'. Analysing research 4.3 explains this classic experiment and how it relates to the four cognitive processes of observational learning in the social-cognitive model of learning.

> **EXAM TIP**
> One way to remember these four processes is to create an acronym and a visual image to represent the information in multiple formats. For example, we can create an acronym using the first letter of each process: **A** (attention), **R** (retention), **R** (reproduction) and **M-R** (motivation/reinforcement). You then have 'ARRM-R', which can be pronounced like the word 'armour'. You can then create a visual image of a shield or coat of arms to represent armour that has four quadrants on it, placing one of the letters of the acronym in each quadrant, as in Figure 4.17. Try it yourself – we will learn more about the power of such memory techniques in Chapter 5.

A4 paper origami sunglasses

ACTIVITY 4.2 OBSERVATIONAL LEARNING

Try it
Go to the weblink *A4 paper origami sunglasses* and watch the tutor make a pair of sunglasses. Now, try this yourself.

Apply it
Identify the four steps of observational learning. The first one has been done for you.

1. **a** What is attention in observational learning?
   This is where the learner directs their focus towards the model/behaviour to be learned, and the consequences of this behaviour.
   b How is attention demonstrated in this activity?
   When you watch the video you are actively focusing on the model and the steps being demonstrated to help you make A4 paper sunglasses.
2. **a** What is retention in observational learning?
   b How is retention demonstrated in this activity?
3. **a** What is reproduction in observational learning?
   b How is reproduction demonstrated in this activity?
4. **a** What is motivation/reinforcement in observational learning?
   b How is motivation/einforcement demonstrated in this activity?

Exam ready

©VCAA 2010 Q35 ADAPTED

Sian has recently begun skateboarding. She watched footage of the skateboarders at the Tokyo Olympics over and over again. She carefully observed the techniques of Keegan Palmer throughout the heats and the final. However, when Sian went to the skate park, she was unable to perform any of the moves she had seen in the videos.

In terms of observational learning, what is the most likely reason for Sian's inability to put into practice what she had learned by watching Keegan Palmer?

A Attention – Sian did not focus on the videos enough.

B Reinforcement – Sian was not given a medal when skateboarding.

C Reproduction – Sian could remember what she had seen in the video but was not skilled enough to replicate the techniques.

D Retention – even though Sian watched the videos over and over again, this was still not enough to help her form a mental representation of the skateboarding techniques.

ANALYSING RESEARCH 4.3

Learned aggression in children through observation of an adult model

In this famous experiment, Bandura (1965) randomly allocated four-year-old children (33 girls and 33 boys) from an American preschool into three groups. Each child participated in the experiment separately. The experiment began with the child watching a video showing an adult male behaving aggressively towards a large inflatable doll known as a Bobo doll. The adult performed or 'modelled' four specific aggressive actions. Each action was accompanied by a specific spoken phrase such as 'Pow – take that!'

» The difference between the three groups of children was in how the film ended.
» Children allocated to the *model-rewarded* condition saw the adult positively reinforced after producing the aggressive actions, receiving praise and treats.
» Children allocated to the *model-punished* condition saw the adult scolded and spanked with positive punishment for their aggressive behaviour (this was 1965!).
» Children allocated to the *no-consequences* group saw the adult receive no consequences for their behaviour.

After viewing the film, the child was taken to a playroom that contained a range of toys, including the Bobo doll and a toy hammer that had been used in the film. The child was asked to wait there for a few minutes until the experimenter returned. The experimenter then watched the child through a one-way mirror to see whether they would reproduce the specific aggressive acts and phrases seen in the video.

When the experimenter returned, she asked the child to show her the behaviours from the video. The child was reassured that it was OK to reproduce the behaviours and they would receive some sweets for doing so. This was done to find out whether the child had learned the behaviours they saw and could reproduce them if asked, separate from whether they had chosen to reproduce them when they had been left alone to play without instructions.

The bar graphs in Figure 4.18 show the results of the study.

The key difference in the behaviour of the groups occurred in the condition when children were first left alone to play in the playroom without instructions (orange bars in Figure 4.18). The results show that the children in the model-punished group were much less likely to reproduce the aggressive behaviours than the children in the model-rewarded and no-consequences groups when first left alone to play, with the children in the model-rewarded and no-consequences groups being equally likely to reproduce the behaviours. This pattern of results was similar for girls and boys even though girls reproduced fewer of the aggressive behaviours than the boys overall. Crucially, the green bars in Figure 4.18 show that children in all groups could reproduce the aggressive behaviours equally well when given permission and an incentive to do so by the experimenter (although girls were still less likely than boys to do so). This shows that the observed consequences affected whether the children would reproduce the behaviours without being

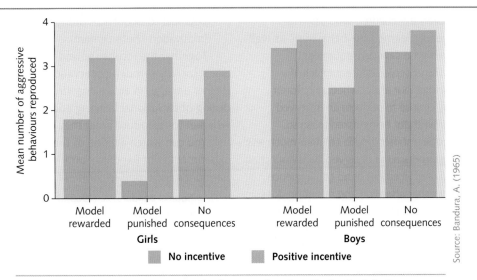

Figure 4.18 Bar graphs showing results from Bandura (1965). On the left, the mean number of aggressive actions reproduced by girls in the three groups when initially left alone to play (orange bars) and when given an incentive to reproduce the actions (green bars). On the right, the data for boys.

directly asked and rewarded to do so, not whether they learned the behaviours. This demonstrates the influence of the motivation-reinforcement component in observational learning.

Figure 4.19 shows still images from one of Bandura's similar experiments in which a female was the adult model. The top row shows the four specific actions 'modelled' by the adult – punching the Bobo doll to the ground, throwing the doll, hitting the doll with a hammer and kicking the doll. The middle and bottom rows show a boy and girl reproducing the aggressive behaviours.

The study was groundbreaking for two reasons.
» First, it showed that children can learn and retain memories of behaviours just by observing (attending to) an adult model, without directly experiencing reinforcement or punishment themselves.

Figure 4.19 These still images are from Bandura's 1965 study of learned aggression in children and show a female model and a preschool-aged boy and girl reproducing the observed behaviours.

In fact, even the children who saw no consequences learned the behaviours as well as children who saw consequences.

» Second, the observed consequences of a model's behaviour affect whether children will freely choose to perform an observed behaviour without being asked to do so. Observed punishment made children much less likely to reproduce the observed behaviour than observed reinforcement or no consequence. This shows that attention and retention does not always result in reproduction – whether children reproduce an observed behaviour depends on their active reasoning about the potential consequences of the behaviour (motivation-reinforcement).

Together the results showed that learning can occur *socially* through observation, and that the learner engages in active *cognitive* processes both during observational learning and when deciding whether to reproduce what they have observed.

You can watch a video of the original experiment with Bandura narrating it at the *Bandura's Bobo doll experiment* weblink.

Source: Bandura, A. (1965). Influence of models' reinforcement contingencies on the acquisition of imitative responses. *Journal of personality and social psychology, 1*(6), 589.

Weblink
Bandura's Bobo doll experiment

4.3.1 OBSERVATIONAL LEARNING

4.3.2 APPLICATION OF RESEARCH METHODS

Questions

1 Identify the independent and dependent variables used in this experiment. Note, there are *three* independent variables in this study – can you identify all of them?
2 This is an example of a *mixed* experiment design. Explain why.
3 Explain how each of the four cognitive processes of attention, retention, reproduction, motivation/reinforcement apply to the children's behaviours in this experiment.
4 Using information from the bar graphs in Figure 4.18, explain how the results of Bandura's study provide evidence that *attention* and *retention* occurred in all children, but that *reproduction* differed between groups of children according to the *motivation/reinforcement* they had observed in the video and the reward the experimenter later offered to reproduce the behaviours.

Current applications of social-cognitive learning

The social-cognitive approach to learning has generated a huge amount of research since Bandura's early work in the 1960s. For example, recent research has uncovered the neural and developmental processes involved in observational learning and has demonstrated that we can learn classically conditioned fear responses through observing others.

Mirror neurons activate during observation of movement

Studies using fMRI have uncovered how the brain represents observed motor sequences during Bandura's attention and retention processes in observational learning. The fMRI images show activation of regions of the motor cortex during perception of action that contain 'mirror neurons' – these are neurons that simulate the behaviour being observed (Hardwick et al., 2018; Rizzolatti et al., 2001).

The fMRI images in Figure 4.20 show the shared brain regions that become active when we

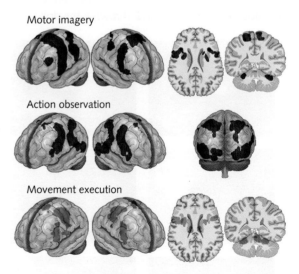

Source: adapted from Hardwick, R. M., Caspers, S., Eickhoff, S. B., & Swinnen, S. P. (2018). Neural correlates of action: Comparing meta-analyses of imagery, observation, and execution. *Neuroscience & Biobehavioral Reviews,* © Elsevier

Figure 4.20 This data from fMRI scans shows brain regions activated when we imagine, observe and execute a movement.

Figure 4.21 The development of learning through observation of others. An Infant follows the gaze of a researcher towards an object.

imagine performing an action, when we observe someone else performing an action, and when we perform an action. Notice that the same cortical regions are involved when we observe an action as when we imagine or perform that action, but that the subcortical regions of basal ganglia and cerebellum are not engaged when we are just observing an action.

The development of observational learning: gaze following and shared-attention

Other recent research has explored the development of observational learning in infants. The research shows that babies are born with an innate attentional bias to follow the gaze of a person they are interacting with (Flom et al., 2017; Stephenson et al., 2021). This shared focus of attention helps children understand how the words their mother is using relate to the things they represent in the world. Figure 4.21 shows a baby 'sharing attention' with the researcher by following her gaze to see what she is referring to.

Observational learning of conditioned fear responses

One of the most adaptive functions of observational learning is that it allows us to learn about dangerous stimuli through observing others. This saves us from needing to experience dangerous or threatening situations first-hand (Debiec & Olsson, 2017).

This research builds on the fear conditioning experiments we explored earlier in this chapter.

In socially learned fear conditioning experiments, participants (observers) watch a model experience electric shock to their forearm when particular stimuli are presented (for example, blue squares on a computer screen) and no shock for other stimuli (for example, yellow squares on the computer screen). The observer watches the model through a window or via a monitor and can see the stimuli and the model's reactions, including their arm responding to the shock and their facial expressions.

Later, when the observer is shown the same stimuli, they produce an elevated skin conductance response (SCR) to the stimuli they had seen associated with shock for the model. This elevated SCR occurs even though the observer never directly experiences a shock for any of the stimuli! Figure 4.22 shows the experimental procedure used in socially conditioned fear studies.

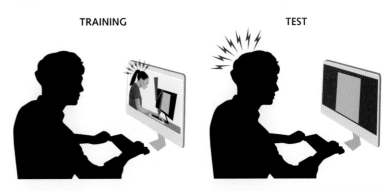

Figure 4.22 Social conditioning of fear. An observer watches a model receive electric shocks associated with particular stimuli and no shock for other stimuli. When the observer is later shown the same stimuli, they demonstrate a strong skin conductance response to the stimuli they observed being associated with shock.

KEY CONCEPTS 4.3

» The social-cognitive approach to learning emphasises the role of observational learning and active cognitive processes in human learning.
» Observational learning is a key process of social learning and involves four cognitive processes of attention, retention, reproduction and motivation/reinforcement.
» Bandura's 1965 study of learned aggression in children showed that children can learn aggressive behaviours through observing the aggressive actions of an adult model but that the observed consequences determine whether children choose to demonstrate their learning through reproduction.

Concept questions 4.3

Remembering

1. Use the table below to name and define the four cognitive processes involved in observational learning. r

Processes of observational learning

Name of process	Definition

Understanding

2. Explain why observational learning is a form of social learning. e

Applying

3. Apply social-cognitive learning theory to explain how young AFL fans who see their role models behave aggressively on the field might be influenced to reproduce the aggressive behaviours. What factors will influence whether the young fans are likely to reproduce the aggressive behaviours? c

HOT Challenge

4. Someone asks you whether it is possible for people to learn operantly conditioned responses through observational learning. What is your response, and what experimental evidence could you provide to support your answer? i

4.4 Aboriginal and Torres Strait Islander peoples' approaches to learning

Content warning

Aboriginal and/or Torres Strait Islander students are advised that this content and the associated weblinks refer to topics that are distressing, including massacres and the forced removal of children. The materials also contain references to the names and images of people who have died.

So far, we have explored Western approaches to understanding learning. In this final section, we explore a fundamentally different approach to understanding learning based on Aboriginal and Torres Strait Islander peoples' knowledge systems.

Situating ourselves as learners

Before we begin it is important to take a step back to reflect on who 'we' are. *We* is the voice that *I* am using to discuss the topics we are learning about – But, who am I? And, what is my relationship to Aboriginal and Torres Strait Islander peoples' ways of knowing? Situating myself in relation to this knowledge is an important part of engaging respectfully with Aboriginal and Torres Strait Islander peoples' ways of knowing.

My name is Meredith McKague. I am a non-Indigenous Australian woman of Anglo-Celtic heritage. I live and work on the country of the Wurundjeri Woi-wurrung people of the Kulin Nation. I am an academic psychologist trained in the methods of Western science and teach and conduct research in the area of cognitive psychology at The University of Melbourne. My cultural heritage and training in Western knowledge systems limits what I can know and teach about Aboriginal and Torres Strait Islander peoples' approaches to learning. My learning in this area comes from my engagement with the Australian Indigenous Psychology Education Project (AIPEP), in which I am a non-Indigenous allied member. AIPEP is a First Nations-led organisation that informs how psychologists are trained to work in ways that are culturally safe and responsive to the needs of Aboriginal and Torres Strait Islander peoples and communities. AIPEP promotes pathways for Aboriginal and Torres Strait Islander students to train as psychologists (see weblink). I have collaborated with Aboriginal psychologists, authors, academics and knowledge holders and highlight their work as the primary sources of knowledge. I am on an ongoing journey of developing my understanding – one that has been one of the most humbling and richly rewarding experiences in my life.

We also includes *you* and your relationship to Aboriginal and Torres Strait Islander peoples' knowledges. Who are you, where are you from, and who are your ancestors? Reflecting deeply on these questions *situates* us in *relationship* with each other and in relationship with the knowledge systems of Aboriginal and Torres Strait Islander peoples. Knowing who we are and where we come from allows us to uncover unconscious cultural biases that influence our attitudes, behaviours, and our relationships with others.

If, like me, you are not an Aboriginal and/or Torres Strait Islander person, then you or your ancestors arrived in Australia sometime within the last ~230 years as part of the process of colonisation (less than 200 years ago for Victorians). In contrast, Aboriginal and Torres Strait Islander peoples have lived on the lands now known as Australia for at least 65,000 years and have maintained, developed and passed down their knowledges over more than 2000 generations (Cane, 2013; Maynard, 2021).

Understanding the shared history of Aboriginal and Torres Strait Islander peoples and non-Indigenous Australians is crucial for how we all approach learning in this space.

The colonisation of the traditional lands of Aboriginal and Torres Strait Islander peoples began in 1788 when Governor Phillip established the colony of New South Wales in the name of the British Crown. This occurred under the false legal claim of '*terra nullius*' – that the land belonged to no one. Over the next century, the expansion of the colonial frontiers occurred by force, without treaties being established with the traditional owners (Reynolds, 2022). By the 1860s, the impacts of the Frontier Wars, massacres, poisonings and spread of introduced infectious diseases reduced the pre-colonial population of Aboriginal and Torres Strait Islander peoples by 90%. In 1901, the federation of the colonial states and territories into the nation of Australia excluded Aboriginal and Torres Strait Islander peoples as citizens. Aboriginal and Torres Strait Islander peoples were removed to missions, stations and reserves under the control of the states. During this time, children were forcibly removed from their parents to be 'assimilated' in Christian educational institutions, raised in white foster families, and/or trained as domestics servants, farm-hands, or labourers in the various industries that have built Australia. Successive government policies have continued to deny Aboriginal and Torres Strait Islander peoples' rights to self-determination and the legitimacy of their knowledges. These historical, cultural, political and social factors continue to negatively impact the social and emotional wellbeing of many Aboriginal and Torres Strait Islander people and communities today, and perpetuate negative stereotypes that shape the attitudes and perceptions of non-Indigenous Australians.

Understanding this shared history opens the way for respectful two-way learning between Aboriginal and Torres Strait Islander peoples and non-Indigenous Australians. Aboriginal and Torres Strait Islander peoples' approaches to learning reflect distinct ways of knowing and offer profound insights for anyone prepared to listen with an open heart and mind. The fact that Aboriginal and Torres Strait Islander peoples' knowledges have survived despite the ongoing impacts of colonisation speaks to the power of their methods of learning, remembering and transmitting knowledge. It is also a testament to the great strength and dignity of their ongoing struggle for self-determination.

Weblink
Australian Indigenous Psychology Education Project (AIPEP)

Learning *with* and *from* Aboriginal and Torres Strait Islander peoples

There is a danger that the need to know and assess this knowledge will produce generalisations that perpetuate racial stereotypes and objectify Aboriginal and Torres Strait Islander peoples, and that distinctive Aboriginal and Torres Strait Islander concepts will be absorbed into familiar Western concepts. To guard against this, I encourage you to adopt an attitude of seeking to learn *with* and *from* Aboriginal and Torres Strait Islander peoples rather than simply learning *about* Aboriginal and Torres Strait Islander peoples. Avoiding stereotypes depends on recognising the great diversity of knowledges and experiences that exist within the hundreds of distinct Aboriginal and Torres Strait Islander nations, clans and language groups. The information boxes provide further information and point you to weblinks to First Nations-led resources to help you develop your understanding.

Weblink
AIATSIS Languages Map

AIATSIS Australia's First Peoples

Torres Strait Island Regional Council

The continent of Australia and it's islands is the traditional home of two broad and distinct groups of First Nations peoples. These are Aboriginal peoples and Torres Strait Islander peoples. The phrase 'Aboriginal and Torres Strait Islander peoples' refers collectively to these two groups of First Nations peoples. The term *peoples* (plural) is used to recognise the many distinct nations, clans and language groups. There is no single Aboriginal and Torres Strait Islander culture or identity.

The term 'Aboriginal' refers broadly to the nations and clans of mainland Australia and Tasmania as well as most of the islands within Australia's maritime borders (e.g., Stradbroke Island, Kangaroo Island, etc). There are over 600 distinct Aboriginal clans and over 250 distinct traditional languages or dialects, only some of which are still spoken today.

The term 'Torres Strait Islander' refers broadly to the peoples of the islands located between the northern tip of Cape York and the south-west coast of Papua New Guinea that are part of the state of Queensland. Seventeen of the approximately 274 islands of the Torres Strait are home to 20 distinct cultural groups. The cultures and histories of Torres Strait Islander peoples are distinct from those of Aboriginal peoples.

The Australian Institute of Aboriginal and Torres Strait Islander Studies (AIATSIS) map shows the rough borders of the over 250 distinct Aboriginal and Torres Strait Islander language groups. To learn more, explore the First Nations-led weblinks provided.

Weblink
The Stolen Generations: Healing Foundation

AIATSIS Explore

Understanding the current experiences and perspectives of Aboriginal and Torres Strait Islander peoples requires developing your understanding of the social, cultural, political and historical factors that have shaped modern Australia. Learn about the Frontier Wars, policies of 'protection' and 'assimilation', the removal of people to missions, reserves and stations, the Stolen Generations, and the proud history of Aboriginal and Torres Strait Islander resistance and activism by exploring the First-Nations-led weblinks provided.

Aboriginal and Torres Strait Islander peoples' ways of knowing

The Study Design phrases the key knowledge for this section as follows:

"Approaches to learning that *situate* the learner within a system, as illustrated by **Aboriginal and Torres Strait Islander ways of knowing** where learning is viewed as being **embedded in relationships** where the learner is part of a **multimodal system** of knowledge patterned on *Country*".

The bolded words indicate the key concepts needed to understand Aboriginal and Torres Strait Islander peoples' approaches to learning. The phrase **Aboriginal and Torres Strait Islander ways of knowing** reflects the diverse approaches to learning used by the many distinct Aboriginal and Torres Strait Islander nations and clans. Although there is no single Aboriginal and/or Torres Strait Islander way of knowing, there are some common features between nation groups that can described, noting that unique expressions will occur in different contexts. The most fundamental concept within this broad approach is the concept of *Country*.

Country

The following information is informed by knowledge shared by the Bawaka Collective. Bawaka is a beautiful region in North-east Arnhem Land in the Northern Territory of Australia, more than 600 kms east of Darwin. The Bawaka Collective are a group of human and 'more-than-human' researchers led by Bawaka *Country* itself. Already you can see a very different way of knowing in which *Country* itself holds and gives knowledge. Bawaka Collective have shared their knowledges in a range of publications that provide deep insights into their unique ways of knowing (Bawaka *Country*, including Burarrwanga, et al., 2022). When reading this content it is important to remember that this account reflects an approach unique to Bawaka, although the broader principles are common to many Aboriginal and Torres Strait Islander nations. Figure 4.23 shows the people of Bawaka Collective on Bawaka Country. You can learn more about the Bawaka Collective in the weblinks.

Country is an Aboriginal English word (see weblink) that has a very different meaning to the standard English meaning of country as land or landscape. In this text we capitalise *Country* and use italics to indicate this culturally specific term. *Country* refers to the living system of all entities that exist in the universe. *Country* was created by the ancestors and those ancestors still live within *Country* in their many forms. People are only one part of *Country* and are **embedded in relationships** with the more-than-human entities within this system. The more-than-human entities include animals, plants, geographical features (rocks, mountains, etc.), waterways, seasons, tides, weather events (e.g., winds, rain), the skies and celestial bodies within them. When people die they return to *Country* in an ancestral form. *Country* also includes entities such as artworks, ceremonial objects and tools, as well as intangible things such as knowledge of laws, philosophies, medicines, astronomy, navigation, ceremonies, songs, stories, and dances. Everything is part of *Country*.

All entities within *Country* are sentient – that is, they live, breathe, think and feel. *Country* is a **multimodal system** because each of these entities has its own way of knowing and of being known. Each entity communicates with its own sensory language that may be observed, heard, smelled, tasted, or felt. This means that all entities within *Country* have the capacity to teach people knowledge. Learning happens through being in relationship with *Country*, attending quietly, and developing the ability to sense the signals that *Country* communicates.

This way of knowing is deeply ethical and spiritual. It gives respect to entities other than people, and acknowledges that people's existence depends on looking after the health and wellbeing of all other entities within the system. People *learn from, with and through* other entities, not just *about* them. Notice how different this is to a Western way of objectively analysing the natural world separate from people (Bawaka Collective, including Suchet-Pearson, et al., 2015).

Learning and knowledge embedded in relationships: Kinship

Each Aboriginal or Torres Strait Islander nation, and clans within nations, is connected to a specific geographical region that defines their identity through their relationship to *Country*. For example, the Wurundjeri people are one of the clans of the Kulin nation of south-central Victoria. They are the traditional owners of the area now known as greater Melbourne and speak Woi-wurrung language. On their website, the Wurundjeri explain that the word 'wurun' is the name for the Manna Gum trees that grow along the 'Birrarung' (Yarra River) and 'djeri' is the name of a grub found near the tree that is crucial to the health of the entire ecosystem. So, the Wurundjeri people are the 'witchetty Grub People'.

Weblink
Bawaka Collective

Bawaka Life

Weblink
Aboriginal English

Weblink
Wurundjeri
Woi-wurrung Council Website

Figure 4.23 Members of Bawaka Collective on Bawaka Country, North-East Arnhem Land.

Source: Matt Webb and the Bawaka Collective

Weblink
Nations, clans and family groups
Moiety
Totems
Skin names

A person's relationship within *Country* is defined within the **kinship system**. In Aboriginal and Torres Strait Islander cultures, the kinship system defines the relationships that people have with each other, the knowledges they are responsible for, and the entities within *Country* that they have a responsibility to care for. This concept of kinship goes beyond Western ideas of blood relationships. Terms like mother, father, auntie, uncle, sister, brother, cousin refer to relationships within and across entire generations of multiple family groups. This means each person has many people they call mother or father, etc. The distribution of knowledge throughout kinship structures weaves the system of knowledge into a living fabric of connections between people and *Country*. The distribution of knowledge and responsibilities ensures that all aspects of *Country* are maintained, and that everyone plays their part within the wider knowledge system. The kinship system is another important sense in which the learner and knowledge are **embedded in relationships** that are **patterned on** *Country*. There is much more to learn about these complex kinship systems and their connections to *Country*, including the role of skin names, moieties, and totems. To learn more, explore the kinship weblinks.

A system of knowledge patterned on *Country*

The survival of Aboriginal and Torres Strait Islander peoples and their knowledges for over 65,000 years has depended on people having a deep knowledge of *Country* that has been passed down more than 2000 generations. This includes knowledge of things such as which animals and plants are good to eat; those that are poisonous or medicinal, as well as when they are available. knowledge of the stars for navigation, methods of land and water management, laws and spiritual knowledge. Because Aboriginal and Torres Strait Islander cultures are **oral cultures**, their knowledges have been traditionally stored and transmitted through methods other than written language. Instead knowledge is 'written' in and on *Country* and is expressed through **Songlines**.

Songlines is an English word that refers to the sung stories that hold knowledge of *Country* (also known as songspirals, song cycles, or Dreaming tracks). Different Aboriginal and Torres Strait Islander nations have their own words for Songlines in each of their languages. Songlines tell the epic stories of the journeys and experiences of the ancestral beings who created *Country* and so are deeply connected to the places described. The stories encode knowledge of places that are significant for finding water, food and medicines, of things that are dangerous and should be avoided, of the consequences of behaviours, and of laws and spiritual knowledge. Knowledge is encoded in layers within Songlines, with deeper levels of understanding relating being revealed at different stages of learning (Bawaka *Country*, including Burarrwanga, et al., 2022; Neale & Kelly, 2020).

Broadly speaking, each entity within *Country* is associated with a story (or stories) that is sung. Songlines are linked sequences of songs that trace a path across *Country* – they both store knowledge of *Country* and provide a 'map' of *Country*. The linking of knowledge to physical locations on *Country* means that those places hold knowledge. When stories are told or sung, they may be accompanied by drawn symbols and/or gestures. For example, Figure 4.24 shows how an Eastern Anmatyerr woman from the central desert region uses sand drawings and gestures to help convey the meaning of a story (Green & Turpin, 2013). Songlines can also be expressed in paintings, dance, and carved or woven objects. These are all examples of the **multimodal system** in which the learner and knowledge are embedded.

Singing the Songlines is used to "enliven" *Country*. The Bawaka Collective say that *Country* and people are in a constant state of "co-becoming" through singing the Songlines. This means that knowledge is constantly evolving. Each time the songs are sung there is an opportunity for new entities to become part of *Country* so that things like smart phones, writing, and even colonisation become part of *Country*. When people "sing up *Country*", the past, present and future co-exist in that place. Notice the contrast to Western ideas of linear time and of time as separate to space (Bawaka *Country*, including Wright, et al. 2016; Bawaka *Country*, including Burarrwanga, et al., 2022).

CHAPTER 4 / Approaches to understand learning

Figure 4.24 An Eastern Anmatyerr woman draws a Songspiral in the sand to convey the structure of the story as it is told/sung.

The Gay'wu Group of Women of Bawaka describe *Country* and its relationship to Songlines (Songspirals), as follows:

> 'Country is the keeper of knowledge … Country gives the knowledge … It guides us and teaches us. Country has awareness, it is not just a backdrop. It knows and is part of us. Country is our homeland … Country is the way humans and non-humans co-become, the way we emerge together, have always emerged together and will always emerge together … And Country is the Songspirals.'
>
> Source: Bawaka *Country*, including Burarrwanga, L. et al., 2019, pp. xxii

The complexity of the system is conveyed in the Gay'wu Group of Women's explanation of the nonlinear relationship between time and space within their songspirals.

> '[the Songspirals] spiral out and spiral in, they go up and down, round and round, forever. They are a line within a cycle. They are infinite. They spiral, connecting and remaking. They twist and turn, they move and loop. This is like all our songs. Our songs are not a straight line. They do not move in one direction through time and space. They are a map we follow through Country as they connect to other clans. Everything is connected, layered with beauty. Each time we sing our Songspirals we learn more, go deeper, spiral in and spiral out.'
>
> Source: Bawaka Collective, Burarrwanga, L. et al., 2019, pp. xvi

These ideas are sophisticated, complex and beautiful. The brief description provided here provides only a glimpse of a fundamentally different way of knowing about who we are and our relationships with each other and all entities in the universe. Songlines are often associated with Aboriginal cultures, but are also part of Torres Strait Islander cultures. For a distinctly Torres Strait Islander perspective on Songlines mapped to the stars, see the weblink. The power of Songlines to encode information in memory is explored further in Chapter 5.

Weblink Torres Strait Islander Songlines

Ways of Knowing: Dadirri (Deep Listening)

Learning from *Country* involves learning how attend to and sense the different languages of *Country*. For the Ngangikurungkurr people of the Daly River in the Northern Territory, the process of attending to *Country* is called **Dadirri**, which is translated in English **deep listening**.

Dr Miriam-Rose Ungunmerr-Baumann, Elder of the Ngangikurungkurr people and Senior Australian of the Year 2021 (Figure 4.25), defines Dadirri as the practice of silent still awareness (Ungunmerr-Baumann, et al., 2022). Practicing Dadirri allows people to notice the relationships between the entities and events that occur within *Country*. For example, noticing that a certain tree coming into flower coincides with fish becoming plentiful in the river makes the tree a seasonal sign that there is good fishing to be had. Observing closely the interrelationships that exist between entities gives people a profound knowledge of the living system of which they are a part. Listen to Dr Ungunmerr-Baumann explain Dadirri in her own words in the weblink and then explore case study 4.3 to deepen your understanding.

Weblink Dadirri – Deep Listening

CASE STUDY 4.3

Dadirri: deep listening with Dr Miriam-Rose Ungunmerr-Baumann

Figure 4.25 Dr Miriam-Rose Ungunmerr-Baumann, Elder of the Ngangikurungkurr people of the Daly River and Senior Australian of the Year 2021

Dadirri recognises the deep spring that is inside us … It is something like what you call 'contemplation' …

A big part of Dadirri is listening. Through the years, we have listened to our stories. They are told and sung, over and over, as the seasons go by. Today we still gather around the campfires and together we hear the sacred stories.

As we grow older, we ourselves become the storytellers. We pass on to the young ones all they must know. The stories and songs sink quietly into our minds and we hold them deep inside …

In our Aboriginal way, we learned to listen from our earliest days. We could not live good and useful lives unless we listened. This was the normal way for us to learn – not by asking questions. We learned by watching and listening, waiting and then acting. Our people have passed on this way of listening for over 40 000 years …

Our Aboriginal culture has taught us to be still and to wait. We do not try to hurry things up. We let them follow their natural course – like the seasons. We watch the moon in each of its phases. We wait for the rain to fill our rivers and water the thirsty earth … We watch the bush foods and wait for them to ripen before we gather them.

We don't like to hurry. There is nothing more important than what we are attending to. There is nothing more urgent that we must hurry away for. We are River people. We cannot hurry the river. We have to move with its current and understand its ways.

My people are used to the struggle, and the long waiting. We still wait for the white people to understand us better. We ourselves had to spend many years learning about the white man's ways. Some of the learning was forced; but in many cases people tried hard over a long time to learn the new ways. We have learned to speak the white man's language. We have listened to what he had to say. This learning and listening should go both ways.

We would like people in Australia to take time to listen to us … To be still brings peace – and it brings understanding. When we are really still in the bush, we concentrate. We are aware of the anthills and the turtles and the water lilies. Our culture is different. We are asking our fellow Australians to take time to know us; to be still and to listen to us …

Source: adapted from Dr. Miriam Rose Ungunmerr-Baumann, (2017). To be listened to in her teaching: Dadirri: Inner deep listening and quiet still awareness. *Earth-Song Journal: Perspectives in Ecology, Spirituality and Education*, 3(4), 14–15

Questions

1 Identify the aspects of deep listening that Dr Ungunmerr-Baumann describes and explain how you think they relate to a systems model of learning.

2 Discuss with your peers how this approach differs from your own experience of learning.

Ways of knowing: Yarning

Dr Tyson Yunkaporta is a member of the Apalech clan from Western Cape York in the far north of Queensland. He is a senior lecturer and researcher at Deakin University in Naarm (Melbourne) where heads the Indigenous Knowledge Systems Lab. Dr Yunkaporta is also a skilled craftsman practising traditional methods of wood carving.

In his book, *Sand Talk* (Yunkaporta, 2019), Dr Yunkaporta explains Aboriginal ways of knowing through engaging in cultural practices that may include drawing, painting or carving visual symbols and the practice of **yarning**. Dr Yunkaporta explains that **yarning** is an Aboriginal cultural practice used for sharing knowledge between people. It takes the form of a conversation in which people seek to their deepen understanding of an idea, experience or process. The structure of a yarn is a free-flowing conversation. It is based on open questioning rather than being analytic. Although it seems informal, yarning is governed by Aboriginal ethical principles of listening and attending deeply and respectfully, and not seeking to dominate with an individual point of view (reciprocity). When yarning, each speaker should build upon what previous speakers have said rather than seeking to defeat them in debate (Bessarab & Ng'Andu, 2010; Frazer & Yunkaporta, 2021; Geia et al., 2013).

One way to deepen understanding of concepts developed through yarning is to engage in a cultural practice that expresses the ideas in a physical symbolic form. This may involve making something through practices like carving, weaving or painting. Making an object or artwork is a way of encoding the knowledge in a form that can be 'read'. The process deepens the learner's understanding of the concepts and ideas as they work. Dr Yunkaporta gives an example of a wooden shield he carved before writing the introductory chapter for his book. The images carved into the shield represent the concepts that are unpacked in each chapter of the book, providing a visual table of contents. Each chapter takes the form of a yarn that unpacks the knowledge contained in the symbol.

Explore Case Study 4.4 to learn more about Dr Yunkaporta's work and explore five 'sand talk' symbols that express five different **ways of knowing**, adapted from Yunkaporta (2019) and Bilton, et al. (2020, pp. 87–92).

Figure 4.26 Dr Tyson Yunkaporta, member of the Apalech clan from Western Cape York and director of the Indigenous Knowledge Systems Lab at Deakin University, author of *Sand talk: How Indigenous thinking can save the world*

Figure 4.27 Drawing of the surface of a shield carved by Dr Yunkaporta showing sand talk symbols as a visual table of contents for the chapters in his book

CASE STUDY 4.4

Sand talk symbols for five ways of knowing

Figure 4.28 shows the symbol for **kinship mind** which is about improving learning and memory through relationships and connectedness. Kinship represents the broad Aboriginal understanding that nothing exists outside of its relationship to something else. For example, if you learned something with or from another person, then knowledge the sits within the relationship between you two. There are no isolated variables – instead, every element must be considered in relation to the other elements in context. The symbol represents the connection between two things – people, ideas, places – connected by the line at the centre. Within the system, each pair is connected to a multitude of other pairs.

Figure 4.28 Sand talk image for kinship mind

Figure 4.29 is the symbol for **story mind**, which is about the role of narrative in memory and knowledge transmission. In oral cultures, narrative is the most powerful tool for remembering, particularly when connected meaningfully to places. Story mind includes yarning as a method of knowledge production and transmission. The symbol represents two people sitting (as if viewed from above) bringing their stories together to share. Yarning provides a way to share and check understanding with others to ensure validity of understanding.

Figure 4.29 Sand talk image for story mind

Figure 4.30 is the image for **Dreaming mind**, which is about using metaphors to understand ideas. The circle on the left of the symbol represents abstract knowledge, and the circle on the right represents tangible, concrete knowledge. The connecting curved lines represent communication between the abstract and the concrete. The upper connection represents knowledge from the Dreaming, which occurs through metaphors. The lower connection represents knowledge from the physical lived world which occurs through actions. The metaphors are the images, dances, songs, languages, objects, and rituals that embody knowledge.

Figure 4.30 Sand talk image for Dreaming mind

The image in Fig. 4.31 is for **ancestor mind**. This involves connecting with a timeless state of mind that is optimal neural state for learning. It is characterised by complete concentration, engagement, immersion and losing track of time – what some Western writers call 'flow'. People can reach this state through engaging in cultural activities and immersing themself in the process. The symbol represents the form of a coolamon (female) or a shield (male). The shield is a reminder to protect knowledge. The coolamon – an object that can hold an infant, ceremonial ochre or food – represents holding cultural knowledge and nourishing it in children's minds.

Figure 4.31 Sand talk image for Ancestor mind

Figure 4.32 is the symbol for **pattern mind**. Dr Yunkaporta explains that this 'is about seeing entire systems and the trends and patterns within them, and using these to make accurate predictions and find solutions to problems within those systems...Pattern mind links back to the beginning, to the first symbol of kinship mind, to the assertion that all elements, people and variables are interconnected. Mastery of Indigenous ways of knowing requires us being able to see beyond the object of study and even the learner and teacher, to seek a viewpoint incorporating the complex, dynamic systems.'

Figure 4.32 Sand talk image for pattern mind

HOT Challenge

Learning through culture, not about culture

To truly begin to understand Aboriginal and Torres Strait Islander ways of knowing requires engaging in learning practices informed by Aboriginal and/or Torres Strait Islander ways of knowing. Dr Yunkaporta calls this *learning through culture*, not *about* culture. To experience this, he encourages you to try drawing each of the sand talk symbols for yourself, carefully and mindfully – on the ground, on paper, or on a screen – encoding the principles you are learning about through the process of drawing. Then, to see how the five ways of knowing come together as a system, you can re-create one integrated symbol, as shown in Figure 4.33. Drawing this combined symbol can produce insights about relationships between symbols within the knowledge system. Try this and share what you learn from the process with your peers.

Figure 4.33 Symbol combining the five ways of knowing in a system

Final reflections

You may notice that there are some aspects of Aboriginal and Torres Strait Islander peoples' approaches to learning that seem similar to Western approaches to learning. For example, there is experimental evidence to show that classically and operantly conditioned associations can be learned through storytelling, without requiring direct experience (Field et al., 2001). Such cultural learning of stimulus-response relationships is crucial for survival as it allows us to learn associations that would otherwise need to be learned through potentially fatal trial and error experiences. However, it is a mistake to equate Aboriginal and Torres Strait Islander peoples' ways of knowing with Western concepts – these are distinct and unique ways of knowing with value beyond superficial similarities with aspects of Western approaches to learning. Each of the approaches to learning we have explored in this chapter has its own strengths and offers its own unique insights that reflect the cultures that have produced them.

4.4.1 THE SITUATED MULTIMODAL SYSTEMS APPROACH TO LEARNING

KEY CONCEPTS 4.4

» There is no single Aboriginal and Torres Strait Islander approach to learning, but there are some broad commonalities in approach between the hundreds of distinct nation and clan groups.

» Aboriginal and Torres Strait Islander peoples learn through distinctive ways of knowing in which the learner and knowledge are embedded (situated) within the system of relationships that comprise *Country*.

- *Country* is the living system of all human and more-than-human entities in which knowledge is held. All entities within *Country* communicate in their own language, making *Country* a multimodal system. All entities are living and have the ability to teach people about *Country*.
- Knowledge is held and expressed through sung narratives, known in English as Songlines. Songlines are one example of knowledge that is patterned on *Country*.
- Kinship structures determine who can hold particular knowledges and people's responsibility to care for *Country*. This is an example of how the learner is embedded in relationships within the multimodal system of *Country*.
- Dadirri, yarning, and sand talk symbols are examples of Aboriginal ways of knowing.

Concept questions 4.4

Remembering
1. Where are the Torres Strait Islands?
2. Approximately how many distinct Aboriginal and Torres Strait Islander language groups exist?
3. For approximately how many generations have Aboriginal and Torres Strait Islander peoples passed down their knowledges?

Understanding
4. Explain why the kinship systems of Aboriginal and Torres Strait Islander peoples are an example of learning being embedded in relationships.
5. Explain why *Country* is a multimodal system of knowledge.
6. Why do we say 'Aboriginal and Torres Strait Islander peoples' approaches to learning' (plural), rather than 'Aboriginal and Torres Strait Islander people's approach to learning'?

Applying
7. Thinking about the meaning of *Country*, describe how the destruction of *Country* and the removal of people from *Country* would affect Aboriginal and Torres Strait Islander peoples' capacity to learn and pass down their knowledges?
8. Go to the Bawaka Life weblink and read the article. Apply your understanding of Aboriginal and Torres Strait Islander peoples' ways of knowing to find examples of each of the key terms below.

Key term	Example
Learner embedded in relationships (situated)	
Multimodal system of knowledge	
Knowledge patterned on *Country*	
(Aboriginal) ways of knowing	

HOT Challenge
9. Use a visual organiser to explain the broad principles of Aboriginal and Torres Strait Islander peoples' approaches to learning. You could construct a concept map, draw a Venn diagram, use a series of concentric circles or a mix of organiser types to organise the information in meaningful ways.

Concept labels to select from are listed below. Use as many as you can. You can include additional relevant terms not listed.

Instructions:
- Write each term on a sticky-note.
- Add a short definition or explanation.
- Arrange the sticky-notes on a large sheet of paper to show the links between the concepts.
- Select a visual organiser that best suits your arrangement.
- Annotate with additional descriptive links from the terms below:

Multimodal	relationships	ways of knowing
Country	learner	knowledge
Dadirri	deep listening	yarning
Ancestors	Patterned	Songline
story	song	carving
painting	weaving	system
more-than-human	entities	sentient

4 Chapter summary

KEY CONCEPTS 4.2a

- Behaviourist approaches to learning are based on behaviourism, an approach to psychology introduced by John B. Watson in the early 20th century that focuses on measuring observable behavioural responses to stimuli.
- Watson's behaviourism was inspired by the work of Russian physiologist Ivan Pavlov, who described the process of classical conditioning through experiments investigating dogs' learned salivation responses to stimuli other than food.
- The three-phase process of classical conditioning involves three stages in the acquisition of a conditioned response:
  1. Before conditioning
  2. During conditioning
  3. After conditioning.
- Classical conditioning is the process of learning an involuntary association between an initially neutral stimulus (NS) and an unconditioned stimulus (UCS) so that the neutral stimulus becomes a conditioned stimulus that can cause a conditioned reflex response in the absence of the unconditioned stimulus.

KEY CONCEPTS 4.2b

- B. F. Skinner was a behaviourist psychologist who discovered and described the principles of operant conditioning.
- Operant conditioning is a form of associative learning in which voluntary behaviours are influenced by the consequences associated with them – behaviours are strengthened through reinforcement and weakened through punishment.
- Consequences for voluntary behaviours can be of four different kinds:
  1. Positive reinforcement (addition of a rewarding stimulus)
  2. Negative reinforcement (removal of an aversive stimulus)
  3. Positive punishment (addition of an aversive stimulus)
  4. Negative punishment (removal of a rewarding stimulus).
- Skinner studied the effects of reinforcement and punishment on the voluntary behaviours of rats and pigeons using a special enclosure called a Skinner box.
- The Skinner box allowed precise recording of behaviours and control of the stimuli and consequences associated with behaviours.

KEY CONCEPTS 4.2c

- The three-phase process of operant conditioning (ABC model) describes how the environmental stimuli that are present when we experience consequences for our behaviours can become cues for us to produce voluntary behaviours to gain rewards or avoid punishment.
- An antecedent is an environmental stimulus that has become associated with the consequences of a voluntary behaviour.
- Antecedent stimuli cue behaviours that are associated with rewarding consequences through the release of dopamine in anticipation of a reward.

KEY CONCEPTS 4.2d

- The three-phase process of operant conditioning can be used to train animals.
- Psychologists use behaviour modification programs based on the ABC model to help people change unwanted behaviours and to promote alternative positive behaviours.
- Behaviour modification steps are: Identify the behaviour to change; Behaviour analysis; Goal setting; Action planning; and Monitoring.
- Behavioural models of classical and operant conditioning can be used in advertising campaigns.

KEY CONCEPTS 4.3

- The social-cognitive approach to learning emphasises the role of observational learning and active cognitive processes in human learning.
- Observational learning is a key process of social learning and involves four cognitive processes of attention, retention, reproduction, and motivation/reinforcement.
- Bandura's 1965 study of learned aggression in children showed that children can learn aggressive behaviours through observing the aggressive actions of an adult model but that the observed consequences determine whether children choose to demonstrate their learning through reproduction.

4 End-of-chapter exam

Section A: Multiple choice

The following information refers to Questions 1 and 2.
A Psychology class completed an activity to see if they could condition an eye-blink response to the sound of a pencil tap on the desk. The students were instructed to tap the pencil several times at irregular intervals without presenting a puff of air. This was done to get the participant used to the tapping sound alone so that they no longer responded by blinking. When the participant showed no sign of blinking to the tapping alone, the students could begin conditioning.

There were 15 trials. In each of trials 1 to 10, the tapping sound was presented first, and immediately followed by a puff of air aimed at the bridge of the participant's nose. In trials 11 to 15, only the tapping sound was presented (no air puff) for the five trials.

1. The unconditioned stimulus (UCS) and the conditioned stimulus (CS) respectively were
   A pencil tap, puff of air.
   B puff of air, pencil tap.
   C pencil tap, eye blink in response to pencil tap.
   D puff of air, eye blink in response to puff of air.

2. The unconditioned response (UCR) and the conditioned response (CR) respectively were
   A blinking in response to pencil tap, no response.
   B no response, blinking in response to pencil tap.
   C blinking in response to pencil tap, blinking in response to puff of air.
   D blinking in response to puff of air, blinking in response to pencil tap.

3. In classical conditioning, both the unconditioned and the conditioned responses
   A are voluntary.
   B are reflexes.
   C are spontaneous.
   D only occur if the learner is aware of the association.

4. Which of the following is not a difference between classical and operant conditioning?

	Classical conditioning	Operant conditioning
A	Learn to associate two previously unrelated stimuli	Learn to associate behaviour with its consequences
B	The learner is passive	The learner is active
C	The unconditioned response and the conditioned response are involuntary	The behaviour is voluntary
D	Cannot learn classical conditioned responses through observational learning	Can learn operantly conditioned responses through observational learning

5. Which statement about behaviourist approaches to learning is incorrect?
   A Behaviourist approaches to learning include operant and classical conditioning.
   B Behaviourist approaches to learning describe the learning of voluntary behaviours.
   C Behaviourist approaches to learning describe learning as a process of making associations between stimuli and responses.
   D Behaviourist approaches to learning were built on the work of Pavlov, who researched the salivation response in dogs.

6. In the Little Albert experiment, John Watson conditioned a phobia in an infant boy. The unconditioned stimulus (UCS) and the conditioned stimulus (CS) respectively were
   A the white rat, the loud sound of the metal bar being struck.
   B the loud sound of the metal bar being struck, the white rat.
   C the white rat, fear of the white rat.
   D fear of the white rat, fear of the loud sound.

7. In the Little Albert experiment, the conditioned response of fear of the white rat was
   A voluntary.
   B a somatic reflex.
   C an autonomic reflex.
   D learned because he initially showed interest in the white rat.

8. Which of the following has not been shown by experiments and fMRI studies of the brain during fear conditioning?
   A People without an amygdala do not acquire conditioned fear responses.
   B Non-phobic people acquire conditioned fear responses to snakes and spiders more quickly than they do to pictures of neutral or pleasant stimuli.
   C The amygdala is active when conditioned stimuli (previously paired with electric shock) are presented without the unconditioned stimulus of the shock.
   D The amygdala is not active when conditioned stimuli (previously paired with electric shock) are presented without the unconditioned stimulus of the shock.

The following information relates to Questions 9–12.

Four-year-old Jamie does not like getting his hair cut since the barber nicked his ear with the scissors. Although his mother now takes him to a different hairdresser, Jamie winces and turns away when the hairdresser tries to trim his fringe. When the hairdresser offers Jamie a lollipop if he can sit still while she cuts his hair, she is able to finish his haircut without him wincing and turning away.

9. Operant and classical conditioning may occur in the same situation. In the case of Jamie, which one of the following responses is evidence of classical conditioning?
   A Jamie sitting still after the offer of a lollipop.
   B Jamie wincing and turning away in response to getting his ear nicked.
   C Jamie associating haircuts with the barber nicking his ear with scissors.
   D Jamie eating the lollipop.

10. In terms of the three-phase process of operant conditioning, if the antecedent in this scenario is Jamie getting his hair cut then the behaviour would be
    A Jamie sitting still.
    B Jamie wincing and turning away when the hairdresser tries to trim his fringe.
    C Jamie associating haircuts with the barber nicking his ear with scissors.
    D Jamie eating the lollipop.

11. In many situations, adults and children can reinforce each other. If the antecedent is Jamie wincing and turning away when the hairdresser tries to trim his fringe, then the behaviour would be
    A the hairdresser offering Jamie a lollipop.
    B Jamie sitting still.
    C Jamie eating the lollipop.
    D Jamie's mother taking him to a different hairdresser.

12. In terms of operant conditioning, the relief experienced by the hairdresser when she can cut Jamie's fringe without him wincing and turning away after he is given the lollipop is an example of
    A negative punishment.
    B positive punishment.
    C positive reinforcement.
    D negative reinforcement.

13. Negative punishment involves the
    A addition of an unpleasant stimulus that weakens behaviour.
    B removal of an unpleasant stimulus that weakens behaviour.
    C removal of a rewarding stimulus that strengthens behaviour.
    D removal of a rewarding stimulus that weakens behaviour.

14. Effective behaviour modification programs based on operant conditioning
    A focus on the behaviour while keeping antecedents and rewards the same.
    B modify antecedents to cue wanted behaviours and use rewards that motivate positive behaviour.
    C apply negative punishment.
    D apply positive punishment.

15 In one of Bandura's Bobo doll experiments, preschool children were randomly allocated to one of three conditions.

Condition 1: The aggressive model was rewarded with lollies, soft drink and praise.

Condition 2: The aggressive model was punished with a spanking and verbal criticisms.

Condition 3: There were no consequences for the aggressor's behaviour – the model was neither rewarded nor punished.

The results of the research were groundbreaking as they demonstrated

  A that learning can occur socially through paying attention to the behaviour of models.

  B that learning through observation can remain 'hidden' unless an appropriate consequence is offered to the observer to reproduce the learned behaviour.

  C that the learner engages in active cognitive processes during observational learning and when deciding to reproduce the behaviour.

  D all the above.

16 The correct order of the cognitive processes in observational learning is

  A attention, retention, reproduction, motivation and reinforcement or punishment.

  B motivation, attention, retention, reproduction, punishment and reinforcement.

  C attention, retention, reproduction, motivation and reinforcement.

  D attention, reproduction, retention, motivation and reinforcement.

17 Learning of conditioned fear responses through storytelling or observing others being classically conditioned

  A is not possible because the learner must have direct experience of the conditioned stimulus paired with the aversive unconditioned stimulus.

  B is not possible because the response in classical conditioning is reflexive.

  C occurs through operant conditioning.

  D is adaptive as it allows us to learn about dangerous stimuli without needing to experience the threatening situation first-hand.

18 Select the option that correctly matches the learning approach with its description.

A	Behaviourist	Explains how people learn through observing and thinking about the behaviour of others
B	Social-Cognitive	Explains how people learn associations between stimuli and behavioural responses
C	Aboriginal and Torres Strait Islander	Explains how people learn through being embedded in relationships as part of the multimodal system of *Country*.
D	Observational learning	Learning through the practice of Dadirri and deep listening

19 For Aboriginal and Torres Strait Islander peoples, the term *Country* refers to:

  A Non-urban landscapes or countryside.

  B The place of their birth.

  C A system of human and more-than-human entities in which all knowledge is contained.

  D All of the above.

20 Which statement best describes Aboriginal and Torres Strait Islander peoples' approach to learning?

  A The learner is situated in relationship with all other entities.

  B The learner learns from, with and through other entities.

  C Learning occurs through multiple modalities patterned on *Country*.

  D All the above.

Section B: Short answer

Question 1 ©VCAA 2020 Q4 ADAPTED

a   Advertisers often use behavioural models of classical and operant conditioning in advertising campaigns. An advertisement for a new soft drink features people having a good time while consuming the product. This is intended to make potential customers experience positive emotions about the soft drink. Apply the three-phase model to the soft drink advertisement described to identify the NS, UCS, CS, UCR and CR. [5 marks]

b   Advertising campaigns for new products such as the soft drink are usually played repeatedly. In terms of neural plasticity, how does long-term potentiation contribute to the learning of the conditioned response? [3 marks]

[Total = 8 marks]

Question 2 ©VCAA 2016 MC Q25 ADAPTED

Simran's four-year-old child Ava regularly throws tantrums when she is not given what she asks for, such as when she asks for chocolate just before dinnertime. Simran sought advice from Ava's kindergarten teacher, who suggested Simran ignore the tantrums, and when Ava is behaving well, Simran should praise Ava and give her a treat.

a   According to operant conditioning, identify the antecedent, behaviour and consequence for Ava when Simran ignored her tantrums. [3 marks]

b   For Ava, the chocolate is a highly valued reward. What is likely to happen to her dopamine levels when she asks for chocolate before dinnertime? [2 marks]

c   Simran has noticed that her two-year-old son, Noah, has recently started throwing tantrums when he is not given what he asks for. With reference to the four cognitive processes of observational learning, explain how Noah might have learned to throw tantrums. [4 marks]

[Total = 9 marks]

Question 3

a   Aboriginal and Torres Strait Islander peoples' approach to learning is said to be 'embedded in relationships.' Explain what this means using two examples. [3 marks]

b   Explain what is meant by saying that knowledge is 'patterned on *Country*'. [3 marks]

[Total = 6 marks]

The psychobiological process of memory

Key knowledge

- the explanatory power of the Atkinson-Shiffrin multi-store model of memory in the encoding, storage and retrieval of stored information in sensory, short-term and long-term memory stores
- the roles of the hippocampus, amygdala, neocortex, basal ganglia and cerebellum in long-term implicit and explicit memories
- the role of episodic and semantic memory in retrieving autobiographical events and in constructing possible imagined futures, including evidence from brain imaging and post-mortem studies of brain lesions in people with Alzheimer's disease and aphantasia as an example of individual differences in the experience of mental imagery
- the use of mnemonics (acronyms, acrostics and the method of loci) by written cultures to increase the encoding, storage and retrieval of information as compared with the use of mnemonics such as sung narrative used by oral cultures, including Aboriginal peoples' use of Songlines

Key science skills

Develop aims and questions, formulate hypotheses and make predictions
- identify independent, dependent and controlled variables in controlled experiments
- formulate hypotheses to focus investigation

Generate, collate and record data
- record and summarise both qualitative and quantitative data, including use of a logbook as an authentication of generated or collated data
- organise and present data in useful and meaningful ways, including tables, bar charts and line graphs
- identify and analyse experimental data qualitatively, applying where appropriate concepts of: accuracy, precision, repeatability, reproducibility and validity of measurements; errors (random and systematic); and certainty in data, including effects of sample size on the quality of data obtained
- evaluate investigation methods and possible sources of error or uncertainty, and suggest improvements to increase validity and to reduce uncertainty

Construct evidence-based arguments and draw conclusions
- evaluate data to determine the degree to which the evidence supports the aim of the investigation, and make recommendations, as appropriate, for modifying or extending the investigation
- evaluate data to determine the degree to which the evidence supports or refutes the initial prediction or hypothesis
- use reasoning to construct scientific arguments, and to draw and justify conclusions consistent with the evidence and relevant to the question under investigation

Source: VCAA VCE Psychology Study Design (2023–2027) page 35 & 12–13

5 The psychobiological process of memory

We usually think of memory as the ability to recall knowledge and experiences from the past. But, have you ever considered how memory also gives us awareness of present and allows us to imagine the future? Or, how Alzheimer's disease affects these processes and its relationship to mental imagery?

5.1 What is memory?
p. 188

Memory is a set of systems and processes that each depend on different neural structures. Atkinson and Shiffrin developed a framework of memory called the multi-store model of memory, which includes three memory stores and the cognitive processes that occur to enable encoding, storage and retrieval of memories.

5.3 The neural basis of explicit and implicit memories
p. 203

Memory is supported by interconnected networks of billions of neurons. Neural networks connecting the neocortex, hippocampus and amygdala support explicit memory of facts and personally experienced events. The basal ganglia and cerebellum support implicit memories of procedures and fine motor-control.

5.2 The structure of long-term memory
p. 198

There are two distinct forms of long-term memory. Explicit memory stores knowledge that is conscious and that we can express in words such as facts for an exam or memories of our experiences. Implicit memory occurs without conscious recollection, such as riding a bike, singing along with a favourite song or walking home from school.

Adobe Stock/Orawan

5.5 Mnemonics
p. 218

Mnemonics are techniques to assist memory. Written cultures use mnemonics such as acrostics, acronyms, and the method of loci. In contrast, oral cultures embed knowledge in sung narratives that are deeply connected to place. The Songlines of Aboriginal peoples interweave the powerful mnemonic effects of context, narrative, song, dance, emotion and imagery to pass down knowledge over thousands of generations. Mnemonics can also be a useful device to use when studying for exams!

5.4 Autobiographical memory
p. 209

ABM is part of our explicit memory. Semantic ABM stores our self-knowledge and episodic ABM allows us to transport ourselves forward and backward in time to remember our past and imagine possible futures. ABM is impaired in Alzheimer's disease and may be related to the absence of vivid voluntary mental imagery in experienced in Aphantasia.

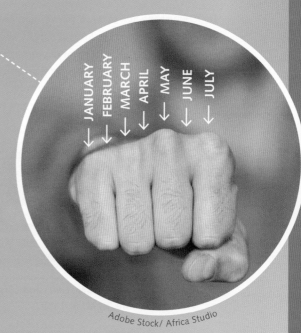
Adobe Stock/ Africa Studio

Memory encapsulates past, present and future. There are ways to enhance your memory that you can use when trying to remember a lot of material, such as for your Psychology exam. Make your memory work for you!

Slideshow
Chapter 5 slideshow

Flashcards
Chapter 5 flashcards

Test
Chapter 5 pre-test

Assessment
- Pre-test
- End-of-chapter exam

Revision
- Chapter map
- Key term flashcards
- Key concept summary
- Slideshow

Investigation
- Investigation: Serial position effect
- Data calculator
- Logbook template: Controlled experiment

Worksheet
- The serial position effect

To access these resources, visit
cengage.com.au/nelsonmindtap

 Nelson MindTap

Know your key terms

- Acronym
- Acrostic
- Alzheimer's disease (AD)
- Amygdala
- Aphantasia
- Autobiographical memory (ABM)
- Basal ganglia
- Brain imaging studies
- Brain lesion study
- Cerebellum
- Cerebral cortex
- Encoding
- Episodic autobiographical memory (EAM)
- Episodic memory
- Explanatory power
- Explicit memory
- Hippocampus
- Implicit memory
- Long-term memory (LTM)
- Mental imagery
- Method of loci
- Mnemonic
- Multi-store model of memory
- Neocortex
- Oral cultures
- Retrieval
- Semantic hubs
- Semantic memory
- Sensory memory
- Short-term memory (STM)
- Songlines
- Storage
- Sung narratives

In 1953, at the age of 27, a young man named Henry Molaison – now known simply as patient H.M. (Figure 5.1) – underwent radical experimental surgery to cure severe epileptic seizures. What followed was absolutely transformative for our understanding of memory and the brain. Dr William Scoville performed the surgery, removing a structure called the hippocampus from deep within both of H.M.'s temporal lobes. Figure 5.2 shows how H.M.'s brain would have looked after the surgery, compared with a normal brain. The surgery

Figure 5.1 Patient H.M. (left) and surgeon William Scoville (right) in 1953

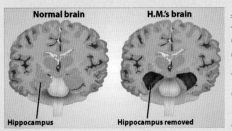

Figure 5.2 Cross-section showing H.M.'s brain compared to an intact brain (as if looking at the brain front-on with front half cut away)

successfully controlled H.M.'s seizures, but left him with a complete inability to remember any of the events that happened to him over the remainder of his life, right up to his death, aged 82, in 2008 (Scoville & Milner, 1957).

H.M. was also unable to remember the events of his life prior to the surgery, including the death of a favourite uncle just three years earlier. In contrast, H.M. could recall the facts of his pre-surgery life – such as his name and address, his parents' names, where he went to school, names of teachers, the floor plan of the home he grew up in, where he had worked and his job. He also retained his pre-surgery knowledge of language and concepts, as well as knowledge of important historical events, famous people, and of music and films.

Psychologist Dr Brenda Milner (Figure 5.3) assessed H.M's memory after the surgery. She and her student, Dr Suzanne Corkin, continued to study H.M. over his entire life until he died in 2008. Despite meeting Dr Milner and Dr Corkin on many occasions over 48 years, H.M. never remembered who they were from one visit to the next, nor any of the testing sessions he attended with them.

H.M. described his life as being like continuously awakening from sleep. Despite his profound inability to remember events or facts, other memory functions were unaffected. For example, H.M. retained a normal ability to recall sequences of numbers immediately after he had heard them. He could also perform the skills he acquired before his surgery, such as how to fix an engine. Strikingly, Milner and Corkin showed that H.M. could also learn a *new* motor skill normally, using a task called 'mirror-tracing'. This task requires you to draw a line between the narrow boundaries of two other lines that form the shape of a star, without drawing outside of the lines. The trick is, you have to do this while viewing your hands and the drawing through a mirror. The graph in Figure 5.4 shows H.M.'s normal acquisition of this skill over three days *even though he could not recall learning the skill and claimed not to know what mirror-tracing was when asked.*

Left: Eva Blue/ Flickr https://creativecommons.org/licenses/by/2.0/deed.en; Right: Alamy Stock Photo/GL Portrait

Figure 5.3 Professors Brenda Milner (left) and Suzanne Corkin (right), who worked with H.M. for 48 years. (At the time of writing in 2022, Brenda Milner is still actively researching, aged 104!)

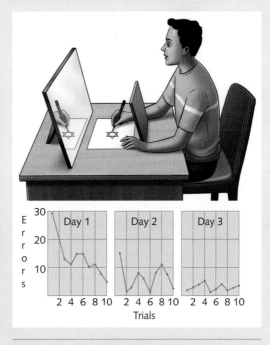

Figure 5.4 The mirror-tracing task. The graphs show Patient H.M.'s improving performance over three days of practising.

Weblink
Dr Corkin recounts H.M.'s death

MRI scan taken of H.M.'s brain

Before his death, H.M. (Figure 5.5) agreed to donate his brain to neuroscience. Dr Corkin recounts the events of the night he died in 2008 (see weblink). The full story of H.M. is in Dr Corkin's book *Permanent Present Tense: The memorable life of patient H.M.*

We owe a great debt of gratitude to H.M. for his lifetime commitment to ensuring others will learn from his loss. We return to H.M.'s story throughout the chapter as we map each of the different kinds of memory we discuss, and their neural basis, back to this extraordinary single case study.

Figure 5.5 Patient H.M. – Henry Molaison, aged 60 in 1986

Henry Molaison, aged 60, at MIT in 1986. Photograph and copyright, Jenni Ogden, author of "Trouble In Mind: Stories from a Neuropsychologist's Casebook" OUP, New York, 2012

5.1 What is memory?

The case study of H.M. demonstrates vividly that memory is not a single system – it is a set of systems and processes that each depend on different neural structures. With this in mind, we can define memory as *the set of psychobiological systems and processes that allow our past experiences to inform how we respond to and interpret our current experiences and to imagine the future*. That's right – memory is as much about the present and future as it is about the past!

Our memories are formed through a continuous cycle that integrates our current actions, perceptions, thoughts and feelings with those from our past, and with our plans for the future.

As you explore this chapter you will learn how memory can function like a built-in time-travelling machine, allowing us to re-experience the past and imagine the future. You will learn that even our experience of the present moment is actually a kind of memory.

The multi-store model of memory

In 1968, two cognitive psychologists, Richard Atkinson and Richard Shiffrin, proposed a framework of memory systems and processes known as the **multi-store model of memory** (Atkinson & Shiffrin, 1968). For over 50 years, the model has guided research to increase our understanding of the storage systems, processes, and brain networks that support memory (Malmberg et al., 2019).

Figure 5.6 shows the components of the original model. Notice that there are three distinct memory *stores* where information can be held, and a set of cognitive *processes* that control the transfer of information between stores. The sensory and short-term stores hold information temporarily so that it can be processed and encoded into the long-term

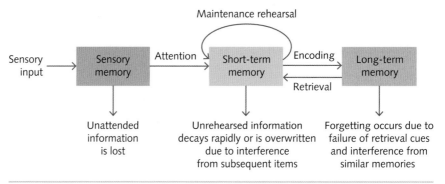

Figure 5.6 The multi-store model of memory (adapted from Atkinson & Shiffrin, 1968)

store, which holds memories permanently. The three memory processes control the processing (encoding), retention (storage), and recall (retrieval) of information. Each store is described below in terms of its **capacity** – the amount of information that can be stored – and its **duration** – the length of time that memories are retained.

Sensory memory

Sensory memory is the first memory store. Although the model shows it as a single box, there is a separate sensory memory system for each of our senses. Their purpose is to hold sensory information briefly so we have just enough time to attend to it. Attending to sensory information transfers it to short-term memory, where it can be processed consciously and given meaning.

Each sensory system transmits information to a specific region of the **cerebral cortex** that stores the information. For example, visual sensations are captured through movements of our eyes. Each movement produces a snapshot of information, with vision being suppressed between eye-movements. These snapshots travel to the primary visual cortex in the occipital lobes. Here, the neural activation from each snapshot persists briefly in visual sensory memory as an **iconic memory** trace. The capacity of the iconic trace is very large, capturing all of the visual input from each eye-movement. However, the duration of neural activation is only very brief, lasting about half-a-second (500 milliseconds). This is just long enough to provide the illusion of a continuous visual experience, rather than a series of static snapshots.

In contrast, auditory sensations (sounds and speech) are registered in the primary auditory cortex, located on the lateral (side) surface of each temporal lobe. These are called **echoic memory** traces. Like visual (iconic) sensory memory, the capacity of auditory (echoic) sensory memory is large. However, the duration of the echoic trace is longer, lasting 3–4 seconds. Echoic memory is particularly important for language. It allows us to capture and connect the rapid sequence of speech sounds as they unfold across time.

The brief persistence of sensory memory allows just enough time to attend to some of the information. However, because our attention is limited most of the information is lost as the sensory memory traces fade (decay), and as new information overwrites earlier information (interference). For this reason, we never become aware of most of the information that registers in sensory memory.

Short-term memory

Attending to information in sensory memory transfers the information to **short-term memory (STM)**, where it is represented consciously. STM provides a mental workspace where we make sense of information and use it to achieve our current goals. For this reason, Atkinson and Shiffrin also referred to short-term memory as a 'working memory', an idea that was developed later by others, such as Baddeley et al. (2019). Another way to think about STM is that it represents your conscious experience of the present moment. Because information in STM has been held in sensory memory previously, 'the present moment' is actually a memory of events that have already occurred!

Encoding and retrieval processes in STM

One of the most important functions of STM is to **transfer** information to long-term memory (LTM). Transfer is not a passive process; it requires us to do something with the information in STM. Active processing of information in STM is called **encoding**. One of the simplest ways to encode information is to mentally repeat it using our 'inner voice'. Atkinson and Shiffrin called this **maintenance rehearsal**. Figure 5.6 shows maintenance rehearsal as a loop connected to the STM store. We use maintenance rehearsal when we want to keep information active in STM to achieve a goal. An everyday example is when we mentally repeat a sequence of numbers like when we want to remember a passcode. Maintenance rehearsal will not produce a strong long-term memory trace, but it is effective for maintaining information long enough for us to use it.

Long-term retention requires information to be encoded meaningfully. To do this, we need to bring relevant information from LTM into STM as part of the encoding process. We do this through the process of **retrieval**. Figure 5.6 shows the retrieval process as an arrow pointing from LTM to STM. Meaningful encoding involves a cycle of interaction between STM and LTM. For example, imagine that you hear someone say the phrase 'albino elephant' and that this is not a phrase you have heard before. Attending to the words in STM encodes the information into LTM.

This causes information about the meaning of each word to become active in LTM and to be retrieved into STM. Now, you can interpret the words and understand that they refer to a particular kind of animal that has a particular inherited condition, causing its normally grey hide to appear white and its eyes to be coloured red. You might even be able to visualise this. Actively interpreting the information in STM by relating it to information retrieved from LTM results in **storage** of your new understanding in LTM.

STM capacity

STM capacity can be measured using the digit-span task, which tests your ability to repeat random sequences of single-digit numbers in the order they were presented (i.e. random sequences of numbers from 1–9). Digit-span is an example of a test of **immediate serial recall**. The sequence of digits is either spoken (i.e. 'six, nine, three'), or is presented visually (6, 9, 3), one at a time at the rate of one digit per second. The test begins with sequences (spans) of two digits. The span then increases by one digit after every two trials of a span that are correctly repeated. Your digit-span score is determined by the longest sequence you can repeat accurately for both trials of a span. For example, if you make an error on both trials at span eight, your digit-span is seven digits. When STM capacity is measured this way, the average capacity of STM is approximately seven units of information with a standard deviation of two (i.e. between five and nine units of information). You can try this for yourself at the weblink *Digit span task*.

The often-quoted capacity of STM as 'seven plus-or-minus two' units of information was first proposed by Miller (1956). However, Miller's most important contribution was to show that the definition of a unit is flexible and depends on our prior knowledge. He described how our STM capacity appears to increase when we process information meaningfully. This happens when our brain recognises a familiar pattern in STM and combines the individual elements into a single meaningful unit called a 'chunk'. **Chunking** in STM allows us to recognise familiar objects within a scene as meaningful wholes, rather than as disconnected visual features that would exceed seven units of information. Similarly, chunking allows us to recognise words and phrases instead of hearing strings of meaningless sounds. Thus, Miller proposed that the capacity of STM is actually seven chunks of information, which explains why STM capacity often feels greater than it actually is (Cowan, 2015).

Chunking can be used as a deliberate encoding strategy to enhance storage and later retrieval from LTM. For example, if you wanted to learn the sequence of 12 numbers in Figure 5.7, you could increase your STM capacity by grouping the numbers into sets of two, three or four numbers each. In the example, the first eight digits become three chunks, freeing up space in STM to encode the remaining digits. Think about how we remember mobile phone numbers or credit card numbers – we tend to recite short sets of three or four digits with a pause between each set. This simple chunking strategy is effective because it taps into our brain's love of rhythm and phrasing.

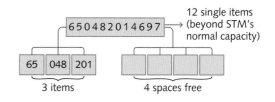

Figure 5.7 Chunking during encoding works by grouping individual items into larger meaningful units, which increases capacity of short-term memory.

STM duration

Information can be held in STM for significantly longer than in sensory memory, but the duration is still very brief. The duration of a STM trace can be measured by asking participants to remember a very short meaningless sequence, such as 'XPW', and varying the length of time that the information must be retained. This period of retention is called the **retention interval**. Studies like this show that if maintenance rehearsal is prevented by filling the retention interval with another task, such as counting aloud backwards, then the STM trace will **decay** (fade) rapidly within 15–30 seconds. In addition to decay, STM traces are very sensitive to **interference** from events that occur during the retention interval that overwrite earlier traces (e.g. the counting backwards task). The rapid decay of STM traces and their sensitivity to interference explain why we so often forget someone's name just moments after we have met them. This happens because the trace decays if we fail to rehearse the person's name before moving on with the conversation, and because the conversation interferes with the earlier information about their name.

5.1.1 ATKINSON AND SHIFFRIN'S MULTI-STORE MODEL OF MEMORY

Weblink
Digit span task

ACTIVITY 5.1 SHORT-TERM MEMORY

Try it

1 For this task, you will be completing a memory test. You will be required to read number sequences in your head; the number sequences are listed from a–h below.

After you have read each number sequence, write down as many of the numbers as possible in the same order they appear here.

List 1: Read each number individually with a one second break between each number.

a 2 5 3 6 7
b 4 6 9 0 1 5
c 5 8 2 4 1 3 2
d 9 0 4 6 2 1 5 3
e 3 4 3 5 6 2 7 0 6
f 2 1 5 3 6 7 8 0 1 0
g 5 3 6 7 8 8 9 2 3 6 1
h 3 5 4 5 3 6 7 9 2 8 9 3

List 2: Read each group of numbers as they are written (i.e. as they are 'chunked').

a 157 79
b 109 985
c 275 7531
d 765 925 34
e 485 975 253
f 907 385 7513
g 382 256 259 76
h 975 297 223 198

A correct response is counted as all numbers in the sequence being remembered in order. Did you perform better on list 1 or list 2?

Apply it

2 Using your knowledge of the capacity of short-term memory, explain why participants would likely have better recall of the items in the second list as opposed to the first list.

Exam ready

3 With regards to the Atkinson-Shiffrin model of memory, which one of the following statements is most correct?

A Sensory memory and short-term memory are both believed to have a limited duration but an unlimited capacity.

B Short-term memory receives information from sensory memory and information from long-term memory.

C Episodic and semantic memories are both types of implicit memories.

D Maintenance rehearsal increases the duration of short-term memory, whereas elaborative rehearsal increases the capacity of short-term memory.

Long-term memory

The third memory store of the multi-store model is **long-term memory (LTM)**. LTM stores all of the knowledge we acquire during our lives. The capacity is unlimited and the duration is permanent (at least within individual lifetimes). However, our ability to retrieve any given LTM trace depends on how it was encoded and the cues that are available when we want to remember it. This was demonstrated in a classic experiment by Craik and Tulving (1975). Participants were first asked to make speeded 'yes or no' judgements about individual words. The judgements were of three kinds:

1 Is the word written in upper case?
2 Does the word rhyme with _____?
3 Does the word fit in this sentence _____?

Crucially, the participants believed that the researchers only wanted to measure their response times for the different kinds of judgements. They were not aware that they would be tested on their memory of the words.

The purpose of the different judgement tasks was that each engaged a different **level of processing** during encoding. The case judgement focused attention on the surface features of the word (upper or lower case print), the rhyme judgement focused on the sound of the word, and the sentence judgement focused on the meaning of the word.

Craik and Tulving's hypothesis was that **deeper levels of encoding** would increase storage strength and make retrieval more likely, even when participants were not intentionally studying the words for a test.

After the encoding tasks, participants completed some filler tasks for 20 minutes. Then, they completed a surprise **recognition memory** test in which they had to indicate which words from a list were words they had seen in the previous judgement tasks. Figure 5.8 shows that participants were much more likely to recognise words from the sentence judgement task (~80% accurate) than from the rhyme task (~50% accurate), and that recognition was worst for words encoded in the case task (~15% accurate). Craik and Tulving concluded that LTM storage was the natural outcome of attending to information, and that meaningful encoding produces much better retention and retrieval than encoding the surface features of information.

Craik and Tulving's findings have been reproduced many times, including with **free recall** rather than recognition. In free recall tests, participants write down all of the words they can remember from the encoding task in any order (i.e. not serial recall). The key difference between free recall and recognition tests is the kind of **retrieval cues** that are available. In recognition tasks, the to-be-remembered items (e.g. words, numbers, images) are included among the items on the test. This provides strong retrieval cues (as in a multiple-choice exam). In contrast, free recall provides very few retrieval cues other than the instructions about what to recall (e.g. 'please recall the words from the list you just studied'; or, 'name the capital of France').

Retrieval cues assist memory through reminding us of past experiences, and are particularly useful when the cues available at retrieval match the cues that were present at encoding. For this reason, tests of recognition are easier than recall. Encoding information meaningfully improves recall by creating richer internalised retrieval cues (links and associations) within the memory system itself. Section 5.5 on mnemonics is all about how to create effective retrieval cues within your own mind.

Analysing research 5.1 explores another encoding factor that affects the storage and retrieval of long-term memories. But first, consider the following scenario: imagine you are studying for an upcoming exam and you have read through your textbook or notes, perhaps highlighting key points. What would you want to do next? Would you (a) go back and restudy either all of the material or parts of it, or (b) try to recall the material without restudying? Which method do you think would produce better long-term learning and why? The results of the experiment reported below might surprise you.

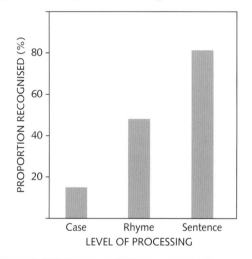

Figure 5.8 Results from Craik and Tulving (1975) showing the effect of depth of processing at encoding on recognition memory for words

ANALYSING RESEARCH 5.1

Practising retrieval improves long-term retention

This classic experiment by Roediger and Karpicke (2006) is one of the most highly cited studies of the last 30 years of psychological science. Its impact comes from challenging our intuitions about how we learn information from texts.

Students from Washington University were randomly allocated to one of three groups. All participants were given a 250-word passage to learn from a science textbook. Students in the first group read the passage repeatedly in four separate 5-minute blocks. Because it was a short passage, participants in this condition were able to read the passage about 14 times in total. This was called the SSSS condition, where S stands for study.

The second group of students read the passage during three 5-minute blocks, reading the passage about 10 times on average. Then, they took a single 5-minute test to recall as much of the passage as possible. This condition was called the SSST condition, where T stands for test.

The third group read the passage during a single 5-minute study period, reading it about

3.5 times on average. Then, they spent the remaining three 5-minute blocks trying to recall as much information from the passage as they could. This was called the STTT condition.

After the learning phase, all participants rated how well they thought they would remember the passage a week later. Then, half of the participants from each group were tested on their ability to recall the text after 5 minutes. The other half were dismissed and asked to return a week later for their final test.

Figure 5.9 shows the results. Notice that when the final test occurred just 5 minutes after learning, recall was better for students who had read the text more often, which means that the SSSS group recalled more than the SSST group, who recalled more than the STTT group. However, when memory was tested one week after learning, the exact opposite pattern held. Recall was much better for the students who had tested their knowledge, rather than rereading (i.e. the STTT group outperformed the other groups). If you are surprised by this result, you are not alone. Roediger and Karpicke reported that students overwhelmingly believed that long-term retention would be best for students who studied the text most often, but the outcome was exactly the opposite!

This 'retrieval–practice effect' has been reproduced by other researchers many times, making it highly reliable (for example, Adesope et al., 2017; McDermott, 2021; Soderstrom et al., 2016). It demonstrates clearly that although 'cramming' (repeated rereading) can lead to good performance on an immediate test, most of the information will be forgotten within a week of learning. In contrast, repeated effortful retrieval of information from LTM into STM strengthens the memory trace through a process of re-encoding. The authors note that their findings affirm what one of psychology's earliest researchers, William James (Figure 5.10), observed about his own memory over 100 years earlier:

'A curious peculiarity of our memory is that things are impressed better by active than by passive repetition. I mean that in learning (by heart, for example), when one almost knows the piece, it pays better to wait and recollect by an effort from within than to look at the book again. If we recover words in the former way, we shall probably know them the next time; if in the latter way, we shall very likely need the book once more.' (James, 1890, p. 646)

Figure 5.10 Pioneering U.S. psychologist William James noted long ago that testing memory results in long-term retention.

Source: Alamy Stock Photos Pictorial Press Ltd

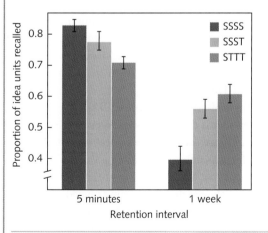

Figure 5.9 This figure, adapted from Roediger and Karpicke (2018), shows the results of their classic 2006 study demonstrating the benefits of retrieving information from LTM into STM for increasing long-term memory retention.

Source: adapted from Roediger III, H. L., & Karpicke, J. D. (2018). Reflections on the resurgence of interest in the testing effect. *Perspectives on Psychological Science*, 13(2), 236–241

> **Questions**
> 1 What kind of research design has been used in this study?
> 2 Identify the dependent variables in this study.
> 3 Interpret the graph in Figure 5.9 and re-express the results in terms of the difference between the three learning groups in the proportion of information that was forgotten after one week.
> 4 Evaluate the internal and external validity of this study and the relevance of the results for teachers and students.

Retrieval and elaborative encoding as protection against forgetting from LTM

Together, the depth of encoding and retrieval-practice strengthen long-term memories. These two processes can supercharge LTM if we use retrieval-practice as an opportunity to re-encode information more deeply. For example, failing to retrieve some information alerts us to what we don't know, which can guide further study to find the missing information. Adding this to your existing knowledge deepens and extends your understanding through a process called **elaborative encoding**.

Elaborative encoding includes asking yourself questions about the information you have retrieved. For example, if you successfully retrieve a fact such as 'the duration of visual sensory memory is about half a second', you can then ask yourself how this relates to short-term memory. Not only will this strengthen the existing memory traces, it will *organise* the information in a way that assists with retrieval and clarifies important distinctions between concepts.

Increasing the organisation and distinctiveness of information in LTM helps protect us from the two major causes of forgetting in LTM. **Retrieval failure** occurs when we are unable to locate the required information. **Interference** occurs when information cannot be distinguished from other similar information.

Unfortunately, the fluent understanding we experience when reading and re-reading information misleads us into believing we know more than we actually do. As noted earlier, *recognition* of familiar information when it is in front of you is very different from being able to *recall* the information using only your own internal memory cues. However, if we encode information through repeated cycles of retrieval-practice and elaboration, we strengthen our memories and organise them in a way that provides effective cues for retrieval.

The explanatory power of the multi-store model

Now that we understand the basic components and processes of the multi-store model, we are in a position to *evaluate* its **explanatory power**. To be useful, a model of memory must explain how we process information from our current sensory experiences, and how we transform these experiences into stored knowledge that can be accessed in the future. It must also explain why we forget information and what factors improve memory.

Multi-store model strengths

One of the multi-store model's greatest strengths is its ability to explain how information is transferred to LTM from sensory memory and STM, and why forgetting occurs. This is demonstrated in the model's explanation of **serial position effects**.

One of the most reproducible effects in memory research is the effect of the serial order of events on our ability to later recall those events (Murdock, 1962; McLeod, 2008). In laboratory studies of serial position effects, participants first study a list of words that exceeds STM capacity (i.e. 10 or more words). Then, memory is tested using immediate or delayed free recall. The question of interest is how our memory for any given word from the study list is affected by the serial order in which it was presented during learning.

Tests of immediate free recall show a typical pattern that is described as the **serial position curve**. Figure 5.11 shows a typical example. Notice how the probability of recalling a word is highest for words that appeared either at the beginning or end of the list during learning (encoding). The advantage for initial words is called the **primacy effect** (*primal* meaning 'first') and the advantage for the final few words is called the **recency effect**. Words from the middle of the list are remembered least well, giving the curve its characteristic U-shape.

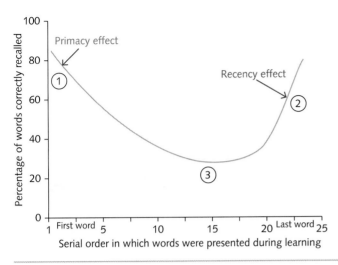

Figure 5.11 The serial position curve showing the effects of primacy and recency on free recall

According to the multi-store model, the primacy effect occurs because the first few words are repeatedly rehearsed, which transfers them to LTM. The recency effect occurs because the last two or three words are still available in sensory memory and STM, at least when the test requires immediate recall. The poor recall of the middle words is explained by a combination of factors. First, these items begin to decay due to a lack of maintenance rehearsal. This is because they occur in positions that exceed STM rehearsal capacity. Second, the words that occur later in the list remain in STM. These items tend to be recalled first, which displaces earlier items from STM and interferes with their retrieval.

If it is true that the recency effect occurs because the last items remain in sensory memory and STM, then introducing a filled retention interval immediately after encoding should eliminate the effect. Glanzer and Cunitz (1966) tested this hypothesis by asking one group of participants to recall immediately, while a second group recalled after 30 seconds of counting backwards in threes (Figure 5.12). As you can see, the recency effect disappeared for the second group, but their primacy effect remained. If the last items were in LTM then this should not have happened.

We can test the explanation of the primacy effect by preventing rehearsal during encoding. This can be done by asking participants to repeat a word aloud (e.g. 'the, the, ...') while encoding, and then testing immediate recall. This eliminates the primacy effect, but the recency effect remains.

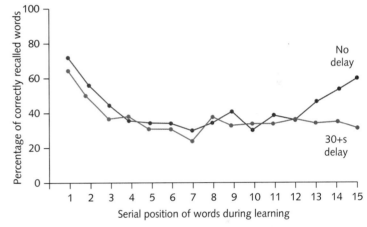

From Simply Scholar Ltd article "Serial Position Effect" By Dr. Saul McLeod, published 2008

Figure 5.12 These results from Glanzer and Cunitz (1966) show elimination of the recency effect after delayed recall.

Multi-store model weaknesses

The multi-store model provides a strong account of the kind of memory that can be tested using lists of words or digits, and about how this information is transferred between different memory stores. However, it does not explain the different forms of long-term memory that were so apparent in the case of H.M. For example, the model does not explain how we encode, store and retrieve information about the events of our lives, how we organise our knowledge of concepts and words, or how we learn and perform skills. The model also fails to explain the neural basis of the different memory stores and processes, and why some forms of memory are affected by hippocampal damage and others are not.

Worksheet
The serial position effect

INVESTIGATION 5.1 SERIAL POSITION EFFECT

Scientific investigation methodology
Controlled experiment

Introduction

Have you ever made a list and then when you wanted to remember what was on the list, you found you could only remember the first and last items? These are known as serial position effects and are described in this chapter on pages 194–195. Serial position effects have been cited as evidence for separate short-term and long-term memory stores in the multi-store model of memory. The key evidence was provided in the experiment by Glanzer and Cunitz (1966).

For this investigation, you will create a controlled experiment that tests a hypothesis related to serial position effects. What interests you about how the order in which we experience events affects our memory for those events?

Remember to use your logbook throughout this investigation. Present your findings as a scientific poster. The poster may be produced electronically or in hard-copy format and should not exceed 600 words. Please see the VCAA Psychology Study Design pages 15 and 16 for further specifications of the poster format.

Logbook template
Controlled experiment

Data calculator

Pre-activity preparation

Before you start you will need to review some literature on the topic. You could start with the brief review of serial position effects in *Psychology Today* (see weblink). You could then try reading the original journal article by Glanzer and Cunitz, which is available via the APA website (see weblink). More recently, researchers have studied applications of serial position effects on things like jury decisions (see weblink). Record your literature review in your logbook. Do not forget to record the full references that you use. You can find tips about appropriate referencing style for psychology from the APA-style website. From your literature review decide what aspect of the serial position effect you want to investigate. Develop a research question, aim and hypothesis for your investigation.

Remember that you need to collect primary quantitative data.

Weblink
Psychology Today

Glanzer & Cunitz, 1966 journal article

Jury decisions

Citation and referencing guides

Method

When designing your controlled experiment, refer to the information in Chapters 1 and 11 about experimental designs. You will need to consider the following issues:

- Population and sample – how big does your sample need to be?
- Variables – dependent, independent, controlled. Are there any extraneous variables you have not accounted for?
- What quantitative data are you going to record and how are you going to record it?
- Are there any ethical or safety issues you have to take into account?

Results

Collect and collate the results for all participants. Process the quantitative data obtained, using appropriate mathematical calculations and units (for example, descriptive statistics such as a mean). Organise, present and interpret your data using an appropriate table and/or graph.

Discussion

1. State whether the hypothesis was supported or refuted.
2. Identify and describe any patterns that occur within the collated data.
3. Provide an explanation of why certain words (or other stimuli) were recalled with greater frequency than others.
4. Identify and provide an explanation of any outliers and/or contradictory or incomplete data.
5. Discuss any implications of the results.
6. Discuss whether it is possible to generalise any of the results to the wider population.
7. Evaluate the investigation methods and possible sources of error or uncertainty.
8. Suggest future improvements to increase validity and to reduce uncertainty.

Conclusion

Write a conclusion to respond to the research question.

KEY CONCEPTS 5.1

» Atkinson and Shiffrin's multi-store model of memory consists of three memory stores: sensory memory, short-term memory (STM) and long-term memory (LTM).
» Information is maintained and transferred between the stores using three processes: encoding, storage and retrieval.
» Sensory memory registers information from our senses and holds it briefly so we can attend to it. The capacity is large, but duration is brief. Visual sensory memory persists as an iconic trace for approximately half a second. Auditory sensory memory persists as an echoic trace for up to 4 seconds.
» Short-term memory (STM) provides a temporary store for information that is the current focus of attention. It integrates attended information from sensory memory with information retrieved from LTM.
» STM capacity is seven plus-or-minus two units when measured using digit-span. Capacity can be increased through chunking multiple items of information into larger meaningful units.
» STM duration is limited by our attention span and by the ability to mentally rehearse information. Without maintenance rehearsal, STM traces decay rapidly within 15–30 seconds.
» Transfer of information from STM to LTM occurs through encoding processes, including maintenance rehearsal and elaborative encoding.
» Long-term memory (LTM) has unlimited capacity and its duration is permanent.
» The depth of encoding in STM affects the strength and organisation of long-term memories. Elaborative encoding and retrieval-practice are two deep encoding strategies that enhance storage and retrieval of long-term memories.
» A strength of the multi-store model is its ability to explain serial position effects of primacy and recency.
» A weakness of the multi-store model is its failure to explain different forms of long-term memory and their basis in the brain and neural processes.

Concept questions 5.1

Remembering

1 Using the table below, name the three stores in the multi-store model and describe the capacity and duration of each. **r**

Name of memory store	Capacity and duration

2 Using the table below, name the three processes used to transfer and maintain information in the multi-store model of memory and describe the function of each. **r**

Process to transfer and maintain information	Function

3 Describe the function of sensory memory and its relationship with STM. **e**

Understanding

4 Explain how the processes of encoding and retrieval interact when we encode information meaningfully. **e**

Applying

5 You notice the dangerous behaviour of another car on a freeway and make a mental note of the number plate as it speeds out of view. As you start to mentally rehearse the information, your fellow traveller comments on how dangerous the behaviour was and you agree. When you then try to recall the number plate you realise that you can't remember it. Use the multi-store model to explain why you forgot the number plate so quickly after encoding it in STM. **c**

> **HOT Challenge**
> 6 Go back to the case study of H.M. at the beginning of the chapter. Identify which aspects of H.M.'s memory abilities and problems can be explained by the multi-store model. Make a list of things that the model cannot explain. **C**

Feature of H.M.'s memory	Multi-store model explanation

5.2 The structure of long-term memory

H.M.'s case alerted memory researchers to two distinct forms of LTM called **explicit memory** and **implicit memory**, and to further subdivisions of each. In this section, we explore these different forms of LTM.

Explicit memory

H.M.'s greatest difficulty was his complete inability to recall any of the events he experienced after his surgery. He was also unable to learn new concepts, facts or words. The inability to store memories for events and facts is called **anterograde amnesia**, with *antero* coming from Latin, meaning 'situated in front'.

H.M. was also unable to recall any of the events of his life from before the surgery. We call this **retrograde amnesia**, with *retro* being Latin for 'backwards'. However, H.M. retained his general knowledge of the words, concepts and facts he had learned before his surgery, including factual knowledge about himself and his pre-surgery experiences (Steinvorth et al., 2005). Figure 5.13 visually represents anterograde and retrograde amnesia in relation to the time-point of the amnesia-causing brain injury.

The kind of long-term memory that is affected by anterograde and retrograde amnesia is called explicit memory (also called declarative memory). **Explicit memory** stores knowledge that we can represent consciously and express in words. H.M.'s case revealed two distinct forms of explicit memory called **episodic memory** and **semantic memory**.

Episodic memory

Episodic memory is the component of explicit LTM that allows us to encode, store and retrieve the personally experienced events (episodes) of our lives. For this reason, it is more accurately referred to as **episodic-autobiographical memory (EAM)**. Retrieval of EAMs is accompanied by the sense of experiencing the event from a first-person perspective, with perceptual and emotional details. For example, if you take your mind back to your first day at high school you may be able to recall how you felt and what you saw, as if reliving the event.

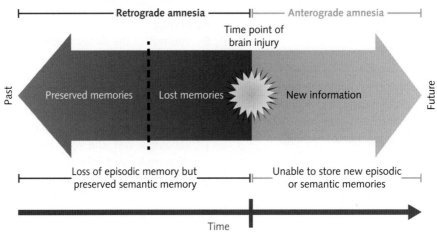

Figure 5.13 Retrograde and anterograde amnesia in relation to the amnesia-causing event

Retrieval of EAMs can occur involuntarily when an environmental cue, such as a song, a scent, or a photograph, reminds you of a previous event and transports you back in time. Or, retrieval can be voluntary, like when we search EAM to respond to a question like 'Where were you at the time of the accident?', or to a self-generated thought like 'Where did I leave my phone?'

It might surprise you to learn that EAM is the kind of memory that is tested in memory experiments when participants study a list of items for later retrieval. The study phase of these experiments is the personally experienced 'episode' to which participants mentally return during testing, and the items (words, numbers or images) are the events and perceptual details that are retrieved. Studies like this have revealed much about the workings of memory, but they lack *external validity*, leaving out much of the richness of our everyday experience of EAM.

EAM also allows us to imagine how we will experience an event in the future from a first-person perspective. This is called **episodic future-thinking**. An example would be imagining what you will do and how you will feel after your last VCE exam, or what it will be like to attend your formal. At first, this might not seem to be memory at all – you might call it imagination instead. However, imagining ourselves in the future is a form of episodic memory because it draws on aspects of our experiences (e.g. celebratory events), and it involves a first-person perspective. We will see later in the chapter that episodic future-thinking also depends on information from semantic memory.

Episodic future-thinking is a critical part of training and preparation for athletes and performers, enabling them to mentally rehearse their performance and to practise regulating their emotions in high-stakes situations. But episodic future-thinking is also something that everyone can use effectively in daily life. For example, we can rehearse what we would like to say and how we would like to behave ahead of an important meeting. Simulating different possible outcomes can help us pre-empt issues and prepare appropriate responses when things don't go to plan (Schacter et al., 2015).

Regardless of whether we are reliving the past or imagining the future, EAM always involves an experiencing 'self' situated in time and place, with a personally experienced past and an imagined future. For this reason, EAM is sometimes referred to as the ability to engage in mental time-travel. EAM is thought to be uniquely human. It gives us the capacity for self-reflective consciousness so that we can reflect on our past experiences and use them to inform our future behaviour (Renoult & Rugg, 2020; Tulving, 2002).

Semantic memory

Semantic memory is the component of explicit LTM we use to encode, store and retrieve general knowledge of words, concepts and facts. Examples include 'Canberra is Australia's capital city', or 'an apple is a kind of fruit', and 'birthdays are celebrated with a party and cake'. It also includes self-related facts like 'I was born in 2004', or 'on Friday night I am going out with friends'. Notice that semantic memories can be about events, such as birthdays. They can also be about ourselves and about the times and places that events occurred or will occur.

So, the difference between semantic memories and episodic memories is not what they are about. The difference is that retrieval of semantic knowledge is accompanied by a sense of 'knowing' rather than experiencing. This is because semantic memories are formed by extracting general information from episodic memories. For example, multiple experiences with coffee across a range of contexts gradually builds general knowledge of its perceptual features (taste, colour, smell, warmth), the contexts in which it occurs (cafés, kitchens), and the objects, actions and people it is associated with (cups, milk, drinking, baristas) (Davis & Yee, 2019).

Our knowledge of concepts and the relationships between them is organised in **semantic networks**. For example, because our experiences with coffee share features with tea experiences, these two concepts can be represented as examples of the higher-order concept of 'hot beverages', which is associated with even more general concepts like 'beverage' and 'food'. An example of a semantic network can be seen in Figure 5.14.

Semantic representations called **schemas** represent our knowledge of familiar situations, and about spatial relationships between concepts. For example, a 'café schema' stores the general knowledge that has been extracted from experiences of going to cafés. The schema includes knowledge of the kinds of things we will experience in a café, and the typical spatial relationships between concepts in the schema (e.g. the barista is behind the coffee machine).

Semantic memories often include mental imagery, but the imagery is more general than in EAM. It represents the typical features of concepts like 'cat', 'mountain' or 'airport'.

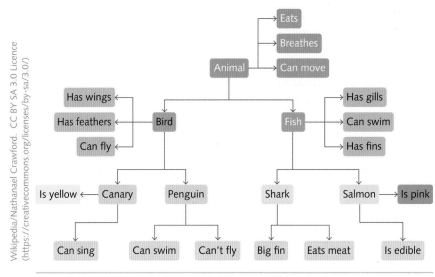

Figure 5.14 This example of a semantic network represents knowledge of categories (Animal) and concepts (Bird) and features of concepts ('can fly').

EXAM TIP
Remember that episodic memory and autobiographical memory are not the same thing. Episodic memory refers to memories of personally experienced events that are vividly re-experienced from a first-person perspective, whereas autobiographical memory includes both episodic memories and semantic knowledge about your life and experiences.

Semantic knowledge of your particular cat's features is stored as a general image of your cat. Semantic knowledge of abstract concepts like 'fun', 'beauty' and 'justice' is less likely to include perceptual imagery, but is often associated with emotional information.

As a general rule, semantic memories are expressed with statements like, 'I know that …', whereas episodic memories are expressed with statements like, 'I remember when …' (Tulving, 2002). Semantic memory can be tested using a category fluency test, in which people generate words associated with a cued category. For example, if the cue was 'animals', then you would list or name as many examples of the category as possible in 1 minute. Other semantic memory tests include word-picture matching and naming famous faces. Figure 5.15 shows how non-verbal semantic memory can be tested.

5.2.1
CONCEPT MAP
OF MEMORY

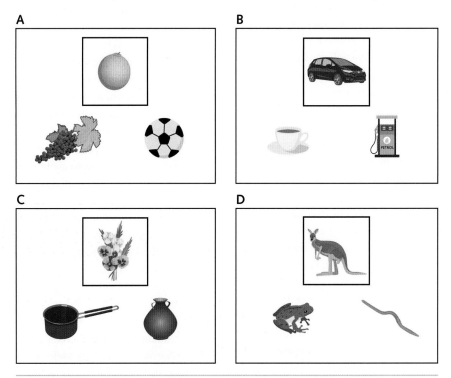

Figure 5.15 Example of a non-verbal test of semantic memory. The participant/patient is asked to point to which of the two lower images is most closely related in meaning to the upper image.

Implicit memory

In stark contrast to H.M.'s severe amnesia for aspects of explicit memory, he retained the skills he acquired before his surgery, and could learn new skills after his surgery. For example, H.M. learned the new skill of mirror-tracing just as well as matched controls, despite not remembering the learning episodes. Performing skills is one example of implicit memory (also known as non-declarative memory). **Implicit memory** is the kind of long-term memory that is demonstrated through changes in behaviour and adaptive responses as a result of repetition or practice, without conscious recollection of the knowledge that underlies the performance. Implicit memory represents our 'know-how', whereas explicit memories relate to knowing 'what, why, when and where'.

The kind of implicit memory that allows us to master skills is called **procedural memory**. Examples include riding a bike, playing an instrument, knowing how to tie your shoelaces or type a text message. Cognitive skills like reading, writing or performing long-division are also forms of procedural memory. All of these skills require coordinated sequences of perceptual, motor and cognitive actions learned through repetition and practice. With practice, such skills become 'automatised' so that they can be initiated and executed without thinking about how we do them. Procedural knowledge is not easily described in words. For example, a gymnast may be able to perform but not describe the exact movements required for a routine.

Another important form of implicit memory is **conditioning**. As you learned in Chapter 4, conditioning allows us to learn associations between stimuli that predict rewarding or punishing outcomes. Learning these kinds of associations is the foundation of both classical and operant forms of conditioning. Although we are sometimes explicitly aware of the associations we have learned (e.g. we know that a tone predicts an electric shock, and recall the conditioning episodes), we do not *need* to be aware of conditioned associations for conditioned responses to occur! Patients like H.M., who have anterograde amnesia for explicit memories, *can* still learn new conditioned associations (Madaboosi et al., 2021). We explore this more when we look at the brain mechanisms involved in implicit memory.

Figure 5.16 shows each of the components of LTM described above. In the next section we will explore the regions of the brain that support explicit and implicit memory.

5.2.2 COMPARING IMPLICIT AND EXPLICIT MEMORIES

EXAM TIP
Remember that implicit memory and procedural memory are not the same. Procedural memory is one form of implicit memory. Conditioning is another form of implicit memory.

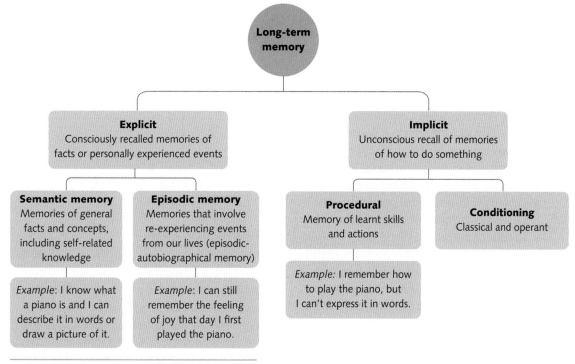

Figure 5.16 Components of long-term memory

KEY CONCEPTS 5.2

- » Long-term memory (LTM) includes explicit and implicit memory systems.
- » Explicit memory is the kind of LTM that allows conscious recall of experiences and facts. It includes episodic memory and semantic memory.
- » Episodic memory (or episodic autobiographical memory) is the kind of explicit memory that allows mental time-travel. It includes retrieval of personally experienced events from a first-person perspective, with details of time, place, perceptions, thoughts and emotions. It also includes the ability to imagine ourselves in the future from a first-person perspective, known as episodic future-thinking.
- » Semantic memory stores our general knowledge of words, concepts and facts. It includes semantic autobiographical knowledge.
- » Semantic networks store our knowledge of concepts and the words that represent them. Schemas store our knowledge of familiar situations and spatial relationships between concepts.
- » Anterograde amnesia and retrograde amnesia affect the ability to store and retrieve explicit memories, respectively.
- » Implicit memory is the kind of LTM that allows us to perform skills and respond to stimuli adaptively without conscious recollection of the knowledge that underlies these abilities. It includes procedural memory and conditioning.
- » Procedural memory is the kind of implicit memory that is demonstrated through skilled performance of a sequence of actions due to practice and repetition. It includes physical and cognitive skills.

Concept questions 5.2

Remembering
1. Name and define the two kinds of explicit memory. **r**
2. Name and define the two kinds of implicit memory. **r**

Understanding
3. Explain the difference between implicit and explicit forms of LTM. **e**
4. Explain the relationship between semantic memories and episodic memories. **e**

Applying
5. In 2005, Professor Susan Corkin visited patient H.M. to conduct further tests of his memory for personally experienced events and knowledge acquired before his surgery in 1953. She found that he was able to describe facts about his life and about public events that occurred before 1953, such as the name of the U.S. president and events of the Second World War. However, he could not recall personal experiences related to these public events, such as where he was and how he felt when he heard that a nuclear bomb had been dropped on Hiroshima. When Professor Corkin visited him again the next day, H.M. did not remember who she was or what they did the previous day, but he did feel a warm familiarity towards her that he could not explain. Use the table below to analyse this scenario and explain which elements relate to explicit episodic and semantic memories, and which relate to implicit memories. Identify which instances of forgetting reflect anterograde and retrograde amnesia. **c**

Element of scenario	Semantic, Episodic, Implicit?	Type of amnesia

6. Using Figure 5.16 as a foundation, create a personal example of each form of LTM from your experiences. **c**

HOT Challenge
7. H.M. gave his informed consent to participate in each episode of research he contributed to over his life, and he consented to donating his brain for further research after his death. Imagine you are on the research ethics panel assessing a research proposal to conduct a study with H.M. Identify the ethical issues that arise and explain how (or if) you think they could be managed. **i**

5.3 The neural basis of explicit and implicit memories

The extraordinary storage capacity of LTM is supported by the 86 billion neurons in our brains and the thousands of synaptic connections each neuron has with other neurons (Hawkins & Ahmad, 2016; Herculano-Houzel, 2009; Rah & Choi, 2022). Although we now know that there are not more neurons in the human brain than there are stars in the galaxy (there's about 100 billion stars), there are still an awful lot! It has been estimated that each adult human has acquired 12.5 million bits of semantic information alone (Mollica & Piantadosi, 2019).

The roles of the neocortex, hippocampus and amygdala in explicit memory

Figure 5.17 shows the major brain regions involved in memory. The structures of the **neocortex** (including prefrontal cortex), **hippocampus** and **amygdala** are each involved in different aspects of encoding, storing and retrieving explicit memories. We will cover the roles of the **basal ganglia** and **cerebellum** when we discuss the brain regions involved in implicit memory.

Hippocampus

Notice how the hippocampus sits deep within the temporal lobe in a region called the medial temporal lobe. The hippocampus gets its name from the Greek word for seahorse (Figure 5.18). It plays a crucial role in **binding** together the different elements of our experiences and **consolidating** these as explicit memories in the neocortex. We explore the role of the hippocampus in more detail shortly.

Amygdala

Just in front of the hippocampus sits the almond-shaped structure of the amygdala (plural amygdalae), which can be seen more clearly in

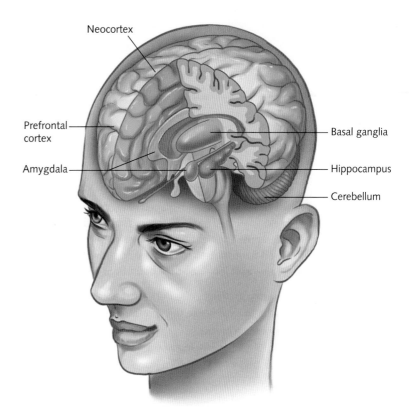

Figure 5.17 Areas of the brain that support long-term implicit and explicit memory

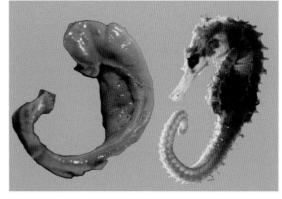

Figure 5.18 The shape of the hippocampus resembles a seahorse.

Figure 5.19. The amygdala plays an important role in the rapid and unconscious processing of emotions (implicit memory) and feeds this information to the hippocampus so that emotional information can be integrated into explicit memories.

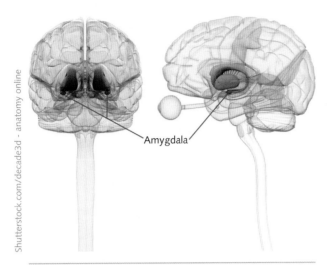
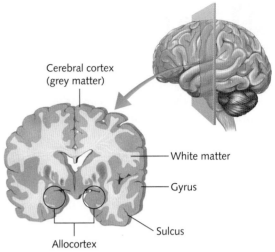

Figure 5.19 The almond-shaped amygdala ('almond' in Greek) is located immediately in front of the hippocampus. The right image shows an internal view of the left hemisphere. The left image shows both amygdalae as if looking at the brain front-on.

Figure 5.20 Coronal section (front-on interior view) of the human cerebral cortex, consisting mostly of the neocortex and the much smaller allocortex, which includes the hippocampus and olfactory bulb.

People with damage to both their amygdalae are unable to experience the emotions associated with episodic memories, and do not acquire implicit conditioned fear responses. Explicit memories associated with strong emotional responses in the amygdala are more memorable than non-emotional memories (LeDoux, 2020; Lindquist et al., 2012; McGaugh, 2013; Feinstein et al., 2013).

Neocortex

The neocortex is shown in Figure 5.17 as the wrinkled outermost layer of the cerebrum, with its characteristic folds of gyri (ridges) and sulci (grooves). *Cortex* comes from the Greek for 'bark', and *neo* means 'new', so the neocortex is the 'new bark' of the brain. Indeed, the neocortex is the most recently evolved area of the brain. It is the largest region of the cortex, including over 80 per cent of the grey matter of the four cerebral lobes in each cerebral hemisphere. It contains approximately 16 billion neurons in six layers of cells (Herculano-Houzel, 2009). The neocortex is distinguished from an evolutionarily much older and much smaller area of cortex called the **allocortex**, which includes the hippocampus and olfactory bulb. Figure 5.20 shows the neocortex, with the allocortex circled.

The neocortex processes the sensory, motor and perceptual information that we become aware of. The **prefrontal cortex** (the most frontal portion of the neocortex) controls the higher cognitive functions of attention, thought and language that allow us to reflect consciously on our experiences. It directs attention to the contents of the different memory stores as we encode information from STM into LTM. Our long-term explicit memories are stored in networks of neurons throughout the rest of the neocortex, in the very same regions that represent our experiences during perception, action and thought.

The neural basis of episodic memory

As we interact with the world, the prefrontal cortex directs attention to the regions of the neocortex that are processing the sensory-motor and perceptual aspects of the experience. The information is then transmitted to the hippocampus via a network of connections between the hippocampus and neocortex. The hippocampus then **binds** the separate sources of information about the episode into an integrated memory trace, which it then feeds back to the neocortex. The feedback between the hippocampus and neocortex strengthens the activation of neocortical neurons involved in processing the experience *and their connections with each other*.

As you learned in Chapter 2, the process of strengthening connections between neurons that are active at the same time is called **long-term potentiation** (LTP). Here, we see that LTP is

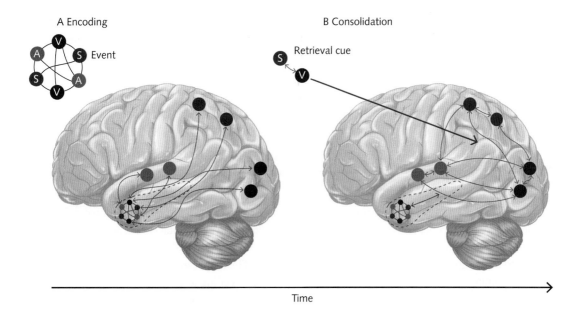

Figure 5.21 Consolidation of episodic memory via hippocampal-neocortical connections. (A) During encoding, an event with auditory (A), visual (V) and spatial (S) information is registered in regions of the neocortex specialised for each kind of information and is communicated to the hippocampus where a bound representation is formed. Feedback between the hippocampus and neocortex strengthens connections between the different regions of the neocortex involved in processing the experience. (B) During consolidation, neural connections between regions of the neocortex are strengthened to form an integrated memory representation of the experience. This memory can be reconstructed in response to a partial cue that shares some features with the original experience. (S-V) via hippocampal-neocortical connections. Notes, dotted outline indicates the hippocampus lying deep within the temporal lobe; double-ended arrows represent feedback connections between neurons.

the mechanism that **consolidates** (strengthens) long-term episodic memories in the neocortex. Figure 5.21 shows the interactions between the hippocampus and neocortex during consolidation of episodic memories. Damage to the hippocampus removes the ability to consolidate our experiences as explicit long-term memories, even though we can still experience events consciously in STM.

Retrieval of episodic memories occurs when the connections between the hippocampus and neocortex are reconstructed in response to a cue. For example, when you hear an old song and it transports you back to an experience associated with it. In this example, the information in the auditory sensory cortex is sent to the hippocampus, which retrieves and binds associated memory traces. This is then projected back to the neocortex where the event is reconstructed and represented consciously. It is this reactivation of neocortical regions that gives episodic remembering its quality of reliving an event. When we retrieve episodic memories intentionally via self-generated cues (thoughts), the prefrontal cortex signals the hippocampus to reconstruct the memory in a similar way to retrieval via an external cue.

The neural basis of semantic memory

Our semantic knowledge is distributed throughout the same regions of the neocortex that process our experiences. So, how do we distinguish semantic from episodic knowledge? The answer to this is that the temporal lobes contain two specialised **semantic hubs** that organise different aspects of our semantic knowledge.

Our knowledge of words, concepts and their meanings (semantic networks) is stored in a hub located at the front of the temporal lobes with connections to the frontal lobes, including a region known as Broca's area. Our knowledge of schemas and how words can be combined to describe knowledge and events (grammar) is located in a hub at the back of the temporal lobes with connections to the parietal lobes (including the region known as Wernicke's area). The two semantic hubs are referred to as the **anterior** (front) **temporal lobe hub** and the **posterior** (back) **temporal lobe hub**, respectively.

Each semantic hub has connections with the neocortical regions that represent the sensory-motor and perceptual information that is associated with concepts. So, when we think of using words, concepts and schemas, these thoughts activate the relevant sensory-motor and perceptual knowledge in the rest of the neocortex. Another way to think about it is that words, concepts and schemas act like pointers to the information that is associated with them. This is how it is possible for us to generate mental imagery associated with a word, concept or schema. And, it is how perceptual experiences activate concepts and words. Damage to the temporal lobe semantic hubs affects our ability to use language, name objects and understand familiar situations (Flinker et al., 2015; Irish & Vatansever, 2020; Patterson & Ralph, 2016).

The consolidation of semantic memories relies on the hippocampus because general knowledge is extracted from episodic memories. However, retrieval of semantic memories occurs through direct activation of the semantic networks themselves, either through perceptual cues – seeing a cat, or hearing the word 'cat' – or via self-generated cues when we think about or imagine cats. This explains why damage to the hippocampus causes anterograde amnesia for episodic and semantic memories, but causes retrograde amnesia only for episodic memories.

Bringing it together: hippocampal-neocortical networks for explicit memory

Although episodic and semantic memory are distinct kinds of LTM, they interact in important ways within an integrated hippocampal-neocortical network. The prefrontal cortex acts as the director of STM encoding and retrieval processes that build LTM. The hippocampus binds episodic information during encoding and consolidates it in the neocortex, including emotional information from the amygdalae. The hippocampus also coordinates the reconstruction of episodic memories during retrieval. The temporal lobe semantic hubs extract information from activated episodic memories to construct general knowledge from our experiences. Semantic memory relies on the hippocampus for consolidation, but not for retrieval. Explore Analysing research 5.2 to gain further insights into how semantic knowledge is represented across the neocortex.

ANALYSING RESEARCH 5.2

Weblink
Brain atlas video

Where words are stored: the brain's semantic map

A team of neuroscientists at the University of California, Berkeley, led by Alexander Huth, has created a comprehensive 'atlas' that shows where different meanings are represented in the human brain. The researchers played two hours of stories from the Moth Radio Hour (a popular podcast) to seven participants while recording their brain activity in a functional MRI scanner. They then analysed the activity across the entire brain, creating detailed maps of where different meanings were represented in each individual. They found that semantic information was distributed over much of the neocortex, extending beyond areas traditionally thought of as language centers. Some areas were found to respond selectively to words related to people, whereas other areas respond to concepts related to places or numbers, and many more besides. The maps were remarkably similar from one participant to the next, though not identical. The researchers developed a statistical tool that enabled them to produce a general semantic 'atlas', by finding areas common to all participants. The map they produced is seen in Figure 5.22.

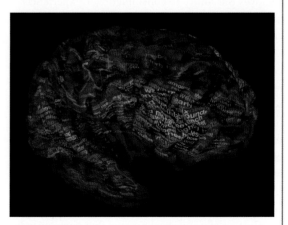

Figure 5.22 This representation of an 'atlas' of meaning in the brain shows where semantic concepts are distributed across regions of the neocortex.

Source: adapted from Makin, S. 'Where Words are Stored: The Brain's Meaning Map' in *Scientific American Mind*, 27, 5, 17 (1 September 2016). doi:10.1038/scientificamericanmind0916-17a; based on original research reported in Huth, A. G., De Heer, W. A., Griffiths, T. L., Theunissen, F. E., & Gallant, J. L. (2016). Natural speech reveals the semantic maps that tile human cerebral cortex. *Nature*, 532(7600), 453–458.

https://www.scientificamerican.com/article/where-words-are-stored-the-brain-s-meaning-map/

Questions
1 Explain why this study relates to semantic memory rather than episodic memory.
2 What does this study tell us about where semantic knowledge is stored in the brain?
3 Explain why this study is an example of a modelling study.

The role of the basal ganglia and cerebellum in implicit memory

Implicit memories are stored in the basal ganglia and the cerebellum. These areas can operate independently of the hippocampus, which is why the abilities to encode, store and retrieve procedural memories and conditioned associations are preserved in people with hippocampal damage.

Basal ganglia

The basal ganglia are a set of structures that lie beneath the cortex. They store procedural memories of sequences of practiced voluntary movements. The basal ganglia work together with the motor cortex and prefrontal cortex to initiate practiced sequences of movements and inhibit unwanted movements. The result is a smooth sequence of movements stored as a procedural memory.

Dopamine-producing cells within the structures of the basal ganglia are important for initiating sequences of movement. Damage to these structures causes the movement disorders seen in Parkinson's disease, including slowed movement, tremor and rigid posture (Foerde & Shohamy, 2011).

Cerebellum

The cerebellum is the major structure of the hindbrain and is densely packed with 70 billion of the brain's 86 billion neurons (Herculano-Houzel, 2009). It controls the movements, balance and coordination needed to master tasks that require constant monitoring and fine motor adjustments, such as threading a needle, or the mirror-tracing task described earlier. The cerebellum also plays a role in cognitive functions through its connections with the basal ganglia and cortex (Bostan & Strick, 2018). It seems that the cerebellum regulates the speed, consistency and appropriateness of mental processes just as it regulates the rate, rhythm, force and accuracy of movements (Schmahmann, 2019). For example, the cerebellum is known to be important in learning how to read (Alvarez & Fiez, 2018).

WB 5.3.1 BRAIN STRUCTURES INVOLVED IN LONG-TERM MEMORY

ACTIVITY 5.2 TWO TRUTHS AND A LIE

Try it
Have you ever played the classic ice-breaker game called 'two truths and a lie'? In this game you need to think of two true statements and one false statement and see if another person can work out what is true and what isn't.
1 For this version of the game, you will need to come up with two true statements and one false statement about the five brain regions, listed below, involved in long-term memory.
- Hippocampus
- Amygdala
- Neocortex
- Basal ganglia
- Cerebellum

Quiz a classmate once you have completed your 'truths' and 'lies'.

Apply it

In a 2011 study, a patient known as S.M. was exposed to a number of situations to provoke fear. S.M. was used as she has a rare condition that has damaged her amygdala in both hemispheres of her brain (Feinstein et al., 2011).

Over the course of the study, S.M. handled live snakes and spiders, toured a haunted house, and was shown emotionally intense movies. However, as the researchers noted, at no point did S.M. exhibit fear.

2. With reference to the role of the amygdala, explain why S.M. is unable to experience fearful memories.

Exam ready

3. Research suggests that semantic memories are encoded by the _____ and are transferred to the _____ for storage.

    A   cerebellum; amygdala
    B   neocortex; hippocampus
    C   hippocampus; cerebellum
    D   hippocampus; neocortex

KEY CONCEPTS 5.3

- The brain regions that encode, store and retrieve explicit memories are the hippocampus, amygdala and neocortex.
- The hippocampus is located deep within each temporal lobe (in the allocortex). It has two-way (feedback) connections with the neocortex. Its role is to bind and consolidate episodic memories in the neocortex. It is also required for retrieval of episodic memories.
- The amygdala is located in front of the hippocampus. Its role is to process emotional information. Implicit emotional information is integrated with explicit episodic memories through connections between the amygdala and hippocampus.
- The neocortex is the largest region of the cortex, including 80 per cent of the grey matter of each of the four lobes of the brain (excluding the allocortex).
- The prefrontal cortex coordinates attention, thought and language during encoding and retrieval of memories.
- Neocortical regions behind the prefrontal cortex process sensory-motor information during perception and action and store semantic and episodic memories.
- The temporal lobes of the neocortex contain two semantic hubs (anterior and posterior) that store semantic knowledge that has been generalised from episodic memories.
- Neural representations of words, concepts and schemas in the semantic hubs are connected with sensory-motor and perceptual representations throughout the neocortex. These connections allow us to imagine the perceptual features of words and concepts.
- The basal ganglia and cerebellum are the brain structures involved in encoding, storing and retrieving implicit memories. These regions can function independently of the hippocampus.
- The basal ganglia is a set of subcortical structures that coordinate the smooth performance of sequences of actions involved in procedural memory.
- Loss of dopamine-producing cells in the basal ganglia causes problems with initiating movement and inhibiting unwanted movement, as seen in the motor symptoms of Parkinson's disease.
- The cerebellum is the major structure of the hindbrain with connections to the basal ganglia and prefrontal cortex. It coordinates fine motor movements and balance when performing skilled actions. It also plays a role in procedural memory of cognitive skills such as reading.

Concept questions 5.3

Remembering
1. What regions of the brain define the neocortex? (r)
2. Where in the brain is the hippocampus found? (r)
3. Describe the role of the basal ganglia in procedural memory. (r)

Understanding
4. Explain the relationship between the hippocampus and amygdala in explicit memory. (e)
5. Explain why the hippocampus is required for storage (consolidation) but not retrieval of explicit semantic memories. (e)

Applying
6. Imagine that you are learning archery. Explain the distinct roles of the basal ganglia and the cerebellum in learning this skill. (c)

HOT Challenge
7. Use the diagram below, which shows the lateral (side) view of the left hemisphere of the brain, to draw in the two temporal lobe semantic hubs as they are described in the text. Then annotate the diagram with feedback arrows to show how words and schemas are connected with sensory-motor and perceptual representations in the neocortex. (d)

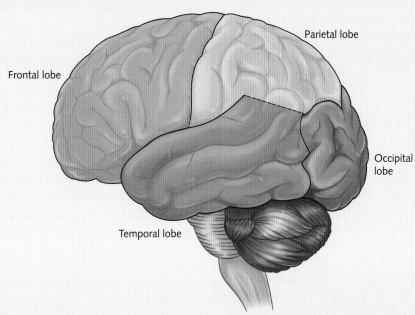

5.4 Autobiographical memory

Autobiographical memory (ABM) is the component of explicit memory that represents our memories of personally experienced events and self-knowledge. ABM includes the whole of episodic memory (aka episodic autobiographical memory, or EAM) as well as the component of semantic memory that stores self-knowledge, called **semantic autobiographical memory (SAM)**. Together, EAM and SAM allow us to represent ourselves as an enduring person situated within a cultural-historical context (e.g. Australia in the 21st century), and to project ourselves forwards and backwards within personally experienced time.

The EAM component of ABM provides the first-person experience of remembering events from the past and imagining ourselves in the future. The SAM component builds a stable core of self-knowledge and identity over time. Like other kinds of semantic knowledge, SAM is generalised from EAM.

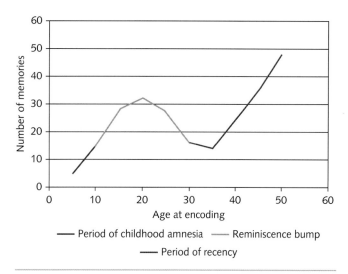

Figure 5.23 The typical distribution of ABMs over the lifespan

For example, we develop a self-schema that represents our relationship to people, places and events, as well as our personal characteristics (like being shy, funny or smart) and our goals for the future (Conway & Loveday, 2015; Levine et al., 2002; Tulving, 2002).

The distribution of ABMs across the lifespan can be seen in Figure 5.23. Notice the period of **childhood amnesia** that occurs between the ages of 0–5 years, from which people report very few ABMs. A **reminiscence bump** occurs between the ages of 10 and 30. The bump is the period of time from which older adults recall most ABMs (Munawar et al., 2018). A strong recency effect occurs for the events experienced in the past few months. The reminiscence bump occurs because the events of adolescence and early adulthood involve emotionally significant transitions that are central to our developing self-concept; like graduating from school, first job, moving out of home, significant relationships, marriage and children.

The roles of semantic and episodic memory in remembering the past and imagining the future

Theories of ABM propose that SAM is organised in a hierarchical structure. Broad life-stages are represented at the highest level (e.g. childhood, adolescence, early adulthood, etc.), followed by general life events (e.g. birthdays, holidays, employment), and finally specific events (e.g. your 16th birthday). In these theories, SAM provides an organising framework, called a 'scaffold', that helps us locate detailed episodic knowledge. For example, when searching for a specific event, we can begin by locating ourselves within the general time period. From there, we can locate general event information, which helps locate specific event details. The search can also occur in reverse. For example, a sensory cue or mental image can activate an episodic memory, which enables the broader semantic knowledge associated with the event to be recovered (Conway & Loveday, 2015; Irish & Piguet, 2013).

Imagining ourselves in a future situation also involves contributions from EAM and SAM. However, the **semantic scaffolding theory** of episodic future-thinking proposes a central role for *general* semantic knowledge in this process (Irish, 2016; Strikwerda-Brown et al., 2022). For example, imagine that you have a job interview next week and you want to mentally rehearse what you are going to say and do in the interview. To construct this imagined scene, you first need to draw on a general 'job interview schema' from semantic memory. This schema includes information extracted from your past experiences of similar situations (if you have any) as well as information you have observed, read or heard about such situations, and whatever the employer has told you to expect. This semantic knowledge 'sets the scene' in which you can then place your imagined future self.

Functional magnetic resonance imaging (fMRI) studies show that the same hippocampal-neocortical network is active during both retrieval of ABMs and episodic future-thinking. There are striking similarities between the two tasks. During both processes, the hippocampus coordinates with attentional networks in the prefrontal cortex to retrieve and consciously represent semantic knowledge from the temporal lobe and episodic information from regions of the parietal cortex (Dickerson & Eichenbaum, 2010).

Individual differences in ABM and episodic future-thinking

Research using self-report questionnaires has found stable individual differences between people in their ABM style. The differences lie on a continuum. At one end are people who experience **highly superior autobiographical memory (HSAM)**, and at the other are people who report **severely deficient autobiographical memory (SDAM)** (McGaugh & LePort, 2014; Palombo et al., 2018; Sheldon et al., 2016).

People with HSAM can recall most of the events of their lives with a high degree of episodic detail. They recall the date and day of their personal experiences and of public events (e.g. what they were doing when Prince Harry and Meghan Markle were married, and the date and day of the week that this occurred). Interestingly, they do not perform differently to age-matched controls on laboratory tests of memory, nor are they superior learners.

In contrast, people who experience SDAM report a lifelong inability to vividly recollect personally experienced events from a first-person perspective. They have only semantic autobiographical memories of their lives, represented as a life story from a third-person perspective. They also find it difficult to imagine themselves in the future (Watkins, 2018).

Between the extremes of HSAM and SDAM, people vary in the relative dominance of EAM and SAM components when retrieving ABMs and imagining the future. Interestingly, EAM-dominant people tend to be involved in the arts, whereas SAM dominant people tend to be involved in engineering and science (Palombo et al., 2018).

Different ABM-styles are related to different patterns of connectivity between the hippocampus and the neocortex during ABM retrieval and episodic future-thinking. EAM-dominant people show greater hippocampal connectivity with the posterior temporal lobe semantic hub and with occipital regions associated with mental imagery. In contrast, SAM-dominant people show greater hippocampal connectivity with the anterior temporal lobe semantic hub associated with language (Petrican et al., 2020; Schacter et al., 2017).

KEY CONCEPTS 5.4a

- Autobiographical memory (ABM) is the component of explicit memory that stores knowledge of personally experienced events and self-knowledge.
- ABM includes episodic-autobiographical memory (EAM) and semantic autobiographical memory (SAM).
- EAM provides the first-person experiential component of ABM. SAM provides our self-knowledge.
- SAM provides an organising 'scaffold' for retrieving EAMs. In contrast, general semantic knowledge is thought to scaffold our ability to imagine ourselves in future events.
- The relationship between ABM retrieval and episodic future-thinking is seen in fMRI studies showing shared hippocampal-neocortical networks for both tasks.
- The relative contributions of EAM and SAM components to ABM vary between people along a continuum. People with highly superior ABM (HSAM) recall rich episodic details for each day of their life. People with severely deficient ABM (SDAM) recall only semantic knowledge about their lives.
- Individual differences in ABM style are reflected in different patterns of hippocampal connectivity with the neocortex. HSAM people show greater connectivity with parietal and occipital regions associated with mental imagery. SDAM people show greater connectivity with the frontal temporal lobe semantic hub (language).

Concept questions 5.4a

Remembering
1. Define autobiographical memory. **r**
2. Define episodic future-thinking. **r**

Understanding
3. Explain how semantic memory is involved in retrieving autobiographical memories and episodic future-thinking. **e**

4. Explain how episodic-autobiographical memory is different from autobiographical memory. **e**

Applying
5. Imagine you have been asked to provide evidence for a cold-case investigation into a robbery that you witnessed over a decade ago

> that is now believed to be related to a series of increasingly violent crimes. The interviewing officer asks what you remember about the episode. Write two accounts of the event, one from the perspective of someone with HSAM and the other from the perspective of someone with SDAM. **C**
>
> and ask them to retrieve a memory of a specific event experienced on a holiday in the past 10 years and one from when they were in their 20s. Code their responses for EAM and SAM components. Consider how and whether the two descriptions differ and explain this using your knowledge of EAM and SAM contributions to ABM over the lifespan, and individual differences in ABM-style. **C**
>
> **HOT Challenge**
>
> 6 Interview a family member or friend who is 50+ years old. Use the cue 'holiday'

The role of mental imagery in ABM and episodic future-thinking

5.4.1 STUDYING APHANTASIA

Individual differences in ABM style may be related to differences in the ability to generate **mental imagery**. Mental imagery refers to perception-like experiences in the absence of sensory input, for example when we 'see something in our mind's eye' or hear a song in our heads (Pearson et al., 2015).

Theories of ABM propose that mental imagery is a crucial component of the ability to retrieve rich first-person ABMs and to imagine the future. This is because mental images provide powerful cues to search the hierarchical structure of ABM (Conway & Pleydell-Pearce, 2000; Rubin, 2005). Studies of the relationship between ABM and visual mental imagery support this idea. They find that people's self-reported 'vividness' of visual imagery is strongly related to whether they experience vivid EAM components during ABM retrieval and episodic future-thinking (Greenberg & Knowlton, 2014; Vannucci et al., 2016).

Visual imagery has two components: object imagery and spatial imagery (Blazhenkova, 2016). **Object imagery** refers to the ability to 'picture' the shape, colour and texture of objects, as well as people, faces, animals and scenes. It is associated with activation in the visual cortex in the occipital lobes. **Spatial imagery** is the ability to imagine relationships between objects, and between parts of objects (e.g. the mechanism in a clock), and their movement in three-dimensional space. Spatial imagery is involved in activities like 'seeing' how to construct IKEA furniture from a flat-packed box, or mentally rotating an object to 'view' it from different angles, and predicting the trajectory of an object like a billiard ball in response to force. Spatial imagery is more abstract than object imagery and is not necessarily accompanied by vivid object imagery. It is associated with activity in the parietal lobes.

Aphantasia: the lack of mental imagery

Some of the strongest evidence for a relationship between mental imagery and ABM style comes from studies of people who experience aphantasia. **Aphantasia** is the lifelong absence of voluntary mental imagery. Aphantasic people represent 2–3% of the general population. It seems that visual object imagery is mostly affected, but some aphantasics also report absent or reduced mental imagery for other senses. Most aphantasics experience involuntary imagery during dreaming, although they report fewer and less vivid dreams. Importantly, aphantasics report strong non-visual spatial imagery and are often exceptional at spatial imagery tasks like those described above (Dawes et al., 2020; Keogh et al., 2021; Zeman et al., 2015).

Consistent with the hypothesis that visual imagery is crucial for the episodic component of ABM, aphantasics often report severely deficient autobiographical memory (SDAM) *and* an inability to imagine the future (Watkins, 2018). And, like SAM-dominant people, aphantasics are attracted to science and engineering more than the arts. At the other end of the mental imagery spectrum are people who experience **hyperphantasia**. Hyperphantasic people report mental imagery 'as vivid as real seeing'. They also commonly report experiencing an involuntary co-activation of sensory imagery (e.g. 'seeing' colours when hearing sounds), known as synaesthesia (Keogh et al., 2021; Zeman et al., 2020).

Where are you on the mental imagery/ABM spectrum?

The vividness of mental imagery and episodic-ABM is assessed using self-report questionnaires. You can try these for yourself at the Aphantasia Network site (see weblink). This site is run by prominent researchers in the field, including Australian cognitive neuroscientist Professor Joel Pearson from The University of New South Wales (UNSW). It includes a forum and resources for people who experience aphantasia. You can experience how the relationship between autobiographical memory and mental imagery is studied by completing a 15-minute questionnaire hosted by the University of Toronto (see weblink).

Weblink
Vividness of Visual Imagery Questionnaire (VVIQ)

Autobiographical memory questionnaire

🔑 KEY CONCEPTS 5.4b

- Researchers have found that individual differences in ABM style are related to individual differences in the ability to generate mental imagery.
- Mental imagery refers to perception-like experiences in the absence of the relevant sensory input.
- People who can generate vivid mental images of objects, people and scenes (object visual imagery) also experience more vivid first-person EAMs during ABM retrieval.
- People with aphantasia experience an absence of voluntary mental imagery. They also report having severely deficient ABM and are unable to imagine themselves in the future. However, they have strengths in non-visual spatial imagery.
- These findings support theories that propose a strong role for visual mental imagery in retrieving ABMs.

Concept questions 5.4b

Remembering
1. Define aphantasia. r

Understanding
2. Explain the difference between object imagery and spatial imagery. e
3. Explain how the two forms of mental imagery are related to autobiographical memory and aphantasia. e

Applying
4. Your friend Max has arranged a coffee date with someone next week and is very nervous about what to say and how to behave. You suggest that she should imagine the scene at the café, how her date will look, and to imagine herself looking relaxed and feeling happy as she greets them. Max complains that she does not know how to do this. Is Max being difficult or is there another explanation for her response? What other things might Max find difficult? c

HOT Challenge
5. Fifty participants each provide a self-reported vividness of mental imagery score using a five-point scale from 1 'No image at all' to 5 'As vivid as real seeing'. The researcher also calculates an episodic-autobiographical memory (EAM) score for each participant as the mean number of EAM details they produce when describing memories of personal events, with scores ranging from 0 to 20. Write a hypothesis for this study in terms of the expected correlation (i.e. relationship) between self-reported vividness of mental imagery and episodic-autobiographical remembering. Use the axes below to draw a graph to show what the hypothetical relationship between the two scores should look like. i

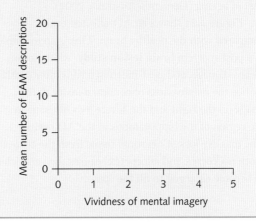

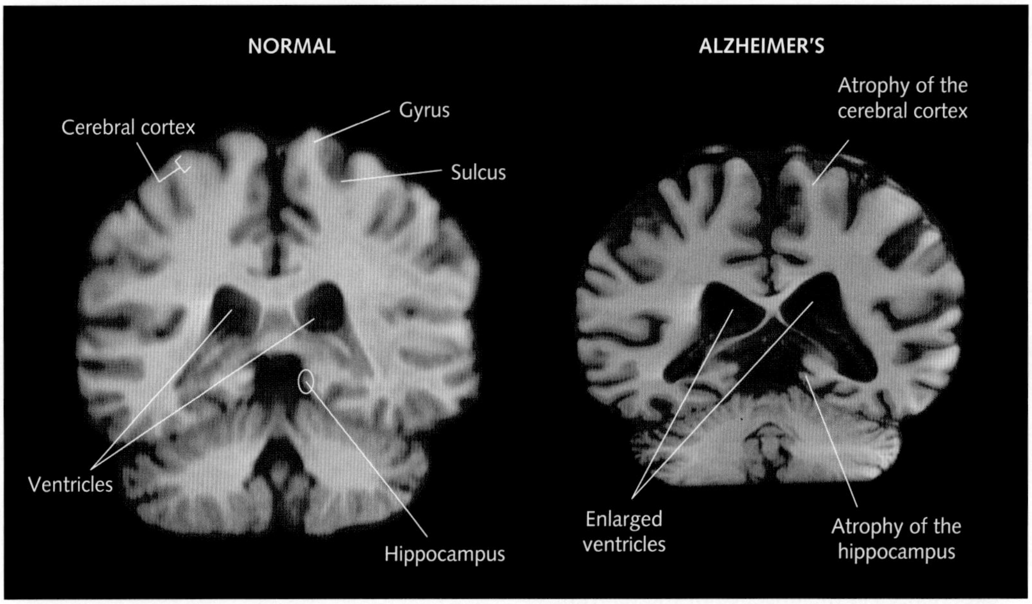

Figure 5.24 Post-mortem lesion study comparing a healthy brain (left) with an AD-affected brain (right). The healthy hippocampal region is circled on the image on the left. The image on the right shows hippocampal and neocortical atrophy in AD. Notice the enlarged ventricles in the AD brain.

ABM in Alzheimer's disease

Alzheimer's disease (AD) is a neurodegenerative disease that causes a progressive loss of brain tissue (atrophy) that is eventually fatal. AD is most common in people over the age of 65, but can also occur in younger adults. It causes symptoms of dementia, including progressive memory loss, confusion, changes in personality, and an inability to perform familiar tasks. AD is the most common form of dementia, accounting for 70 per cent of all cases. As of January 2022, one in ten Australians over the age of 65 has AD (approximately 340 000 people). Although common in ageing, AD is not a normal part of ageing (Dementia Australia, 2002).

Figure 5.24 shows an AD-affected brain compared with the healthy brain. This is an example of a post-mortem **brain lesion study**, where lesion means damage, and post-mortem means after death. Notice the massive loss of cortical density in the image on the right and the severe loss of tissue in the region of the hippocampus in the AD brain. The massive loss of neuronal tissue from the cortex (both neocortex and allocortex) causes the ventricles to become enlarged to fill the space.

The cause of AD is the focus of ongoing research, but it seems certain to be related to an abnormal build up and overactivity of beta-amyloid and tau proteins in the brain. In AD, beta-amyloid collects into clumps called **amyloid plaques** that build up between neurons and disrupt their functioning. Tau forms **neurofibrillary tangles** inside the neuron cell body, which interrupts synaptic transmission. Figure 5.25 shows amyloid plaques as green clumps between the cells and neurofibrillary tangles are shown in brown within the body of the neuron.

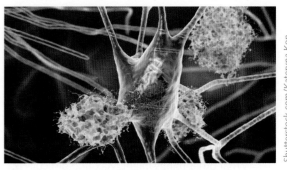

Figure 5.25 The build-up of amyloid plaques (green clumps) and neurofibrillary tangles (brown) are the key biological markers of Alzheimer's disease brain pathology.

Until recently, the presence of plaques and tangles could only be confirmed after death. However, new techniques are making diagnosis more precise in the early stages of AD while people are still alive. For example, it is now possible to measure concentrations of beta-amyloid and tau in cerebrospinal fluid using a lumbar puncture. A dye injected into the bloodstream, called a tracer, can also be used to stain tau so that neurofibrillary tangles are visible in positron emission tomography (PET) images (Bayram et al., 2018).

Brain imaging studies of AD use MRI and fMRI to study the structural and functional brain changes that occur as the disease progresses. The earliest structural and functional changes in AD occur in brain regions involved in consolidating and retrieving explicit autobiographical memories, making these regions a key focus of research to identify biomarkers for early detection of AD.

Autobiographical memory in AD

One of the earliest signs of AD is severe anterograde amnesia for episodic and semantic memories. That is, people with AD are unable to consolidate memories about recent events, nor can they remember facts learned during those events. For example, they may have difficulty remembering important upcoming events, like forgetting that today is their grandson's first birthday party. Or, they may forget where they are and why they are there, or forget what has just been said in a conversation. They may repeat stories or repeat the same question after just hearing the answer. These difficulties are much more severe than the normal forgetfulness associated with healthy ageing. The Alzheimer's Association provides an excellent guide to the 10 key memory problems associated with AD (see weblink).

As AD progresses, retrograde amnesia develops for episodic-autobiographical memories, with memories for more recent events being affected before remote memories. This is why it can seem that people with AD are 'living in the past'. Indeed, it is the life events from early in the reminiscence bump, from late childhood and adolescence, that seem most resistant to AD. However, these memories tend to lack episodic detail, being more semanticised (Kirk & Berntsen, 2018).

People with AD also find it difficult to imagine themselves in future situations (Addis et al., 2009; Irish & Piguet, 2013). Recent research shows that the early loss of episodic future-thinking is accompanied by a loss of voluntary visual imagery, particularly object imagery. This is consistent with what we know about the relationship between EAM and visual object imagery in healthy brains (El Haj et al., 2019).

As AD progresses, other areas of memory become affected. Loss of general semantic knowledge becomes increasingly evident, so that objects, people and familiar situations are no longer recognised. Finally, procedural memory is affected as people lose the ability to perform even the most highly practiced tasks.

Evidence from brain imaging studies

Brain imaging studies of people with early-stage AD show a loss of volume in the hippocampus and abnormal activation of the hippocampus during episodic memory tasks, including episodic future-thinking (Addis et al., 2009; Irish & Piguet, 2013). Early hippocampal damage explains why anterograde amnesia appears first in AD, followed by retrograde amnesia for EAMs. The damage to the hippocampus prevents the binding and consolidation of recent events and of semantic knowledge associated with those events. It also explains why retrograde amnesia affects retrieval of recent EAMs, and the sparing of the more semanticised remote memories.

As the disease spreads to the prefrontal cortex and temporal lobes, people lose the ability to retrieve semantic memories, including voluntary generation of visual mental imagery. They have difficulty finding words and naming familiar objects. Spatial memory becomes affected as the disease spreads to the parietal lobes, causing people to become lost even in previously familiar surroundings. Finally, damage in the occipital lobes at the back of the brain prevents recognition of familiar objects and people (Bayram et al., 2018).

Weblink
10 signs of Alzheimer's disease

Figure 5.26 shows the results of an fMRI study that compared brain activity during a recognition memory task between early-stage AD patients and healthy age-matched controls (Kivistö et al., 2014). Participants viewed pictures of faces paired with names that were repeated throughout the experiment. Then, they had to indicate when they recognised a correct face–name pairing from earlier in the experiment. This is an episodic memory task that requires the hippocampus to bind the face-name information from each episode (trial) and consolidate it in LTM. The researchers found that the AD participants showed *increased* activation in hippocampal and neocortical regions when viewing repeated pairs, whereas the healthy controls showed *decreased* activation for repeated pairs compared to non-repeated pairs. This finding indicates that the AD patients had greater difficulty remembering which face-name pairs they had just seen.

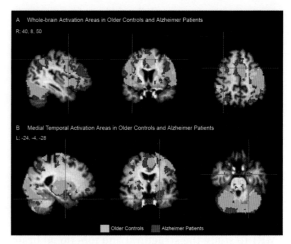

Source: Kivistö, J., Soininen, H., & Pihlajamaki, M. (2014). Functional MRI in Alzheimer's disease. In *Advanced brain neuroimaging topics in health and disease – methods and applications*. IntechOpen. CC BY 3.0 Licence (https://creativecommons.org/licenses/by/3.0/)

Figure 5.26 Brain-imaging study (fMRI) showing increased activation in neocortical (A) and hippocampal (medial-temporal) (B) brain regions during processing of repeated face–name pairs in patients with early AD (in red) relative to healthy control subjects (in yellow).

ACTIVITY 5.3 AUTOBIOGRAPHICAL MEMORY, APHANTASIA AND ALZHEIMER'S DISEASE

Try it

1. **Part A.** Spend a few minutes jotting down as many pieces of information as you can remember from the first concert you ever attended. Include information such as: who you saw, who you were with, and your impressions of the concert in general.

    Part B. Once you have written down as many points as you can remember from the concert, classify each piece of information as either a semantic memory or an episodic memory.

Apply it

2. Based on your understanding, explain whether a person with aphantasia would struggle to recall their first concert.

Exam ready

3. Which of the following statements is most correct about Alzheimer's disease?

    A. The first region of the brain affected in a person with Alzheimer's disease is the amygdala.

    B. Neurofibrillary tangles are found between neurons, while amyloid plaques are found within the neuron.

    C. The first region of the brain affected in a person with Alzheimer's disease is the neocortex.

    D. Neurofibrillary tangles are found within the neuron, while amyloid plaques are found between neurons.

🔑 KEY CONCEPTS 5.4c

» Alzheimer's disease (AD) is a progressive and fatal neurodegenerative disease causing symptoms of dementia, especially explicit memory loss.
» AD is associated with abnormal levels of beta-amyloid and tau proteins in the brain that form amyloid plaques and neurofibrillary tangles, respectively.
» The first symptom of AD is anterograde amnesia for episodic-autobiographical memories and semantic knowledge. People are unable to remember recent personally experienced events, nor facts associated with those events.
» Early symptoms also include difficulty with episodic future-thinking and generating voluntary mental imagery.
» These problems are consistent with damage to the hippocampus in the early stages of AD, which affects binding and consolidation of explicit memories.
» As the disease progresses, people experience retrograde amnesia for the events of their lives, with more recent memories being lost first. Remote memories from early in the reminiscence bump are the last to be affected, as they are more semanticised.
» Semantic memory deficits follow as areas of temporal, parietal and occipital lobes become progressively affected.
» Post-mortem lesion studies confirm diagnosis, identifying the characteristic plaques, tangles and atrophy.
» Functional MRI brain-imaging studies aim to detect changes in brain activity associated with early episodic memory decline before significant structural changes are evident.
» These studies reveal abnormal activity in the hippocampus and its connections with the neocortex during episodic memory tasks and offer hope for early detection and treatment.

Concept questions 5.4c

Remembering
1 Define Alzheimer's disease, including both the psychological and biological symptoms that affect autobiographical memory. r

Understanding
2 Explain the relationship between visual imagery and autobiographical memory loss in Alzheimer's disease. e

Applying
3 Zihan's aunt complains that she is occasionally forgetting upcoming appointments and often forgets where she left her keys, but is then able to locate them by retracing her movements. She also gets confused about which day of the week it is, but is then able to remember. Use the Alzheimer's Association checklist on their website (see weblink on page 211) to work out whether you think Zihan should be concerned about his aunt. c

HOT Challenge
4 Imagine that you are a neuropsychologist and that you want to develop a test to measure the set of inter related cognitive processes that are thought to be early markers specific to AD. Use the table below to list the key cognitive functions that should be affected in early AD and those that should be preserved in early AD. Describe a task that would provide a measure of each function. c

Function showing cognitive decline in early AD	Task to measure functioning	Function preserved in early AD	Task to measure functioning

5.5 Mnemonics

In this section, we explore memory techniques under the expert guidance of Dr Lynne Kelly, author of *The Memory Code*, and co-author of *Songlines: The power and promise*.

Mnemonic techniques are methods to help us encode information in a way that makes it more memorable. The word *mnemonic* comes from Mnemosyne, the goddess of memory in Greek mythology. To pronounce mnemonic, just ignore the first 'm' and say 'nemonic'.

In this section, we'll look at four mnemonic techniques that increase in complexity. You'll see that the key to all mnemonics is the meaningful links you create between the knowledge you want to learn and things that are naturally memorable like imagery, rhyme, emotions, locations, objects, stories and songs. The deep encoding involved in using mnemonics strengthens the storage of information in long-term memory. Retrieval is facilitated by the powerful internal and external cues that are created. The most powerful mnemonics interweave multiple strategies and have been used to encode, store and retrieve knowledge in **oral cultures** over thousands of generations.

Mnemonics based on written language

The invention of writing systems occurred in several cultures within the last 5000 years. Writing provides a visual code of spoken (oral) languages that allows speech to be recorded in a stable visual form. This makes the written characters of alphabets excellent retrieval cues for words. Acronyms and acrostics are both examples of how we can use letters to help us remember information.

5.5.1 MNEMONICS: ACRONYMS AND ACROSTICS

Acronyms

An **acronym** is an abbreviation formed by using the first letter of each word in a phrase or title we want to remember to form a unit (i.e. chunk) that can be pronounced as a word. For example, Anzac Day commemorates the 25th of April 1915 during the First World War when the **A**ustralian and **N**ew **Z**ealand **A**rmy **C**orps (ANZAC) set out for the Gallipoli peninsula. ANZAC is always pronounced as a word, so much so that Anzac Day is often written using lower case. Popular internet acronyms include things like LOL and FOMO.

Acronyms can also be abbreviations that use the first letters of the words you want to remember to form a unit that is not pronounceable, such as VCE (Victorian Certificate of Education) or NBN (National Broadband Network). Technically, these are **initialisms**, but can be included under a broad definition of acronyms.

Some abbreviations can cross over the two forms. For example, Australia's national science agency, CSIRO (Commonwealth Scientific and Industrial Research Organisation) is sometimes an acronym, pronounced as a word ('sigh-roe'), and is sometimes said by spelling out the letters as an initialism: 'C-S-I-R-O'.

Acronyms are also excellent for remembering a sequence of actions. For example, you can remember the first-aid procedure with ABC – first clear the **a**irway, then check **b**reathing and then perform **C**PR (cardiopulmonary resuscitation).

Acrostics

Unlike acronyms, which use the actual words to be remembered in the memory aid, **acrostics** use different words to help remember some kind of sequence. Acrostics use the first letter of each word of the to-be-remembered information to create memorable phrases or sentences from words that begin with those letters. Acrostics are traditionally known from poems where the first letter of each line spells out a word. A famous acrostic poem was written by Lewis Carroll in *Through the Looking-Glass*. The whole poem is an acrostic for the real Alice's name: Alice Pleasance Liddell. It starts:

> **A** boat, beneath a sunny sky
> **L**ingering onward dreamily
> **I**n an evening of July –
> **C**hildren three that nestle near,
> **E**ager eye and willing ear,
> **P**leased a simple tale to hear –

Acrostic poems and texts were originally designed to help memory by relating the acrostic to the theme of the text. The embedded word or phrase provides a set of cues to recall the first word of each line, focusing on the meaning of the text to produce deeper encoding. The definition of acrostic includes mnemonic phrases. In these acrostics, the first letter of each word in a catchy phrase or rhyme matches the first letter of each word in a sequence you want to remember.

For example, the phrase *kids pick candy over fancy green salads* is an acrostic for the original classification of living organisms: *kingdom, phylum, class, order, family, genus, species*. The modern classification includes **l**ife and **d**omain in front of **k**ingdom. It would be easy to add two words to the front, such as **L**ike **d**ogs, *kids pick candy over fancy green salads*.

Another commonly used acrostic is for the order of the planets from the Sun (Figure 5.27). The planets are *Mercury, Venus, Earth, Mars, Jupiter, Saturn, Uranus* and *Neptune*. A commonly used acrostic used to be *My Very Educated Mother Just Served Us Nine Pizzas*. In 2006, Pluto was reclassified as a dwarf planet, and so the Pizzas had to go. Maybe she could serve us **N**oodles or **N**achos or even **N**othing.

The method of loci

5.5.2 MNEMONICS: METHOD OF LOCI

Acronyms and acrostics are great memory aids for brief sequences, but you can't encode any further knowledge to the items represented. For example, it would be very difficult to add more information about Mercury, say, to the planet acrostic. For that reason, there are much more sophisticated memory methods.

The most common of the more complex methods is called the **method of loci**. The word 'loci' is the Greek for locations, the singular being 'locus'. As the name implies, the method is dependent on using physical locations to help you remember information. The set of locations is also known as a memory palace, mental walk, mind palace, memory space, memory trail and the journey method.

Figure 5.27 The solar system with the planets and dwarf planets – a good target for creating an acrostic to remember the planets in order.

Each of these titles are describing the same thing: a set of locations that you can mentally or physically walk through and recall a great deal of information in some kind of structured order.

Imagine using your home as a memory palace for the planets instead of the acrostic. At the front door, you place the first planet, Mercury. You might imagine the Greek god (Mercury) (Figure 5.28) is pouring silvery mercury down on you as you enter. The more ridiculous, bizarre, vulgar, violent and vivid the scenes and characters you use, the better you will make the connection.

Figure 5.28 The Greek God Mercury, who you might imagine at your first locus for the method of loci approach to remembering information about the solar system.

To add the other planets, you would then move around the house from location to location in order, placing some memory aid for each planet at each position. To recall them, you would imagine yourself walking around the house and remembering the image you have created at each place.

The beauty of the method is that you can add more information to each location. For example, to add that Mercury is the smallest planet, you could mentally reduce the size of the door to a tiny space for you to squeeze through as Mercury pours the mercury all over you.

It is amazing how effective this method can be and how fast you will learn to use it. You can adapt the method of loci to any sort of information that you want to remember during your formal education and throughout life.

The method of loci and the human nervous system

In this exercise we will use your home as a memory palace to recall the structure of the nervous system as described in Chapter 2. Look back at Figure 2.4: the divisions of the nervous system.

There are two main divisions of the nervous system: the central nervous system (CNS) and the peripheral nervous system (PNS). Let's put the CNS in your living room, thinking of it as *central* to home life. We'll put the PNS in the kitchen, because it is *peripheral*, somewhere on the side of the living room. Although sample rooms are shown in Figures 5.29 and 5.30, you should use your own home.

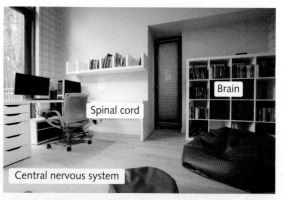

Figure 5.29 Example of how you could use locations in your living room to encode information about the central nervous system using the method of loci. The living room is the location of the central nervous system in the house. The bookshelf is the brain. The power cord to the computer is the spinal cord.

Next, we need to encode the two major components of the CNS – the brain and the spinal cord – in the living room. What will you use for the brain? The first thing that comes to mind will usually be the most effective. Let's say you chose a bookcase for the brain because it contains a lot of information like your brain.

For the spinal cord, think about what device in your living room has a cord attached. That location can be the starting point for everything you want to remember about the spinal cord.

Let's move to the kitchen for the PNS. There are two key subdivisions – the somatic nervous system (SNS) and the autonomic nervous system (ANS). We need two places in the kitchen to represent these systems.

The SNS is the voluntary part of the PNS – so make this a place you go to voluntarily. The refrigerator would be perfect to represent the SNS. What is in there? Is there *some* ice-cream and *some* milk and *some* fruit juice and *some* butter and *some*…? Think of all the things there is *some* of in your refrigerator and you have the link to 'somatic' through a pun on the sound of 'some'.

We'll put the ANS around the stove. Autonomic? Auto? Car. Nomic? Name. Let's link the image you create to a car name. What is cooking on the stove? Is it a (Volkswagen) 'beetle' or a (Ford) 'falcon' or even a Jaguar? Choose any car name and start cooking it. That gives you *auto* and *nom*.

You now need to encode the three subdivisions of the ANS. Choose three aspects of the stove for the sympathetic, enteric and parasympathetic nervous systems. Is there a *sympathetic* person at the stove? Is something *enter*ing the oven? Whatever occurs to you first will probably be the best mnemonic for you. Everyone will be different.

By visualising the living room and kitchen, you should be able to recall all the parts of Figure 2.4. You can now add any amount of information to your memory palace by using stories, puns and images. The more you imagine vivid stories and act them out, the more the information will stick. You can add layer upon layer of information over time. The basis of the memory palace will be there for you forever.

You can use a memory palace to memorise anything that can be organised into some kind of structure.

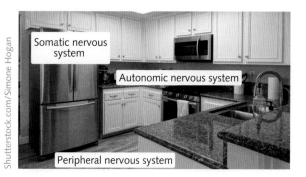

Figure 5.30 Example of how you could use locations in your kitchen to encode information about the peripheral nervous system using the method of loci.

ACTIVITY 5.4 METHOD OF LOCI

Try it
The method of loci is a mnemonic technique in which items that are to be remembered are visualised and placed in familiar locations, for example your bedroom or your street.
1 Use the method of loci to attempt to remember the 10 words below.

   1 Ball
   2 Cup
   3 Pen
   4 Dog
   5 Cat
   6 Wheel
   7 Fork
   8 Phone
   9 Wallet
   10 Shoe

Apply it
2 Would you expect the average person to remember all the words from the list above using only maintenance rehearsal?

Exam ready
3 Steve stops into the supermarket on the way home to buy ingredients for dinner. Steve needs to buy bread, eggs, avocado, mustard and salami.

   As Steve walks around the supermarket, he repeats the word BEAMS. What mnemonic technique is Steve using?

A Method of loci, as Steve is walking around the supermarket to get the items from particular locations

B Acrostic, as Steve has taken the first letter of each dinner item to make a single word

C Acronym, as Steve is walking around the supermarket to get the items from particular locations

D Acronym, as Steve has taken the first letter of each dinner item to make a single word.

The history of the method of loci

The method of loci is usually attributed to an ancient Greek, Simonides of Ceos, who lived around 500 BCE. Simonides was chanting a poem as the dinner entertainment at the home of a wealthy nobleman when he was called outside. The roof of the dining hall collapsed, killing everyone inside. To identify the badly crushed bodies for burial, Simonides recalled each name by their position at the table. This gave him the idea of using physical locations to remember information. The method retains not only the information but the order in which it was encoded.

The method of loci became so popular that it was documented by famous Roman orators such as Cicero (106–43 BCE) and Quintillian (35–100 CE). To prepare for their long speeches, the orators would place each point of the speech in a location in a building or streetscape (Figure 5.31). When giving the speech, they would imagine themselves walking their memory palace recalling each point as they went. This method works so well that it is still used by public speakers today.

All contemporary memory champions use the method of loci. In competitions, they temporarily store hundreds of images of cards from a shuffled deck, long lists of numbers and words, names for faces and many other items in their memory palaces. Brain imaging studies show that memory champions engage the hippocampal-parietal lobe network associated with visual-spatial memory and navigation when encoding with the method of loci (Dresler et al., 2017).

Figure 5.31 Cicero giving a long speech from memory

But a much more sophisticated version of the method of loci has been used by indigenous cultures for thousands of years before Simonides and is still used today.

The sung narratives of oral cultures

It is here that I need to respect cultural protocols as I write about Indigenous memory systems. As a white Anglo-Australian woman, I can only understand and experience these methods at a superficial level. Despite this limitation, I am constantly astounded by how effectively I am able to implement what I have learnt from indigenous peoples, both Australian and international. I have worked closely with Australian Aboriginal colleagues, including Margo Neale (Figure 5.32), my co-author on the book *Songlines: the power and promise* (Neale & Kelly, 2020). I live and work on the traditional lands of the DjaDja Wurrung people. I acknowledge the traditional owners and pay my respects to their Elders – past and present.

Without writing, cultures around the world use alternative methods to store the vast amount of information on which their survival depends physically and culturally. Research shows that oral

Figure 5.32 Professor Margo Neale, Senior Indigenous Curator & Principal Advisor to the Director at the National Museum of Australia. Margo is of Aboriginal and Irish descent, from Kulin nation with Gumbayngirr clan connections, and is co-author with Dr Lynne Kelly of Neale, M., & Kelly, L. (2020). **Songlines: The power and promise**. Thames & Hudson Australia.

memory methods date back many thousands of years before the ancient Greek and Roman orators (Kelly, 2015).

In oral cultures, Elders memorise knowledge of all the animals and plants. Their classifications involve hundreds if not thousands of species. They store knowledge of land management and health including a vast array of medicines. They record astronomy, weather, climate, season, geology, genealogies, laws, ethics, expectations, relationships and obligations … the list goes on and on. How could they possibly remember so much information?

You are probably perfectly capable of singing many songs because music makes information so much more memorable. For oral cultures, the songs tell of the animals, plants and all the other genres of information listed above. The songs tell stories handed down by the Ancestors teaching all the knowledge needed for the community. Most, if not all of the songs will be linked to one or more physical locations within on Country.

Oral cultures know their landscape intimately and use it as a vast, sophisticated method of loci. In Australian Aboriginal terms, physical features of the landscape are considered to be sacred places that hold cultural knowledge. The landscape is not just geography. When understood in terms of the encyclopedia encoded there, the landscape is Country. All of the sacred locations are linked by a **sung narrative** creating pathways across Country, as laid down by the ancestors when the land was formed. Because these lines of connection are sung, Europeans called them **Songlines**. Every one of the hundreds of Aboriginal cultures has words for 'Songlines' in their own language.

First Nations cultures around the world used sung narratives to map their knowledge systems. Native American First Nations people call them pilgrimage trails. Pacific Islanders use ceremonial roads. For the Incas, the sung narrative paths are *ceques*. For a thousand years, the Incas ruled a vast empire in South America. The invading Spanish recorded their use of *ceques* before totally destroying the Inca empire in only 50 years. Explore the weblinks to learn how ancient Aboriginal Songlines have been found to record geological and astronomical events that occurred thousands of years ago and how this information has been passed down over thousands of generations.

Weblink
Ancient Aboriginal stories preserve history of a rise in sea level

Aboriginal legends an untapped record of natural history written in the stars

CASE STUDY 5.1

The Seven Sisters Songline

The Seven Sisters Songline is almost certainly the largest sung narrative trail in the world. It crosses the entire Australian continent, covering about 500 000 square kilometres as it tells the story of an Ancestral Being who pursues seven sisters in an attempt to possess them (Figure 5.33). Encoded in this story is an enormous amount of information that is critical for survival on this vast dry continent. During the pursuit, the country was created, leaving significant features which record the plight of the sisters. Sometimes the Songline goes underground. Sometimes it leaps into the sky; astronomy is part of the knowledge being stored.

As with all memorable stories, the pursuit of the seven sisters involves high drama.

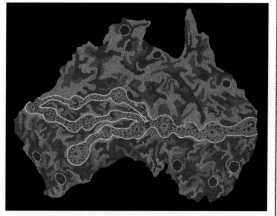

Figure 5.33 *Seven Sisters Songline* 1994 by Josephine Mick, Ninuku Arts, © the artist/Copyright Agency 2020

Margo Neale, Senior Indigenous Curator at the National Museum of Australia, explains:

It is a tale of tragedy and comedy, obsession and trickery, desire and loss, solidarity and sorrow that touches on life's moral dimensions: how to live with each other on this earth in a sustainable way; how to care for each other and share resources equitably. It also instructs on gender relations, kinship, marriage rules and other codes of behaviour. These lessons are embodied in compelling tales of intrigue, drama and passion that connect people and places across time. In this way, the story has been easily remembered and willingly retold to each generation for millennia. It is a saga of mythological dimensions and meanings.

Source: Neale and Kelly, 2020, p. viii

Songlines are unique and sacred Aboriginal knowledge systems that are deeply connected to Country – they cannot be reduced to a collection of simple mnemonic techniques. These distinctly Aboriginal Australian ways of knowing offer great insights to Western psychobiological understandings of memory. For example, the power of the Method of Loci can be increased by incorporating lessons from Songlines. Try the following exercise to see how linking a sung narrative to a series of locations along a journey can help you remember information.

The history of psychology as a sung narrative

You can set up a sung narrative in the form of a timeline for the history of psychology by starting with just a few locations. You can then continually add new events and people or more detail to those you have already encoded. The more vivid your story or song and your actions and interactions at the location, the more memorable you will find it.

1 Choose a portion of the school grounds.
2 Aboriginal people always learn with their community, so try this when setting up your Songline. Working with a group is very effective. You will find the interaction with others will make the knowledge more memorable.
3 The start of your sung narrative is to represent 1900. Mark out every ten years until 2020 by finding something significant – a tree, building, corner, door, gate … whatever stands out. Make each of your locations at least 10 metres apart. That allows plenty of room to add more events in each decade.
4 At about the location of 1906, encode Ivan Pavlov publishing his findings on classical conditioning.
5 At 1920, add Watson and Rayner publishing their research on the classical conditioning of fear with their subject, Little Albert.
6 At 1957, add Brenda Milner, Patient H.M. and the role of the hippocampus in explicit memory.
7 At 1968, add Atkinson and Shiffrin's multi-store model of memory.
8 For 2008, add the death of Patient H.M., who left his brain to science.

Once you have used your sung narrative for a while, you will find that you become very attached to it. You are just glimpsing the extraordinary memory techniques used by Aboriginal people.

Sung narratives are reflected in many other mnemonic devices used by Indigenous cultures. Aboriginal artworks, for example, encode knowledge from sung narratives, often making reference to Country. Many cultures around the world create miniature versions of their storied landscapes, encoding sung knowledge to handheld devices. The African lukasa is a memory device of the Luba people, indigenous to the Democratic Republic of the Congo (Figure 5.34). Together, the landscape, songs, stories, art and cultural objects form the 'library' of oral cultures.

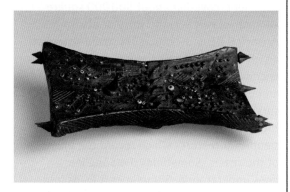

Luba. Lukasa Memory Board, late 19th or early 20th century. Wood, metal, beads, 10 × 5 3/4 × 2 1/4 in. (25.4 × 14.6 × 5.7 cm). Brooklyn Museum, Gift of Marcia and John Friede, 76.20.4. Creative Commons-BY (Photo: Brooklyn Museum, 76.20.4_view2_PS9.jpg)

Figure 5.34 Lukasa memory board of the African Luba people in the Brooklyn Museum, USA

Final reflections

Why would anyone want to use these ancient memory methods today when we have everything at our fingertips on our devices? Isn't using memory systems a lot more work because you need to remember more information – the system and the information you are encoding?

The reason mnemonic methods have been used throughout time and are still used by all memory champions today is that they are far more efficient and effective than rote learning. It only seems like more work at first. Because the knowledge is more firmly grounded in the system, you need less repetition and it will stay there much more firmly over time.

Creativity requires seeing knowledge in new ways, making new connections. When we search online for something, we can only find the knowledge the way someone else has interpreted it. The more connections you can make between things you know and the things you want to learn, the greater the depth of understanding you will gain.

5.5.3 RESEARCH INTO MEMORY

KEY CONCEPTS 5.5

» Mnemonics are techniques you can use to improve encoding, storage and retrieval of explicit long-term memories.
» They work through elaborative, multi-modal methods of encoding, which strengthen and organise the information stored in memory, and create effective cues for retrieval.
» Mnemonics based on written language include acronyms and acrostics. Both methods use the first letter of words as memory cues, but in different ways.
» The method of loci was described by the ancient Greeks. It uses sequences of locations on a familiar route or in a building as cues for memory.
» The method of loci has its origins in the much older traditions of oral cultures who used the landscape with song, stories, dance and art to encode their knowledge.
» Aboriginal Australians have practised oral memory techniques for tens of thousands of years. The Songlines of Aboriginal peoples encode knowledge in sung narratives (stories) and performance (e.g. dance, ritual), and these are associated with locations in the landscape and night-sky.
» Oral memory can also be encoded onto portable objects that serve as retrieval cues for knowledge.

Concept questions 5.5

Remembering
1 What do acronyms and initialisms have in common and how do they differ? Use examples to explain your answer. e
2 What did the ancient Greek orators perform using the method of loci? r
3 What is the name of the Songline that crosses the breadth of Australia? What is the basis of the story it encodes? r

Understanding
4 What are the acronyms for 'self-contained underwater breathing apparatus', 'radio detection and ranging' and 'National Aeronautics and Space Administration'? c
5 What is a limitation of acronyms and acrostics, and how can that be solved by using the method of loci? Explain using the 'salads' in the acrostic 'Like dogs, kids pick candy over fancy green salads', as your example. c
6 Socrates, who did not use writing, was quoted by Plato as saying about writing:
'For this invention will produce forgetfulness in the minds of those who learn to use it, because they will not practise their memory.'
How does the use of Songlines in Aboriginal cultures demonstrate a different approach to remembering knowledge from that taught in Western education? c

Applying
7 In your living room, you have a location for the spinal cord. It is often described as 'the body's information highway'. Revise how the

spinal cord acts as a communication highway for the brain. Then, add to your memory palace for the nervous system by encoding more information about the function of the spinal cord. Act out the way messages are sent from the brain through the spinal cord to the peripheral nervous system (PNS). Describe what you did and whether it helped you remember the information. d

HOT Challenge

8 Some memory champions experience aphantasia, yet they still claim to use the method of loci as their preferred mnemonic technique. What does this suggest to you about the role of mental imagery in the method of loci? e

5 Chapter summary

KEY CONCEPTS 5.1

- » Atkinson and Shiffrin's multi-store model of memory consists of three memory stores: sensory memory, short-term memory (STM) and long-term memory (LTM).
- » Information is maintained and transferred between the stores using three processes: encoding, storage and retrieval.
- » Sensory memory registers information from our senses and holds it briefly so we can attend to it. The capacity is large, but duration is brief. Visual sensory memory persists as an iconic trace for approximately half a second. Auditory sensory memory persists as an echoic trace for up to 4 seconds.
- » Short-term memory (STM) provides a temporary store for information that is the current focus of attention. It integrates attended information from sensory memory with information retrieved from LTM.
- » STM capacity is seven plus-or-minus two units when measured using digit-span. Capacity can be increased through chunking multiple items of information into larger meaningful units.
- » STM duration is limited by our attention span and by the ability to mentally rehearse information. Without maintenance rehearsal, STM traces decay rapidly within 15–30 seconds.
- » Transfer of information from STM to LTM occurs through encoding processes, including maintenance rehearsal and elaborative encoding.
- » Long-term memory (LTM) has unlimited capacity and its duration is permanent.
- » The depth of encoding in STM affects the strength and organisation of long-term memories. Elaborative encoding and retrieval-practice are two deep encoding strategies that enhance storage and retrieval of long-term memories.
- » A strength of the multi-store model is its ability to explain serial position effects of primacy and recency.
- » A weakness of the multi-store model is its failure to explain different forms of long-term memory and their basis in the brain and neural processes.

KEY CONCEPT BOX 5.2

- » Long-term memory (LTM) includes explicit and implicit memory systems.
- » Explicit memory is the kind of LTM that allows conscious recall of experiences and facts. It includes episodic-autobiographical memory (EAM) and semantic memory.
- » EAM is the kind of explicit memory that allows mental time-travel. It includes retrieval of personally experienced events from a first-person perspective, with details of time, place, perceptions, thoughts and emotions. It also includes the ability to imagine ourselves in the future from a first-person perspective, known as episodic future-thinking.
- » Semantic memory stores our general knowledge of words, concepts and facts. It includes semantic autobiographical knowledge.
- » Semantic networks store our knowledge of concepts and the words that represent them. Schemas store our knowledge of familiar situations and spatial relationships between concepts.
- » Anterograde amnesia and retrograde amnesia affect the ability to store and retrieve explicit memories, respectively.
- » Implicit memory is the kind of LTM that allows us to perform skills and respond to stimuli adaptively without conscious recollection of the knowledge that underlies these abilities. It includes procedural memory and conditioning.
- » Procedural memory is the kind of implicit memory that is demonstrated through skilled performance of a sequence of actions due to practice and repetition. It includes physical and cognitive skills.

KEY CONCEPT BOX 5.3

- » The brain regions that encode, store and retrieve explicit memories are the hippocampus, amygdala and neocortex.
- » The hippocampus is located deep within each temporal lobe (in the allocortex). It has two-way (feedback) connections with the neocortex. Its role is to bind and consolidate episodic memories in the neocortex. It is also required for retrieval of episodic memories.
- » The amygdala is located in front of the hippocampus. Its role is to process emotional information. Implicit emotional information is integrated with explicit episodic memories through connections between the amygdala and hippocampus.
- » The neocortex is the largest region of the cortex, including 80 per cent of the grey matter of each of the four lobes of the brain (excluding the allocortex).
- » The prefrontal cortex coordinates attention, thought and language during encoding and retrieval of memories.
- » Neocortical regions behind the prefrontal cortex process sensory-motor information during perception and action and store semantic and episodic memories.
- » The temporal lobes of the neocortex contain two semantic hubs (anterior and posterior) that store semantic knowledge that has been generalised from episodic memories.
- » Neural representations of words, concepts and schemas in the semantic hubs are connected with sensory-motor and perceptual representations throughout the neocortex. These connections allow us to imagine the perceptual features of words and concepts.
- » The basal ganglia and cerebellum are the brain structures involved in encoding, storing and retrieving implicit memories. These regions can function independently of the hippocampus.
- » The basal ganglia are a set of subcortical structures that coordinate the smooth performance of sequences of actions involved in procedural memory.
- » Loss of dopamine-producing cells in the basal ganglia causes problems with initiating movement and inhibiting unwanted movement, as seen in the motor symptoms of Parkinson's disease.
- » The cerebellum is the major structure of the hindbrain with connections to the basal ganglia and prefrontal cortex. It coordinates fine motor movements and balance when performing skilled actions. It also plays a role in procedural memory of cognitive skills such as reading.

KEY CONCEPT BOX 5.4a

- » Autobiographical memory (ABM) is the component of explicit memory that stores knowledge of personally experienced events and self-knowledge.
- » ABM includes episodic-autobiographical memory (EAM) and semantic autobiographical memory (SAM).
- » EAM provides the first-person experiential component of ABM. SAM provides our self-knowledge.
- » SAM provides an organising 'scaffold' for retrieving EAMs. In contrast, general semantic knowledge is thought to scaffold our ability to imagine ourselves in future events.
- » Studies of ABM show that older people tend to report more 'semanticised' ABMs.
- » The relationship between ABM retrieval and episodic future-thinking is seen in fMRI studies showing shared hippocampal-neocortical networks for both tasks.
- » The relative contributions of EAM and SAM components to ABM vary between people along a continuum. People with highly superior ABM (HSAM) recall rich episodic details for each day of their life. People with severely deficient ABM (SDAM) recall only semantic knowledge about their lives.
- » Individual differences in ABM-style are reflected in different patterns of hippocampal connectivity with the neocortex. HSAM people show greater connectivity with parietal and occipital regions associated with mental imagery. SDAM people show greater connectivity with the frontal temporal lobe semantic hub (language).

KEY CONCEPT BOX 5.4b

- » Researchers have found that individual differences in ABM-style are related to individual differences in the ability to generate mental imagery.
- » Mental imagery refers to perception-like experiences in the absence of the relevant sensory input.
- » People who can generate vivid mental images of objects, people and scenes (object visual imagery) also experience more vivid first-person EAMs during ABM retrieval.
- » People with aphantasia experience an absence of voluntary mental imagery. They also report having severely deficient ABM and are unable to imagine themselves in the future. However, they have strengths in non-visual spatial imagery.
- » These findings support theories that propose a strong role for visual mental imagery in retrieving ABMs.

KEY CONCEPT BOX 5.4c

- » Alzheimer's disease (AD) is a progressive and fatal neurodegenerative disease causing symptoms of dementia, especially explicit memory loss.
- » AD is associated with abnormal levels of beta-amyloid and tau proteins in the brain that form amyloid plaques and neurofibrillary tangles, respectively.
- » The first symptom of AD is anterograde amnesia for episodic-autobiographical memories and semantic knowledge. People are unable to remember recent personally experienced events, nor facts associated with those events.
- » Early symptoms also include difficulty with episodic future-thinking and generating voluntary mental imagery.
- » These problems are consistent with damage to the hippocampus in the early stages of AD, which affects binding and consolidation of explicit memories.
- » As the disease progresses, people experience retrograde amnesia for the events of their lives, with more recent memories being lost first. Remote memories from early in the reminiscence bump are the last to be affected, as they are more semanticised.
- » Semantic memory deficits follow as areas of temporal, parietal, and occipital lobes become progressively affected.
- » Post-mortem lesion studies confirm diagnosis, identifying the characteristic plaques, tangles and atrophy.
- » Functional MRI brain-imaging studies aim to detect changes in brain activity associated with early episodic memory decline, before significant structural changes are evident.
- » These studies reveal abnormal activity in the hippocampus and its connections with the neocortex during episodic memory tasks and offer hope for early detection and treatment.

KEY CONCEPT BOX 5.5

- » Mnemonics are techniques you can use to improve encoding, storage and retrieval of explicit long-term memories.
- » They work through elaborative, multi-modal methods of encoding, which strengthen and organise the information stored in memory, and create effective cues for retrieval.
- » Mnemonics based on written language include acronyms and acrostics. Both methods use the first letter of words as memory cues, but in different ways.
- » The method of loci was described by the ancient Greeks. It uses sequences of locations on a familiar route or in a building as cues for memory.
- » The method of loci has its origins in the much older traditions of oral cultures who used the landscape with song, stories, dance and art to encode their knowledge.
- » Aboriginal Australians have practised oral memory techniques for tens of thousands of years. The Songlines of Aboriginal peoples encode knowledge in sung narratives (stories) and performance (e.g. dance, ritual), and these are associated with locations in the landscape and night-sky.
- » Oral memory can also be encoded onto portable objects that serve as retrieval cues for knowledge.

5 End-of-chapter exam

Section A: Multiple choice

1. Which option correctly lists the three memory storage systems of the multi-store model?
   A encoding, storage, retrieval
   B semantic, episodic, autobiographical
   C sensory memory, short-term memory, long-term memory
   D implicit memory, procedural memory, explicit memory

2. Which option correctly lists the three memory processes in the multi-store model?
   A binding, consolidation, reconsolidation
   B encoding, storage, retrieval
   C maintenance rehearsal, elaborative rehearsal, retrieval-practice
   D recall, recognition, cued recall

3. Which option correctly states the capacity and duration of visual sensory memory?
   A large capacity; half-a-second duration
   B three plus-or-minus one units; 3–4 seconds
   C large capacity; 3–4 seconds
   D seven plus-or-minus two units; 15–30 seconds

4. Which option correctly states the capacity of verbal STM?
   A three plus-or-minus one unit
   B seven plus-or-minus two units
   C unlimited
   D none of the above

5. How would the multi-store model explain your failure to remember the name of someone you have just been introduced to and have been chatting with for 30 seconds?
   A Forgetting occurs due to lack of rehearsal and interference from the events that follow hearing their name.
   B Forgetting occurs due to a lack of retrieval cues.
   C Forgetting occurs due to trace decay after 30 seconds.
   D Forgetting occurs due to a failure to encode the name.

6. Asking yourself questions about how one fact relates to another fact is an example of what kind of encoding?
   A retrieval-practice
   B elaborative encoding
   C maintenance rehearsal
   D consolidation

7. An experimenter has a hypothesis that the recency effect occurs in tests of immediate free recall because items at the end of the studied list are still available in sensory and STM, rather than being in LTM. Which experiment and finding would allow you to determine if this is true?
   A Introduce an unfilled 30-second retention interval after the last studied item to find if the recency effect remains.
   B Prevent participants from rehearsing items during encoding by having them repeat a word aloud during encoding and find that the primacy effect is lost.
   C Reproduce the experiment in multiple different labs and see if you get the same result.
   D Introduce a 30-second filled retention interval after the last studied item to find if the effect of recency is lost.

8. Which one of the following is a weakness of the multi-store model of memory?
   A It does not distinguish between different kinds of LTM.
   B It does not explain the neural mechanisms involved in memory.
   C It does not explain how we maintain and work with visual-spatial information in STM.
   D All of the above.

9 Which one of the following statements is correct about episodic (episodic-autobiographical) memory?
    A It is the kind of long-term explicit memory that allows us to consciously retrieve factual information about ourselves and the events we have experienced.
    B It is the kind of long-term explicit memory that allows us to consciously re-experience the events of our lives from a first-person perspective.
    C It is the kind of long-term explicit memory that allows us to consciously recall the events of our lives.
    D It is the kind of long-term explicit memory that allows us to imagine what the outcome of an upcoming election might be.

10 Which one of the statements below is an example of episodic-autobiographical memory?
    A My birthday will be on a Saturday this year.
    B I had a lovely holiday in Bali last month.
    C I can still see how her face looked when we surprised her.
    D All of the above.

11 Which one of the following statements is NOT true about implicit memory?
    A We cannot consciously recall how we learned implicit memories.
    B Implicit memory reveals itself in skilled behaviour without conscious awareness of the knowledge that underlies the performance.
    C Implicit memories can be encoded, stored and retrieved even when both hippocampi have been surgically removed.
    D Procedural memory is a form of implicit memory.

12 Which brain region coordinates fine motor control?
    A basal ganglia
    B cerebellum
    C hippocampus
    D prefrontal cortex

13 Which one of the following statements is correct about the role of the hippocampus in memory?
    A Signals from the hippocampus to the neocortex strengthen connections between neurons active during an episode through long-term potentiation.
    B The hippocampus sends feedback about the associations between elements of an episode to the neocortex where it is consolidated.
    C The hippocampus binds neocortical sources of information associated with an episode into integrated memory traces.
    D All of the above.

14 Which one of the following statements is correct about how explicit semantic memories are stored in the brain?
    A Semantic memory is stored in networks of connections between two semantic hubs in the temporal lobes and regions of the neocortex that represent the sensory-motor and perceptual features of concepts.
    B Semantic memory is stored in a network of connections between the hippocampus and regions of the neocortex involved in sensation, perception and action.
    C Semantic memory is stored in connections that link the cerebellum with the basal ganglia and prefrontal cortex.
    D All of the above.

15 Which statement is true about the role of episodic memory in autobiographical memory (ABM)?
    A Episodic memory and autobiographical memory are different terms for the same kind of memory.
    B Episodic memory provides an organising framework, or 'scaffold' for autobiographical memory.
    C Episodic memory is a component of autobiographical memory that enables first-person remembering of events.
    D Episodic memory operates independently of autobiographical memory.

16 Which one of the following statements is correct about the relationship between episodic future-thinking and retrieval of autobiographical memories?
    A Episodic future-thinking and retrieval of autobiographical memories are carried out by overlapping hippocampal-neocortical networks in the brain.
    B Episodic future-thinking and retrieval of autobiographical memories are both scaffolded by semantic autobiographical memory.
    C Episodic future-thinking and retrieval of autobiographical memories are both associated with a first-person experience of mental time-travel.
    D All of the above.

17 What is the relationship between aphantasia and autobiographical memory?
   A Aphantasia is associated with the experience of highly superior autobiographical memory (HSAM); aphantasics can only retrieve episodic-autobiographical memories.
   B Aphantasia is associated with the experience of highly superior autobiographical memory (HSAM); aphantasics can only retrieve semantic-autobiographical memories.
   C Aphantasia is associated with the experience of severely deficient autobiographical memory (SDAM); aphantasics can only retrieve semantic-autobiographical memories.
   D Aphantasia is associated with the experience of severely deficient autobiographical memory (SDAM); aphantasics can only retrieve episodic-autobiographical memories.

18 Which statement is true about the relationship between Alzheimer's disease and autobiographical memory?
   A People in the early stages of Alzheimer's disease are unable to consolidate episodic-autobiographical memories but remain able to consolidate semantic autobiographical memories.
   B People in the most advanced stage of Alzheimer's disease can still retrieve semanticised autobiographical memories.
   C People in the early stages of Alzheimer's disease are unable to consolidate episodic-autobiographical memories and semantic autobiographical memories.
   D As Alzheimer's disease progresses from the early stage, people lose the ability to retrieve their most remote episodic-autobiographical memories before their more recent ones.

19 How is mental imagery affected in Alzheimer's disease?
   A The ability to generate mental imagery is preserved in the early stages of Alzheimer's disease.
   B The ability to generate mental imagery in Alzheimer's disease is only affected once the occipital lobes begin to atrophy.
   C Involuntary mental imagery associated with hallucinations occurs in the early stages of Alzheimer's disease.
   D The ability to generate mental imagery during episodic future-thinking is affected early in Alzheimer's disease.

20 Which option correctly shows an example of an acronym and an acrostic?

	Acronym	Acrostic
A	Every good boy deserves fruit.	**A** boat, beneath a sunny sky **L**ingering onward dreamily **I**n an evening of July – **C**hildren three that nestle near, **E**ager eye and willing ear,
B	ANZAC	**E**very **g**ood **b**oy **d**eserves **f**ruit.
C	**A** boat, beneath a sunny sky **L**ingering onward dreamily **I**n an evening of July – **C**hildren three that nestle near, **E**ager eye and willing ear,	ANZAC
D	ANZAC	CSIRO

Section B: Short answer

1 a Describe how the multi-store model of memory explains the transfer of information from short-term memory (STM) to long-term memory (LTM). [2 marks]
   b With reference to serial position effects, explain the evidence that supports the multi-store model account of transfer from STM to LTM. [3 marks]
   [Total = 5 marks]

2 Eliza is looking forward to attending her school formal at the end of year 12. Describe how semantic memory is involved in Eliza imagining herself attending the formal. [3 marks]

3 Describe two ways in which Australian Aboriginal peoples' use of Songlines is different from the method of loci. [2 marks]

Unit 3, Area of Study 2 review

Section A: Multiple choice

Question 1 ©VCAA 2021 Q10 ADAPTED MEDIUM
Which of the following is a similarity between classical conditioning and operant conditioning?
A Both require active attention.
B Both consist of a three-phase model, which can be abbreviated to ABC.
C Both are forms of learning through associations between stimuli.
D Both involve reinforcers and punishers to associate two stimuli together.

Use the following information to answer Questions 2 and 3.
Oliver, aged five, adores his big brother Luca, aged ten. Luca recently learned how to play the piano.

Question 2 ©VCAA 2021 Q11 ADAPTED MEDIUM
Oliver is likely to learn to play the piano due to which stage of observational learning and associated reason?

	Stage of observational learning	Associated reason
A	attention	Oliver is actively watching his older brother play the piano.
B	retention	Oliver is able to complete the action of playing the piano.
C	motivation	Oliver makes a mental representation of playing the piano.
D	reproduction	Oliver is being reinforced to play through praise of Luca.

Question 3 ©VCAA 2021 Q12 ADAPTED MEDIUM
For Luca, learning how to play the piano
A developed neural changes in his cerebellum.
B caused a reflexive response in his nervous system.
C was a conscious response that involved his somatic nervous system.
D led to the frequent activation of sensory neurons, followed by motor neurons.

Question 4 ©VCAA 2021 Q13 ADAPTED EASY
The use of Songlines in Aboriginal cultures is an example of
A the ability of written cultures to explain their surrounds.
B how oral cultures sing their experiences.
C how oral cultures use song to better store and retrieve information.
D how Aboriginal cultures share their experiences.

Use the following information to answer Questions 5–7.
Doug is speaking to a friend who he has not seen in a while. They exchange phone numbers and Doug tries to memorise it without writing it down.

Question 5 ©VCAA 2021 Q14 ADAPTED MEDIUM
According to the Atkinson-Shiffrin multi-store model of memory, how many numbers will Doug be able to store in his short-term memory?
A between 15 and 30
B between 5 and 9
C between 7 and 9
D between 9 and 12

Question 6 ©VCAA 2021 Q15 ADAPTED MEDIUM
Which of the following methods is most likely to assist Doug with keeping the phone number in his short-term memory long enough to add to his phone and why?

	Method of transferring to short-term memory	Why
A	Repeat the number to himself out loud	The repeated auditory exposure will help keep it in his working memory.
B	Try to have a nap	The consolidation process is aided by sleep.
C	Relearn the number in the car where he first heard it	The environment in the car will act as a memory cue.
D	Use reconstruction	The order of the numbers will assist with recall.

Question 7 ©VCAA 2021 Q16 ADAPTED MEDIUM

If Doug was to commit this number to long-term memory, what region of the brain will this memory be temporarily stored in?

A amygdala
B neocortex
C basal ganglia
D hippocampus

Question 8 ©VCAA 2002 EXAM 2 Q6 ADAPTED MEDIUM

James conducted an experiment in which there were two groups of 30 people. Group 1 was given a list of nonsense words to remember, whereas Group 2 was asked to remember a group of real words. The results showed that the participants in Group 2 recalled more because they were able to group individual items together on the basis of some shared characteristic. This finding provides support for the importance of

A rehearsing information to ensure it has been learned properly.
B chunking information to retain it in long-term memory.
C organising information during encoding to help with later retrieval.
D the recency effect.

Question 9 ©VCAA 2021 Q29 ADAPTED HARD

Olga is playing netball. At every game, her mum buys her a bottle of her favourite energy drink. Now each time she enters the netball stadium she feels thirsty.

Which type of conditioning is this likely a result of and why?

	Type of conditioning	Why
A	operant	Olga is being rewarded with her favourite drink, encouraging her to play.
B	operant	Olga feels excited to play netball.
C	classical	Olga is thirsty as a result of associating the energy drink with the game.
D	classical	Olga has been conditioned to want to play netball due to getting her favourite drink.

Question 10 ©VCAA 2020 Q12 ADAPTED HARD

Khoi is teaching her dog Mia to sit when she instructs her to do so. She rewards Mia by giving her a treat every time she sits when Khoi says the word 'sit'. According to the three-phase process of operant conditioning, which is the antecedent in this situation?

A sitting when given a treat
B the treat
C sitting in response to the word 'sit'
D the word 'sit'

Question 11 ©VCAA 2020 Q13 ADAPTED HARD

Joel was about to sit on a bench at the beach when he noticed that he felt happy and had been unconsciously smiling. He realised that this feeling was probably associated with his girlfriend, whom he often brings to this bench to watch the sunset. This is because of

A classical conditioning, as he has associated his girlfriend with the positive feelings of the beach.
B operant conditioning, as he has been rewarded with quality time with his girlfriend, making him more likely to go to the beach.
C observational learning, as he has viewed other happy couples and learned that the beach is a happy place for him and his girlfriend.
D classical conditioning, because of the feeling of happiness when he is with his girlfriend, which he has associated with the bench on the beach.

Use the following information to answer Questions 12 and 13.

Elton is constantly leaving his room a mess and his mum Linda has tried everything to get him to keep it clean. In one last attempt to get Elton to clean his room she tells him that he cannot go to the movies on the weekend with his friends unless it is cleaned.

Question 12 ©VCAA 2020 Q14 ADAPTED MEDIUM

If Elton does not clean his room and cannot go out with his friends, which principle of operant conditioning is being demonstrated?

A Positive reinforcement, because Elton is upset by not being able to see his friends.
B Negative reinforcement, because of the threat of not being able to see friends.
C Negative punishment, because something Elton was excited to do was taken away from him.
D Positive punishment, because he is being forced to stay home when he was looking forward to seeing friends.

Question 13 ©VCAA 2020 Q15 ADAPTED MEDIUM

Which one of the following actions is most likely to help Elton's mum negatively reinforce Elton to clean his room?

A Stops nagging him.
B Ignoring it and not asking him anymore.
C Confiscating his Xbox.
D Leading by example by having her own room clean.

Use the following information to answer Questions 14–17.

Mr Stevens is an experienced psychology teacher. He has been teaching for 30 years but still remembers his first day of teaching so vividly. He can remember the names and faces of his students, the day of the week, the cologne he wore on the day, what the classroom looked like and the weather forecast. He also recalls feeling equally scared and excited at the same time. He remembers thinking he would be late but showing up 1 hour before classes even started. He also remembers thinking about how his life would look in 30 years and pictured himself still teaching.

Question 14 ©VCAA 2020 Q13 ADAPTED EASY

What type of memory is Mr Stevens constructing and in which region was it primarily stored?

A episodic, hippocampus
B semantic, hippocampus
C episodic, neocortex
D semantic, basal ganglia

Question 15 ©VCAA 2020 Q13 ADAPTED HARD

What would occur if Mr Stevens suffered from aphantasia?

A He would be unable to recall the feelings of being stressed/excited.
B His memories of his first day of teaching would eventually be forgotten due to lesions on the brain.
C He would be unable to make a mental representation of possible future scenarios.
D He would be unable to visualise the students' faces and the classroom.

Question 16 ©VCAA 2020 Q13 ADAPTED HARD

In terms of the Atkinson and Shiffrin multi-store model of memory, explain the method of retrieval of Mr Stevens' memories of his first day of teaching as he recalls it years later.

A When Mr Stevens actively pays attention it is retrieved from long-term memory into short-term memory.
B Mr Stevens commits the memory from sensory memory into short-term memory.
C Mr Stevens commits the memory from short-term to long-term storage.
D It is actively thought of directly from long-term memory.

Question 17 ©VCAA 2020 Q13 ADAPTED MEDIUM

Mr Stevens learned to associate the school bell as beginning lunch, ultimately getting hungry each time he heard the bell. Which of the following is true in regard to classical conditioning?

A Eating food is the conditioned response.
B The bell represents the conditioned stimulus.
C Hunger represents the neutral stimulus.
D The bell represents the unconditioned response.

Question 18 ©VCAA 2020 Q22 ADAPTED HARD

Which one of the following statements about memory is incorrect?

A The capacity of short-term memory can be increased by chunking.
B The greatest number of items that can be held in short-term memory is fifteen.
C The duration of short-term memory can be increased by using maintenance rehearsal.
D Short-term memory receives information from both sensory memory and long-term memory.

Question 19 ©VCAA 2020 Q24 ADAPTED HARD

Which one of the following is not a characteristic of Alzheimer's disease?

A The hippocampus is the first area of the brain to be affected.
B It initially affects short-term memory more than long-term memory.
C It is caused by an increase in the level of the neurotransmitter dopamine.
D It involves neurofibrillary tangles, a build-up of abnormal protein within the neurons in the brain.

Question 20 ©VCAA 2019 Q14 ADAPTED HARD

Memory of how to ride a bike will be

A consolidated and stored in the hippocampus.
B processed in the hippocampus and stored in the amygdala.
C consolidated by the hippocampus and stored in the cerebral cortex.
D consolidated by the hippocampus and stored in the cerebellum.

Question 21 ©VCAA 2003 Q 12 EXAM 2 ADAPTED

Susie was involved in a car accident in which she hit her head. As a result she was unable to remember anything about the few days before the accident. Susie is suffering from

A anterograde amnesia.
B Alzheimer's disease.
C retrograde amnesia.
D aphantasia.

Question 22 ©VCAA 2019 Q19 ADAPTED HARD

Renee's teacher, Mrs Louis, wanted to test Renee's memory ability. She showed Renee a series of 11 photographs. After her presentation, Mrs Louis asked Renee to recall what she saw in all of the photographs. Renee could only remember seven of the photographs that she saw. Why is this the case?

A She wasn't paying direct attention to each item, thus the sensory memory was not committed to short-term memory.
B The photos were not all committed to long-term memory because the duration was too short.
C The photos exceeded the duration of sensory memory.
D There were too many photos to remember, thus exceeding the capacity of short-term memory.

Use the following information to answer Questions 23–25.

Kalvin conducted a matched participant experiment because he was interested in testing whether memory triggers would lead to better recall. One group was taught a lesson and given an acronym to remember the main points. The other group was given the same lesson without any acronyms to remember it by. One week after the lesson, each group completed a test on the information.

Question 23 ©VCAA 2019 Q23 ADAPTED MEDIUM

Which one of the following identifies an independent variable in this experiment?

A results on the test
B the lesson being taught
C being exposed to the acronym or not
D the participants' knowledge prior to learning

Question 24 ©VCAA 2019 Q24 ADAPTED HARD

The results that would be expected from this experiment are that

A the group without the acronym will perform better on the test.
B the group exposed to the acronym will perform better on the test.
C both groups will perform the same on the test.
D individuals who do not go home and revise for the test will not get any of the questions right.

Question 25 ©VCAA 2019 Q25 ADAPTED HARD

What would be a potential extraneous variable in this study?

A Amount of time between the learning and the test, as some individuals may have studied.
B Sample size, as it may lead to an unrepresentative data set.
C Conditions being unequal between the two groups.
D Participants were not aware of whether they were in the experimental or control group.

Use the following information to answer Questions 26 and 27.

Layla is two years old and is learning to potty train. Every time she uses the potty her mum rewards her by giving her a treat. Now every time Layla uses the potty she salivates.

Question 26 ©VCAA 2018 Q15 ADAPTED EASY

In terms of classical conditioning, the conditioned stimulus in this scenario is

A Layla seeing her mum carrying the potty.
B Layla salivating when she uses the potty.
C Layla's mum giving her a treat.
D Layla using the potty.

Question 27 ©VCAA 2018 Q16 ADAPTED EASY

Layla's mum using a treat to encourage Layla to use the potty is also an example of operant conditioning. Which of the following types of reinforcement/punishment was used?

A positive punishment
B positive reinforcement
C negative reinforcement
D negative punishment

Use the following information to answer Questions 28 and 29.

Billy paid careful attention to his soccer coach, who was showing the team how to do a new trick. He was very attentive and when he did the trick in soccer practice the following week, he got lots of praise for his skill from his coach, which made him feel proud of himself.

Question 28 ©VCAA 2018 Q16 ADAPTED EASY

In terms of observational learning, which of the following is NOT true?

A Billy paid close attention to the demonstrations his coach gave.
B Billy had the physical ability to replicate the skill.
C Billy's motivator was being able to show off his skills to his teammates.
D Billy was able to make a mental representation of the skill demonstrated by his coach.

Question 29 ©VCAA 2018 Q16 ADAPTED MEDIUM
In terms of operant conditioning the praise received by Billy for showing off his newly learned skill is an example of
A negative punishment.
B positive punishment.
C positive reinforcement.
D negative reinforcement.

Question 30 ©VCAA 2018 Q16 ADAPTED EASY
Which one of the following is NOT an example of a mnemonic device used by written cultures?
A acronyms
B Songlines
C methods of loci
D acrostics

Section B: Short answer

Question 1 ©VCAA 2016 SECTION B Q5 ADAPTED
Mila learned to ride a bike when she was five years old. She is now 30 and has not ridden a bike in 25 years. Despite this, when Mila and her boyfriend decided to ride a bike ride around Amsterdam on their European holiday, it only took a few minutes, and she was riding it as well as when she was a child.
a Identify the specific type of implicit memory that this event is for Mila. Justify your response. [2 marks]
b Identify the specific areas of Mila's brain that stored her memory of how to ride a bike. Explain the role of each area in memory formation. [2 marks]
[Total = 4 marks]

Question 2
Ashita was diagnosed with early-stage Alzheimer's disease. Which region of the brain would be affected first? Justify your response, giving examples of what the first onset of symptoms would be. [3 marks]

Question 3 ©VCAA 2020 SECTION B Q4 ADAPTED
In schools, teachers typically provide students with rewards for doing the right thing. Ms Lewis offers Michael a sticker for doing all of his work on time in class.
a What type of conditioning is used to generate positive emotions towards Michael completing all his work? Give two reasons to justify your response. [3 marks]
b Lucas watches Michael get rewarded for doing his work and the following day also does all of his work and also gets a sticker. Identify two different processes involved in observational learning that helped Lucas to do all of his work. Justify your response for each process. [4 marks]
[Total = 7 marks]

Question 4 ©VCAA 2019 SECTION B Q3 ADAPTED
Allison is trying to encourage people who attend her bowling centre to also purchase a meal. She ensures her café staff always have the oven on and items cooking with a fan to blow the smell around the centre. She also decided to introduce a '2-for-1 deal' so that if you order one meal you get the second one free. She found that when her regulars were coming in, they were mentioning how hungry bowling makes them.
a In terms of classical conditioning, describe how Allison creates an association between the neutral stimulus and the unconditioned stimulus to develop a conditioned response. [3 marks]
b Operant conditioning was also used in this training. Name the antecedent, the subsequent behaviour and the type of consequence in the training sessions. [3 marks]
[Total = 6 marks]

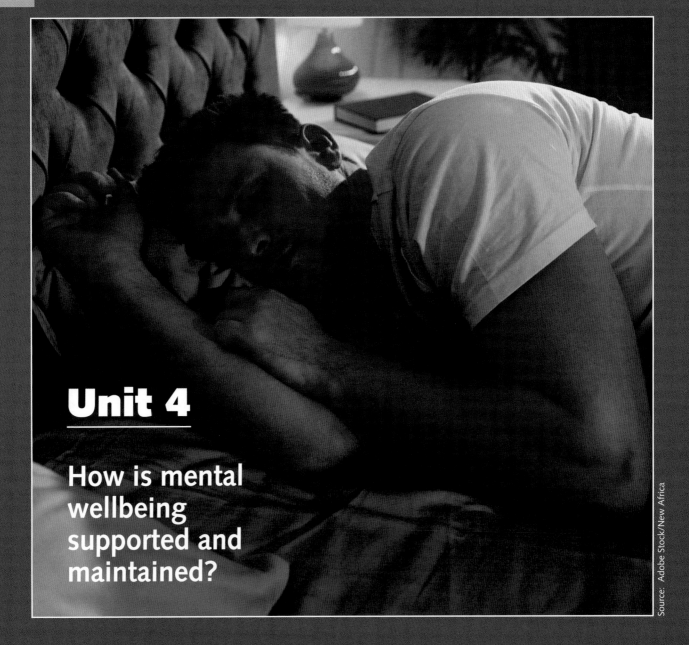

Unit 4

How is mental wellbeing supported and maintained?

Area of study 1: How does sleep affect mental processes and behaviour?

Area of study 2: What influences mental wellbeing?

Area of study 3: How is scientific inquiry used to investigate mental processes and psychological functioning?

The demand for sleep 6

Key knowledge

» sleep as a psychological construct that is broadly categorised as a naturally occurring altered state of consciousness and is further categorised into REM and NREM sleep, and the measurement of physiological responses associated with sleep, through electroencephalography (EEG), electromyography (EMG), electro-oculography (EOG), sleep diaries and video monitoring

» regulation of sleep–wake patterns by internal biological mechanisms, with reference to circadian rhythm, ultradian rhythms of REM and NREM Stages 1–3, the suprachiasmatic nucleus and melatonin

» differences in, and explanations for, the demands for sleep across the life span, with reference to total amount of sleep and changes in a typical pattern of sleep (proportion of REM and NREM)

Key science skills

Develop aims and questions, formulate hypotheses and make predictions
» formulate hypotheses to focus investigation

Comply with safety and ethical guidelines
» demonstrate ethical conduct and apply ethical guidelines when undertaking and reporting investigations

Generate, collate and record data
» systematically generate and record primary data, and collate secondary data, appropriate to the investigation
» record and summarise both qualitative and quantitative data, including use of a logbook as an authentication of generated or collated data
» organise and present data in useful and meaningful ways, including tables, bar charts and line graphs

Analyse and evaluate data and investigation methods
» evaluate investigation methods and possible sources of error or uncertainty, and suggest improvements to increase validity and to reduce uncertainty

Construct evidence-based arguments and draw conclusions
» evaluate data to determine the degree to which the evidence supports the aim of the investigation, and make recommendations, as appropriate, for modifying or extending the investigation
» evaluate data to determine the degree to which the evidence supports or refutes the initial prediction or hypothesis
» use reasoning to construct scientific arguments, and to draw and justify conclusions consistent with evidence base and relevant to the question under investigation
» discuss the implications of research findings and proposals, including appropriateness and application of data to different cultural groups and cultural biases in data and conclusions

Source: VCE Psychology Study Design (2023–2027) pages 39 and 12–13

6 The demand for sleep

We all know what sleep is. Let's face it, we've been doing it since we were born. We are sleep gurus! But, you guessed it, there is a lot more to sleep than closing your eyes and counting sheep.

6.2 Regulation of sleep–wake patterns

p. 258

Do you go to bed at around the same time every night? Do you wake up around the same time every morning? With some variation your body is able to regulate your sleep and wakefulness in two ways –circadian rhythms and sleep–wake homeostasis.

6.1 What is consciousness?

p. 243

Consciousness is your awareness of yourself and your environment. This changes throughout the day and night from total awareness in your Psychology class to almost a complete lack of awareness when you are in a deep sleep. Your sleep is made up of REM and NREM sleep and there are lots of ways each of these can be measured.

Adobe Stock/Jacob Lund

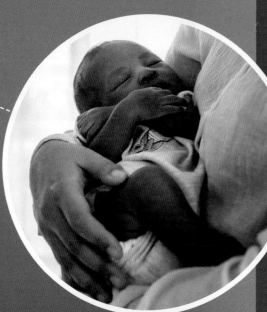

6.3
The changes in sleep over the life span

p. 265

We spend about one third of our life asleep, most of it in NREM sleep. When we are babies we need a lot of sleep but as we age the amount of REM sleep gets less. VCE students need about 9 hours of sleep every night, about 1 hour more than adults need.

Sleep is very important to your mental wellbeing. You need to make sure you get enough sleep so you can function at your optimal level.

Slideshow
Chapter 6 slideshow

Flashcards
Chapter 6 flashcards

Test
Chapter 6 pre-test

Assessment
- Pre-test
- End-of-chapter exam

Revision
- Chapter map
- Key term flashcards
- Key concept summary
- Slideshow

Investigation
- Investigation: Sleep patterns across the life span
- Data calculator
- Logbook template: Controlled experiment

Worksheet
- Are children and adolescents getting enough sleep?

To access these resources, visit
cengage.com.au/nelsonmindtap

Nelson MindTap

Know your key terms

Alpha waves
Altered states of consciousness (ASC)
Beta waves
Biological rhythm
Circadian rhythm
Consciousness
Delta waves
Dreams
Hypnogogic state
K-complex
Melatonin
Non-rapid eye movement (NREM) sleep
Normal waking consciousness (NWC)
Paradoxical sleep
Pineal gland
Rapid eye movement (REM) sleep
Self-report
Sleep
Sleep demand
Sleep diary
Sleep laboratory
Sleep spindles
Slow-wave sleep (SWS)
Suprachiasmatic nucleus (SCN)
Theta waves
Ultradian rhythm
Zeitgeber

Fifteen people have emerged from a cave in south-west France after spending 40 days underground in an experiment called the 'Deep Time Project' (Figure 6.1). The aim of this project was to see how the absence of clocks and daylight with no communications would affect their sense of time.

Scientists at the Human Adaptation Institute say the experiment will help them understand better how people adapt to drastic changes in living conditions and environments.

The group were all volunteers and they lived in and explored the cave for the 40 days without any natural light, the temperature was 10°C and the relative humidity was 100 per cent. They had no contact with the outside world, no updates on the pandemic nor any communications with friends or family. They had to find ways to occupy their time; one participant ran 10 000-metre circles in the cave to stay fit.

In partnership with laboratories in France and Switzerland, scientists monitored the participants' sleep patterns, social interactions and behavioural reactions via sensors. One sensor was a tiny thermometer inside a capsule that participants swallowed like a pill. It measured body temperature and transmitted data to a computer until it was expelled naturally.

The team members followed their biological clocks to know when to wake up, go to sleep and eat. They counted their days not in hours but in sleep cycles.

As expected, those in the cave lost their sense of time. One team member estimated they had been underground for 23 days.

Figure 6.1 The participants in the Deep Time Project lived together in a cave in France for 40 days.

6.1 What is consciousness?

Before we investigate sleep, we need to set the scene and talk about a couple of things first to put sleep into context. Sleep is a 'state of consciousness'. **Consciousness** is everything you experience; it is your awareness of yourself and your environment. It consists of the ever-changing array of thoughts, feelings, sensations, perceptions and memories that you are aware of at any given moment. Throughout the day we move through various states of consciousness. Your state of consciousness refers to your level of awareness of stimuli, both internal and external. For example, we may go from being highly alert and aware of our external environment, to daydreaming and focusing on our internal environment (our thoughts), and at some time in our day we are likely to experience sleep, which is a very different state of consciousness.

There are no distinct boundaries to indicate where one state of consciousness ends and another begins. One way to describe the different states of consciousness is to place them on a continuum (range) from complete lack of awareness (unconscious) to total awareness (focused attention). There are many different states of consciousness between the two extremes of the continuum. Figure 6.2 shows the continuum from complete lack of awareness to total awareness.

Consciousness can be divided into two distinct categories: normal waking consciousness and altered states of consciousness.

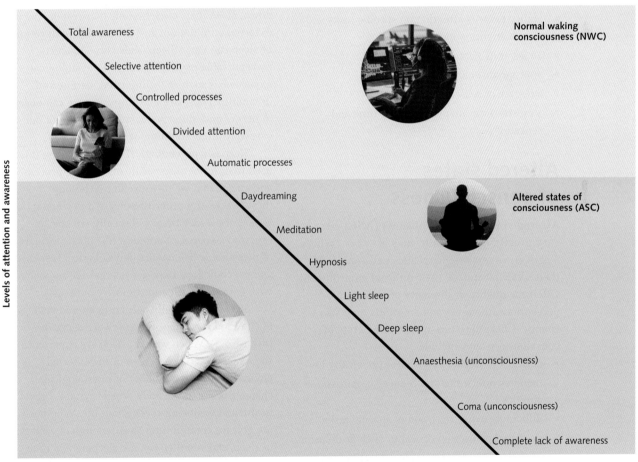

Figure 6.2 The different states of consciousness can be represented on a continuum, from total awareness at one end, to complete lack of awareness at the other. In between there are various states of consciousness that can be broadly separated into two categories: normal waking consciousness and altered states of consciousness.

Normal waking consciousness

Each state of consciousness brings with it a different level of awareness of our internal and external environments. We spend most of our lives in **normal waking consciousness (NWC)**, which is a state of mostly clear, organised alertness to internal and external stimuli. This state of consciousness is at the high end of the continuum, where we mostly perceive time, places and events as real, meaningful and familiar. Although everyone has an individual consciousness that is personal and unique, there are several common psychological characteristics of an individual when they are experiencing normal waking consciousness. These include:

» moderate to high levels of awareness
» good memory and cognitive abilities
» ability to focus attention on specific tasks and to switch attention between tasks
» an accurate perception of reality
» appropriate emotions
» a degree of self-control
» a mostly accurate perception of time and sensations.

Altered states of consciousness

Altered states of consciousness (ASC) can vary greatly, but generally, they are states where a person will experience reduced awareness of their external and/or their internal environment. Often the quality or intensity of sensations, perceptions, thoughts and emotions may also change. Characteristics of an altered state of consciousness may include:

» changes to levels of awareness (generally lower)
» memory difficulties and reduced cognitive abilities
» difficulty paying attention to specific tasks
» distorted perception of reality, such as delusions
» inappropriate or uncharacteristic emotions
» a lack of self-control
» distortions when perceiving time
» distortions of sensations.

There are many different altered states of consciousness, and they can occur naturally, or they can be induced. Induced ASC involve physiological and psychological changes that have been intentionally produced. For example, a person decides to drink alcohol or to take drugs and therefore will experience physiological and psychological changes because of this consumption.

Naturally occurring ASC involve physiological and psychological changes that occur automatically and are produced spontaneously beyond our conscious control. Naturally occurring ASC include dreaming, daydreaming, some psychoses and sleep.

Sleep as a psychological construct

Sleep is a readily observed behaviour that we all experience every day. When people sleep they lay down, close their eyes, become non-responsive to external stimulation, and the depth and duration of sleep can vary. Sleep is also accompanied by reportable psychological and physiological experiences such as feeling tired, fatigue and dreaming. These are all things that we can observe in others and/describe ourselves as part of our everyday understanding of the term 'sleep'.

Sleep is a psychological construct when it is used to describe the specific underlying processes and stages that we measure in psychology (brainwaves and other responses) and when it is part of a wider theory. Basically, we take an everyday experience and describe it as a formal set of measurable behaviours, physiological responses and psychological experiences that are associated with various health and cognitive outcomes.

Sleep is a naturally occurring altered state of consciousness that we experience every day (Figure 6.3). When we are asleep, we experience the suspension of awareness of the external environment, and a number of physiological changes occur to the body. Sleep is vital for good health and wellbeing throughout life. Whether you are getting enough total sleep, enough of each type of sleep, and whether you are sleeping at a time when your body is prepared and ready to sleep will all affect your ability to function and feel well while you are awake. Sleep can help to protect mental health, physical health and overall quality of life.

Each of us has experience with sleep and over the course of our lifetime, we will spend

Figure 6.3 Sleep is a naturally occurring altered state of consciousness.

approximately one third of our lives asleep. Up until around the 1950s, sleep was considered as a kind of 'turning off' of the brain and was of little interest to scientists. It was accepted that sleep was necessary, but there was little understanding about what occurs during sleep besides an awareness that humans need to sleep to 'recharge' in some way. Since then, it has been discovered that there is more to sleep than simply closing your eyes and drifting off into slumber, that in fact sleep is a dynamic process. With the invention of technologies, we now know that the brain and body undergo changes in activity throughout the course of a sleep period, and that these changes occur in patterns and can be recorded using various devices, which we will investigate soon. There are still many unsolved questions when it comes to what is happening to us when we sleep and why. We are going to explore sleep architecture, which is the basic structural organisation of normal sleep, and we will look at some of the methods used, including technologies to record what is happening when in a sleep state.

The two categories of sleep: REM and NREM

Not all sleep is the same. In a typical night of sleep, a healthy adult will experience about five cycles of sleep that will consist of different sleep stages (some deep and some lighter sleep). These sleep cycles will last around 90 minutes and change as the night progresses. Within these cycles there are two distinct types of sleep: **rapid eye movement (REM)** and **non-rapid eye movement (NREM) sleep**. From birth to old age, sleep patterns change, but every person no matter their age, if they are experiencing healthy sleep, will experience both REM and NREM sleep.

Just like we need to breathe and consume water and food, sleep is essential. At times, it is possible to cope without sleep, but the longer we are awake the stronger the urge to sleep becomes. The exact role and function of sleep has been a hot debate for researchers, but most agree that sleep serves a restorative purpose, both for our bodies and for our minds. There is evidence that deep NREM sleep (stage 3) is mostly involved with restoring the body and physical energy, while REM sleep is most important for restoring brain function such as memory and concentration. Sleep is important for general physical health, restoring energy, recovering from injuries or illness, growth, psychological wellbeing and mood, concentration, memory, work performance, getting along with others and helps us to regulate our emotions. Much of the knowledge and understanding we have about the purpose of sleep comes from studies of animals and humans, where they have been deprived of sleep and the resulting changes in behaviour have been documented. A large body of research is also emerging from studies of what is happening in our brain cells, hormones, neurotransmitters and other systems in our bodies when we sleep. There appears to be a complex array of mechanisms involved in the process of sleep and it is not yet fully understood.

NREM sleep

When we first go to sleep, we experience the first stage of non-rapid eye movement sleep, which we will call NREM sleep from now on. NREM sleep is divided into three distinct stages: stages 1, 2 and 3. In each stage, the sleep experienced becomes progressively deeper, with stage 3 NREM sleep often called slow-wave sleep (SWS), the deepest stage of NREM sleep.

NREM sleep is thought to be important for the restoration and repair of the body. It is thought that biological processes, such as restoring hormone levels depleted by daytime activity, repair of muscles and tissues that may have been damaged as a result of daytime activity and detoxifying muscles, occur during NREM sleep. We need NREM sleep (particularly stage 3) in order to feel refreshed and non-fatigued in the morning. Physiological activities such as heart rate, muscle tension and respiration all slow down during NREM sleep, indicating that sleep is the time when the body gains the vital rest it needs in order to function at its best.

The amount of NREM sleep we get in a full night of sleep changes as we age, but typically from about childhood onwards, the majority of sleep will be made up of NREM sleep. As adults we spend approximately 80 per cent of the night in NREM sleep, although at times, such as after physical exertion, the amount of time spent in NREM sleep may increase, which suggests that NREM sleep helps us to recover from fatigue incurred during the day. Sleep also supports healthy growth and development. Deep sleep (stage 3 NREM) triggers the body to release growth hormone that promotes normal growth in children, teens and adults. This hormone also boosts muscle mass and helps to repair cells and tissues. Elderly people experience less stage 3 NREM sleep. The pituitary gland, located at the base of the brain, releases growth hormones when you are in deep sleep, so prolonged lack of sleep, particularly NREM sleep, may cause physical growth processes to be interrupted, especially in children and teenagers. During sleep, our bodies also secrete hormones that help to control appetite, energy metabolism and glucose processing. It appears that NREM sleep is also involved in consolidating new memories and skills. There is evidence to show that specific patterns during NREM sleep produce better memory and learning, particularly in relation to verbal fluency, motor learning and word retrieval (National Sleep Foundation, 2020).

REM sleep

The other distinct type of sleep is rapid eye movement sleep (REM). If you watch a person who is asleep, you may be able to determine which type of sleep they are experiencing by noticing whether their eyes occasionally move under their eyelids. These movements, known as rapid eye movements, are indicative of a sleeper experiencing REM sleep, hence the name (Figure 6.4). The first period of REM sleep occurs at the end of the first sleep cycle, approximately 90 minutes after falling asleep for most age groups. Along with rapid eye movements, REM sleep involves heightened brain activity. In fact, the brain is so active during REM sleep that a device used to measure brain activity (an EEG, page 250) will indicate a reading that looks as if the person is awake (Hobson et al., 1998).

Rapid eye movements are strongly associated with dreaming. Roughly 85 per cent of the time, people woken during REM sleep report vivid **dreams**.

Although the brainwaves resemble alertness, REM sleep is actually considered quite a deep sleep

Figure 6.4 When we are dreaming during REM sleep, our eyes are moving but our body is motionless. These rapid eye movements indicate that we are likely dreaming. If you observe a baby's eyelids to be flickering, you know they are likely to be dreaming.

because it is difficult to wake an individual in this stage. This is one reason that REM sleep is referred to as **paradoxical sleep**. The word 'paradoxical' means 'something that contradicts itself'. For example, during REM sleep typically the heart beats faster, breathing is more rapid and irregular, similar to our awake state, blood pressure varies and the eyes dart around in their sockets. Yet with all of this internal activity going on, the body appears totally relaxed – in fact, the skeletal muscles are in a state of atonia, or paralysis, which is probably a good thing as it might be what stops you acting out your dreams!

It is suggested that REM sleep is important for brain health and development and that it is vital for keeping the brain functioning properly (Hobson, 1999). Research suggests that REM sleep helps us sort memories formed during the day and consolidate them into stable long-term memories (Ackermann & Rasch, 2014). REM sleep is important to your sleep cycle because it stimulates the areas of your brain that are essential in learning and making or retaining memories. According to the National Institute of Neurological Disorders and Stroke, a study depriving rats of REM sleep significantly shortened their life span—instead of living for 2 to 3 years, they lived for 5 weeks. Rats deprived of all sleep cycles lived only 3 weeks. The importance of REM sleep is attributed to the fact that during this phase of sleep, your brain exercises important neural connections that are key to mental and overall wellbeing and health. REM sleep stimulates the brain regions used in learning. This may be important for normal brain development during infancy, which would explain why infants spend much more time in REM sleep than adults (Figure 6.4). While there is still debate over the relative roles of REM and NREM sleep in synaptic plasticity (Chapter 2) and information processing, studies show that both are important for learning (Chapter 4) and memory (Chapter 5), although their relative roles remain a topic of debate.

REM sleep may increase dramatically when there is some sort of emotionally charged event in a person's life, such as a death in the family, trouble at school or family conflict. Daytime stress tends to prompt an increase in REM sleep. However, on average, REM sleep totals about 90 minutes per night or 20 per cent of an adult's total night's sleep. The amount of REM sleep increases as the night progresses. The first period of REM sleep may be as brief as a few minutes, and the final period of REM sleep may last up to one hour.

Because of the distinct differences in NREM and REM sleep, measurement of physiological responses can indicate which stage of sleep an individual is experiencing (Figure 6.5). We know that REM sleep is particularly essential, as when we go without sleep (including REM) the next time we go to sleep, we experience REM rebound, which is the increase in the proportion of REM sleep compared to a 'normal' period of sleep. It appears our body has a way of 'making up for' lost REM.

Sleep cycles

As stated previously, sleep is not one static state; it consists of cycles of sleep each lasting around 90 minutes and each cycle consisting of the various stages of sleep (Figure 6.6a). What does a cycle of sleep consist of?

The sleeper initially enters stage 1, and this is usually the only time throughout the sleep cycle in which they will enter stage 1 NREM sleep (as shown in the hypnogram in Figure 6.6b). After stage 1 NREM, the sleeper progresses through the subsequent stages (2 and 3), gradually entering deeper sleep. Once they have reached stage 3, the sleeper then goes back to stage 2 in reverse order. Instead of entering stage 1 NREM sleep again, the sleeper moves from stage 2 into their first period of REM sleep. The cycle then begins again: stage 2 NREM, followed by stage 3 NREM, followed by stage 2 NREM, followed by REM.

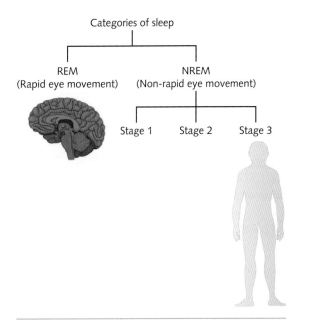

Figure 6.5 There are two categories of sleep: REM and NREM sleep. REM sleep is thought to be important to restore brain function and NREM is thought to be important to restore the body's functions after a day of activity.

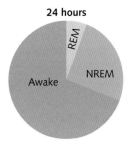

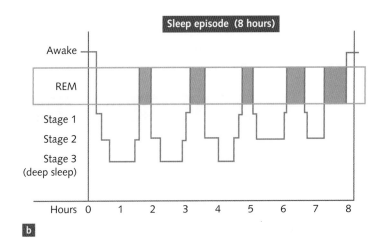

Figure 6.6 a This piechart shows the average proportion of time an adult spends in REM and NREM sleep during a 24-hour period. **b** This hypnogram shows typical changes in sleep stages over one night.

Every sleep cycle lasts approximately 90 minutes. As the night progresses, the time we spend in NREM sleep becomes shorter while the time we spend in REM sleep lengthens. Close to the end of a night's sleep we can be spending up to an hour in REM sleep during one cycle.

To identify the types and stages of sleep and the corresponding physiological variations that accompany each stage, researchers primarily use the EEG to measure the electrical activity of brain neurons.

Stage 1 NREM sleep

As you start to relax, you enter stage 1 NREM sleep. Your nervous system begins to slow, your heart rate slows, your breathing becomes irregular and your muscles relax. This physiological slowing down during the transition between wakefulness and the first stage of NREM sleep is called a **hypnogogic state** and it can last for several minutes. The hypnogogic state is characterised by slow, rolling eye movements. People in a hypnogogic state sometimes see vivid mental images or flashes of light and colour, which are known as hypnogogic images. Although they are somewhat dream-like, these images are usually not associated with dreaming. The hypnogogic state may also trigger a hypnic jerk. A hypnic jerk (also known as a myoclonic jerk or a sleep start) is a reflex muscle contraction or twitch throughout the body that may momentarily jerk a person awake. Devices used to record muscle movement during the sleep stages indicate that they only occur in stage 1 NREM sleep, often just as someone is falling asleep. Hypnic jerks are common, completely normal experiences and people often describe them as a falling sensation or being like a mild electric shock. What causes hypnic jerks is unknown, but one theory suggests they may be the result of the muscles relaxing. Another theory suggests they may be a reflex used to keep the body functioning when your brain misperceives the slower rate of internal activity as a sign that you are falling, so it alerts your arms and legs to stay upright. Stage 1 lasts for approximately 2 to 10 minutes. People wakened during this stage may or may not feel as if they have been sleeping.

Stage 2 NREM sleep

As sleep deepens, body temperature drops, and physiological responses such as heart and respiratory rate continue to slow down. In stage 2 sleep, the sleeper is still receptive to external stimuli, such as loud noises. Stage 2 sleep lasts for approximately 20 to 30 minutes.

Stage 3 NREM sleep

In stage 3 sleep, physiological responses begin to steady, the sleeper is in deep sleep and is more difficult to wake than previous stages. This stage is the start of what is known as **slow-wave sleep (SWS)**. Deep sleep is reached after about an hour of entering the sleep cycle. Brainwaves become almost pure delta waves (page 251). If a sleeper is wakened from stage 3 sleep, for example by a very loud noise, they will wake in confusion and may not recall hearing the noise. They may also take a few minutes to orientate themselves. Stage 3 can last for approximately 20 to 40 minutes but decreases in length as the night progresses. It is during stage 3 sleep that we typically see the appearance of sleep problems such as sleepwalking or bedwetting.

Weblink
Understanding our sleep cycles

ACTIVITY 6.1 CONSCIOUSNESS FLOWCHART

Try it

1. Put the following terms into a flowchart similar to the one on page 249.

   Altered states of consciousness (This term has been added for you.)

   Consciousness

   Induced altered state of consciousness

   Naturally occurring altered state of consciousness

   Normal waking consciousness

   NREM sleep

   REM sleep

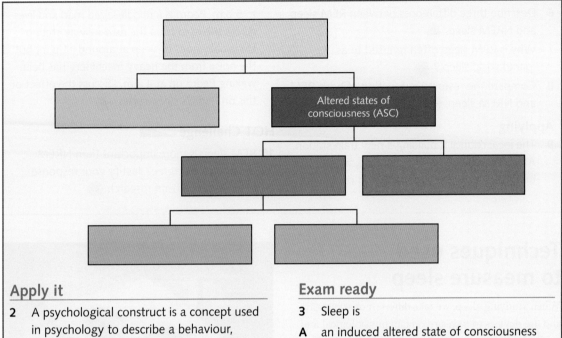

Apply it

2. A psychological construct is a concept used in psychology to describe a behaviour, psychological state, trait or process that is part of a psychological theory or model. Modify this definition of a psychological construct to show how it applies to sleep.

Exam ready

3. Sleep is
   A an induced altered state of consciousness categorised into REM and NREM sleep.
   B a naturally occurring altered state of consciousness categorised into REM and NREM sleep.
   C a subset of normal waking consciousness categorised into REM and NREM sleep.
   D a state of consciousness that cannot be measured.

🔑 KEY CONCEPTS 6.1a

» Consciousness refers to your level of awareness of stimuli, both internal and external.
» There are two broad categories of consciousness: normal waking consciousness (NWC) and altered state of consciousness (ASC).
» NWC is indicated by higher awareness than ASC.
» Sleep is a naturally occurring ASC described as the suspension of awareness of the external environment.
» There are two categories of sleep: non-rapid eye movement (NREM) and rapid eye movement (REM) sleep.
» NREM sleep is thought to be important for restoring body functions.
» REM sleep is thought to be important for restoring brain functions.

Concept questions 6.1a

Remembering
1. What are the two broad categories of consciousness? r
2. Identify the two categories of sleep. r
3. List the stages of sleep. r

Understanding
4. Why is sleep called a naturally occurring ASC? e
5. Describe three characteristics of most ASC. e

6 Describe three differences between REM sleep and NREM sleep.
7 Why is REM sleep often referred to as 'paradoxical sleep'?
8 Compare the restorative function of both REM and NREM sleep.

Applying

9 The local council is building a new train station. As construction is behind schedule, works begin at 4 a.m. every morning and continue to 5 p.m. Pedro is a middle-aged man and lives across the road from the new railway station. He usually goes to sleep at around 11 p.m. but the noise from the heavy machinery has been waking Pedro up at 4 a.m. Discuss the effect of this on Pedro's sleep cycles.

HOT Challenge

10 REM sleep is more important than NREM sleep. Do you agree? Justify your response with evidence from research.

Techniques used to measure sleep

When studying sleep, we take different observations and use devices to measure certain physiological and psychological changes that occur. From this data, we infer the state of the individual as being asleep, awake or in a particular stage of sleep.

Sleep behaviour can be measured objectively or subjectively. Subjective data are measurements collected through personal observations or personal reports of behaviour. These are often influenced by researcher or observational biases or may be influenced by the participants' biased view of their own behaviour. Objective data are measurements of behaviour collected under controlled conditions. They minimise many biases encountered in research and represent a more scientific, accurate and reliable method of data collection that allows experiments to be replicated independently.

Sleep scientists have come to understand some of the physiological changes that occur during the sleep cycle, how this cycle changes across the life span and also how to treat people who may be suffering from sleep disturbances. Most sleep research takes place in a **sleep laboratory** (Figure 6.7), which is a controlled environment that enables the electronic recording and measurement of sleep patterns.

A sleep laboratory also allows researchers to watch people's behaviour over an extended period as they sleep. Sleep laboratories are usually designed to look like a normal bedroom so that sleep is more natural and comfortable for the

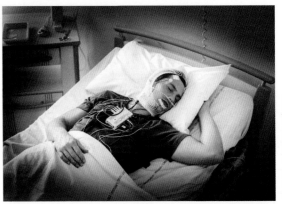

Figure 6.7 In a sleep laboratory, sleep scientists study the physiological responses that occur during the sleep cycle.

person. Common reported sleep problems often bring individuals to visit a sleep laboratory. These sleep problems include frequent waking during the night, excessive sleepiness during the day, snoring or difficulty breathing while sleeping, night terrors and sleepwalking. The use of a laboratory allows researchers to monitor a variety of responses that may explain some of these problems.

Electroencephalograph (EEG)

The electroencephalograph (EEG) is a non-invasive device used by researchers to detect, amplify and record the brain's electrical activity, measured in the form of brainwaves. Our brain is made up of billions of neurons that use electricity to communicate and send signals to each other. As signals are sent throughout the brain, they generate electrical activity, which is recorded by the EEG through electrodes placed on the scalp. It is a non-invasive device.

A patient may be attached to an EEG to record brainwave activity during sleep. The electrical activity recorded will give researchers an indication of the person's likely stage of sleep by looking at the frequency and amplitude of the brainwave pattern. The frequency of activity refers to the number of brainwaves per second. A pattern of high frequency waves indicates greater brain activity, meaning more brainwaves per second. A pattern of reduced frequency refers to slow brain activity, meaning fewer brainwaves per second.

Brain activity is also measured by amplitude, or intensity. Amplitude is measured by the size of the peaks and troughs in brain activity compared to a baseline of zero activity, as displayed on the EEG machine. High amplitude brainwaves have larger peaks and troughs, and low amplitude brainwaves have small peaks and troughs. Different combinations of frequency and amplitude in an EEG recording indicate different types of brainwave activity and thus varying stages of sleep.

There are four types of brainwave patterns, known as beta, alpha, theta and delta waves. All have different combinations of frequency and amplitude. When beta brainwaves are present, a person's brain is alert and active, which usually indicates the person is in NWC and thus, awake. Any other brainwave pattern generally indicates a person is in an ASC including that they may be in a certain stage of sleep (Table 6.1).

When you are awake and alert, a pattern of **beta waves** indicates that the brain is active. Beta waves are characteristic of normal waking consciousness, and they have a high frequency and low amplitude. The frequency of a wave is how close the waves are together. The amplitude of a wave is the height of the wave. When we start to relax, beta waves shift to a pattern of larger and slower waves called **alpha waves**. Alpha waves are experienced during a deeply relaxed state and they have a medium–high frequency and low–medium amplitude. These waves occur when you are about to fall asleep and your thoughts begin to drift.

6.1.2 BRAINWAVES

Table 6.1 Overview of brainwave patterns detected, amplified and recorded by an EEG

Brainwave	EEG recording	Amplitude on EEG	Frequency on EEG	When does it occur?
Beta		Low	High	» Normal waking consciousness (e.g. awake and alert) » Beta-like waves can be experienced during REM sleep, which is an altered state of consciousness
Alpha		Low–medium (higher than beta waves)	Medium–high (lower than beta waves)	Deeply relaxed or meditative state
Theta		Medium–high (higher than alpha waves)	Low–medium (lower than alpha waves)	Early or light sleep
Delta		High (highest of all brainwave types)	Low (lowest of all brainwave types)	Deep sleep

Stage 1 sleep, also known as light sleep, is marked by low–medium amplitude and medium–high frequency alpha brainwaves. As you progress through stage 1, **theta waves** begin to appear. Theta waves are a mix of medium–high amplitude and low–medium frequency brainwaves that indicate that the electrical activity in your brain is slowing and you are moving further away from consciousness. So, although the brainwaves in stage 1 sleep are irregular, they become higher in amplitude and lower in frequency as you move through this stage. The EEG indicates that there is a high prevalence of theta-wave activity at this time. The EEG recording also begins to show **sleep spindles,** which are bursts of distinctive brainwave activity, indicated by a short burst of high-frequency brainwaves that last for approximately one second. Sleep scientists suggest that sleep spindles may mark the true boundary of where sleep begins.

The EEG also shows that stage 2 sleep is characterised by a phenomenon known as a **K-complex**. A K-complex is a single, large burst of high-amplitude brainwaves. Scientists don't know why we have sleep spindles and K-complexes. One theory suggests that sleep spindles occur because the brain is attempting to block signals from external stimuli (for example, the sound of a crying baby) from disturbing sleep. It has also been suggested that K-complexes help suppress arousal and aid in memory consolidation.

In stage 3, **delta waves** are high in amplitude and low in frequency, and their presence signals deeper sleep and a further loss of consciousness. Finally, in REM sleep, the brain becomes 'active' in sleep, and 'beta-like' waves of high frequency and low amplitude will be produced.

Electro-oculograph (EOG)

An electro-oculograph (EOG) is a device that detects, amplifies and records the electrical activity of the muscles surrounding the eyes that move or rotate them in their sockets. It records the activity through small electrodes that are attached to the skin around the eyes. When we are awake and alert during NWC, we are able to track moving objects; that is, move our eyes in all directions and either converge or fixate them. The speed of our eye movements and, therefore, the electrical activity generated by the muscles controlling their movement, will vary according to the activity we are engaged in at the time. However, during an ASC our ability to control eye movement and the variations in the speed of eye movement do not occur to the same degree. Therefore, by demonstrating the level of electrical activity of muscles that control eye movement, the EOG can help determine a person's state of consciousness and their stage of sleep to a degree.

Electro-oculography readings are helpful in determining whether someone is awake, in NREM sleep or in REM sleep. When in REM sleep, the EOG will detect a high amount of electrical activity because the eyes are moving rapidly beneath the eyelids. When in NREM sleep, the EOG will detect low electrical activity because the muscles surrounding the eyes will have little to no movement.

Electromyograph (EMG)

An electromyograph (EMG) is a device that detects, amplifies and records the electrical activity created by active skeletal muscles. During normal waking consciousness our skeletal muscles are tense in order to maintain posture and are activated to move our bodies. When we are in NWC, an EMG will show a pattern of electrical activity that is moderate to high. However, as we enter into sleep, this pattern changes as skeletal muscles gradually relax and our ability to stay upright or to control voluntary movement diminishes. Because an EMG detects and amplifies the electrical activity created by active muscles, it is able to record the degree of tension or relaxation in the muscles. When this information is combined with EEG and EOG readings, we can gather fairly accurate information about the stage of sleep a person is experiencing.

When an EMG is used, electrodes are attached to the skin's surface, overlying muscle. EMG electrodes are typically placed under the chin, and on the arms and legs, because muscle tension

Weblink
Stages of sleep day

EXAM TIP
When answering questions regarding EOG, be careful to refer to the device measuring electrical activity of the muscles surrounding the eye which are involved in moving the eye, and not to say that it detects 'eye movement'. Indirectly we can infer that the eyes are moving if there are high levels of electrical activity, but eye movement is not what the EOG measures.

in these areas changes as we move through the various stages of sleep. EMG measurements show a gradual decrease in muscle tension as we enter the sleep cycle beginning at stage 1 NREM sleep and decreasing in the deeper stage of NREM sleep until atonia (muscle paralysis) is present during REM sleep.

These three devices combined produce a polysomnogram, a moving chart that displays data collected from an EEG, EOG and EMG simultaneously, which is a valuable, non-invasive method to objectively determine sleep patterns (Figure 6.8). A hypnogram is a type of polysomnogram. It uses the data from an EEG to show brainwave activity. A sleep scientist trained in the ability to analyse the data produced from these devices can look for patterns in sleep and compare them to the 'norm', to see whether sleep is normal and healthy or not. A summary of these methods is given in Table 6.2.

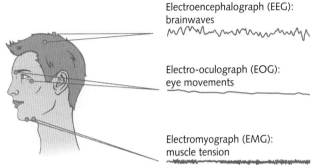

Figure 6.8 EEG, EMG and EOG recordings are made simultaneously on a continuously moving chart as part of a polysomnogram.

6.1.3 PHYSIOLOGICAL CHANGES AND STAGES OF SLEEP

Table 6.2 Summary of the objective methods used to measure physiological changes when in an ASC such as sleep

Device	What it measures	How might it demonstrate state of consciousness?
Electroencephalograph (EEG)	Electrical activity of the brain (brainwaves)	Brainwaves (alpha, theta and delta waves) can indicate an altered state of consciousness. Alpha waves may indicate a meditative or deeply relaxed state, whereas beta waves indicate a high level of alertness.
Electro-oculograph (EOG)	Electrical activity of the muscles surrounding the eyes	The EOG gives us an indication of which stage of sleep a person is in and therefore their state of consciousness. When in REM sleep, the EOG will detect a high amount of activity because the eyes are moving rapidly beneath the eyelids. When in NREM sleep, the EOG will detect low activity because the muscles surrounding the eyes will display little to no movement.
Electromyograph (EMG)	Electrical activity produced by active muscles	The EMG gives us an indication of which stage of sleep a person is in and therefore their state of consciousness. When in REM sleep, a person's muscles will not be moving at all. When in a light stage of sleep, muscles may twitch.

Sleep diaries

In conjunction with these physiological measures, a **self-report** can be useful to collect data about sleep activity that physiological devices will not detect. Subjective reporting, or self-reports, involve an individual keeping a record of their own subjective experiences (thoughts, feelings and behaviours). An advantage of using self-reports to measure sleep is that they provide researchers with some insight into covert thoughts. An example of a self-report is the Interactive Epworth Sleepiness scale (see weblink), which involves an individual rating their experiences of 'sleepiness' doing everyday things. If the sleepiness rating is high, they may require medical intervention. A self-report may provide information, for example, about an individual's thoughts or subjective experiences while they are attempting to get to sleep, maintain sleep and how they are feeling upon waking from sleep (Figure 6.9).

Weblink
Interactive Epworth Sleepiness Scale

How sleepy are you?

Name: _____ Date: _____

Age (years): _____

Consider the activities below.

Based on your experiences from the last month, answer the following multiple-choice questions. Please answer each question even if you haven't done the activity recently.

A = I have never fallen asleep this way.

B = Very occasionally I may drop off.

C = I often fall asleep this way.

D = I regularly fall asleep doing this.

How likely is it that you would fall asleep whilst carrying out the following activities? Mark A, B, C or D.

Activities	Chance of falling asleep			
	A	B	C	D
1 Sitting down whilst reading	☐	☐	☐	☐
2 Eating lunch alone	☐	☐	☐	☐
3 Being a passenger in either a car or on public transport	☐	☐	☐	☐
4 Chatting to somebody	☐	☐	☐	☐
5 Eating lunch alone	☐	☐	☐	☐
6 Lying on the couch and watching a film or series	☐	☐	☐	☐

Scoring
Tally your score based on the system below.
A=0 B=1 C=2 D=3
A score 7 or more: suggests excessive daytime sleepiness
A score greater than 10: suggests a high level of excessive daytime sleepiness

Figure 6.9 This scale will help determine how likely it is that you would fall asleep in certain situations.

Another example of a self-report is a **sleep diary**. To establish a person's sleep patterns over time, sleep scientists will often ask people who are experiencing sleep disturbances to keep a sleep diary (Figure 6.10). A sleep diary is a log of subjective behavioural and psychological experiences surrounding a person's sleep and daytime activities. Some examples of information gathered through a sleep diary may include general activities before bedtime, any consumption of food and drinks taken including the time consumed, the amount of time it took to fall asleep, any sleep disturbances throughout the night, how many times they woke during the night, dreams they might remember and their feelings on waking in the morning. Sleep diaries can gather both qualitative and quantitative data, as some of the information may be descriptions in words of experiences (qualitative) and some may be the numerical data such as the number of hours slept (quantitative).

Sleep diaries are inexpensive and are a simple way for researchers to gain insight into when

Sleep diary Morning							
Start date: Day of week: (write the day of the week):	______	______	______	______	______	______	______
Time I went to bed:	PM / AM	PM / AM	PM / AM	PM / AM	PM / AM	PM / AM	PM / AM
Time I got up this morning:	AM / PM	AM / PM	AM / PM	AM / PM	AM / PM	AM / PM	AM / PM
Last night:							
I fell asleep easily I fell asleep after some time It was difficult to fall asleep I woke up __ of times I was awake for __ minutes	☐ ☐ ☐	☐ ☐ ☐	☐ ☐ ☐	☐ ☐ ☐	☐ ☐ ☐	☐ ☐ ☐	☐ ☐ ☐
Last night I slept for:	Hours	Hours	Hours	Hours	Hours	Hours	Hours
My sleep was affected by: (List mental or physical factors including noise, lights, pets, allergies, temperature, discomfort, stress, etc.)							
When I woke up this morning, I felt:							
Refreshed A little bit refreshed Tired	☐ ☐ ☐	☐ ☐ ☐	☐ ☐ ☐	☐ ☐ ☐	☐ ☐ ☐	☐ ☐ ☐	☐ ☐ ☐
Note: add any other factors that may have affected your sleep (i.e.getting up for work or feeling ill).							

Evening							
Day of week: (write the day of the week):	______	______	______	______	______	______	______
I consumed drinks with caffeine in the:							
Morning/Afternoon/ Evening/Not Applicable Note how many drinks	______	______	______	______	______	______	______
I did exercise for at least 20 minutes in the:							
Morning/Afternoon /Evening/None							
List of medications I took:							
Took a nap? (circle)	Yes No	Yes No	Yes No	Yes No	Yes No	Yes No	Yes No
If Yes, write how long.							
How likely was I to doze off while performing daily activities:							
Moderate Chance High Chance Slight Chance No Chance							
My mood was:							
Very Pleasant Pleasant Unpleasant Very Unpleasant							
Within 2-3 hours of going to bed, I consumed:							
Alcohol A heavy meal Caffeine None	☐ ☐ ☐ ☐	☐ ☐ ☐ ☐	☐ ☐ ☐ ☐	☐ ☐ ☐ ☐	☐ ☐ ☐ ☐	☐ ☐ ☐ ☐	☐ ☐ ☐ ☐
My bedtime routine included: List activities (for example, reading a book, using your mobile phone, taking a bath, doing breathing exercises, etc.)							

Based on sleepfoundation.org

Figure 6.10 A sleep diary is a subjective method used to provide information regarding lifestyle habits and sleep experience from the individual's perspective.

sleep patterns are uncharacteristic, the degree to which they are affecting an individual's daily routine, and some of the psychological and behavioural factors that can contribute to sleep disturbances. A sleep diary gives participants the opportunity to express how they are feeling, which is helpful particularly if technological devices may be indicating a normal pattern of sleep experienced; however, if someone is feeling tired during the day and unfulfilled by their sleep, this may be highlighted in the diary.

Subjective measurements such as a sleep diary have some limitations. One is in the accuracy of recording. Often people will not remember all that they did during the day, or they forget to write details in the diary. When recording experiences in the morning after sleep (an altered state of consciousness), it can also be difficult to accurately recall details, such as the length of time taken to fall asleep, meaning estimates may be unreliable. People may also neglect to record things out of fear of embarrassment – they may be worried about how it will be interpreted–and may only record what they believe the researcher will want to read.

Video monitoring

Video monitoring involves using infrared cameras to video record a person while they are sleeping so any observable disturbances in their sleep can be analysed (Figure 6.11). People who experience sleep disturbances or problems will often have their sleep monitored in a sleep laboratory so that their sleeping behaviour can be monitored using the array of devices mentioned earlier such as the EEG, EOG and EMG, as well as being video recorded. A video monitor of their sleep may show frequent waking, restless legs, nightmares, sleepwalking and other behaviour.

Some advantages of using video monitoring to observe sleep behaviour include that the footage is recorded and therefore can be viewed at any time after a period of sleep. It can also be given to several people to interpret, and it can be shown to the participant (or the sleeper) to increase their awareness and understanding of their sleep behaviour. Limitations of using video monitoring include that the results may be inconclusive, interpretation may be subjective, and if the recording has taken place in a sleep laboratory, the artificial environment and attachment of extensive monitoring equipment may affect the person's ability to sleep normally. Video monitoring alone will only give a small part of the overall sleep experience, as it can only monitor or record overt or observable behaviours. Used in conjunction with other devices and methods, it can give an insight into observable sleeping behaviour to create a more informed overall picture of what is happening during sleep.

Figure 6.11 Sleep scientists can analyse video monitoring data of a participant in a sleep laboratory in real-time.

Weblink
What happens during an in-lab sleep study?

Weblink
The Quantified Scientist: Apple Watch Pillow app vs EEG

ACTIVITY 6.2 SLEEP MONITORING

Try it

Smartphones now have a sleep app in them and there are now wearables that claim to monitor your sleep.

1. Watch the video in the weblink in which the Quantified Scientist, Rob, compares the Pillow Apple Watch app with an actual EEG. He wore both devices for 10 nights and then evaluated the Pillow app's ability to detect REM and NREM. He also used a night vision camera to detect his movements while he slept.

   From the graph below, during which times was Rob experiencing REM sleep?

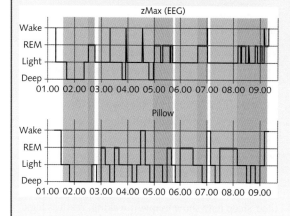

Apply it

The technician in a sleep lab can spend about 45 to 60 minutes setting you up for your sleep study. One of the tests that will take place while you sleep is an EEG. In addition to these electrodes, several of the following may be part of your sleep study set-up:
» a flat, plastic microphone taped to your neck to record snoring
» stretchy cloth belts that go across the chest and stomach to measure breathing
» sticky pads or electrodes applied to the shins or forearms for EMG
» an EOG, which uses electrodes placed near the eye.

All of these wires will be connected to a small, portable box that you can easily carry with you if you need to get out of bed (for a trip to the bathroom, for example). The technician will help you into bed and connect the wire box to a computer so that they can monitor you from another room. There will likely be a small infrared camera and a two-way speaker in the room. If you need to get up during the night, this is how you will call for help to do that.

2. Describe the role of the EEG, EMG and EOG in the sleep lab in distinguishing between REM and NREM sleep.

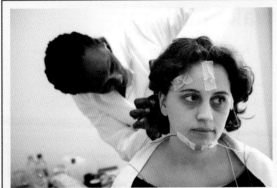

Exam ready

3 Which type of data is provided by a sleep diary and an EOG?

A objective data

B aubjective data

C objective and subjective respectively

D subjective and objective respectively

🔑 KEY CONCEPTS 6.1b

» Sleep is a psychological construct as it is something we believe to exist, as we can measure its effects, but we cannot directly observe or measure it.
» Devices used to measure the physiological activity in the brain, skeletal muscles and muscles that move the eyes during sleep include the EEG, EMG and EOG, which enable scientists to determine the stages of sleep.
» An electroencephalograph (EEG) detects, amplifies and records the electrical activity of the brainwave patterns from beta, alpha, theta and delta waves.
» An electromyograph (EMG) detects, amplifies and records the electrical activity of the skeletal muscles in the body (high levels of electrical activity indicate more muscle tension).
» An electro-oculograph (EOG) detects, amplifies and records the electrical activity of the muscles surrounding the eye (high levels of electrical activity occur when awake and during REM sleep).
» Sleep diaries are a useful subjective measurement of sleep behaviour and other lifestyle habits that may affect sleep.
» Video monitoring uses an infrared camera to record sleep behaviour unobtrusively, which can be used to indicate sleep behaviour such as movement and snoring (sounds).

Concept questions 6.1b

Remembering

1 What are the four different brainwave patterns as indicated on an EEG?
2 Name three devices used to study sleep.

Understanding

3 Describe the level of alertness and state of consciousness as indicated by the four different brainwave patterns.
4 Without referring to your textbook, explain how an EMG can be used to determine whether a person is asleep or in NWC.
5 Explain why sleep is considered to be a psychological construct.
6 Explain why it would be unlikely for someone to sleepwalk during a dream.
7 Why are sleep diaries an example of a subjective measure of reporting consciousness?

Applying

8 Evaluate which device used to measure sleep would be the most reliable measure to assess the stage of sleep that a person is experiencing. Justify your response.

HOT Challenge

9 Sleepwalking is an undesirable behaviour that occurs when an individual walks or performs other complex behaviours while asleep. Janine is a 45-year-old woman attending a sleep clinic to investigate her problem sleepwalking. Explain why a video recording, rather than an electromyograph (EMG), would be a more appropriate method for monitoring Janine in the sleep clinic.

6.2 Regulation of sleep–wake patterns

How do you know when it is time to go to sleep and when to wake up? The human body regulates sleep and wakefulness in two ways – through the interaction between circadian rhythms and sleep – wake homeostasis (**sleep demand**). Circadian rhythms are controlled by a biological clock in the brain which responds to the light/dark cues in the environment, as well as other cues. This master 'clock' or pacemaker activates systems in our body so that we are likely to go to sleep or wake up. The 'clock' also sets the timing for other biological rhythms that regulate our body and behaviour, such as when some hormones are released, regulating body temperature, as well as eating and digestion.

Our sleep and wake cycles are also driven by sleep–wake homeostasis, which is the pressure that builds the longer we stay awake, until we get to the point where we have the urge to sleep. The pressure to sleep gets stronger the longer we stay awake and decreases during sleep, reaching a low after a full night of quality sleep. The homeostatic process begins to build again after we awaken. Our sleep demand will be higher after completing tasks that are more cognitively challenging or physically demanding, as well as when our immune system is fighting infection, which generally increases our sleep demand. As a result, our sleep may be longer and deeper after those experiences.

Biological rhythms are the cyclical natural rhythms our body follows in order to perform a variety of functions (Figure 6.12). Examples of biological rhythms include the sleep–wake cycle, body temperature changes, endocrine activity, the menstrual cycle and levels of alertness. Biological rhythms can be influenced by internal factors such as our internal body clock or external factors such as light, noise and other environmental stimuli. Our body experiences a number of biological rhythms, including circadian rhythms and ultradian rhythms.

Circadian rhythms

Circadian rhythms are biological processes that roughly follow a 24-hour cycle. The term 'circadian' is derived from Latin, meaning 'about a day'. Circadian rhythms are largely endogenous, which means they originate and are controlled by internal biological processes. However, circadian rhythms are also influenced by **zeitgebers** (a German word for 'time givers'), which are external cues that can influence cyclical changes, such as light, external temperature, noise and food. These circadian rhythms influence daily cycles, like when people get hungry, when the body releases certain hormones, and when a person feels sleepy or alert.

For humans, the dominant circadian rhythm is the sleep–wake cycle. Throughout the 24-hour day, there are periods of time when we naturally feel awake, active and alert, and periods of time when we are inactive, asleep and resting. This biological circadian rhythm is largely controlled by our internal body clock and other biological processes. Our internal biological clock (or circadian clock) is found in the hypothalamus and is called the **suprachiasmatic nucleus (SCN)** (Figure 6.13). The SCN is a cluster of 20 000 neurons found deep within the brain located just above the optic chiasm, which is where the optic nerves from the two eyes cross over. The SCN receives information from the optic nerve about light, which assists its function. At night, the lack of light stimulation activates the SCN to trigger another brain region, the **pineal gland**, to release a hormone called **melatonin**. The pineal gland is located in the centre of the brain, between the two hemispheres, and helps to regulate body rhythms and sleep cycles. Melatonin levels in the bloodstream respond to cycles of light and dark by rising at

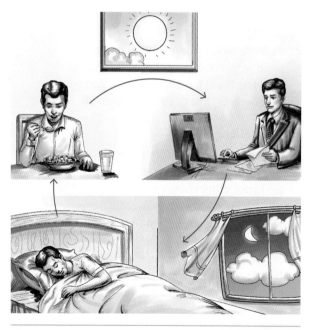

Figure 6.12 Many of our bodily processes and daily activities are controlled by innate biological rhythms.

dusk when it begins to get dark in most places and peaking around midnight. This increased release of melatonin into the bloodstream makes us feel drowsy, which helps us fall asleep. Melatonin levels then fall again as morning approaches, making it easier for us to wake up. As the sun rises, your body releases cortisol. This hormone naturally prepares your body to wake up by increasing arousal levels.

Although circadian rhythms are influenced by zeitgebers, they are largely endogenous because research has found that even in the absence of light, humans still follow a similar sleep–wake cycle, although this circadian cycle seems to extend to roughly 25 hours in total darkness. Research has also found that when the SCN of animals has been damaged, their sleep–wake cycles become totally disrupted and disorganised.

Other disruptions to our circadian rhythms can be caused by flying across time zones, where individuals may experience jet lag, or working in shifts, with workers forced to stay awake at night and sleep during the day. These changes to the normal light/dark cycle can put the body clock out of synchronisation. We will examine the impact of such events in Chapter 7.

There are other factors that will affect sleep–wake patterns, such as stress, hunger, intake of caffeine and other drugs, and exposure to things like light from electronic devices at certain times of the day or night, as well as lifestyle choices. Other regions besides the SCN in the brain can influence the sleep system too. Messages from the brainstem (the base of the brain, top of the

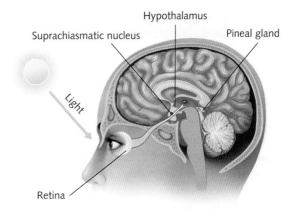

Figure 6.13 The suprachiasmatic nucleus (SCN) and pineal gland in the human brain respond to light. With 'dark' conditions, the pineal gland will secrete melatonin and with 'light' conditions the SCN will signal the release of cortisol and inhibit the release of melatonin.

spinal cord) that relay information about the state of the body such as 'having a full stomach' are influential in enabling us to fall asleep (Colten & Altevogt, 2006).

As the amount of time spent awake increases throughout the day, the pressure to sleep increases. As the day progresses into night, the 'dark' signals to the suprachiasmatic nucleus to secrete melatonin, which prepares us for sleep by making us drowsy and promoting sleep onset. We tend to go to sleep when the pressure to sleep is high and the SCN signals the release of melatonin. In a sleep episode, there will be about five cycles consisting of both NREM and REM sleep (Figure 6.14). We tend to wake in the morning during REM sleep.

6.2.1 CIRCADIAN RHYTHMS

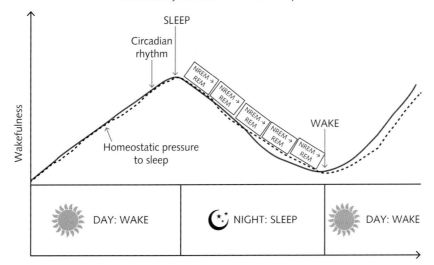

Figure 6.14 The circadian rhythm and the homeostatic sleep demand (pressure to sleep) are synchronised.

ANALYSING RESEARCH 6.1

Even dim light before bedtime may disrupt a preschooler's sleep

Lisa Marshall

Funded by the National Institutes of Health, this report examines how the central body clock of young children is unique. It suggests that preschoolers are highly susceptible to the physiological impacts of light at night, and some children may be even more sensitive than others.

'Our previous work showed that one, fairly high intensity of bright light before bedtime dampens melatonin levels by about 90 per cent in young children,' said first author Lauren Hartstein. 'With this study, we were very surprised to find high melatonin suppression across all intensities of light, even dim ones.'

Light: the body's strongest time cue

Light is the body's primary time cue, influencing circadian rhythms that regulate everything from when we feel tired or hungry to what our body temperature is throughout the day.

When light hits the retina, a signal transmits to a part of the brain called the suprachiasmatic nucleus, which coordinates rhythms throughout the body, including nightly production of melatonin. If this exposure happens in the evening as melatonin is naturally increasing, it can slow or halt it, delaying the body's ability to transition into biological night-time.

Because children's eyes have larger pupils and more transparent lenses than adults, light streams into them more freely. (One recent study showed that the transmission of blue light through a 9-year-old's eye is 1.2 times higher than that of an adult.)

'Kids are not just little adults,' said senior author Monique LeBourgeois, an associate professor of Integrative Physiology and one of the few researchers in the world to study the circadian biology of young children. 'This heightened sensitivity to light may make them even more susceptible to dysregulation of sleep and the circadian system.'

Research in a 'cave'

To quantify how susceptible they are, the researchers collaborated with Colorado School of Mines mathematician Cecilia Diniz Behn for a new study.

They enlisted 36 healthy children, ages 3 to 5 years, for a nine-day protocol in which they wore a wrist monitor that tracked their sleep and light exposure. For seven days, parents kept the children on a stable sleep schedule to normalise their body clocks and settle them into a pattern in which their melatonin levels rose at about the same time each evening.

On the eighth day, researchers transformed the children's home into what they playfully described as a 'cave' – with black plastic on the windows and lights dimmed – and took saliva samples every half hour starting in the early afternoon until after bedtime. This enabled the scientists to get a baseline of when the children's biological night naturally began and what their melatonin levels were.

On the last day of the study, the young study subjects were asked to play games on a light table in the hour before bedtime, a posture similar to a person looking at a glowing phone or tablet. Light intensity varied between individual children, ranging from 5 lux to 5000 lux. (One lux is defined as the light from a candle 1 metre away.)

When compared to the previous night with minimal light, melatonin was suppressed anywhere from 70 per cent to 99 per cent after light exposure. Surprisingly, the researchers found little to no relationship between how bright the light was and how much the key sleep hormone fell. In adults, this intensity-dependent response has been well documented.

Even in response to light measured at 5 to 40 lux, which is much dimmer than typical room light, melatonin fell an average of 78 per cent. And even 50 minutes after the light extinguished, melatonin did not rebound in more than half of children tested.

'Together, our findings indicate that in preschool-aged children, exposure to light before bedtime, even at low intensities, results in robust and sustained melatonin suppression,' said Hartstein.

What parents can do

This does not necessarily mean that parents must throw away the nightlight and keep children in absolute darkness before bedtime. But at a time when half of children use screen media before bed, the research serves as a reminder to all parents to shut off the gadgets and keep light to a minimum to foster good sleep habits in their kids. Notably, a tablet at full brightness held 30 centimetres from the eyes in a dark room measures as much as 100 lux.

And for those children who already have sleep problems? 'They may be more sensitive to light than other children,' said LeBourgeois, noting that genes – along with daytime light exposure – can influence light sensitivity. 'In that case, it's even more important for parents to pay attention to their child's evening light exposure.'

Source: University of Colorado Boulder

Questions

1. What are some of the physiological differences mentioned in the article between preschoolers and adults?
2. Why was it important for researchers to establish a baseline of when the children's biological night naturally began and what their melatonin levels were? How did they achieve this?
3. What was the research design used in the experiment?
4. What were the independent and dependent variables in this experiment?
5. Identify at least two possible extraneous variables.
6. How was data collected in the experiment? Evaluate the data collection technique.

HOT Challenge

7. Compare the findings from this study to what you know about sleep. Are there any similarities or differences?

Ultradian rhythms

During sleep, an **ultradian rhythm** determines the timing and duration of our sleep states. Ultradian rhythms are another example of biological rhythms, although they differ from circadian rhythms in the length of time before biological cyclical changes occur. While circadian rhythms roughly follow a 24-hour cycle, ultradian rhythms are biological rhythms that follow a cycle of less than 24 hours; for example, eye blinks, heartbeats and sleep cycles. In sleep, the ultradian rhythm refers to the alternation of two distinct types of sleep, NREM and REM sleep, throughout the sleep period.

Ultradian rhythms can last 1 minute. For example, humans blink approximately 15 times in 1 minute and our heart beats approximately 60–100 times per minute. The cyclical stages of sleep are performed in an ultradian rhythm, where each cycle lasts approximately 90 minutes. As discussed previously, when we sleep, a number of physiological changes occur at different stages of our sleep cycle. For example, our eye movements, muscle tension and brainwaves all change cyclically throughout a typical night's sleep (Figure 6.15). The first sleep cycle, which involves stages of NREM sleep followed by REM, makes up one ultradian cycle that lasts approximately 90 minutes. As the night progresses, this sleep cycle changes slightly, with periods of stage 1 NREM no longer occurring; stages 2 and 3 NREM getting shorter; and stages of REM sleep getting longer. However, each cycle still lasts approximately 90 minutes, so we will typically experience five ultradian sleep cycles each night.

6.2.2 STAGES OF SLEEP

6.2.3 EVALUATION OF RESEARCH

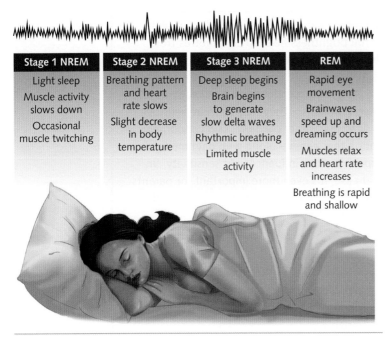

Figure 6.15 On average, an adult will experience five cycles of sleep during one sleep episode. Each cycle will repeat every 90 minutes or so, making the cycles of sleep an ultradian rhythm.

Some researchers also argue that ultradian rhythms can be seen when we are awake, through our varying levels of alertness. Our levels of alertness, including our attention span and concentration levels, are also believed to occur in 90-minute cycles. These cycles are then followed by a period of drowsiness. This is why researchers argue that to improve our work efficiency and productivity we should have a 15-20-minute break every 90 minutes. After 90 minutes of high alertness, we have a period of drowsiness, and if we try to push through and work through these periods this can lead to low performance. You might be able to relate to these findings depending on the length of each period at your school. During a typical lesson at your school, when does your attention span tend to wear out? Circadian and ultradian rhythms are compared in Table 6.3.

Table 6.3 A comparison of circadian and ultradian rhythms

	Circadian rhythm	Ultradian rhythm
Definition	A biological process that roughly follows a 24-hour cycle	A biological rhythm that follows a cycle of less than 24 hours
Examples	The sleep–wake cycle	Sleep cycle, eye blinks and heartbeats
Factors affecting this rhythm	» Internal factors such as our internal body clock through the release of hormones » External factors such as light, noise and other environmental stimuli	

ACTIVITY 6.3 LIGHT AND MELATONIN

Try it

1. Put these elements (right) into a flowchart to demonstrate the effect of light on melatonin levels. You may add more elements to your flowchart.

Melatonin molecule

Suprachiasmatic nucleus

Apply it

Refer to the chart below to answer Question 2.

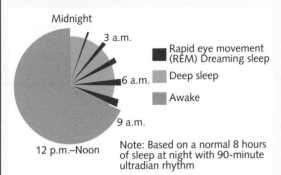

Note: Based on a normal 8 hours of sleep at night with 90-minute ultradian rhythm

2 a How many times will the sleep–wake cycle repeat in a 24-hour period?
   b How many times will the REM/NREM cycle repeat in a 24-hour period?

Exam ready

3 Which of the following statements is correct?

A The release of melatonin by the suprachiasmatic nucleus follows an ultradian rhythm, as its cycle repeats once per day.

B The release of melatonin by the suprachiasmatic nucleus follows a circadian rhythm, as its cycle repeats once per day.

C The release of melatonin by the suprachiasmatic nucleus follows a circadian rhythm, as its cycle repeats many times per day.

D The release of melatonin by the suprachiasmatic nucleus follows an ultradian rhythm, as its cycle repeats many times per day.

ANALYSING RESEARCH 6.2

Disruption of circadian rhythms due to chronic constant light leads to depressive and anxiety-like behaviours in rats

Research has shown that mood disorders such as depression are closely related to the circadian system, particularly when the circadian system is disrupted (Tapia-Osorio et al., 2013). Long periods of artificial light, shift work and jet lag can all disrupt the circadian system as it affects the suprachiasmatic nucleus (SCN) (internal biological clock). People suffering from mood disorders such as depression also often report changes in sleep patterns.

A team of researchers (Tapia-Osorio et al., 2013) was interested in determining whether circadian disruption, caused by long-term constant light, would lead to depressive symptoms in rats and whether these symptoms are related to altered activity of the SCN. To carry out this study, researchers used a group of rats and separated them into three groups. One group of rats was exposed to regular light and dark conditions across a 24-hour cycle, one group was exposed to constant darkness for 24 hours a day, and one group was exposed to constant artificial light for 24 hours. All conditions lasted for 8 weeks.

Baseline behaviours of the rats were recorded before the experimental condition took place and were then compared with behaviours at the conclusion of 8 weeks. Following the 8 weeks, rats were tested on their general behavioural activities, levels of pleasure and the neuronal activity of the SCN. Results found that rats in the constant light group experienced anhedonia (inability to feel pleasure), which was measured using a sucrose consumption test where rats showed a significantly reduced consumption of sucrose indicating a depressive-like state. These rats were also prone to increased grooming, suggesting an anxiety response. Higher levels of corticosterone (stress hormone in animals) and lower body weight also indicated a depressive-like state. When comparing the circadian rhythms of all three groups, rats in the control group's cycle remained rhythmic, while rats in the dark group did not express any major disruption in rhythmic activity, with a cycle generally lasting longer than 24 hours. Rats in the light group, however, reported a major disruption in rhythmic activity, with rats following an ultradian rhythm of activity approximately every 4 hours. Researchers concluded this was due to the changes in neural activity in the SCN.

Questions

1. What was the aim of this study?
2. Identify the control group and the experimental groups in this study.
3. Which research design was used in the experiment? Outline a benefit and a limitation when using this particular design.
4. How were the rats' circadian rhythms affected by the different exposure to light?
5. Write a conclusion for this study.

KEY CONCEPTS 6.2

- Regulation of the sleep–wake pattern is driven in part by a circadian rhythm.
- A circadian rhythm is a biological rhythm (pattern) that occurs once every 24 hours and is endogenous in humans.
- The circadian rhythm of sleep–wake is influenced by external factors such as when there is a lack of daylight (evening), when the absence of light signals received by the eyes stimulates the suprachiasmatic nucleus (master clock) to signal to the pineal gland to release melatonin.
- Melatonin is a hormone that promotes drowsiness and readies us for sleep.
- When light conditions increase, cortisol is released, which 'wakes' us up and makes us more alert.
- In the absence of light/dark cues in the environment, the circadian rhythm extends slightly, but still operates.
- Ultradian rhythms are biological rhythms that occur more frequently, following a cycle of less than 24 hours.
- Healthy adults experience about five cycles of sleep in a night, each lasting about 90 minutes, and are an example of an ultradian rhythm.

Concept questions 6.2

Remembering
1. Identify the biological rhythm that repeats every 24 hours and regulates our sleep–wake cycle. r
2. Where in the brain is the 'body clock' located? r
3. What are the names of the hormones involved in promoting sleep and wakefulness? r
4. Which part of the brain is responsible for releasing melatonin? r

Understanding
5. Compare ultradian and circadian rhythms and give an example of each related to sleep. e
6. Explain the role of melatonin in regulating sleep. e

Applying
7. Explain how the circadian rhythm and sleep–wake homeostasis work together to regulate sleep. e

HOT Challenge
8. Sixteen-year-old Hannah is experiencing problems getting to sleep. Explain to Hannah how her brain uses 'light' to regulate sleep. What advice would you give her to improve her ability to get to sleep? c

6.3 The changes in sleep over the life span

By now, you may be asking why we sleep? We will spend about a third of our lifetime sleeping, and we also become vulnerable when we are asleep in an ASC, because we lack awareness of our external environment. Whatever the purpose of sleep, it must be important. We know that sleep is essential, it is innate (or inborn) in humans, but why?

The exact purpose of sleep is not clear cut. It appears that there may not be just one purpose of sleep. We know that sleep is vital for waking cognition. We need to sleep to have the ability to think clearly, to be alert and vigilant, and to maintain concentration and attention when we are awake. We know that sleep is vital for memory consolidation, and for emotional regulation. Researchers have investigated what happens when humans and other animals don't sleep, and they have studied the animal kingdom looking for similarities and differences between species in the way they sleep and the patterns of sleep. Why we sleep is still not well understood.

Some theories of why we sleep include the restoration theory, which states that the function of sleep is to repair and replenish cell components and functions that have been depleted during the day, including muscle repair, tissue growth and protein synthesis. Evidence to support this theory is that in studies, animals deprived of sleep entirely lose all immune function and die in a relatively short period of time. Prolonged sleep deprivation has been shown to lead to death in rats, dogs, fruit flies and humans. It is believed that in these cases, death has resulted from damage to cells, caused by metabolic processes, which would normally be repaired by getting rid of wastes and restoring our systems—work that occurs primarily during sleep (Cao et al., 2020). The evolutionary theory alternatively proposes that the human sleep–wake cycle is based on the idea that night-time sleep is adaptive due to the lower risk from predators as we don't function well in the dark (we can't see!). A more recent explanation for why we sleep is linked to brain plasticity. This is based on research findings that sleep corresponds to changes in the structure and organisation of the brain. Evidence to support the role of sleep in brain plasticity is suggested by the vital role sleep plays in the brain development of infants and children, and adults in optimising the ability to learn and perform a variety of tasks. This development is significantly reduced when there is a lack of sleep. Why we sleep is still not fully explained, and it may be that there is no one explanation, that sleep in fact serves many purposes.

When we look at the changing patterns in sleep over the life span, some of these functions of sleep may offer some explanation of the differences in sleep as a person ages.

ANALYSING RESEARCH 6.3

Screen time and sleep among school-aged children and adolescents

Screen time refers to the amount of time spent in front of an electronic device with a 'screen'. Many electronic devices use LED back-light technology to help enhance screen brightness and clarity. These LEDs emit very strong blue light waves. Mobile phones, computers, tablets and flat-screen televisions are just among a few of the devices that use this technology. Because of their widespread use and increasing popularity, we are gradually being exposed to more and more sources of blue light and for longer periods of time.

The Australian guideline for the amount of screen time for children aged 2–18 years is no more than 2 hours a day. Children who spend less than 2 hours in front of a screen each day may be less likely to be overweight, engage in more positive social interactions, and have improved school performance and sleep patterns. The Australian Health Survey on Physical Activity in 2011–12 revealed that children and adolescents (aged 5–17 years) spent more than 2 hours a day on a screen-based activity, with physical activity decreasing and screen-based activities increasing as age increased (Australian Bureau of Statistics, 2013).

One study conducted by Hale and Guan (2015) compared 67 studies conducted between 2009 and 2014, investigating the relationship between sleep outcomes and screen time among adolescents and school-aged children.

The analysis revealed that screen time was adversely related to sleep outcomes. In 90 per cent of the studies, increased screen time led to poorer sleep patterns, including delayed sleep time and shortened duration of sleep.

Researchers concluded that young people should be limiting their screen time, especially near their bedtime, to prevent a disruption to their sleep patterns and general wellbeing.

Questions

1. What was the aim of this study?
2. What scientific investigation methodology was used in this study?
3. 'In 90 per cent of the studies, increased screen time led to poorer sleep patterns, including delayed sleep time and shortened duration of sleep.' Using what you know about sleep, offer an explanation for these results.
4. What did researchers conclude based on the findings of this study?
5. Reflection: How many hours per day do you spend in front of a screen? Is this above or below the recommended time per day?

Demand for sleep: how much sleep is enough?

To answer this question, we need to first look at age as a factor affecting sleep requirements. We will explore how sleep changes over the life span in terms of total amount of sleep required for optimal performance during the day, REM and NREM proportions, changes in patterns and type of sleep required and discuss some explanations for the changes in sleep across the life span.

Many experts would agree that there is a kind of 'sweet spot' that most people should aim for when it comes to how much sleep is required for each stage of life. The amount of sleep that is needed across the life-span and patterns of sleep vary with age. Generally, the demand for sleep changes so that people need more sleep early in life, when they are growing and developing, and the older we get the less sleep we require and the less time we spend in REM sleep. Although recommended hours of sleep are suggested for each age group, the demand for sleep will depend on individual differences and may vary slightly (but not significantly) among two people of the same age.

Newborns

It should come as no surprise that newborns do not have established sleep–wake rhythms. They will sleep both during the day and the night. On average, newborn babies sleep for approximately 16 hours a day, and 50 per cent of their sleep time is spent in REM sleep and 50 per cent in NREM sleep. The high proportion of REM sleep experienced at this age may provide supportive evidence for the restorative theory, which would suggest that babies need proportionately more REM sleep than older age groups to help replenish the mental processes they exhaust when learning information during the day. When we are babies, the world is full of new and exciting things. There are things to discover and explore and, as a result, our brain is making many new meaningful connections. Another explanation as to why babies need so much REM sleep is because it is believed that newborn babies lack the capacity for long, deep NREM sleep; a capacity that only develops with brain maturation during childhood and adolescence (Hobson, 2001). Newborns experience 50 per cent NREM and while this is proportionately less than older age groups, they still have more NREM and REM in total, as they are sleeping about 16 hours a day. This is equivalent to experiencing about 8 hours in REM and 8 hours in NREM sleep. The high amount of both categories of sleep may be explained by the need for sleep to both support the growth and replenishment of the body (NREM), as it is in deep sleep that the body physically repairs itself, boosts the immune system and restores bones, muscles and tissue, and it supports the brain changes due to healthy brain development in the first years of life (REM).

Newborn babies require sleep to be broken up throughout the 24-hour period as they have not developed a circadian rhythm, and because in order to feed and tend to other bodily functions, like changing nappies, they wake regularly. This makes sleep during the newborn stage of life polyphasic, meaning sleep is not achieved in one block of time, but instead is divided into several periods of time throughout a 24-hour period.

EXAM TIP
It is important to learn the figures for each age group in terms of the total amount of sleep, proportion of REM and NREM sleep and to be able to describe the changes to typical patterns of sleep as a person ages. You may need to state these figures for any age group or to compare different age groups.

Milk is the main source of nutrition in the early months, and it needs to be ingested regularly to meet the growing needs of the baby. Newborn babies will typically feed every 3–4 hours in the early months, some more regularly than this, so they will wake for a feed, change, then return to sleep, spending most of a 24-hour period asleep. Sleep cycles consist of active sleep and quiet sleep. During active sleep, the newborn may move, groan, open their eyes, cry out or breathe noisily or irregularly. During quiet sleep, they will lie relatively still, and their breathing will be more regular. They sleep in short sleep cycles usually lasting around 20–50 minutes. Newborn sleep can also start with an initial active REM stage, different to adults who start in stage 1 of NREM and experience REM sleep at the end of each sleep cycle.

Infants

As the newborn moves into infancy, the demand for sleep drops to about 14 hours, with the proportion of sleep in REM also dropping. Although REM sleep begins to reduce to about 35 per cent of sleep time, it is still experienced at a higher proportion than adults, in order for infant's to continue growing and developing. At the age of between 4 and 6 months the circadian rhythm starts to influence sleep and at this point the infants sleep architecture will most likely resemble that of an adult, with a longer period of sleep at night (monophasic), but they will most likely differ from adults, in that they will take naps during the day (biphasic). The development of the circadian rhythm results in greater durations of wakefulness during the day and longer periods of sleep at night (Sheldon, 2002). By 3 months of age, sleep cycles become more regular: sleep onset now begins with NREM, REM sleep decreases and shifts to the later part of the sleep cycle, and the total NREM and REM sleep cycle is typically 50 minutes (Anders et al., 1995; Jenni & Carskadon, 2000). Sleep cycles also change because of the greater responsiveness to external social cues (such as feeding and bedtime routines). As sleep cycles mature, the typical skeletal muscle paralysis of REM sleep replaces the movement behaviour in what was called 'active sleep' as a newborn. By 12 months old, the infant typically sleeps around 14 hours per day with the majority of sleep consolidated in the evening and during one to two naps during the day.

Children

Children demand on average 11–12 hours of sleep, with the proportion spent in REM dropping to approximately 20–25 per cent. It is normal for children to continue napping in the day until about the age of three to five. If a child still requires daytime naps after the age of five, this may indicate they are not getting enough sleep at night. Young children tend to sleep more in the early evening. The amount of slow-wave sleep (a stage of NREM sleep) peaks in early childhood and then drops sharply after puberty. It continues to decline as people age.

During childhood we are still developing in both brain and body, which may explain the high amount of total sleep this age group experiences. Even though REM sleep has continued to decline proportionately to between 20–25 per cent of sleep, that accounts for about 2–3 hours of REM sleep in a night. Children are now having more NREM sleep proportionately, about 75–80 per cent, which is equivalent to roughly 9–10 hours of NREM sleep in a night. As a child approaches adolescence, sleep timing changes, from children sleeping earlier in the evening and waking early in the morning, shifting forward to a later sleep time in the evening and later wake time in the morning (if they don't have school early!).

Adolescents

During adolescence total sleep required for optimal functioning is approximately 9 hours. Sleep consists of 20–25 per cent REM sleep and 75–80 per cent, NREM sleep. The architecture of sleep will show more stage 3 NREM with brief periods of REM at the end of each cycle in the first half of sleep and shift in the second half of the night so that longer periods of REM are experienced. Many adolescents do not always get the required 9 hours of sleep a night, and this may be due to their busy lives, trying to juggle school, work and social commitments. It could also be due to a natural sleep–wake cycle shift as the rhythm and timing of the body clock changes, which prevents adolescents from feeling tired until late in the evening.

Adolescents fall asleep later at night than younger children and adults. One reason for this is because melatonin is released and peaks later in the 24-hour cycle for teens.

Worksheet
Are children and adolescents getting enough sleep?

As a result, it is natural for many teens to prefer later bedtimes and to sleep later in the morning than adults. This will cause a delay in the time when sleep is initiated and delay in the time of waking, so the whole sleep period shifts later by 1–2 hours. This shift is discussed in more detail in Chapter 7. Adolescence is a time of growth and development and therefore requires sleep to support the developing brain and growing body.

Adults

6.3.1 INVESTIGATING SLEEP

The demand for sleep in adults is approximately 7–8 hours a night, with 20 per cent of this spent in REM sleep and 80 per cent in NREM. Following adolescence, the amount of time spent in REM and NREM stabilises into old age. Adults often report not always getting the required hours of sleep because of their busy lifestyles.

The timing of sleep changes again, this time due to the earlier release of melatonin (1–2 hours before an adolescent will typically release the hormone), making adults feel drowsy and sleepy at an earlier time in the evening than during adolescence. This, in turn, shifts wake times for adults to earlier in the morning. There will be approximately five cycles of sleep in a night, each lasting around 90 minutes (Figure 6.16). Sleep will consist of more stage 3 deep sleep in the first half of the night with brief periods of REM, to longer periods of REM in the latter part of the night.

Elderly

Older adults tend to go to bed earlier and wake up earlier. As people enter old age, they often sleep less than they did earlier in life; however, it is recommended that they get approximately 7–8 hours on average. About 20 per cent of their sleep time is spent in REM sleep and 80 per cent in NREM. As people move into old age, they will also report waking more frequently throughout the night and they spend less time in stage 3 (deep sleep) and nap more frequently during the day. Some of the changes in sleep as people get older may be explained by an ageing SCN and a deterioration in other brain areas and physical systems in the body. A deterioration in the SCN may disrupt the circadian rhythms, which affects when a person is going to feel tired as well as when they are going to feel more alert and awake (Mattis & Sehgal, 2016).

Lifestyle changes that reduce the amount of daylight the elderly are exposed to may further impact the circadian rhythm and this may be more restricted for those living in nursing home care. Some research has shown that many elderly people are averaging around one hour of exposure to daylight each day (Stepnowsky & Ancoli-Israel, 2008).

Older adults tend to have sleepiness earlier in the evening and wake up earlier in the morning than desired. This earlier sleep timing in older adults may be due to the age-related phase advance in their circadian rhythm. This phase advance is seen not only in the sleep–wake cycle, but also in the body temperature rhythm, and in the timing of secretion of melatonin and cortisol, all of which are about one hour advanced in older people compared to young adults (Tranah et al., 2017).

The changes in the sleep patterns of the elderly may also contribute to changes in production of hormones, such as melatonin and cortisol, which may disrupt sleep. As people age, the body secretes less melatonin, which is normally produced in response to darkness to help promote sleep by synchronising circadian rhythms (Newsom & DeBanto, 2020). When less melatonin is produced, the feeling of tiredness when the brain receives the 'dark' signal is not as effective and the reduced level of cortisol makes it less likely that the person will wake up feeling alert.

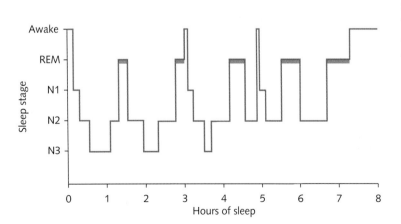

Figure 6.16 A typical hypnogram for a healthy adult. When interpreting a hypnogram, note the amount of time spent asleep is indicated on the *x*-axis and the stages of sleep are on the *y*-axis. In the first half of a sleep episode more time will be spent in NREM sleep, particularly stage 3 (deep sleep). The second half of the sleep episode will have an increase in the amount of time spent in REM sleep and a reduction in stage 3 NREM.

There are many medical conditions that occur more in the elderly that can make sleep difficult. Changes in lifestyle in old age, such as retirement and being less active during the day, can also impact on sleep demand. Older people also tend to sleep lightly. They wake up more often and spend less time in deep, refreshing sleep (National Sleep Foundation, 2020). There is a change in sleep patterns so that the elderly experience less stage 3 NREM (deep slow-wave sleep) compared to more stages 1 and 2 NREM sleep, resulting in lighter sleep, which leads to people in old age being more easily woken. This decline in deep sleep corresponds to changes in hormone levels, including a reduction in the release of growth hormone (Copinschi & Caufriez, 2013).

If you are unsure whether you are getting the required amounts of sleep each night, you should consider how you feel and function throughout the day. If you feel well rested, energetic and alert during the day, then the sleep you are getting is likely to be adequate, even if it is slightly below the recommended hours of sleep for your age group (Table 6.4; Figure 6.17). Alternatively, if you are experiencing fatigue, irritability or concentration problems, you may want to revise your sleep habits to improve your sleep quality. Sleeping the number of recommended hours regularly is associated with better health outcomes, including improved attention, behaviour, learning, memory, emotional regulation, quality of life, and mental and physical health.

Table 6.4 Sleep recommendations across the life span

Age	Hours of sleep
Newborns	16 hours
Infant (4–11 months)	12–15 hours
Toddler (1–2 years)	11–14 hours
Preschooler (3–5 years)	10–13 hours
School-aged child	9–11 hours
Teen (14–17 years)	8–10 hours
Young adult (18–25 years)	7–9 hours
Adult (26–64 years)	7–9 hours
Older adult (65+ years)	7–8 hours

Source: National Sleep Foundation

> **EXAM TIP**
> If asked to compare the sleep patterns of different age groups, you must include both groups in the comparison. For example, when comparing infants to adults: infants typically nap during the day, *whereas* adults do not.

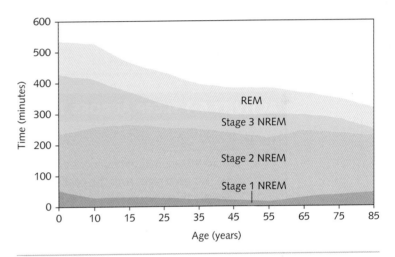

Figure 6.17 This graph shows the typical changes in the amount of NREM and REM sleep across the life span.

INVESTIGATION 6.1 SLEEP PATTERNS ACROSS THE LIFE SPAN

Scientific investigation methodology
Controlled experiment

Aim
To compare the quantity and/or quality of sleep across different age groups.

Introduction
This scientific investigation utilises the sleep scale shown in Figure 6.9 to obtain data on sleep across different age groups.

You will investigate the relationship between age and a measure of sleepiness. The task requires you to access a small number of participants of different ages (for example an infant, a child, an adolescent, an adult and an elderly person). You will be required to collect data and use evidence from your findings to explain any relationship.

You must maintain a logbook for this practical activity for recording and authentication, and particularly if this activity is to be used for assessment purposes. Record in your logbook all elements of your investigation planning, comprising identification and management of relevant risks, recording of raw data, and preliminary analysis and evaluation of results, including identification of outliers and their subsequent treatment.

Logbook template
Controlled experiment

Data calculator

Procedure
Develop a research question, state an aim, formulate a hypothesis, and plan an appropriate method to answer the research question, while complying with safety and ethical guidelines. You will design the data collection method using the sleep scale shown in Figure 6.9.

Results
Collate the results for all participants and process the quantitative data obtained. Organise, present and interpret data using an appropriate table and/or graph.

Discussion
1. State whether the hypothesis was supported or refuted.
2. Describe how sleep changes across the life span.
3. Discuss any implications of the results.
4. State if it is possible to generalise any of the results to the wider population.
5. Discuss any confounding variables and how they may have affected the data.
6. Suggest future improvements to address the confounding variables if the investigation was to be repeated.

Conclusion
Write an appropriate conclusion for your investigation.

ACTIVITY 6.4 SLEEP ACROSS THE LIFE SPAN

Try it
1. a Record the number of hours of sleep you have had each night over the past week.
   b Calculate the mean duration of one night's sleep.
   c Refer to Table 6.4: How do you compare to the number of hours of sleep recommended for your age group? Are you getting enough sleep? What age group is your sleep pattern most like?

Apply it
2. Why do babies sleep more and why do they have more REM sleep than adults?

Exam ready
3. There are changes to our sleep patterns over the life span. Which statement best describes differences in sleep across the life span?
   A The proportion of NREM sleep increases as we age.
   B The total amount of NREM sleep increases as we age.
   C The proportion of REM sleep increases as we age.
   D The total amount of REM sleep increases as we age.

ANALYSING RESEARCH 6.4

A sound night's sleep grows more elusive as people get older. But what some call insomnia may actually be an age-old survival mechanism, researchers report.
A study of modern hunter-gatherers in Tanzania finds that, for people who live in groups, differences in sleep patterns commonly associated with age help ensure that at least one person is awake at all times.

The research suggests that mismatched sleep schedules and restless nights may be an evolutionary leftover from a time many, many years ago, when a lion lurking in the shadows might try to eat you at 2 a.m.

'The idea that there is a benefit to living with grandparents has been around for a while, but this study extends that idea to vigilance during night-time sleep,' said study co-author David

Samson, who was a postdoctoral fellow at Duke University at the time of the study.

The Hadza people of northern Tanzania live by hunting and gathering their food, following the rhythms of day and night just as humans did for hundreds of thousands of years before people started growing crops and herding livestock.

The Hadza live and sleep in groups of 20 to 30 people. During the day, men and women go their separate ways to forage for food in the savanna woodlands in Tanzania. Then each night they reunite in the same place, where young and old alike sleep outside next to their hearth, or together in huts made of woven grass and branches.

'They are as modern as you and me. But they do tell an important part of the human evolutionary story because they live a lifestyle that is the most similar to our hunting and gathering past. They sleep on the ground, and have no synthetic lighting or controlled climate – traits that characterised the ancestral sleeping environment for early humans,' said co-author Alyssa Crittenden, associate professor of anthropology at the University of Nevada, Las Vegas.

As part of the study, 33 healthy men and women aged 20 to 60 agreed to wear a small watch-like device on their wrists for 20 days, that recorded their night-time movements from one minute to the next.

Hadza sleep patterns were rarely in sync, the researchers found. On average, the participants went to bed shortly after 10 p.m. and woke up around 7 a.m. But some tended to retire as early as 8 p.m. and wake up by 6 a.m., while others stayed up past 11 p.m. and snoozed until after 8 a.m.

In between, they roused from slumber several times during the night, tossing and turning or getting up to smoke, tend to a crying baby, or relieve themselves before nodding off again.

As a result, moments when everyone was out cold at once were rare. Out of more than 220 total hours of observation, the researchers were surprised to find only 18 minutes when all adults were sound asleep simultaneously. On average, more than a third of the group was alert, or dozing very lightly, at any given time.

'And that's just out of the healthy adults; it doesn't include children, or people who were injured or sick,' said Samson, now an assistant professor of anthropology at the University of Toronto, Mississauga.

Yet the participants didn't complain of sleep problems, Samson said.

The findings may help explain why Hadza generally don't post sentinels to keep watch throughout the night – they don't need to, the researchers say. Their natural variation in sleep patterns, coupled with light or restless sleep in older adults, is enough to ensure that at least one person is on guard at all times.

Previous studies have found similar patterns in birds, mice and other animals, but this is the first time the phenomenon has been tested in humans, Samson said.

The researchers found that the misaligned sleep schedules were a byproduct of changing sleep patterns common with age.

Older participants in their 50s and 60s generally went to bed earlier, and woke up earlier than those in their 20s and 30s.

They call their theory the 'poorly sleeping grandparent hypothesis'. The basic idea is that, for much of human history, living and sleeping in mixed-age groups of people with different sleep habits helped our ancestors keep a watchful eye and make it through the night.

'Any time you have a mixed-age group population, some go to bed early, some later,' Nunn said. 'If you're older you're more of a morning lark. If you're younger you're more of a night owl.'

The researchers hope the findings will shift our understanding of age-related sleep disorders.

Duke University. Live-in grandparents helped human ancestors get a safer night's sleep: sleep changes common with age may have helped our ancestors survive the night. *ScienceDaily*, 11 July 2017. http://www.sciencedaily.com/releases/2017/07/170711215825.htm

6.3.3 CASE STUDIES

Questions

1. Considering the method used to collect data in the study performed using the Hadza people, does the study have validity? Justify your response.
2. Outline the scientific investigation methodology used and identify the research design.
3. What explanation do the researchers give for the changes in sleep patterns across the life span?
4. Can the results from the Hadza people be generalised to other populations of people? Justify your response.

KEY CONCEPTS 6.3

- The amount of total sleep required decreases over the life span, remaining stable from adulthood (newborns 16 h, infants 14 h, childhood 12 h, adolescence 9 h, adults 7–9 h, elderly 7–8 h).
- The proportion of REM sleep decreases over the life span, stabilising somewhat in adulthood (newborns 50%, infants 35%, childhood 20–25%, adolescence 20–25%, adults 20%, elderly 20%).
- The proportion of NREM sleep increases over the life span, stablising somewhat in adulthood (newborns 50%, infants 65%, childhood 75–80%, adolescence 75–80%, adults 80%, elderly 80%).
- Newborns have an irregular sleep pattern due to an under-developed circadian rhythm.
- The high amount of NREM sleep in the early years of life may be explained by the need for sleep to support the growth and the replenishment of the body, as it is in deep sleep that the body physically repairs itself, boosting the immune system and restoring bones, muscles and tissue.
- The high amount of REM sleep in the early years may be explained by the need to support the brain changes needed for healthy brain development in the first years of life.

Concept questions 6.3

Remembering

1. What is the recommended total amount of sleep for newborns, infants, children, adolescents, adults and the elderly? **r**
2. Fill in the gaps: Newborns will experience ___ per cent REM and ___ per cent NREM sleep, while an adult will experience ___ per cent REM and in old age ___ per cent NREM. **r**

Understanding

3. Compare the sleep patterns of adolescents to adults. **e**
4. How are the sleep patterns of the elderly different from any other age group? **e**

Applying

5. Thomas is a 16-year-old student who is having arguments with his parent over his bedtime. He wants to go to bed later than his parent, as he is finding it difficult falling asleep at 9.30 p.m. when his parent is going to bed. Outline an argument based on this stage of the life span showing that Thomas may have a reason for wanting a later bedtime. **c**
6. Give explanations of a social factor and a biological factor for why people in the elderly stage of life may require naps during the day. **e**

HOT Challenge

7. Describe the informed consent procedures that should be used by researchers investigating children or the elderly with sleep problems or disorders. **e**

6 Chapter summary

KEY CONCEPTS 6.1a

- Consciousness refers to your level of awareness of stimuli, both internal and external.
- There are two broad categories of consciousness: normal waking consciousness (NWC) and altered states of consciousness (ASC).
- NWC is indicated by higher awareness than ASC.
- Sleep is a naturally occurring ASC described as the suspension of awareness of the external environment.
- There are two categories of sleep: non-rapid eye movement (NREM) and rapid eye movement (REM) sleep.
- NREM sleep is thought to be important for restoring body functions.
- REM sleep is thought to be important for restoring brain functions.

KEY CONCEPT BOX 6.1b

- Sleep is a psychological construct as it is something we believe to exist, as we can measure its effects, but we cannot directly observe or measure it.
- Devices used to measure the physiological activity in the brain, skeletal muscles and muscles that move the eyes during sleep include the EEG, EMG and EOG, which enable scientists to determine the stages of sleep.
- An electroencephalograph (EEG) detects, amplifies and records the electrical activity of the brain (brainwave patterns from beta, alpha, theta and delta).
- An electromyograph (EMG) detects, amplifies and records the electrical activity of the skeletal muscles in the body (high levels of electrical activity indicate more muscle tension).
- An electro-oculograph (EOG) detects, amplifies and records the electrical activity of the muscles surrounding the eye (high levels of electrical activity occur when awake and during REM).
- Sleep diaries are a useful subjective measurement of sleep behaviour and other lifestyle habits that may impact sleep.
- Video monitoring uses an infrared camera to record sleep behaviour unobtrusively, which can be used to indicate sleep behaviour such as movement and snoring (sounds).

KEY CONCEPT BOX 6.2

- Regulation of the sleep–wake pattern is driven in part by a circadian rhythm.
- A circadian rhythm is a biological rhythm (pattern) that occurs once every 24 hours and is endogenous in humans.
- The circadian rhythm of sleep–wake is influenced by external factors such as the absence of light signals received by the eyes when there is a lack of daylight (evening), which stimulates the suprachiasmatic nucleus (master clock) to signal to the pineal gland to release melatonin.
- Melatonin is a hormone that promotes drowsiness and readies us for sleep.
- When light conditions increase, cortisol is released, which wakes us up and makes us more alert.
- In the absence of light/dark cues in the environment, the circadian rhythm extends slightly, but still operates.
- Ultradian rhythms are biological rhythms that occur more frequently, following a cycle of less than 24 hours.
- Healthy adults experience about five cycles of sleep in a night, each lasting about 90 minutes, which is an example of an ultradian rhythm.

KEY CONCEPT BOX 6.3

- The amount of total sleep required decreases over the life span, remaining stable from adulthood (newborns 16 h, infants 14 h, childhood 12 h, adolescence 9 h, adults 7–9 h, elderly 7–8 h).
- The proportion of REM sleep decreases over the life span, stabilising somewhat in adulthood (newborns 50%, infants 35%, childhood 20–25%, adolescence 20–25%, adults 20%, elderly 20%).
- The proportion of NREM sleep increases over the life span, stablising somewhat in adulthood (newborns 50%, infants 65%, childhood 75–80%, adolescence 75–80%, adults 80%, elderly 80%).
- Newborns have an irregular sleep pattern due to an under-developed circadian rhythm.
- The high amount of NREM sleep in the early years of life may be explained by the need for sleep to support the growth and the replenishment of the body, as it is in deep sleep that the body physically repairs itself, boosting the immune system and restoring bones, muscles and tissue.
- The high amount of REM sleep in the early years may be explained by the need to support the brain changes needed for healthy brain development in the first years of life.

6 End-of-chapter exam

Section A: Multiple choice

1 The sleep–wake cycle we experience every 24 hours is a biological rhythm called _____ rhythm.
   A an ultradian
   B a circadian
   C an infradian
   D a sleep

2 Which of the following statements best describes the noticeable changes that occur to an individual's sleep as they age?
   A The proportion of time spent in NREM sleep increases as we age.
   B The proportion of time spent in REM sleep significantly decreases from infancy and then remains relatively stable as we continue to get older.
   C As we go from being an infant to old age, the proportion of time spent in stages 1 and 2 NREM significantly decreases.
   D The time spent sleeping overall increases as we age.

3 The changes that occur to sleep as we progress through an episode of sleep are typically that
   A more time is spent in NREM sleep and less time is spent in REM sleep.
   B more time is spent in REM sleep and less time is spent in NREM sleep.
   C more time is spent in slow-wave sleep and less time is spent in light sleep.
   D time spent dreaming decreases.

4 Which one of the following is true of a sleep cycle?
   A It is an ultradian rhythm that lasts about 7–8 hours.
   B It is a circadian rhythm that occurs throughout the night.
   C It is a circadian rhythm that is synchronised to the core body temperature rhythm.
   D It is an ultradian rhythm within a circadian rhythm.

5 The role of the EOG is to detect, amplify and record the electrical activity of the
   A eyeballs.
   B eyes.
   C muscles surrounding the eye that enable the eye to rotate and move.
   D brain.

6 Which one of the following is the sequence of events that help regulate sleep?
   A The hypothalamus signals the optic chiasm, which in turn sends signals to the pineal gland.
   B The hypothalamus signals the optic chiasm to release or suppress the release of melatonin.
   C The suprachiasmatic nucleus signals the pineal gland to increase or decrease secretion of melatonin.
   D The pineal gland sends signals to the thalamus, which stimulates the secretion of melatonin.

7 The suprachiasmatic nucleus is located where in the brain?
   A the pineal gland
   B the hippocampus
   C the hypothalamus
   D the optic chiasm

8 A difference between adult and elderly sleep is
   A adults will experience more periods of night-time awakenings than elderly people.
   B there will be a decrease in melatonin release in adults before sleep compared to elderly people.
   C adult sleep begins with a period of REM sleep followed by NREM sleep, while in the elderly this is reversed: NREM is first, followed by REM.
   D adults experience more stage 3 NREM sleep than elderly people.

9 Which of the following is an example of a naturally occurring altered state of consciousness?
   A anaesthetised
   B focused attention
   C sleep
   D hypnotised

10. The first ultradian rhythm of sleep is different from the last ultradian rhythm as the last involves
    A more time spent in NREM sleep.
    B more time spent in REM sleep.
    C only stages 1 and 2 NREM sleep.
    D frequent and brief awakenings.

11. Which is true of altered states of consciousness (ASC)?
    A ASC result in reduced perception of time, making the passage of time difficult to judge accurately.
    B ASC result in an increase in the level of awareness compared to NWC.
    C ASC cannot occur naturally.
    D ASC cannot be induced.

12. Which of the following is an example of a subjective measure of sleep?
    A an EEG
    B an EOG
    C a sleep diary
    D an EMG

13. An EMG and EOG respectively are examples of
    A a physiological method of recording the electrical activity of the skeletal muscles; a psychological method of recording the electrical activity of the brain.
    B a psychological method of recording the electrical activity of the skeletal muscles; a psychological method of recording the electrical activity of the brain.
    C a physiological method of recording the electrical activity of the skeletal muscles; a psychological method of recording the electrical activity of the brain.
    D a physiological method of recording the electrical activity of the skeletal muscles; a physiological method of recording the electrical activity of the muscles surrounding the eyes responsible for eye movement.

14. Pedro stayed up all night studying for his English SAC the next day. It is probable that he will not perform to the best of his ability on this cognitive task due to missing out on what type of sleep?
    A REM
    B NREM stage 1
    C NREM stage 2
    D NREM stage 3

15. During REM sleep
    A skeletal muscles increase in tone and tension.
    B the eyes move slowly and steadily.
    C skeletal muscles become atonic.
    D brain activity decreases.

16. Emilia has joined a representative club for basketball. She trains every second day and has games on Friday nights. After the last game, she attended her best friend's birthday and didn't get much sleep. In the morning she felt like her body had not recovered as well as it usually would. In terms of her sleep, what does Emilia need to do in the future to improve her recovery?
    A She should get more REM sleep.
    B She should try to keep her sleep light in order to get more stages 1 and 2 NREM sleep.
    C She should not bother going to bed but go for a run to loosen her muscles instead.
    D She should get more stage 3 NREM sleep.

17. Which of the following ways to study sleep is considered an objective measure?
    A an electroencephalograph recording the electrical activity of participants' skeletal muscles
    B a sleep diary recording how a participant felt after each sleep episode
    C a questionnaire with a rating scale asking participants to rate their level of tiredness before sleep and after sleep
    D video monitoring which records the time and duration of participants' awakenings

18. An ultradian rhythm
    A occurs more frequently than a circadian rhythm.
    B must occur every hour.
    C includes the sleep–wake cycle.
    D will occur once in a 24-hour period.

19. Billy, a 16-year-old student, lives with his elderly grandfather. In terms of their sleep patterns, which of the following is more typical for people of Billy's age and his grandfather's age respectively?
    A Billy will go to sleep earlier; his grandfather will go to bed later but rise earlier.
    B Billy will sleep for longer; his grandfather will get more REM.
    C Billy will go to bed later and rise later; his grandfather will rise earlier.
    D Billy will get more stage 3 NREM sleep; his grandfather will get more total sleep.

20 Which of the following are sleep–wake shifts that could be observed in the sleep patterns of a healthy adolescent compared with those of an adult?
   A Adolescents go to sleep before an adult, but their wake times are usually after an adult.
   B There are no differences between sleep patterns in adolescents and adults.
   C Adults go to sleep earlier and wake earlier than adolescents.
   D Adults go to sleep later and wake earlier than adolescents.

Section B: Short answer

1 Compare the sleep patterns of healthy infants to the sleep patterns of healthy adults. Identify four ways that sleep differs between the two groups. Give a reason for one of these differences. [4 marks]

2 A sleep scientist wants to determine if a participant is experiencing REM or stage 3 NREM sleep in a sleep study. Explain which device would be useful in making this judgement. [3 marks]

3 Describe the role of the suprachiasmatic nucleus in regulating sleep. [5 marks]

Importance of sleep to mental wellbeing

7

Key knowledge

- the effects of partial sleep deprivation (inadequate sleep either in quantity or quality) on a person's affective, behavioural and cognitive functioning, and the affective and cognitive effects of one night of full sleep deprivation as a comparison to blood alcohol concentration readings of 0.05 and 0.10
- changes to a person's sleep–wake cycle that cause circadian rhythm sleep disorders (Delayed Sleep Phase Syndrome [DSPS], Advanced Sleep Phase Disorder [ASPD] and shift work) and the treatments of circadian rhythm sleep disorders through bright light therapy
- improving sleep hygiene and adaptation to zeitgebers to improve sleep–wake patterns and mental wellbeing, with reference to daylight and blue light, temperature, and eating and drinking patterns

Key science skills

Develop aims and questions, formulate hypotheses and make predictions
- predict possible outcomes of investigations

Generate, collate and record data
- systematically generate and record primary data, and collate secondary data, appropriate to the investigation
- record and summarise both qualitative and quantitative data, including use of a logbook as an authentication of generated or collated data
- organise and present data in useful and meaningful ways, including tables, bar charts and line graphs

Analyse and evaluate data and investigation methods
- identify and analyse experimental data qualitatively, applying where appropriate concepts of: accuracy, precision, repeatability, reproducibility and validity; errors; and certainty in data, including effects of sample size on the quality of data obtained
- construct evidence-based arguments and draw conclusions
- evaluate data to determine the degree to which the evidence supports or refutes the initial prediction or hypothesis
- identify, describe and explain the limitations of conclusions, including identification of further evidence required
- discuss the implications of research findings and proposals, including appropriateness and application of data to different cultural groups and cultural biases in data and conclusions

Analyse, evaluate and communicate scientific ideas
- use appropriate psychological terminology, representations and conventions, including standard abbreviations, graphing conventions and units of measurement
- discuss relevant psychological information, ideas, concepts, theories and models and the connections between them

Source: VCE Psychology Study Design (2023–2027) + pages 39, 12–13

7 Importance of sleep to mental wellbeing

In Chapter 6 you learnt about the mechanics of what happens when you sleep, but what happens when you don't get enough sleep? You have no doubt had many nights when you could not sleep properly. How did you feel the next day? Not on top of your game? Don't despair, there are ways you can improve your sleep–wake patterns.

7.2 Sleep disorders
p. 293

Sometimes not getting enough sleep is not because you chose to party all night. There is a group of sleep disorders that result from disruption to our circadian rhythm – also known as our internal body clock.

7.1 Partial sleep deprivation
p. 283

Even partial sleep deprivation can affect your emotions, thinking and behaviour… but if you do not sleep at all over a 24-hour period then this can affect you in the same way as having a blood alcohol level of 0.10 per cent.

Adobe Stock/franz12

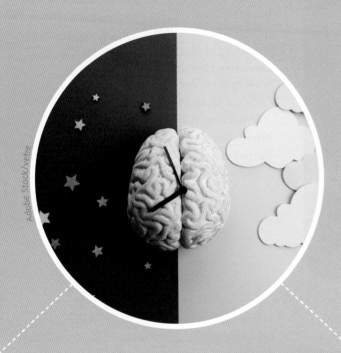

p. 303

**7.3
Improving sleep–wake patterns and mental wellbeing**

You can improve your sleep patterns by being aware of zeitgebers (environmental cues) and making changes to your sleep hygiene practices. Exposure to light, body temperature and eating patterns are all things that can affect how well you sleep.

Sleep can literally make or break us. You perform at your best when you have had enough sleep. There are some routines you can start to develop now to improve your sleep quality and quantity. Sweet dreams!

Slideshow
Chapter 7 slideshow

Flashcards
Chapter 7 flashcards

Test
Chapter 7 pre-test

Assessment
- Pre-test
- End-of-chapter exam

Revision
- Chapter map
- Key term flashcards
- Key concept summary
- Slideshow

Investigation
- Investigation: Investigating circadian rhythms
- Data calculator
- Logbook template: Correlational study

Worksheet
- Delayed sleep disorder and bright light therapy

To access these resources, visit
cengage.com.au/nelsonmindtap

Nelson MindTap

Know your key terms

- Advanced Sleep Phase Disorder (ASPD)
- Affective functioning
- Blood alcohol concentration (BAC)
- Bright light therapy
- Circadian rhythm sleep disorder
- Delayed Sleep Phase Syndrome (DSPS)
- Microsleeps
- Partial sleep deprivation
- REM rebound
- Shift work
- Shift work disorder
- Sleep debt
- Sleep deprivation
- Sleep-deprivation psychosis
- Sleep disorder
- Sleep hygiene
- Sleep–wake cycle
- Total sleep deprivation
- Zeitgeber

On 24 March 1989, the oil tanker *Exxon Valdez*, full of crude oil, had just entered Alaska's Prince William Sound. The ship never made it through. While attempting to avoid an iceberg, the ship ended up running aground on a reef, spewing more than 40 million litres of oil into the pristine Alaskan coastline. The initial response was ineffective in containing much of the spill, and a storm blew in soon after, spreading the oil widely. Eventually, more than 2100 kilometres of coastline were polluted, and hundreds of thousands of animals were killed or left severely affected by the oil.

Fatigue and poor management were blamed at least in part for the accident. The oil tanker was piloted at the time by the Third mate, who allegedly had not slept more than 2 hours in the 16 hours leading up to the accident, leaving him severely sleep deprived. Added to that, the Third mate was not qualified to take control of the vessel. The captain was initially reported to have been intoxicated in his bunk at the time of the accident, and was in charge of a ship with insufficient crew to cover shifts, so crew were working without sufficient sleep and were performing tasks they were not qualified to perform.

Exxon ended up paying billions in clean-up costs and fines and remains tied up in court cases to this day. The captain was acquitted of being intoxicated while at the helm, but convicted on a misdemeanour charge of negligent discharge of oil, fined $50 000, and sentenced to 1000 hours of community service. Though the oil has mostly disappeared from view, many Alaskan beaches remain polluted to this day, with crude oil buried just centimetres below the surface.

The *Exxon Valdez* oil spill is not the only disaster at least in part caused by lack of sleep. The Chernobyl disaster in the Ukraine, when a nuclear power plant exploded in 1986, causing radioactive material to be dispersed into the surrounding environment, was also reportedly caused by poor training and workers experiencing sleep deprivation. Disasters such as these have inspired several research investigations into the effects of sleep deprivation, aimed at improving public health and safety.

However, despite this, inadequate sleep due to sleep disorders, work schedules and chaotic lifestyles continues to threaten health, mental health and safety.

We all know that we often feel 'better' after a good night's sleep. We have learned a lot about the mechanisms of sleep and what happens to our body when we experience sleep from the previous chapter.

So, what are the benefits of sleep for our mental health and why is sleep so important? The relationship between sleep and mental health is bidirectional. This means that while poor sleep can increase mental health problems and challenges, stress and poor mental health can also disrupt sleep.

This works the other way around too: getting good sleep is linked to positive mental health outcomes and when there is the absence of mental health challenges, good sleep is more likely. In this chapter we will learn about the importance of sleep for mental wellbeing by investigating some of the effects of not getting enough sleep or sleep being disrupted by changes to our circadian rhythm (our body clock that helps regulate when we sleep). We will also study ways to improve sleep by creating good habits and being aware of cues that will promote sleep.

Figure 7.1 *Exxon Valdez* oil spill workers hose Quayle Beach, Smith Island in Prince William Sound, Alaska, US

7.1 Partial sleep deprivation

When an individual does not get adequate amounts of sleep or quality of sleep, they experience **sleep deprivation** and functioning in their awake state may become difficult. Sleep deprivation may result from a variety of factors, including work or study demands, social and family responsibilities, the sleeping environment, poor sleep hygiene, medical conditions and sleep disorders.

Regardless of the cause of sleep deprivation, the effects are the same. Going without adequate sleep or without any sleep carries with it both short- and long-term psychological and physiological consequences.

The severity of the effects of sleep deprivation will be determined by the type of sleep loss experienced and the length of time the person was sleep deprived. If the person has some sleep in a 24-hour period but less than they normally require to perform at optimal level during the day, they experience **partial sleep deprivation**. If they have no sleep at all in a 24-hour period or more, they will experience **total sleep deprivation**.

How long might a person go without sleep? With few exceptions, four days or more without sleep becomes unbearable for anyone. Nevertheless, longer periods without sleep are possible. The world record is held by Californian Randy Gardner, who, at 17 years of age in 1964, went 268 hours (11 days) without sleep (Figure 7.2). At various times, Gardner experienced irritability, memory lapses, difficulty concentrating and difficulty in naming common objects. Surprisingly, Gardner needed only 14 hours of sleep to recover (Coren, 1996).

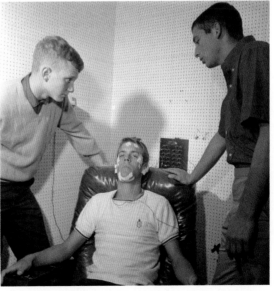

Figure 7.2 Randy Gardner, who at 17 years of age went without sleep for 11 days

As Gardner's experience demonstrates, after a period of sleep deprivation it is not necessary to completely replace lost sleep.

Weblink
Randy Gardner on 'To tell the truth' Randy Gardner study and sleep deprivation

Most symptoms and side effects of sleep deprivation are reversed by a single night of uninterrupted sleep. People who have been sleep deprived may report sleeping for longer than normal in the first few nights following deprivation, but there is generally no need to replace the total amount of sleep that has been lost.

Effects of partial sleep deprivation

Partial sleep deprivation can result when a person gets some sleep but not enough (in quantity or quality) to support their normal waking alertness, performance and health. This can be due to disruptions to the normal progression and sequencing of sleep stages, which leads to fragmented sleep (decreased quality). In the short-term, partial sleep deprivation can affect judgement, mood, and ability to learn and retain information. Although short-term sleep deprivation is usually short and temporary, it still increases the risk of serious accidents and injury during times when a person is awake. In the long term, partial sleep deprivation may lead to a host of health problems including obesity, diabetes, hypertension, high blood pressure and cardiovascular disease. Early mortality may also result from chronic total sleep deprivation (Dinges, 1995).

It is not necessary to go completely without sleep to feel the effects of sleep loss. One third of all adults and most teenagers do not get enough sleep each night. Sleep deprivation studies repeatedly show a variable (negative) impact on mood, cognitive performance and behavioural (motor) function (Durmer & Dinges, 2005).

Affective functioning: amplified emotional responses

Sleep helps us regulate our emotions. Sleep deprivation is detrimental to mood and emotional stability because it amplifies emotional responses. For example, research suggests that the brain's prefrontal cortex, the area of the brain responsible for executive functioning and emotional regulation, is particularly sensitive to sleep deprivation and that sleep loss causes deficits in our ability to regulate our emotions (Durmer & Dinges, 2005; Dahl & Lewin, 2002). Although the effects of sleep deprivation depend on the individual, the amount of sleep they lose, the type of sleep lost and how long they have been deprived of adequate sleep, many sleep-deprived people report a change in their **affective functioning** (emotional control and responses) as the first sign that they are sleep deprived.

Sleep deprivation has been shown to greatly influence the ability to process emotional information, put it into context and produce controlled, appropriate responses (Kahn, et al., 2013). More specifically, frequently disrupted and restricted sleep seems to increase the tendency of the individuals to experience negative emotions and develop mood disorders such as depression and anxiety (Babson & Feldner, 2010; Belenky et al., 2003; Chang et al., 1997; Riemann & Voderholzer, 2003; Van Dongen, Maislin, et al., 2003). Often, sleep-deprived people report feeling uncharacteristically anxious, irritable, angry, aggressive and unmotivated. Studies have shown that sleep loss affects the ability to recognise and categorise other people's emotions, particularly from facial expressions (Minkel et al., 2010; Tempesta et al., 2010; Van Der Helm et al., 2010). This reduces the individual's self-perceived emotional intelligence. As they are less able to understand another person's mental state in terms of emotions, feelings and thoughts, sleep-deprived people become less empathetic towards others (Killgore et al., 2006).

Lack of adequate sleep may also lessen the ability to cope with stress. When sleep deprived, people may feel overwhelmed by routine activities, such as completing homework, doing shopping, keeping up an exercise regime or waiting at traffic lights. Inadequate sleep not only affects mood, but mood and mental states can also affect sleep. For example, anxiety increases agitation and arousal, which makes it more difficult to sleep.

Cognitive functioning: disruption of thinking processes

Sleep is essential for cognitive (thinking) performance ranging from simple attention and alertness to higher-order executive functions (Maquet, 2001; Stickgold, 2005).

When sleep deprived, a person's mental abilities become impaired so their ability to perform cognitive tasks declines, particularly higher order tasks involving vigilant attention. For example, the person's spatial orientation may deteriorate and they may begin to think irrationally and illogically. Executive functioning (through the use of the prefrontal cortex) is compromised and the ability to plan, coordinate, implement and evaluate deliberate actions is disrupted. The person may have difficulty controlling attention and maintaining concentration for an extended time, so their ability to perform tasks to a set standard may be reduced (Durmer & Dinges, 2005). Their ability to make decisions and problem solve may suffer. They may make uncharacteristic errors in judgement because their ability to assess risk, assimilate changing information and revise strategies may be reduced.

Sleep deprivation also impairs memory and learning. People who do not have adequate sleep find it difficult to concentrate on information long enough for it to be processed in short-term memory and transferred to long-term memory for relatively permanent storage. They also find it more difficult to access and retrieve information stored in long-term memory. At a cellular level, changes in synaptic efficacy and membrane excitability are thought to be critical for the formation of memories, and there have been some suggestions that these neuronal properties could be altered during sleep and by the lack of sleep (Bliss & Collingridge, 1993; Graves et al., 2001). Sleep research suggests that the nerve connections that make our memories are formed and strengthened during sleep. For example, neuronal connections that were active during a previous learning experience are reactivated during sleep, suggesting that some memory consolidation may occur during sleep (Louie & Wilson, 2001). If sleep time is cut short or disrupted, it interferes with this process. For example, the percentage of time spent in rapid eye movement (REM) sleep is increased after certain learning tasks (Mandai et al., 1989; Smith & Lapp, 1986), and if REM sleep is prevented after the training, the subsequent performance of this task is impaired (Karni et al., 1994; Smith, 1995).

The sleep-deprived person's ability to successfully complete simple, monotonous or repetitive tasks, particularly those that require reaction speed or vigilance, declines. They struggle to simultaneously focus on several different related tasks. They do not have the speed or creative abilities they normally would have to make quick logical decisions. They find it more difficult to react quickly to unpredicted rapid changes in their circumstance. As a result, there are significant variations in their performance and they are prone to making a number of errors (Doran et al., 2001). These effects make many activities, such as driving a car, playing contact sports or using power tools, very dangerous when you are sleep deprived. This shows the close interaction between the cognitive and behavioural effects of sleep deprivation.

Figure 7.3 When an individual is sleep deprived, it is difficult to maintain attention when undertaking the monotonous task of a long car drive.

Behavioural functioning: impaired actions and reactions

The behavioural effects (observable actions) of sleep deprivation vary depending on a person's age and on how long or how much they have been deprived of sleep. With less sleep, less glucose is metabolised, so muscle strength, speed of movement and endurance are reduced. The ability to perform fine motor functions requiring coordination of the eyes and hands (such as handwriting, computer skills or operating machinery) is also impaired, causing an increase in clumsiness and accidents or injuries. Sleep deprivation leads to slower reaction times, especially on motor tasks.

During the day if we are awake but sleep deprived, the body reacts with brief periods of sleep known as microsleeps. **Microsleeps** are short periods (a few seconds) where the individual appears to be awake – their eyes may even be open – however, EEG recordings of brain activity show brainwaves similar to those shown in the first stage of NREM sleep. This indicates that they are asleep but the individual may have no recollection of what occurred while they were having a microsleep. Microsleeps become more prevalent as a person accumulates a sleep debt, so the more sleep deprived a person is, the greater the chance a microsleep episode will occur. Microsleeps are especially prevalent when people who are sleep deprived complete monotonous or unstimulating tasks.

Figure 7.4 When driving a car, microsleeps can be potentially fatal because they cause the driver to lose awareness of their surroundings.

Table 7.1 The typical effects of partial sleep deprivation

Aspect of functioning	Effects of sleep deprivation
Affective functioning	» Mood swings » Increase in 　– negative emotions 　– irritability » Reduced motivation » Easily bored » Reduced empathy towards others » Inability to cope with stress
Cognitive functioning	» Memory lapses » Difficulty maintaining attention and concentration » Difficulty processing information » Difficulty thinking logically and problem-solving » Reduced creativity » Distorted perceptions » Poor decision-making » Reduced spatial awareness
Behavioural functioning	» Difficulty completing routine tasks » Increase in risk-taking behaviour » Reduced ability to perform fine motor tasks and an increase in clumsiness » Slowed reflexes » Lack of energy (lethargy) » Trembling hands

Although microsleeps may be one way the body attempts to cope with sleep deprivation, the effects of losing awareness for even a couple of seconds can be dangerous and devastating. For example, when driving a car, a person may not notice a sudden change in their external environment. Microsleeps intrude into wakefulness and disrupt attention, and they are regularly implicated in fatal accidents. For this reason, drivers who feel tired are encouraged to take a break or a 15-minute powernap.

As people continue to live busy lifestyles and many industries operate 24/7, requiring workers to do shift work, and students go to school and participate in sports and part-time jobs while juggling a social life, sleep is often ranked low on the list of priorities, therefore sleep deprivation is affecting many people. Research is telling us about the importance of sleep for our safety and mental and physical wellbeing. We know that it is important to get enough sleep, and that sleep has more restorative effects if a person can go through an entire sleep episode with the appropriate sequence of NREM and REM sleep at night, when the body is programmed to sleep by our circadian rhythm. As we will see later in this chapter, when sleep is fragmented like that of shift workers, or the circadian rhythm is not in alignment with environmental cues, this can have a negative effect on waking functions as the person experiences sleep deprivation.

EXAM TIP
It is important to be able to use the terms 'cognitive', 'affective' and 'behavioural' when referring to the various effects of sleep deprivation even though you may learn them more informally as 'thinking', 'emotions' and 'actions'.

ANALYSING RESEARCH 7.1

Chronically sleep deprived? You can't make up for lost sleep

Researchers at Northwestern University in Illinois, United States, have discovered that when rats are partially sleep deprived over consecutive days they do not attempt to catch up on sleep, despite an accumulating sleep deficit. Their study is the first to show that repeated partial sleep loss negatively affects an animal's ability to compensate for lost sleep.

We've all experienced occasional partial sleep deprivation, perhaps by staying out too late at a party on a weeknight or studying into the early hours for a morning exam. It is well established that our bodies try to catch up by making us sleep more and/or more deeply the following night in order to maintain a homeostatic balance between sleep and wakefulness. But what happens when this sleep deprivation becomes chronic, when we lose a little bit of sleep over a period of days, months or even years?

In the study by sleep researchers at Northwestern University, rats were subjected to sleep deprivation, which involved restricting their sleep to 4 hours in a 24-hour period for five days consecutively (so the rats were only able to sleep for 4 hours a day). Normally a rat would sleep for about 15 hours in a day, so this restriction reduced the amount of sleep the rats were having substantially. The team monitored brainwave and muscle activity patterns in order to precisely quantify sleep–wake patterns. On day 1, when the rats were deprived of sleep for 20 hours and then had the opportunity to sleep, the sleep was more intense than usual. The animals exhibited more delta waves (NREM) and more REM sleep compared to the baseline, when they were not sleep deprived. On days two to five, the rats did not show any compensatory response, in that they did not exhibit more delta waves or an increase in NREM or REM sleep. After missing out on 35 hours of sleep over the five days of sleep restriction, the rats failed to regain their lost sleep, even when they were given three days to recover with no sleep restriction. This suggests that the rats were able to compensate for 'acute' sleep deprivation as their sleep was restricted at day 1, but after this when the sleep deprivation lasted longer, they did not show any changes to overcome the loss of sleep. 'We now know that chronic lack of sleep has an effect on how an animal sleeps,' said Fred W. Turek, director of Northwestern's Center for Sleep and Circadian Biology and an author of the paper. 'The animals are getting by on less sleep but they do not try and catch up.'

The findings support what other scientists have discovered in recent experimental studies in humans. Chronic partial sleep loss of even two to three hours per night was found to have detrimental effects on the body, leading to impairments in cognitive performance, as well as cardiovascular, immune and endocrine functions. Sleep-restricted people also reported not feeling sleepy even though their performance on tasks declined. The Northwestern team's results suggest that animals may undergo a change in their need for sleep in situations where normal sleep time is prohibited or where sleep could be detrimental for survival. An extreme but realistic example of this, says Turek, would be how animals respond to catastrophic environmental conditions. No matter how sleep deprived an animal or human may be, it would not be adaptive for the sleep homeostat to kick in and to make the animal fall asleep when it is in the midst of a flood or forest fire. Therefore, the body undergoes some change that allows it to counter its homeostatic need for sleep and to stay awake to avoid danger. Turek and his team propose that this change in the sleep regulatory system is reflective of an allostatic response (the process of achieving stability or homeostasis through physiological or behavioural change). In the short term, allostatic responses are adaptive, but when sustained on a chronic basis, such as in their study, an allostatic load will develop and lead to negative health outcomes. The allostatic load resulting from the accumulating sleep debt loops back to the sleep regulatory system itself and alters it.

Source: adapted from Northwestern University. (2007, July 2). ScienceDaily.

> **Questions**
> 1 Identify the aim of this study.
> 2 Identify the population of research interest.
> 3 Identify the independent variable(s) and dependent variable(s) in this study.
> 4 Identify the participants in this study.
> 5 What were the results of this study?
> 6 What conclusion(s) can be drawn from this study?

7.1.1 SLEEP DEPRIVATION

Total sleep deprivation

If a person is totally sleep deprived, the effects of partial sleep deprivation are magnified. Severe total sleep loss can cause a temporary psychological problem known as sleep-deprivation psychosis. **Sleep-deprivation psychosis** is a major disruption of mental and emotional functioning brought on by lack of sleep, and has symptoms such as confusion and disorientation. Psychologically, they may feel paranoid, or they may become extremely anxious or depressed. If their sleep deprivation lasts for approximately 72 hours they may suffer from perception problems such as hallucinations. Hallucinations may be visual, such as seeing yourself wearing a 'coat of furry worms'; or tactile, such as feeling cobwebs on your face.

In general, people who have been totally sleep deprived for two or three days show little impairment on relatively interesting or complex mental tasks (Binks et al., 1999). Simple and routine tasks, however, seem to be very difficult for the sleep deprived, and in these tasks most people experience problems with attention and concentration. This is particularly important to note for vehicle drivers, pilots or machine operators as, for these people, making a mistake when conducting simple or routine tasks may prove fatal. Sleep deprivation has been shown to produce psychomotor impairments equivalent to those induced by alcohol consumption at or above the legal limit (Dawson & Reid, 1997). Sleep deprivation poses a risk to safely operating all modes of transport and performing other safety-sensitive activities. If a task is monotonous (such as factory work), no amount of sleep deprivation is safe (Gillberg & Akerstedt, 1998).

Recovering from partial sleep deprivation

Fortunately, the effects of short-term partial sleep deprivation tend to be minor and temporary and, once the person resumes their normal uninterrupted sleep pattern and their sleep debt is repaid, they quickly recover. Generally, they fall asleep quicker and sleep longer than normal on their first full night of sleep after being sleep deprived. The person does not need to fully compensate for the amount of sleep lost; a few hours of extra sleep over the next few nights is usually enough for them to recover.

If, during their period of sleep deprivation, the person has been deprived of REM sleep, once they sleep normally, they usually compensate for their REM deprivation by having extra amounts of REM sleep. This is known as **REM rebound**. Later experiments have shown that missing any sleep stage can cause a rebound for that stage.

ANALYSING RESEARCH 7.2

Fatigue trial test a wake-up to drowsy drivers

The Victorian Government has completed an Australian-first, revolutionary new technology trial to detect drowsy drivers. The Government announced an on-track trial of pupil scanning technology has successfully detected excessively fatigued drivers and found drivers with only three hours sleep are ten times more likely to be involved in a crash.

In the Road Safety Victoria and TAC-led trial, participants were kept awake for up to 32 hours before conducting a two-hour drive on a controlled track in Kilsyth while supervised by a qualified instructor in a dual controlled vehicle. Participants also undertook three additional drive tests – with 3 hours sleep and 5 hours sleep in a 24-hour period, and again when they were well-rested after 8 hours sleep. Drivers were tested before and after their drive with technology that measures involuntary movement of their pupils (Figure 7.5) which has shown strong links with increased levels of sleep deprivation.

A range of behavioural, physiological and driver performance data was also collected including brain electrical activity, lane deviations, speed variations and changes in reaction time. The project will examine the accuracy of the technology, with potential to conduct future trials of roadside testing for excessive fatigue – in a similar way to current roadside alcohol and drug testing.

Figure 7.5 Participant wearing technology that measures involuntary movement of their pupils.

Current figures show fatigued drivers are involved in up to 20 per cent of crashes and 11 per cent of fatalities on Victorian roads.

Source: Department of Transport, Victoria.

Questions
1. What is the aim of the study by the TAC?
2. Which research design was used in the study?
3. Outline one advantage and one disadvantage of the method used to collect the data.
4. What ethical concepts and/or guidelines would the researcher need to follow to ensure the wellbeing of the participants?

Comparison between the effects of sleep deprivation and alcohol consumption

Depressants are drugs that calm neural activity and slow down bodily functions. Alcohol is a depressant; it slows down the messages between the brain and the body. Alcohol enters the bloodstream through the stomach and intestines. Its effects slow down if a person has a full stomach; however, someone who has just eaten will still experience the effects of the drug. Some of the effects of moderate alcohol consumption on consciousness include reduced inhibitions and feeling relaxed, calmer and more confident. People under the influence of alcohol may also experience a loss of self-control, impaired mobility and coordination and slower reaction times. Excessive alcohol consumption or 'binge drinking' can lead to more severe effects on consciousness, including blurred vision, nausea and vomiting, aggression, confusion, memory loss, and unconsciousness (passing out, which can lead to coma and even death). The amount of alcohol in our bloodstream is measured as a **blood alcohol concentration (BAC)**. The only way our BAC drops is when our liver metabolises the alcohol, which takes time. Drinking black coffee, vomiting, taking a powernap or having a cold shower do not speed up this metabolic process despite what some people may believe.

Alcohol affects the brain and nervous system, however every individual's response to the drug may differ slightly. Factors such as gender, body size and whether there is food in the stomach will affect alcohol tolerance and metabolism. Generally, however, the more alcohol consumed, the more severe the effects.

7.1.2 A COMPARISON BETWEEN THE EFFECTS OF SLEEP DEPRIVATION AND ALCOHOL CONSUMPTION

EXAM TIP
The comparisons between a BAC 0.05 per cent and 17 hours of no sleep and a BAC 0.10 per cent and 24 hours of no sleep are important figures to remember. It is important that you report them carefully, as a decimal point in the wrong place will cause the comparison to be incorrect. BAC should be reported as a percentage.

As the BAC approaches 0.05 per cent the effects are such that performing tasks such as driving become dangerous. The cognitive effects of alcohol include a reduced ability to problem solve and make decisions, impaired logic and reasoning, and impaired memory. Alcohol may also reduce the ability to assess risks and can impair judgement. Affective (emotions) effects of consuming alcohol include amplified emotions, inappropriate emotional expression, and fluctuation in mood with little control over the mood changes. When consuming alcohol some people may experience a range of moods and emotions, from extreme happiness to extreme sadness.

When we look at the effects of alcohol and compare these effects to a person who is experiencing sleep deprivation, we begin to see some startling similarities. Research has shown that there are similarities in the decrease in cognitive performance and changes to affective functioning and behavioural responses when comparing participants in studies who have been deprived of sleep to those consuming alcohol. Once a person has been awake for 17 hours (waking at 7 a.m. and staying awake until midnight) this is equivalent to the effects of consuming alcohol to a BAC of 0.05 per cent. At 24 hours of sleep deprivation (waking at 7 a.m. and still going at 7 a.m. the next day) the effects are the equivalent to a BAC of 0.10 per cent. The behavioural responses when sleep deprived compared to when alcohol is consumed are also similar. They both generally reduce speed of movement, reduce coordination of eyes and hands, and increase clumsiness. The cognitive deficits experienced with sleep deprivation are like those experienced when consuming alcohol and the affective effects are also similar. However, a difference may be that for most people, sleep deprivation results in a negative mood (feeling irritated and grumpy), whereas when consuming alcohol, mood can be affected positively or negatively and may fluctuate (feeling angry and then happy).

It is deemed safe to drive a car if you are fully licensed in Victoria, with a BAC of less than 0.05 per cent. A BAC of 0.10 per cent would put you at double the legal limit for a fully licensed driver. Driving a car safely is a task that requires a high level of conscious awareness. When you are behind the wheel you need to make important decisions on the road that require total concentration, rapid reflexes, good coordination and the ability to make good judgements. However, when a person is under the influence of alcohol their consciousness is impaired. In Victoria, the legal BAC when driving is under 0.05 per cent for fully licensed drivers and 0.00 per cent for probationary and learner drivers as well as professional drivers such as heavy truck drivers, taxi and bus drivers. The Transport Accident Commission (TAC) reported that one in five drivers and riders killed had a BAC greater than 0.05 per cent (TAC, 2021). People in other jobs such as pilots, machine operators and police officers are required to have a BAC of 0.00 per cent when working. Even though there may be some difference in the way alcohol affects some people, it has been established that alcohol impairs functioning.

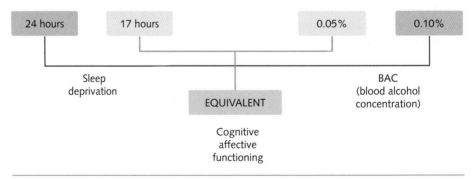

Figure 7.6 Sleep deprivation can have similar effects on cognitive functioning to consuming alcohol.

ACTIVITY 7.1 SLEEPINESS SCALE

Try it

For each of the statements in the table below, give yourself a score from 0–3 based on the sleepiness self-report scale below.

Rating scale:
0 = I have **never** fallen asleep this way.
1 = **Very occasionally** I may drop off.
2 = I **often** fall asleep this way.
3 = I **regularly** fall asleep doing this.

1. Over the past month, how often have you fallen asleep while doing these things?

Activity	Chance of falling asleep (0–3)
Sitting down whilst reading	
Eating lunch alone	
Being a passenger in either a car or on public transport	
Chatting to somebody	
Lying on the couch and watching a film or series	

Scoring

A score of 7 or more: suggests excessive daytime sleepiness

A score greater than 10: suggests a high level of excessive daytime sleepiness

Apply it

Research by the Australian Institute for Family Studies (2019) has estimated that almost half of Australian teenagers aged 16–17 are not getting enough sleep.

2. Identify one affective, one behavioural and one cognitive effect of sleep deprivation.

Exam ready

3. Ron is a taxi driver who works long shifts. During particularly busy times at work, Ron goes more than 24 hours without sleep. It is likely that during busy times at work:

A. Ron would be experiencing total sleep deprivation and would have the same reaction times as a person who has a blood alcohol concentration of 0.05 per cent.

B. Ron is experiencing partial sleep deprivation and would have the same reaction times as a person who has a blood alcohol concentration of 0.10 per cent.

C. Ron is experiencing total sleep deprivation and would have the reaction times of a person with a blood alcohol concentration of 0.10 per cent.

D. Ron is experiencing partial sleep deprivation and would have the reaction times of a person with a blood alcohol concentration of 0.05 per cent.

ANALYSING RESEARCH 7.3

At what point does fatigue start to really affect performance?

Setting safety standards for fatigue due to sleep deprivation would require the ability to compare performance after a known number of waking hours against performance resulting from another factor that is known to decrease performance.

The impact of alcohol is a suitable benchmark for what constitutes safe performance. The effect of alcohol has been measured and used to establish limits on alcohol consumption based on its predicted impact on a persons' ability to drive. Different countries have placed restrictions on how much alcohol is allowed to be consumed while driving, and these restrictions are based on laboratory, simulator, and real-world on-road performance measures of accuracy and speed.

These criteria serve as a reference point for performance deficiencies due to sickness, injury, or in this instance, fatigue resulting from sleep deprivation.

It is widely accepted that a blood alcohol concentration (BAC) of 0.05% is considered to be 'hazardous'.

Therefore it is possible to use the change in performance experienced at a BAC of 0.05% to identify the amount of sleep deprivation required to exhibit the same type of behaviour.

A study was conducted to compare the effects on performance of alcohol consumption and sleep deprivation. The study included 39 participants, 30 of which where employees of the transport industry and the remaining 9 were from the army. Of the 39 participants, 37 were male and 2 were female. Participants were studied over 28 hours of sleep deprivation and during a period in which they measured a BAC of approximately 0.10%. The study was conducted in the laboratory using a within-subject design with counterbalancing. As such, all participants were subjected to both conditions (sleep deprivation and alcohol consumption) where half did the sleep deprivation first and the other the half alcohol consumption first.

At the start of each test session, participants completed questionnaires to report on their sleep the previous night, food and drug intake since waking up and were also asked to complete the Epworth sleepiness scale. Based on the questionnaire results, no participant showed evidence of sleep disorders.

The study found that sleep deprivation of 17–19 hours resulted in a performance on some tests that was equivalent to or worse than having a BAC of 0.05%. Longer periods of sleep deprivation showed performance levels equivalent to a BAC of 0.10% (the maximum amount of alcohol given to participants). These results support the evidence that fatigue due to sleep deprivation may be a contributing factor to deceased performance of accuracy and speed required to be safe on the road, as well as in other industrial environments.

In some communities, being awake for 16 to 17 hours can be considered normal therefore it could be argued that sleep plays a major role in making sure everyone is safe. Being awake beyond 16–17 hours could result in impaired performance and a greater risk of injury. For example, driving home after a long day of work may increase your risk of an accident. Drivers who stay awake for more than 17–18 hours are likely to have a slower reaction time and may miss important information, especially as they stay awake for longer.

The study only looked at the effect of sleep deprivation under day worker conditions and participants were allowed to rest after a certain number of hours sleep the previous night. Today different lifestyles and work schedules often require individuals to be awake for more than 18 hours, therefore shortening the time available for sleep. This type of sleep-wake cycle can often occur over long periods of time.

The implications of this study are clear. Using the legal alcohol limit while driving as a standard for comparison, the findings indicate that after 17–19 hours of being awake, performance on tests dropped to a similar level as those found driving at the legal alcohol limit. As such, countries that set a limit on the amount of alcohol allowed when driving should also consider creating a similar standard for fatigue to make sure that those who have been awake for 18 hours or more are kept from risky behaviours such as operating machinery, flying an aircraft and/or driving.

Source: Adapted from BMJ article 'Moderate sleep deprivation produces impairments in cognitive and motor performance equivalent to legally prescribed levels of alcohol intoxication'

Questions
1. Why did the researchers use BAC as a comparison for fatigue?
2. Who were the sample in this study?
3. Identify the research design used and explain why they used counterbalancing.
4. Identify two ethical guidelines the researchers needed to adhere to.
5. Identify two extraneous variables from the study and explain the possible effect on the study.

KEY CONCEPTS 7.1

» Partial sleep deprivation involves not getting the required quantity of sleep or quality of sleep, which can affect function when awake.
» Partial sleep deprivation can occur because of reduced total sleep time (decreased quantity) or because of disruptions to the normal progression and sequencing of sleep stages, which leads to fragmented sleep (decreased quality).
» Partial sleep deprivation has a negative effect on affective functioning (emotions) as it tends to result in an increase in the tendency to experience negative emotions.
» Partial sleep deprivation has a negative effect on cognitive functioning (thinking) so that mental abilities become impaired.
» Partial sleep deprivation has a negative effect on behavioural functioning (actions), resulting in higher risk of accidents and injury due to poor coordination and slower responses.
» A night of full sleep deprivation (24 hours) produces similar results in an individual's cognitive performance and affective functioning as consuming alcohol to a BAC of 0.10 per cent.
» Sleep deprivation of 17 hours produces similar results in an individual's cognitive performance and affective functioning as consuming alcohol to a BAC of 0.05 per cent.

Concept questions 7.1

Remembering
1. Sleep deprivation refers only to the amount of sleep loss a person experiences. True or false? **r**
2. What is cognitive, affective and behavioural functioning? **r**
3. A full night of sleep deprivation is equivalent to a BAC of ____ per cent. **r**
4. Going without sleep for 17 hours is equivalent to a BAC of ____ per cent. **r**

Understanding
5. Identify a cognitive, affective and behavioural change in functioning caused by partial sleep deprivation. **r**
6. Explain how sleep deprivation can refer to lack of sleep quality or sleep quantity. **e**

Applying
7. David got home late in the evening after basketball training and stayed up watching a movie. He went to bed at 2 a.m. Pedro went to bed at 10 p.m. They both woke for school at 7.30 a.m. At school the next day, who would likely be functioning better? Justify your response. **c**
8. Why might David complain of aches and pains in his muscles the next day at school due to his lack of sleep? **e**

HOT Challenge
9. Design a method to test cognitive, affective and behavioural functioning to compare the effects of sleep deprivation to those of alcohol consumption.

7.2 Sleep disorders

Most people have difficulty sleeping at some time. This is a normal response to a range of factors such as stress and illness. Once these factors have disappeared, most people return to their typical sleeping pattern, which is characterised by normal NREM–REM sleep cycles. However, if the sleeping pattern is regularly disturbed over a prolonged period of time, the sleep–wake cycle may become unbalanced. If this imbalance causes the person distress or interferes with their ability to carry out their normal daily activities, their sleep problem may have developed into a **sleep disorder**. Sleep disorders are a group of syndromes characterised by a disturbance in the amount, quality or timing of sleep, or in behaviours or physiological conditions associated with sleep. Sleep disorders interfere with normal physical, mental and emotional functioning because they disrupt normal restorative sleep.

To be diagnosed with a sleep disorder, the sleep problem must be persistent and cause the person significant emotional distress as well as interfere with their social or occupational functioning.

Circadian rhythm sleep disorders

As we saw in Chapter 6, the sleep–wake cycle refers to the rhythmic biological pattern of alternating sleep with wakefulness over a 24-hour period. For human adults, this cycle typically equates to 8 hours of night-time sleep and inactivity, and 16 hours of daytime wakefulness and activity. The sleep–wake cycle, like many human biological processes, is controlled by circadian rhythms and the main environmental stimulus that synchronises these rhythms to a 24-hour day is light.

Our suprachiasmatic nucleus tells us when to wake up and when to sleep, and it has to be set daily. This is because the length of a day (or period of daylight) changes over the course of a year as the seasons change.

Circadian rhythms influence when, how much and how well people sleep. People with normal circadian systems wake in the morning in time to complete their daily activities and fall asleep at night in time to get enough sleep before having to get up. They can sleep and wake up at approximately the same time every day if they want to. If they start a new routine that requires them to wake earlier than usual, they are generally able to fall asleep at night earlier within a few days. However, if a person's circadian rhythms are disrupted on a regular basis, they may develop a **circadian rhythm sleep disorder**, and they may experience the debilitating effects of sleep deprivation.

Circadian rhythm sleep disorders are a group of disorders that essentially affect the timing of sleep. They occur when the **sleep–wake cycle** operates out of alignment with rhythms in the external environment, particularly the natural day–night cycle.

These disorders can be caused by **intrinsic factors** (caused by the body itself), such as medical conditions and age-related natural shifts in the sleep–wake cycle. They can also be caused by **extrinsic factors** (caused by the environment or external behavioural factors), such as the experience of shift work or jet lag.

Circadian rhythm sleep disorders can affect people in a number of ways, but one key feature shared by these disorders is a persistent disruption to the sleep–wake cycle. This means that the

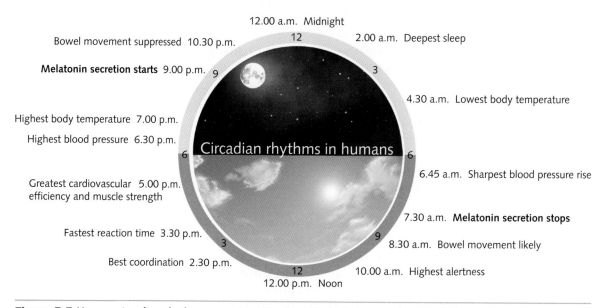

Figure 7.7 Human circadian rhythms are automatic physiological changes that regularly occur during a 24-hour cycle. These rhythms regulate a number of our body processes, including chemical and hormonal production, metabolism and our sleep–wake cycle.

timing of sleep and the ability to sleep and wake for the periods of time necessary to maintain good health and wellbeing are disturbed. People with circadian rhythm sleep disorder are generally able to get enough sleep if allowed to sleep and wake at the times dictated by their body clocks. Unless they have another sleep disorder, their sleep is of normal quality. People with circadian rhythm sleep disorders may be unable to sleep and wake at the times required for normal work, school and social needs, which can severely impact their quality of life. Examples of circadian rhythm sleep disorders include both **Delayed Sleep Phase Syndrome (DSPS)**, which involves a shift in the circadian rhythm so that sufferers are unable to fall asleep until very late at night and do not wake up until much later in the morning, and **Advanced Sleep Phase Disorder (ASPD)**, which involves a shift in the circadian rhythm that leads to a need to fall asleep in the early evening and wake up very early in the morning.

Symptoms of circadian rhythm sleep disorder include difficulty falling asleep at the desired time, chronic tiredness during waking hours, difficulty trying to follow a daytime schedule, and not feeling refreshed or energised when waking. If left untreated, the person can suffer from severe sleep deprivation, which may cause serious problems, for example, depression, impaired work performance, disruption of social schedules and stressed relationships. However, the person is generally able to recover if they get enough uninterrupted sleep and are allowed to sleep and wake at the times dictated by their body clocks.

Delayed Sleep Phase Syndrome (DSPS)

Delayed Sleep Phase Syndrome (DSPS) is a sleep disorder that occurs when a person's circadian rhythm is delayed. People with DSPS have a natural tendency to go to bed later and wake up later than what is typically considered conventional or normal. For example, someone with DSPS may go to sleep at 2 a.m and if left to sleep and wake up 'naturally' (without an alarm or someone waking them) they will wake up much later in the morning, for example 11 a.m. People who suffer from DSPS are unable to easily change their sleep pattern in line with what is considered normal, or even if they do, they are not able to keep it up. It is most common in adolescents, with about 7–16 per cent of adolescents, experiencing DSPS.

DSPS can be a problem if this routine of sleep – going to bed late and waking late–is in conflict with work or school schedules that mean they must get up earlier. Getting up to attend school, for instance, or go to work, would mean the individual would not experience sufficient quantity and quality of sleep. Not only would the total amount of sleep be reduced, but the amount of NREM and REM experienced would also reduce. We know that NREM is important to restore body functions and REM is important for restoring brain function. Limiting the total amount of sleep and the different types of sleep will have a negative effect on a person's ability to thrive.

This leads to experiencing excessive daytime sleepiness and difficulties with the symptoms of sleep deprivation. DSPS is the most prevalent of all circadian rhythm sleep disorders and is most common in adolescents. People who have DSPS often compensate by taking naps during the day, or 'catching up' on sleep on the weekends to overcome the deprived sleep during the week. This can lead to temporary relief but perpetuates the delayed phase cycle, as in the evening they will not feel tired or feel the urge to sleep until later.

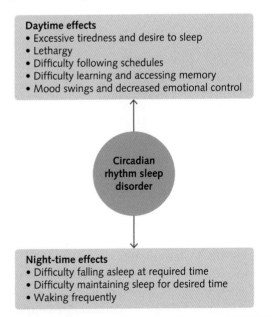

Figure 7.8 Typical effects of circadian rhythm sleep disorder.

Individuals diagnosed with DSPS may be experiencing the negative effects of sleep deprivation, however their body maintains the inclination to go to bed at the usual late time, making it difficult to fall asleep earlier. People with DSPS may be perceived as lazy and unmotivated due to their sleep behaviour. DSPS is a disorder, and it is important to understand that this is not a deliberate behaviour.

Symptoms of DSPS can get worse if the person is not exposed to enough light in the morning, or if they are exposed to too much light at night. The cause of DSPS is not known. It could be caused by an inability to reset the sleep–wake cycle and respond to environmental time cues, or perhaps people suffering from DSPS have biological clocks with longer cycles. A possible cause is that there may be an imbalance in hormones such as melatonin that help maintain the sleep–wake cycle.

7.2.1 CIRCADIAN PHASE DISORDERS AND SHIFTS IN THE ADOLESCENT SLEEP–WAKE CYCLE

7.2.2 SLEEP-WAKE SHIFTS IN ADOLESCENCE

Sleep–wake shifts in adolescence

The circadian rhythms that regulate the sleep–wake cycle naturally shift to a later sleep and wake time as children move into puberty and adolescence. This sleep–wake cycle shift results in a sleep phase delay. The shift in their sleep–wake pattern towards the evening causes a delay in *sleep onset*, or the transition period between wakefulness and sleep. The adolescent is unable to fall asleep until very late at night (or into the early hours of the morning) so during this period individuals may experience symptoms of DSPS. Delayed sleep onset then causes them to have difficulty waking at the time required by their daily work, school or social commitments. Because they have to rise at a normal time, adolescents get less sleep each night and they suffer the effects of sleep deprivation. The automatic shift in the sleep–wake cycle during adolescence, and the sleep deprivation that results, has been observed across various cultures throughout the world and even across several mammalian species (Hagenauer et al., 2009). Researchers studying the sleep periods of adolescents have found that under controlled conditions (for example, with no clocks or lighting cues), adolescents typically sleep 9 hours a night (Carskadon, 2002). However, most adolescents fall short of this sleep time.

The timing of melatonin and cortisol release

The shift in the sleep–wake cycle of adolescents (and the sleep deprivation it brings) is thought to occur because of puberty, when there is a hormonally induced shift of the body clock forward by 1–2 hours. This occurs because melatonin, the sleep hormone, is released approximately 1–2 hours later than in an adult. As a result, the adolescent does not feel sleepy until late at night, usually around 11 p.m. (or later). However, because adolescents need approximately 9 hours of sleep per night, if they have not fallen asleep until around 11 p.m. they are not ready to wake and be alert before 8 a.m. Unfortunately, the demands of their daily school or work schedules don't allow many of them to sleep until this time, so their total amount of sleep is reduced. As a result, they incur a **sleep debt**, which can lead to chronic sleep deprivation. A sleep debt is the amount of sleep loss accumulated from an inadequate amount of sleep, regardless of the cause. According to Melbourne adolescent psychologist Michael Carr-Gregg (2007), by being forced to get up after only 7 hours of sleep rather than 9, teenagers are building up a sleep debt, losing up to 10 hours or more of sleep every school week.

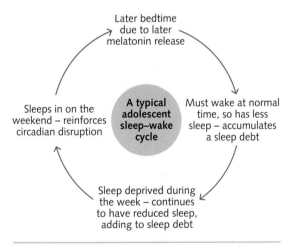

Figure 7.9 A typical adolescent sleep–wake cycle

For the adolescent, this sleep debt is exacerbated by other demands on their time such as homework, sport, part-time work and social commitments. Leisure activities enjoyed during night-time also contribute to them staying awake until late. Watching television or using mobile phones and computers at night exposes the adolescent brain to light cues, which can prevent adequate production of melatonin and keep their brain aroused so they are less able to fall asleep (Figure 7.10). To compensate for their sleep debt, adolescents typically sleep very late into the morning or afternoon on weekends. If this problem is persistent, it may develop into DSPS.

Figure 7.10 Many already sleep-deprived teenagers use their mobile phone when they are in bed. The artificial light emitted by the phone stimulates their brain and slows the release of melatonin, making it difficult to fall asleep at a reasonable time. Because they have to wake up at their normal time, their reduced sleep time adds to their sleep debt.

Advanced Sleep Phase Disorder (ASPD)

Advanced Sleep Phase Disorder involves a problem with the circadian rhythm where the timing is too early. This causes problems staying awake in the evening, which means the individual falls asleep early (for example, falling asleep at 8 p.m.) and then they become alert and wake extremely early in the morning (for example, at 4 a.m.). These unconventional sleep times may impact on social engagements, work and family life (particularly in the evening). ASPD occurs more commonly in the elderly age group and is generally not seen as a problem if sleeping at these times suits the individual. If the individual cannot follow this early to sleep and early to rise sleep–wake cycle, then going to bed later results in still waking early, therefore reducing the amount of sleep the individual will experience. Sleep deprivation is the result. Waking in the early hours of the morning (when most people are asleep) can feel isolating and people may perceive that they are not getting

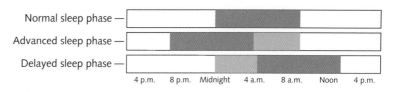

Figure 7.11 Changes to the sleep–wake cycles comparing normal sleep, Delayed Sleep Phase Syndrome and Advanced Sleep Phase Disorder. Dark bar shows when asleep, light bar shows when awake but would normally be asleep.

sufficient sleep – it is still dark at that time! How and why the circadian rhythm shifts in this way is not fully understood.

Shift work: working out of sync with circadian rhythms

Humans are not nocturnal animals. They are programmed to be active and alert during the daytime and inactive, or asleep, during the night. However, shift work requires people to be active and alert at a time when their natural circadian rhythms would prepare them for sleep. **Shift work** involves hours of paid employment that are outside the period of a normal working day and may follow a different pattern in consecutive periods of weeks. Shift-work schedules include night shifts, early-morning shifts and rotating shifts.

Shift work disrupts natural circadian rhythms and, consequently, sleep and waking cycles. People who work a night shift need to be active and alert during the night when melatonin levels are naturally higher and adrenaline and cortisol (hormones that have an arousing and energising effect) levels are lower. Then, they need to sleep during the day when melatonin levels are naturally lower and adrenaline and cortisol levels are higher. So, to succeed at performing their work tasks efficiently, shift workers must try to override their body's natural circadian rhythms. If their attempt to sleep is out of alignment with their body clock and this continues for an extended period, the person may develop shift work disorder.

Shift work disorder occurs when a person's work hours are scheduled during the normal sleep period (at night), causing their circadian rhythms to be out of step with their work schedule. People with shift work disorder are forced to be awake when their circadian rhythms dictate that they should be sleeping, and they may sleep up to 2 hours less than the average worker.

Shift work disorder is an extrinsic circadian rhythm sleep disorder because it is caused by external behavioural factors. It is characterised by insomnia, chronic sleep deprivation and excessive sleepiness when awake. Working night shift leads to loss of synchronisation between the homeostatic sleep pressure drive (the urge to sleep we get in the evening) and the circadian rhythm. If a night-shift worker's circadian system has not adjusted to working at night and sleeping during the day, the sleep pressure drive and circadian rhythms for wakefulness will not be synchronised and, as a result, will not work together (Figure 7.12).

Figure 7.12 Night-shift workers need to be alert and active at a time when their body's natural circadian rhythms prepare them for sleep.

7.2.3 THE EFFECTS OF SHIFT WORK ON THE SLEEP–WAKE CYCLE

Effects of shift work

When their circadian rhythms are no longer synchronised to patterns in the external environment, shift workers often feel sleepy during their work shift. One reason this occurs is because the natural tendency to fall asleep and stay asleep occurs between midnight and 4 a.m., when body temperature drops. Sleep and core body temperature are linked, as core body temperature drops in the evening, which helps to promote sleep onset. As core body temperature rises, it is more difficult to stay asleep. This is one of the reasons why night workers who try to fall asleep at 8 a.m. often struggle to go to sleep and remain asleep through the day.

In particular, people who work night shift find it difficult to adjust to an inverted night–day schedule of activity. Night-shift workers often drive home in morning daylight, which makes it harder to reset their biological clocks. On days off they often resume a normal day–night schedule to spend time with family, and this also disrupts any adjustments their circadian rhythms may have made. It takes about 10 days for the body to adjust to night-shift work. Reverting to daytime routines for a day or two during days off tends to make the circadian rhythm unstable.

Problems with circadian adjustment are also increased by the fact that many night-shift workers work indoors, where the artificial light is too weak to shift their circadian rhythms towards a night–day schedule. However, circadian adjustments can be increased by having very bright indoor lighting at the workplace, keeping bedrooms dark and quiet, and maintaining a schedule of daytime sleep even during days off.

Research into shift work suggests that shift work can lead to partial sleep deprivation, as daytime sleep is typically 1–2 hours shorter than night-time sleep (Tilley et al., 1982). The effects of this reduced quantity of sleep are intensified if the daytime sleep occurs in two split periods. If a shift worker sleeps for a few hours in the morning and then an hour or so before going to work, progression through the sleep stages is disrupted – and this fragmented sleep decreases the quality of their sleep. Environmental factors also make it harder for shift workers to sleep during the day. For example, during the day temperature is generally higher than at night, there is more natural light and there are more disturbances by the activity of those around them. Because their sleep is easily disturbed, shift workers don't spend enough time in REM sleep, the period in the sleep cycle that revitalises and restores mental processes. They also miss out on NREM sleep, the sleep cycle stages when the body repairs and regrows tissues, builds bone and muscle, and strengthens the immune system. If this decreased and fragmented sleep pattern becomes the norm, their functioning is impaired. For example, the risk of diabetes mellitus type 2 is increased in shift workers, especially in men and people working rotating shifts (Yong, 2014). Gastrointestinal disorders and ulcers, obesity, cardiovascular problems, depression, anxiety and other mood disorders are also more common in shift workers.

The problems associated with shift work are exacerbated if the person is working on a rotating shift because their circadian rhythms don't have a chance to adjust to the frequent changes. A rotating shift schedule is a job schedule in which employees work one set of hours for a period and then rotate, or move, to a different set of hours. Rotating shift schedules typically divide the work day into three

8-hour periods. For example, 7 a.m. to 3 p.m., 3 p.m. to 11 p.m. and 11 p.m. to 7 a.m. With a rotating shift schedule, each employee would work one shift for a certain number of weeks and then move onto a different shift. The shifts rotate at regular intervals so that each employee works each shift for the same amount of time. Frequent changes in a work schedule make it difficult for a person's circadian rhythms to adjust, and regular disruption to circadian rhythms can lead to chronic fatigue and other health problems (Figure 7.13).

Figure 7.13 Possible effects of shift work

Sleep research also suggests that, as a result of the effects of sleep deprivation caused by disruption to their circadian rhythms, shift workers increase their risk of having an accident or making a workplace error, particularly in the early hours of the morning, when body temperature is at its lowest point (Colquhoun, 1976; Folkard & Monk, 1979; Richardson et al., 1989).

Reducing the effects of shift work

To alleviate the negative health impacts of shift work, the shift worker should try to stay on one shift as long as possible so their circadian rhythms can adjust. If they have to rotate shifts, they should try to get successively later shifts rather than successively earlier shifts. That is, they should try to move their shift forwards in time so each new shift begins later. Because the human internal clock is slightly longer than 24 hours, it is easier to adjust to lengthening a day than shortening a day by the same amount of time. In other words, moving forwards disrupts the rhythms less than moving backwards. It is also important that shift workers maintain a regular sleep pattern that allows for 7–8 hours of uninterrupted sleep per day and that they take measures to relax and unwind following night-shift work. Shift workers should also try to eat regular meals and a healthy and balanced diet. They should avoid hard-to-digest foods before bed, avoid caffeine and stimulants for at least 6 hours before they plan to sleep, and they should try to exercise regularly during free time. They should also ensure that their sleeping environment is conducive to sleep. For example, they could darken their bedroom and keep it cool, use earplugs and an eye mask to block out sound and light, and have a regular bedtime routine. For people who work night shift, avoiding morning light (for example, by wearing dark goggles or glasses during travel home in the morning) may help their body clock to adapt (Crowley et al., 2004).

Treatments for circadian rhythm sleep disorder

Treatment for circadian rhythm sleep disorder is aimed at 'resynchronising' a person's circadian rhythms to their desired sleep schedule, so properly timed light exposure is essential. Treatment is based on the type of disorder. It can include teaching the person proper sleep hygiene techniques and external stimulus therapy, such as exposing a person to bright light in the morning, avoiding bright light in the evening, and taking melatonin supplements. When combined, these therapies may produce significant results in people with circadian rhythm sleep disorders; however, treatment does not work for everyone. Sleeping pills are rarely effective when used in conflict with the body's natural cycle because they do not correct the underlying circadian abnormality.

Bright light therapy

Bright light therapy exposes people to intense but safe amounts of artificial light for a specific and regular length of time to help synchronise their sleep–wake cycle.

7.2.4 BRIGHT LIGHT THERAPY AND HOW TO SLEEP BETTER

Bright light therapy is a physiological treatment aimed at reducing the symptoms of sleep disorders, particularly circadian rhythm sleep disorders caused by abnormal timing of circadian rhythms. The objective of bright light therapy is to adjust the person's circadian rhythm so their sleep–wake pattern is in sync with environmental shifts in natural light that occur during an external day–night cycle. Several hours of daily exposure to 'light' can shift circadian rhythms by as much as 2–3 hours per day (Shanahan et al., 1999).

During bright light therapy, a person is exposed to a specific level of artificial light for brief periods during strategic times of the day. This is intended to mimic natural daylight involved in regulating the body's sleeping and waking cycles. The person sits near a light box that consists of 2500- to 10 000-lux fluorescent bulbs with a diffusing screen, or they may wear specifically designed bright light glasses. Usually sessions last about 30 minutes, but a session can last between 15 minutes and 2 hours. During a session, people engage in normal daily activities such as reading, eating, using a computer or watching television.

Bright light therapy has proved beneficial for people suffering from various forms of circadian rhythm sleep disorder. For people with Delayed Sleep Phase Syndrome, who have a later-timed circadian rhythm, bright light therapy early in the morning can help to advance or time circadian rhythms earlier, so they should find it easier to sleep at a normal time. Timing the light exposure for early in the morning essentially extends the time the individual will be awake. This will hopefully result in the individual feeling 'sleepy' earlier in the evening, as they have been awake for longer. These people should avoid bright light at night before they go to bed and expose themselves to light early in the morning. If they find themselves becoming sleepy during the day, exposure to bright light can help to increase their alertness. This may be combined with gradually adjusting the time the individual goes to bed, so bedtime is slightly earlier each night. The adjustments would continue gradually until a desirable bedtime is achieved. It is vital to maintain this routine every night, otherwise the circadian rhythm may revert to the misaligned time again.

Bright light therapy can be used to treat Advanced Sleep Phase Disorder, but it is used at different times than DSPS. For ASPD, the bright light exposure should be done in the early evening when the individual would ordinarily want to go to sleep, due to their disorder. This will help to prevent the early release of melatonin and send signals to the suprachiasmatic nucleus (SCN) that it is not time to sleep. The bright light can be removed just before the individual desires to go to sleep. This therapy may take a week to several weeks to achieve a later bedtime and it should be a gradual process of slowly shifting the time later. To experience the amount of sleep required for optimal functioning, the bright light will hopefully push the bedtime to a later time and therefore the wake time to a later time in the morning.

Bright light therapy can be used by shift workers to help them to adjust their sleep–wake cycle to meet their work and sleep needs. When on shift at night-time, wearing bright light glasses or having exposure to a light box may prevent drowsiness and promote alertness. When attempting to sleep during the day, avoidance of light will promote sleep. Bright light may also be used to either adjust sleep times to later or earlier. Bright light exposure in the evening tends to delay sleep times, and bright light in the morning tends to advance sleep times.

Figure 7.14 Bright light therapy exposes people to intense but safe amounts of artificial light so the circadian rhythms that control their sleep–wake cycle can realign to the natural cycle of night and day.

ACTIVITY 7.2 SLEEP DISORDERS

Try it

Infographics convey a great deal of information in a simple, yet visually appealing manner. Figure 7.15 shows an infographic on how to sleep better.

Figure 7.15 Infographic on how to sleep better

1. Create an infographic that explains Delayed Sleep Phase Syndrome, Advanced Sleep Phase Disorder or shift work. Some key points to include are:
   » a definition of the disorder
   » who the disorder affects
   » the causes of the disorder
   » the symptoms of the disorder
   » treatment options for the disorder.

Apply it

2. Outline two differences between Advanced Sleep Phase Disorder (ASPD) and Delayed Sleep Phase Syndrome (DSPS).

Exam ready

3. People who experience Advanced Sleep Phase Disorder (ASPD) tend to go to sleep between 6 p.m. and 9 p.m. and wake between 2 a.m. and 5 a.m.

   Which statement below best describes a treatment that would be effective for ASPD?

   A taking melatonin supplements in the evening to delay the onset of sleep
   B using bright light therapy in the evening to encourage the release of melatonin
   C taking melatonin supplements in the evening to encourage the release of melatonin
   D using bright light therapy in the evening to delay the release of melatonin

KEY CONCEPTS 7.2

» Circadian rhythm sleep disorders are a group of disorders that affect the timing of sleep, resulting from disruptions to a person's circadian rhythms that cause them to operate out of alignment with rhythms in the external environment, particularly the natural day–night cycle.
» Delayed Sleep Phase Syndrome is when the timing of sleep is delayed so that a person cannot fall asleep until much later and will wake later than conventional times of sleep.
» Advanced Sleep Phase Disorder is when the timing of sleep is advanced so that a person falls asleep much earlier and wakes much earlier than conventional times of sleep.
» Shift work involves working at times when the circadian rhythm is programmed to sleep (at night), and therefore sleeping at times when we should be awake. This causes the sleep–wake cycle to be out of alignment with the environmental cues.
» Treatment for circadian rhythm sleep disorders involves timed exposure to bright light therapy to adjust the timing of sleep. Light exposure in the evening delays sleep (fall asleep later) and light exposure in the morning advances sleep (fall asleep earlier).

Concept questions 7.2

Remembering
1. What is a circadian rhythm sleep disorder? r
2. Identify the main treatment for circadian rhythm sleep disorders. r
3. The delayed release of melatonin may occur in which circadian rhythm sleep disorder? r
4. Which of the circadian rhythm sleep disorders is more likely to affect adolescents and which affects the elderly more? r

Understanding
5. Explain the major differences between DSPS and ASPD. e
6. Explain how DSPS and ASPD are similar. e
7. Describe how shift work can affect the circadian rhythms. e

Applying
8. Nadia is finding it difficult to get to sleep at night. She is often still awake at 2 a.m. and finds it difficult to wake up to get to school by 8.30 a.m. How could bright light therapy be used to treat Nadia? c

HOT Challenge
9. You are giving a presentation to nursing staff working shift work in a hospital. Develop a set of notes to outline information the nurses should know about shift work and the effects of sleep deprivation. e

7.3 Improving sleep–wake patterns and mental wellbeing

Sleep hygiene is a term used to describe healthy sleep habits or behaviours to optimise getting to sleep when desired and achieving the quantity and quality of sleep required for good mental health and wellbeing. Sleep hygiene involves changing basic lifestyle habits and the sleep environment that influence sleep (Figure 7.16).

Zeitgebers (German word for 'time givers') are cues in the environment that provide signals to our brains to do things at certain times. An example of a zeitgeber we have explored in detail is light. When our eyes receive lower levels of light, such as when it is getting dark in the environment, the SCN signals to the pineal gland to release melatonin and this makes us feel drowsy and promotes the onset of sleep. So, while our circadian rhythms dictate the timing of daily reoccurring activities such as sleep, energy levels, hunger and temperature, these activities also represent time cues that the circadian system uses to keep our bodies in sync with the environment. If we think about it, many of the good sleep hygiene behaviours are providing zeitgebers to promote sleep (Table 7.2). For example, if you brush your teeth, put pyjamas on and read a book just prior to sleep every night, this routine will be a signal to your brain that it is time for sleep. Our sleep–wake patterns can be improved when an individual is exposed to the correct zeitgebers at the appropriate times to optimise sleep.

As the circadian system is reset every day, the cues present in the environment help to synchronise the SCN and keep the body on a 24-hour schedule. Besides zeitgebers such as light and dark, there are other cues such as temperature and when food is consumed that assist to keep our circadian rhythm on schedule and hence our sleep. When the zeitgebers are conflicting, for example when a person attends shift work at night, or flies across time zones or does exercise late at night, increasing adrenaline and body temperature when usually those are lowering at night, then this can disrupt the circadian rhythm of sleep. Sleep can be improved if entrainment occurs, which is the process of activating or providing a timing cue (zeitgeber) that promotes sleep when it is appropriate. This may involve adapting or making lifestyle choices so that the cues or signals you are giving your body are going to optimise sleep.

Daylight and blue light

Sunlight is the main source of 'blue light' (light on the visible light spectrum with a short wavelength), so exposure to sunlight during the day will provide the signal to the SCN to make us alert and awake. As the sun goes down at the end of the day, this zeitgeber will cue the SCN to signal the pineal gland to release melatonin and we will begin to feel drowsy. However, blue light is not only emitted by the sun: there are also man-made sources including fluorescent and LED lighting on devices such as tablet computers, phones and television screens.

7.3.1 IMPROVING SLEEP HYGIENE

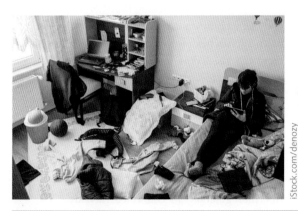

Figure 7.16 To improve sleep–wake patterns, it is ideal to provide an environment that includes calming stimuli, aiding sleep.

Table 7.2 Strategies to improve sleep hygiene

Set a schedule for sleep	» Go to sleep and wake up at about the same time every day, giving yourself enough time to get the amount of sleep you require. » Try to keep the same sleep and wake times on weeknights and weekends. Staying up late and sleeping in late on weekends can disrupt your sleep–wake pattern. » If you want to change your sleep time, do it gradually. For example, if you are not getting enough sleep and currently go to sleep at midnight, then attempting to sleep at 9.30 p.m. will be difficult. Go to bed half an hour earlier and slowly adjust to the earlier times until you reach your target sleep time.
Have a routine to prepare for bed	» Develop a bedtime routine including behaviours like putting on your pyjamas and brushing your teeth; these can reinforce that it's bedtime. » Avoid bright light so that melatonin can be released; this includes avoiding devices that generate blue light (phone, tablet computer, laptop, computer). » If sleep is not achieved in about 30 minutes, don't lay in bed frustrated. Get up and do something relaxing such as reading a book, stretching or meditation in low light before trying to fall asleep again. » Use your bedroom and bed to sleep, not to do homework or watch your favourite program.
Daytime activities and habits	» Get outside and expose yourself to natural light during the day (safely – be sun smart). » Exercise or be physically active for the multitude of health benefits that brings, but also so that you feel tired at the end of the day. » Avoid strenuous exercise before bed (do something calming). » Avoid smoking and reduce consumption of alcohol and caffeine. These drugs can disrupt sleep. If you can't avoid them completely, then avoid taking them in the evening before bed. » Eat dinner early and try to eat a healthy balanced diet. » Avoid napping during the day unless you are a toddler as this will reduce the likelihood of feeling tired at bedtime.
Create a relaxing bedroom environment	» If possible, ensure your mattress and pillow are comfortable, with bedding that is going to offer you the right level of warmth. » The temperature in the room should not be too warm. » Use heavy curtains to block out light. » If your household is noisy, then ear plugs may be useful. » Some scents, like lavender, can be used to promote calmness (if you like them).

Exposure to this form of blue light can act as a zeitgeber perhaps unintentionally, so that the SCN receives a conflicting signal that it should be bedtime, but there is light present in the environment. By limiting your exposure to 'blue light' in the 1 to 2 hours before bed, sleep onset will be easier to achieve (Figure 7.17). If you can't avoid exposure, perhaps due to school or work requirements, consider buying some glasses that block 'blue light', or change the settings on your screens to warmer tones. Many devices have setting options to automatically change the screen brightness at a certain time.

Temperature

The hypothalamus (including the SCN) regulates body temperature automatically so that it rises as dawn approaches, peaks during the day when we feel more energised, dips in early afternoon, and drops before we go to sleep at night. This regulation of physiological activity is important for maintaining homeostasis because it keeps our body working and resting when it needs it (Figure 7.18). The onset of sleep is not only associated with a decrease in body temperature, but also an increase in the release of melatonin. Sleep and control of temperature (thermoregulation) are connected. As body temperature decreases, humans often perform 'nesting' behaviours where they will seek to be warm

7.3.2 ANALYSIS OF A RESEARCH INVESTIGATION

Figure 7.17 Limit your exposure to blue light for 1–2 hours before bedtime.

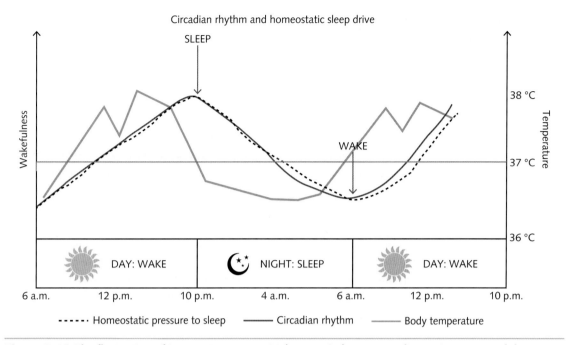

Figure 7.18 The fluctuation of temperature over a 24-hour period, corresponding to increases and decreases in wakefulness due to the circadian rhythm and homeostatic sleep drive

and comfortable – grab a blanket. If there is any major variation in body temperature, this may lead to problems getting to sleep. Body temperature does not vary significantly, so a small change in temperature can have a large effect. Sleep deprivation can lead to changes in the pattern our body temperature usually follows, making sleep onset difficult.

Temperature variations during sleep can also have a significant effect on sleep quality. During NREM sleep your body and brain both cool down; however, in REM sleep your ability to regulate body temperature is impaired. If the ambient temperature becomes too 'hot' or too 'cold', this can change your sleep patterns, causing you to wake up because your body will need to work to cool down or to warm you up.

To be able to get to sleep, the temperature in the room should be considered. If it is an extremely cold evening, perhaps more blankets or sleep wear may be needed to maintain your body's temperature overnight during sleep. It is suggested a good zeitgeber for temperature, if you want to promote the onset of sleep, is to have a warm bath a little before bedtime. This may sound counterintuitive as previously we have said that body temperature drops at the onset of sleep. However, the warm water will warm your skin and this will stimulate vasodilation (enlarging or widening of the blood vessels).

To regulate temperature, the body will react by trying to cool you down, constricting the blood vessels. This will cause a decrease in temperature, which mimics the temperature decrease experienced at night to promote the onset of sleep.

Eating and drinking patterns

Sleep–wake patterns can be affected by not only what you eat, but also when you eat. The circadian system prepares the body to be more efficient at digesting, absorbing and metabolising food earlier in the day, when we are becoming active. For instance, insulin sensitivity (needed to regulate blood sugar) is greater in the morning. For this reason, larger meals are processed better when eaten in the first half of the day. And since melatonin (released at night) reduces insulin release, the body is not able to process sugars (glucose) properly when you eat late at night or very early in the morning, when melatonin is high. When food and drink is consumed outside of the 'optimal' times, say you have a big meal late at night, this may delay sleep. Therefore, eating larger meals earlier in the day and avoiding food for a few hours prior to bedtime may have health and sleep benefits (Manoogian et al., 2019).

Sleep helps to balance and maintain hormones that make you feel hungry (ghrelin) or full (leptin).

7.3.3 ZEITGEBERS

7.3.4 THE IMPORTANCE OF SLEEP TO MENTAL WELLBEING – KEY TERMS

When you are sleep deprived the level of these hormones becomes unbalanced: ghrelin levels go up and leptin levels go down. This can cause you to feel hungrier when sleep deprived than when you have had healthy sleep. Therefore, sleep assists our bodies to balance appetite. Research has shown that people who don't get enough good quality sleep are more likely to consume foods that are high in fat and sugar, be overweight or obese, and develop conditions like diabetes. This may be because when we are tired, we tend to eat foods to boost our mood and energy levels. Other hormones such as insulin and cortisol are usually released in the morning at the end of a healthy sleep. These signal to our bodies to eat breakfast and to wake up and become more alert. When sleep deprivation results in a higher than normal blood sugar level as the release of insulin increases, our body cannot metabolise and properly use the nutrients in the food we eat. This may put people suffering from sleep deprivation at higher risk of diabetes.

How can eating and drinking patterns promote good sleep–wake patterns? Eat bigger meals earlier in the day, stay away from large meals before bedtime. Avoid foods and drinks that contain large amounts of sugar, caffeine and alcohol before bedtime as these will delay sleep onset or disrupt sleep patterns (alcohol).

When our circadian rhythms are out of sync with the environment, sleep and mental wellbeing can be negatively affected. For example, we know that the body expects to do certain activities and eat certain kinds of fuel (fat and sugar) at specific times of the day. Your body is best at doing these activities at certain times, such as digesting food and drinks when you are active and light is present, and sleeping when it is dark. Thus, doing certain activities outside of these times can disrupt this system and compromise health and mental wellbeing. In contrast, a consistent daily cycle using zeitgebers may nurture a healthy circadian clock and optimise health and mental wellbeing.

ACTIVITY 7.3 ZEITGEBERS MIND MAP

Try it

1. Work with a partner to create a mind map that explains how our sleep–wake patterns are influenced by the following:
   » daylight
   » blue light
   » temperature
   » eating and drinking patterns.

Apply it

Jeremy is an otherwise healthy adult who has recently started working as a high school teacher. He has found his first term of teaching rather difficult as he has found it hard to get into an adequate sleep pattern. He usually leaves work at around 4.30 p.m. and is home by 5 p.m. Exhausted from the day, he often has a nap once he gets home. He usually has a coffee after his nap and then has dinner.

Jeremy likes to look after himself, so he spends an hour and a half at the gym. He gets to the gym at around 9.30 p.m. to 10 p.m. as it is quieter. Once home from the gym, he has a banana and a protein shake. Once in bed, he then checks his work emails.

He finds that he gets more frustrated as he waits to fall asleep. He lays there wondering when he will fall asleep and thinks he is just a 'bad sleeper'. He continually checks the time on his phone and calculates how long it will be before his alarm goes off in the morning.

2. Identify three issues with Jeremy's sleep habits and explain how he could address these issues.

Exam ready

3. ©VCAA 2014 Q35 ADAPTED

Body temperature can be used to study sleep and consciousness because it follows a predictable 24-hour cycle with

A our lowest temperatures experienced in the early hours of the morning.

B our lowest temperatures experienced when using selective attention.

C the greatest fluctuation in temperature occurring during normal waking consciousness.

D less fluctuation during sleep.

ANALYSING RESEARCH 7.4

Working night shift burns less energy, increases risk of weight gain

Shift work can make you tired. You have to change your routine and work different shifts, some going against the body's natural inclination to sleep at night and be alert and active during the day. It appears that it can also increase the risk of weight gain and obesity. A shift worker will typically burn less energy and eat at times of the 'night' when their bodies are biologically wired to be asleep. One theory for why weight gain may occur is that the differing shifts a worker may experience mean that they are unlikely to adapt to and synchronise their circadian rhythm, which results in a sluggish, slower metabolism. Shift workers are more likely to be suffering the effects of sleep deprivation and as such may also eat and desire different foods to compensate for low energy levels as they feel tired.

A study at the University of Colorado Hospital's Clinical and Translational Research Centre using 14 healthy adult participants consisted of monitoring their sleep and eating for five nights. For the first two days, participants were asked to follow a normal schedule, where they slept at night and stayed awake during the day. They then transitioned to a three-day shift-work schedule where their routines were reversed, so they were sleeping during the day and staying awake at night.

Over the course of the experiment, the participants were given meals that were carefully controlled so that they consisted of the same amount of food they would normally eat at home to maintain their current weight. When the participants transitioned to the shift work schedule, the timing of their meals changed but the total amount of calories remained the same. When it came time to sleep (either during the night or the day phase of the experiment) the participants were given the same eight-hour sleep opportunity.

What did the results of the study show? The total daily energy used by participants decreased in the later stages of the experiment when they were put on a shift-work schedule. This could be due to the mismatch between the participants' activities and their circadian rhythm. Circadian rhythms can shift over time, if enough time is given to entrain the body to follow new cues in the environment. When people fly overseas and change time zones, their body will eventually learn to adapt to a new light/dark schedule. As we experience small changes to our circadian rhythms during daylight savings transitions – we 'wind' the clocks backwards or forwards, and adapt to the change in time and new cues reasonably quickly. The problem with shift work is that shifts typically rotate and change frequently, so the body may find it difficult to adapt to a new routine. And when shift workers have days off work, they typically switch back to a daytime schedule and sleep at night.

The researchers found that participants burned more fat when they slept during the day compared to when they slept at night. It is not clear why this happens, but the extra fat burning is possibly triggered by the transition day between a daytime schedule and a night-time schedule. On that day, shift workers often take an afternoon nap to prepare for the first night shift, but in total, they are typically awake more hours than usual and, therefore, burn more energy. The need to meet the extra demand for energy may cause the body to begin burning fat.

More research is needed to determine if the fat burning experienced by the participants would happen among actual shift workers, whose diet is not being strictly controlled. In this study diet and eating habits were strictly controlled, while the habits of real shift workers may entail eating more calories on the transition day – an option not available to study participants – which could eliminate the need for the body to start burning fat. Still, the findings suggest that shift workers may be prone not only to gaining weight, but also to a changing composition of fat and muscle mass in their bodies. The researchers were quick to highlight that even though initially the participants burned more fat, this would not lead to weight loss, as the energy expended during the three days of daytime sleeping and wakefulness at night was lower than when participants slept at night and were awake during the day.

Source: Adapted from University of Colorado at Boulder. (2014, November 17). Working night shift burns less energy, increases risk of weight gain. ScienceDaily

> **Questions**
> 1 Identify the independent variable(s) and dependent variable(s) in this study.
> 2 State the results of this study.
> 3 What conclusions can be drawn from these results?
> 4 Discuss the validity of this study and whether the results can be generalised.
> 5 Identify two ethical guidelines the researchers would need to follow. Describe these guidelines and explain how the researchers would implement them.

INVESTIGATION 7.1 INVESTIGATING CIRCADIAN RHYTHMS

Scientific investigation methodology

Correlational study

Introduction

Circadian rhythms are automatic physiological changes that regularly occur during a 24-hour cycle. These rhythms regulate a number of our body processes, including chemical and hormonal production, metabolism and our sleep–wake cycle. Circadian rhythms have a direct influence on energy levels and a number of fundamental cognitive functions throughout the day including motivation, attention and memory ability.

Logbook template
Correlational study

This correlational study investigates the relationship between these cognitive functions and time of day. This will be investigated by completing a self-assessment of energy levels, motivation, attention and memory ability throughout the day for three school days. Before you collect the data, make a prediction for each of the measures of when during the day you will experience higher and lower levels.

Data Calculator

Use the following self-assessment recording table (or similar) to collect the data. Collate and graph your and the class data. Respond to the correlational study questions on page 303.

Self-assessment

Make a judgement out of 10 (10 being highest to 0 being lowest) of your energy levels, motivation, ability to pay attention and memory ability for each of the time periods over three days. Record your results in Table 7.3.

Table 7.3 Self-assessment recording table

	6.00 – 9.00 a.m.	9.00 a.m. – 12.00 p.m.	12.00 – 3.00 p.m.	3.00 – 6.00 p.m.	6.00 – 9.00 p.m.
Energy levels	Day 1: _____ Day 2: _____ Day 3: _____	Day 1: _____ Day 2: _____ Day 3: _____	Day 1: _____ Day 2: _____ Day 3: _____	Day 1: _____ Day 2: _____ Day 3: _____	Day 1: _____ Day 2: _____ Day 3: _____
Motivation	Day 1: _____ Day 2: _____ Day 3: _____	Day 1: _____ Day 2: _____ Day 3: _____	Day 1: _____ Day 2: _____ Day 3: _____	Day 1: _____ Day 2: _____ Day 3: _____	Day 1: _____ Day 2: _____ Day 3: _____
Attention	Day 1: _____ Day 2: _____ Day 3: _____	Day 1: _____ Day 2: _____ Day 3: _____	Day 1: _____ Day 2: _____ Day 3: _____	Day 1: _____ Day 2: _____ Day 3: _____	Day 1: _____ Day 2: _____ Day 3: _____
Memory ability	Day 1: _____ Day 2: _____ Day 3: _____	Day 1: _____ Day 2: _____ Day 3: _____	Day 1: _____ Day 2: _____ Day 3: _____	Day 1: _____ Day 2: _____ Day 3: _____	Day 1: _____ Day 2: _____ Day 3: _____

Correlational study questions

1. Describe the results. Are there any patterns evident within the graphed data over a 24-hour period?
2. Describe the similarities and/or differences in your individual and collated class results.
3. What factors may have impacted on energy, motivation, attention and memory over a 24-hour period?
4. Identify any outliers, contradictory or incomplete data. Suggest reasons for these.
5. Were there any problems encountered in collecting the data? If so, how did these influence the results?
6. Suggest strategies to improve the data collection to increase validity and to reduce uncertainty.
7. What implications do the findings have for your VCE studies?
8. Create a personalised weekly study schedule based on your individual results to maximise learning.

KEY CONCEPTS 7.3

- Sleep–wake patterns can be improved by making changes to sleep hygiene practices which involve changes to bedtime routines, the sleep environment and setting good sleep habits.
- Zeitgebers (time givers) are cues that help identify when we should sleep and when we should be awake.
- Zeitgebers are used at particular times to improve sleep–wake patterns and they include light, temperature, and eating and drinking patterns.
- Light can promote alertness and absence of light will promote drowsiness/sleep at different times of the day.
- Body temperature drops at night (during sleep) and rises in the morning. Changes to body temperature due to activity, illness and ambient temperature can disrupt sleep–wake patterns.
- Sleep affects eating and drinking patterns and keeps our appetite in balance. Ideally, we should not eat for a few hours before bedtime to promote sleep. Eating large meals late at night can disrupt sleep.

Concept questions 7.3

Remembering
1. Define 'zeitgeber'. r
2. Identify three zeitgebers that can be used to improve sleep–wake patterns. r

Understanding
3. How do zeitgebers affect sleep–wake patterns? r
4. Explain how light can be used to adapt to better sleep–wake patterns or contribute to poor sleep–wake patterns. e
5. Identify two ways sleep hygiene might improve sleep–wake patterns. r
6. Why is temperature alone not a reliable indicator of sleep and changes to sleep patterns? r

Applying
7. Explain the effects of blue light used in devices such as PCs and tablet computers on the brain, and how this may affect sleep. e
8. Rhonda has started working at the local service station doing the night shift. Her work begins at 9 p.m. and ends at 5 a.m. She will work this shift for 4 days and then have 3 days off. When Rhonda starts her shift, what should she do to reduce the possible effects of sleep deprivation? c

HOT Challenge
9. Reflect on your own sleep–wake cycle. Design a 5-step routine you could follow to optimise your sleep. Explain the reason for each part of the routine and demonstrate an understanding of zeitgebers and sleep hygiene. c

7 Chapter summary

KEY CONCEPTS 7.1

- Partial sleep deprivation involves not getting the required quantity of sleep or quality of sleep, which can affect function when awake.
- Partial sleep deprivation can occur because of reduced total sleep time (decreased quantity) or because of disruptions to the normal progression and sequencing of sleep stages, which leads to fragmented sleep (decreased quality).
- Partial sleep deprivation has a negative effect on affective functioning (emotions) as it tends to result in an increase in the tendency to experience negative emotions.
- Partial sleep deprivation has a negative effect on cognitive functioning (thinking) so that mental abilities become impaired.
- Partial sleep deprivation has a negative effect on behavioural functioning (actions) resulting in higher risk of accidents and injury due to poor coordination and slower responses.
- A night of full sleep deprivation (24 hours) produces similar results in an individual's cognitive performance and affective functioning to consuming alcohol to a BAC of 0.10 per cent.
- Sleep deprivation of 17 hours produces similar results in an individual's cognitive performance and affective functioning to consuming alcohol to a BAC of 0.05 per cent.

KEY CONCEPTS 7.2

- Circadian rhythm sleep disorders are a group of disorders that affect the timing of sleep, resulting from disruptions to a person's circadian rhythms that cause them to operate out of alignment with rhythms in the external environment, particularly the natural day–night cycle.
- Delayed Sleep Phase Syndrome is when the timing of sleep is delayed so that a person cannot fall asleep until much later and will wake later than conventional times of sleep.
- Advanced Sleep Phase Disorder is when the timing of sleep is advanced so that a person falls asleep much earlier and wakes much earlier than conventional times of sleep.
- Shift work involves working at times when the circadian rhythm is programmed to sleep (at night) and therefore sleeping at times when we should be awake. This causes the sleep–wake cycle to be out of alignment with the environmental cues.
- Treatment for circadian rhythm sleep disorders involves timed exposure to bright light (therapy) to adjust the timing of sleep. Light exposure in the evening delays sleep (fall asleep later) and light exposure in the morning advances sleep (fall asleep earlier).

KEY CONCEPTS 7.3

- Sleep–wake patterns can be improved by making changes to sleep hygiene practices which involve changes to bedtime routines, the sleep environment and setting good sleep habits.
- Zeitgebers (time givers) are cues that help identify when we should sleep and when we should be awake.
- Zeitgebers are used at particular times to improve sleep–wake patterns and they include light, temperature, and eating and drinking patterns.
- Light stimulates the brain to increase alertness. Light can promote alertness and absence of light will promote drowsiness/sleep at different times of the day.
- Body temperature drops at night (during sleep) and rises in the morning. Changes to body temperature due to activity, illness and ambient temperature can disrupt sleep–wake patterns.
- Sleep affects eating and drinking patterns and keeps our appetite in balance. Ideally, we should not eat a few hours before bedtime to promote sleep. Eating large meals late at night can disrupt sleep.

7 End-of-chapter exam

Section A: Multiple choice

1. To treat Advanced Sleep Phase Disorder, when would it be best to expose a patient to bright light therapy?
   A early morning
   B during the day
   C early evening
   D just before midnight

2. Eva and her friends spent the entire night studying together at her house for a Psychology SAC. The next day at school, what behavioural effects may Eva and her friends experience due to being sleep deprived?
   A a lack of interest in making conversation with each other
   B feeling hungry and wanting to visit the school canteen for hotdogs and chips
   C being unable to remember the content for the SAC
   D an inability to sit still while in the morning assembly at school

3. At one point Eva was unable to remember what subject she had after lunch. Sleep deprivation is likely to contribute to her poor memory because
   A behavioural function is compromised by sleep deprivation.
   B affective function is compromised by sleep deprivation.
   C cognitive function is compromised by sleep deprivation.
   D Eva didn't enjoy the subject after lunch.

4. After school that day, Eva went to her friend's house for dinner. Her mum drove over to pick her up at 10 p.m. On the way home, her mum suggested that Eva practise her night driving and offered to let her drive the car. Eva was concerned that she would not be able to concentrate. What information regarding a full night of sleep deprivation would suggest that Eva should not drive the car?
   A A full night of sleep deprivation is equivalent to a BAC of 0.05 per cent.
   B A full night of sleep deprivation is equivalent to a BAC of 0.50 per cent.
   C A full night of sleep deprivation is equivalent to a BAC of 0.10 per cent.
   D A full night of sleep deprivation is equivalent to a BAC of 0.01 per cent.

5. Bruno is suffering from sleep deprivation. Which one of the following is an example of a change to his affective functioning that may be contributing to his lack of sleep?
   A reduced ability to remember things he just learned
   B difficulty completing simple tasks
   C difficulty completing complex tasks
   D decreased ability to control emotions

6. The delay in the onset of sleep resulting in a shift to the sleep–wake cycle of adolescents may be caused by
   A increased release of adrenaline at night.
   B increased release of melatonin at night.
   C delayed release of melatonin at night.
   D delayed release of cortisol at night.

7. Which one of the following changes in the body is an indication that a person is progressing into deeper NREM sleep?
   A The brain becomes more active.
   B The muscles become more active.
   C Body temperature decreases.
   D Cortisol is released.

8. The main distinction between DSPS and ASPD is that DSPS _____ the onset of sleep, whereas ASPD _____ the onset of sleep.
   A delays; advances
   B advances; delays
   C delays; lengthens
   D shortens; delays

9. Which one of the following is not an example of a zeitgeber that could be used to promote sleep at conventional times?
   A eating a big dinner at 9 p.m. just before bedtime
   B brushing teeth before bedtime
   C taking a warm bath before bedtime
   D texting your friends on your phone during the day

10. A treatment for circadian rhythm sleep disorders involves using bright light therapy to help align sleep–wake patterns with conventional bedtimes. It works best when used in the
    A early morning to advance the sleep time.
    B evening to advance the sleep time.
    C morning to delay the sleep time.
    D evening to delay the wake time.

11 Sandra has a sleep disorder that makes it difficult for her to stay awake at night, and she wakes _____ in the morning. This is called _____.
   A early; Delayed Sleep Phase Syndrome
   B late; Delayed Sleep Phase Syndrome
   C early; Advanced Sleep Phase Disorder
   D late; Advanced Sleep Phase Disorder

12 Shift workers often report difficulty getting good sleep due to the rotating shifts they work. The most likely reason for their sleep disturbance problem is a(n) _____ rhythm sleep disorder.
   A ultradian
   B circadian
   C variable
   D delayed

13 A behavioural effect of sleep deprivation is
   A a decrease in the ability to consolidate memory.
   B reduced emotional responses.
   C microsleeps.
   D irrational thoughts.

14 John is a policeman and works night shift. After night shift he has planned a week of holidays. How might John use bright light therapy to help him adjust to 'normal' hours again?
   A He should use bright light in the morning after he finishes his shift to remain awake during the day and then sleep the next night.
   B He should go to sleep after finishing his shift and then use the bright light that night.
   C He should use it only during his shift to stay awake.
   D It should not be used at all as it could potentially be harmful.

15 The association between the bedroom and the use of calming scents and candles may promote calm and sleep through classical conditioning where the bedroom becomes the _____.
   A neutral stimulus
   B unconditioned stimulus
   C conditioned stimulus
   D conditioned response

16 Seventeen hours without sleep results in similar cognitive and affective changes as _____ per cent BAC.
   A 0.05
   B 0.50
   C 0.10
   D 0.01

17 During adolescence the delayed release of which hormone may be responsible for the shift in the sleep–wake cycle of this age group?
   A cortisol
   B melatonin
   C GABA
   D testosterone

18 A cognitive effect of sleep deprivation is
   A an inability to assess risk and consequences.
   B an inability to sit still.
   C clumsiness.
   D increased aggression.

19 Circadian rhythm sleep disorders include
   A sleep quality.
   B sleep quantity.
   C timing of sleep and wakefulness.
   D all of the above.

20 When DSPS is experienced, it is common for the individual to want to follow their body's abnormal sleep rhythm on the weekend. Which statement is correct?
   A They will make up their sleep debt and feel worse because they get too much sleep.
   B This will cause the sleep–wake cycle to advance further.
   C This will cause the sleep–wake cycle to delay further.
   D This will not be a problem as it will allow them to catch up on their sleep debt.

Section B: Short answer

1. Draw a table outlining the cognitive, affective and behavioural changes to functioning caused by partial sleep deprivation. Include at least two changes for each.

   [6 marks]

2. Violet is in Year 12 at school. She is finding it difficult to fit everything into her busy schedule, and often finds herself working until 1 a.m. She doesn't find it difficult to stay awake until this time but finds it difficult to wake for school in the morning. Her younger sister Hannah is in Year 4. She begins to fall asleep at 9 p.m. and consistently wakes at 7 a.m. during the week. She has no problem getting up for her 8.30 a.m. netball game on the weekend. With reference to the scenario, explain why Violet and Hannah have different sleep patterns.

   [6 marks]

3. Compare the similarities and differences between DSPS and ASPD. Indicate how bright light therapy may be used to treat each disorder.

   [6 marks]

Unit 4, Area of Study 1 review

Section A: Multiple-choice

Use the following information to answer Questions 1 and 2.

Jessica is sitting in class and finds her mind drifting to what she is going to order for lunch at the canteen.

Question 1 ©VCAA 2021 Q25 ADAPTED HARD
While sitting in class Jessica is likely to be closer to
A normal waking consciousness with perceptual distortions.
B normal waking consciousness with high levels of awareness.
C an altered state of consciousness with automatic processing.
D an altered state of consciousness with altered time orientation.

Question 2 ©VCAA 2021 Q26 ADAPTED EASY
What state would you classify Jessica to be in while thinking of what to order for lunch in class?
A controlled processing
B divided attention
C daydreaming
D automatic processing

Question 3 ©VCAA 2021 Q30 ADAPTED HARD
Which one of the following is a subjective measure that could be used to investigate possible treatments that would reduce sleep disturbances?
A electromyograph recording participants' brain wave patterns.
B video monitor recording the time and duration of participants' awakenings.
C electromyograph recording patients' movement during their sleep.
D a sleep diary in which the participants would record how they felt after each night's sleep.

Question 4 ©VCAA 2020 Q27 ADAPTED MEDIUM
Normal waking consciousness is
A individual and observable as a lack of consciousness.
B a state of consciousness including focused attention and daydreaming.
C a state in which the person is awake and aware of thoughts, feelings and perceptions.
D a state of consciousness where individuals are always highly focused and aware and cannot multitask.

Question 5 ©VCAA 2020 Q28 ADAPTED EASY
Eric is daydreaming in class. Which one of the following characteristics demonstrates that Eric's state of consciousness is closer to an altered state of consciousness than normal waking consciousness?
A He is lacking awareness of anything happening in the external environment.
B He is able to repeat back instructions given by his teacher.
C He is highly focused on his internal thoughts.
D His mind shifts between his thoughts and his environment.

Question 6 ©VCAA 2019 Q34 ADAPTED HARD
Which one of the following is the best measure of whether an individual is in NREM sleep?
A EEG showing slow wave brain patterns.
B EOG showing highly rapid eye movements.
C EMG showing movement within the muscles.
D video monitoring showing tossing and turning during this stage of sleep.

Question 7 ©VCAA 2021 Q35 ADAPTED MEDIUM
Which one of the following options names and explains the type of rhythm shown in the graph below?

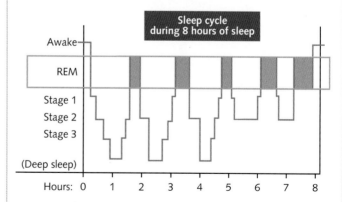

A ultradian rhythm, because it is shorter than a 24-hour cycle.
B ultradian rhythm, because it follows a 24-hour pattern.
C circadian rhythm, because it is over a 24-hour cycle.
D circadian rhythm, which is shorter than 24 hours.

Question 8 ©VCAA 2019 Q26 ADAPTED HARD

Which of the following is true regarding the suprachiasmatic nucleus?

A It is located within the thalamus.
B It regulates all circadian rhythms within the body.
C It releases melatonin in response to circadian rhythms.
D It uses light as an external cue to regulate the sleep–wake cycle.

Use the following information to answer Questions 9–12.

Flynn is an adolescent who finds himself not falling asleep until 1 a.m. This is hard for him as he needs to wake up at 7 a.m. for school and finds himself constantly sleepy at school.

Question 9 ©VCAA 2019 Q29 ADAPTED HARD

Which of the following is true of Flynn's melatonin release?

A advanced release of melatonin leading to inability to stay asleep.
B delayed release of melatonin leading to falling asleep later at night.
C no melatonin release leading to a lack of cue to sleep.
D too much melatonin being released leading to an inability to fall asleep.

Question 10 ©VCAA 2019 Q30 ADAPTED EASY

What should a typical night's sleep for Flynn look like?

A 7 hours with 20% REM 80% NREM
B 8 hours with 20% NREM 80% REM
C 9 hours with 20% REM 80% NREM
D 9 hours with 20% NREM 80% REM

Question 11 ©VCAA 2019 Q31 ADAPTED HARD

Which one of the following could be best used to improve Flynn's sleep hygiene?

A ensuring exposure to daylight first thing in the morning to suppress melatonin.
B eating later in the day so he isn't hungry at night.
C exposure to blue light before bed to induce melatonin release.
D forcing himself to go to sleep earlier in the night.

Question 12 ©VCAA 2019 Q32 ADAPTED MEDIUM

What is likely to be true of Flynn's ultradian rhythms due to his current sleep deprivation?

A less total sleep but similar proportions of REM and NREM to a typical night's sleep.
B same total sleep but a reduction in the amount of NREM sleep.
C less total sleep with a reduction in stage 3 NREM sleep.
D less total sleep with a decrease in REM sleep.

Question 13 ©VCAA 2019 Q33 ADAPTED MEDIUM

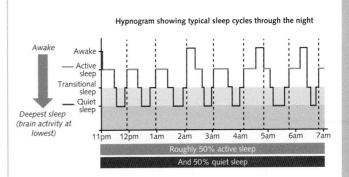

Dr Schiller is a sleep researcher. He has collected data from five healthy participants: an infant, a child, an adolescent, an adult and an elderly person. Dr Schiller forgot to label the hypnograms, so he decided to try to identify them by considering the typical sleep patterns for each life stage.

The hypnogram shown above is likely to belong to which participant?

A the infant, because they spend 50 per cent of sleep in REM.
B the adult, because adults have four to five sleep cycles per night.
C the child, because children sleep for 16 hours.
D the elderly person, because elderly people wake frequently during the night.

Question 14 ©VCAA 2020 Q32 ADAPTED HARD

Sleep changes as we age. Which one of the following statements does not describe a change that occurs?

A The time spent sleeping overall decreases as we age.
B As we age, we spend less time in stage 3 NREM sleep.
C As we progress over time from infancy to old age, the proportion of time spent in stages 1 and 2 of NREM sleep increases.
D The proportion of time spent in REM sleep significantly decreases as we age.

Use the following information to answer Questions 15–19.

Dr Singh compared the effects of consumption of alcohol on number of errors in a driving simulator compared to individuals who were sleep deprived. His random sample included participants aged 18 to 70 years. In the repeated measures experiment, participants consumed one standard drink of alcohol at half-hourly intervals until they reached 0.10 per cent blood alcohol concentration (BAC). Participants completed a series of computer-based driving simulators for total errors at BACs of 0.00 per cent, 0.05 per cent and 0.10 per cent.

Additionally, one week later Dr Singh then placed participants in an overnight study where they went a full 24 hours without sleep, completing the same driving simulator at 17 hours sleep deprivation and 24 hours. The graph below represents the number of errors made in a driving simulator for each of the experimental conditions.

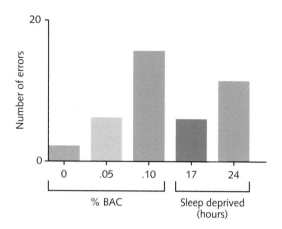

Question 15 ©VCAA 2020 Q33 ADAPTED MEDIUM
The graph above demonstrates that
- A driving errors increase significantly as both BAC and hours sleep deprived increase.
- B the higher the BAC the greater the driving errors, but the higher the sleep deprivation the lower the errors.
- C sleep deprivation results in more errors than alcohol consumption.
- D an altered state of consciousness is achieved.

Question 16 ©VCAA 2020 Q34 ADAPTED MEDIUM
Which of the following includes both an independent variable and a dependent variable for Dr Singh's study?

	Independent variable	Dependent variable
A	Amount of sleep deprivation	BAC
B	Errors in the driving simulator	BAC
C	Cognitive performance	Amount of alcohol consumed
D	Amount of alcohol consumed	Errors in the driving simulator

Question 17 ©VCAA 2020 Q35 ADAPTED EASY
Dr Singh believed that when the participants were in the 0.10 per cent BAC condition, they achieved an altered state of consciousness at the end of the study. The least likely indication of participants being in an altered state of consciousness at the end of the study would be if they
- A had a lack of emotional regulation.
- B struggled with completing automatic processes such as walking in a straight line.
- C remembered the conversations of passers-by outside the laboratory.
- D estimated the study going for a shorter or longer time than it did.

Question 18 ©VCAA 2020 Q36 ADAPTED HARD
Which one of the following accurately describes a limitation of Dr Singh's study?
- A experimenter effects
- B placebo effect
- C order effects
- D individual participant differences

Question 19 ©VCAA 2020 Q37 ADAPTED MEDIUM
What would the results of this study mean in terms of applying its findings to society?
- A Driving with a BAC of 0.05 per cent is equally dangerous to driving after a full night's sleep deprivation.
- B High BAC is more dangerous when driving than sleep deprivation.
- C Driving sleep deprived can help individuals focus their attention and improves driving performance.
- D Driving sleep deprived for 24 hours is equally dangerous to driving with a BAC of 0.10 per cent.

Question 20 ©VCAA 2020 Q38 ADAPTED MEDIUM
Which one of the following represents a qualitative measure of sleep that would indicate whether someone has moved during their sleep?
- A an electroencephalograph, which indicates a deep sleep.
- B an electromyograph, which indicates changes in muscle tone.
- C a video recording, which provides visual evidence of movement.
- D an electro-oculograph, which records changes in eye movements indicating stages of sleep.

Use the following information to answer Questions 21–24.

Chloe is a 60-year-old who feels tired and goes to bed at 7 p.m. Due to going to sleep so early Chloe wakes at 12 a.m. and cannot get back to sleep until 6 in the morning. Despite having the required sleep for her age group Chloe is always tired.

Question 21 ©VCAA 2021 Q31 ADAPTED MEDIUM

Which of the following best describes Chloe's sleep problems?

	Name of sleep disorder	Characterised by
A	Sleep–wake shift	Melatonin secretion peaking later in the day
B	Advanced Sleep Phase Disorder	Melatonin secretion peaking earlier in the day
C	Delayed Sleep Phase Syndrome	Cortisol being released earlier in the morning to assist with waking up
D	Circadian rhythm sleep disorder	The inability to sleep due to melatonin reuptake

Question 22 ©VCAA 2021 Q32 ADAPTED MEDIUM

Chloe is always tired. What could be the likely cause of this?

A not enough NREM stages 1 and 2 sleep due to having a reduced sleep time.
B not enough total hours of both REM and NREM sleep due to broken sleep cycles.
C not enough REM sleep due to falling asleep earlier.
D having broken sleep cycles overall leading to poor quality sleep and an increase in stages 1 and 2 NREM.

Question 23 ©VCAA 2021 Q33 ADAPTED MEDIUM

To cope with the demands of spending time with her partner, who doesn't get home until 6 p.m., Chloe has been forcing herself to stay up until 9 p.m., however finds herself still waking up at 12 a.m. and experiencing even more tiredness the next day. What is a likely affective effect on Chloe's functioning?

A inability to concentrate during her workday.
B reduction in emotional regulation.
C struggling with staying awake during the day.
D slowed reaction time.

Question 24 ©VCAA 2021 Q34 ADAPTED MEDIUM

Chloe's doctor recommends bright light therapy to address Chloe's sleep issues. She provides Chloe with a bright light therapy box. Chloe's doctor is likely to recommend that Chloe administer the therapy

A after lunch to feel more energised during the day.
B in the morning to shift her ultradian rhythm backwards.
C early in the morning to advance her circadian rhythm forward.
D before she goes to bed to resynchronise her sleep–wake cycle.

Question 25 ©VCAA 2020 Q29 ADAPTED HARD

Amanda has a new job as a nurse and is struggling to adapt to working overnight shifts. Which one of the following best indicates the struggles she is facing with her sleep cycle?

A Delayed Sleep Phase Syndrome from going to sleep later than usual because of work.
B Advanced Sleep Phase Disorder as she is waking up earlier than needed.
C shift work disorder as she is struggling with needing to be awake at times we would typically sleep.
D a circadian rhythm sleep disorder as she is experiencing changes to her sleep–wake cycle.

Question 26 ©VCAA 2020 Q31 ADAPTED EASY

Which of the following is not an example of a way to improve sleep hygiene?

A avoiding blue light in the hours leading up to sleep.
B avoiding eating right before bedtime.
C ensuring the room is at an optimal temperature – not too hot or cold.
D ensuring you drink water directly before bed so you don't wake dehydrated.

Use the following information to answer Questions 27–30.

Coral, a university student, conducts an experiment in a classroom to test the effectiveness of avoiding blue light 2 hours before bed on adolescent boys who struggle to fall asleep at night. She recruits nine 16-year-old boys from a suburban boys' school to participate in her experiment. Coral measures quality of sleep every morning for three days prior to avoiding blue light using self-reporting of a rating scale out of 5, ranging from extremely poor to good-quality sleep. On the fourth day, Coral asks the boys to avoid using their phones or electronic devices 2 hours prior to sleep. After one week of avoiding blue light, Coral measures the adolescents' quality of sleep using the same scale.

Question 27 ©VCAA 2019 Q36 ADAPTED MEDIUM

Which one of the following is a possible hypothesis for this experiment?

A Participants will be able to fall asleep quicker when avoiding blue light prior to falling asleep.
B Participants will show an improvement in the symptoms of struggling to fall asleep after avoiding blue light.
C Participants will show lower quality of sleep after avoiding blue light before bed.
D Participants will have higher overall quality of sleep after avoiding blue light 2 hours prior to sleep.

Question 28 ©VCAA 2019 Q37 ADAPTED MEDIUM

What is an ethical guideline that needs to be upheld in this experiment?

A consent obtained from the adolescents.
B not allowing participants to have the option to remain anonymous.
C participants must have the right to withdraw from the experiment with no reason needed.
D coral is able to inflict deception in the experiment.

Question 29 ©VCAA 2019 Q38 ADAPTED MEDIUM

The graph below represents Coral's results.

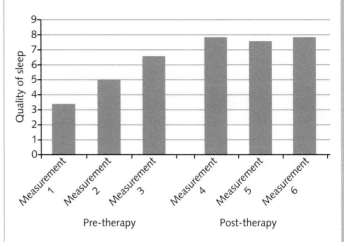

Based on these results, is avoidance of blue light before bed likely to be an intervention to individuals struggling to fall asleep?

A No, because avoidance of blue light had no effect on participants' quality of sleep.
B Yes, because avoidance of blue light had an immediate effect on participants' quality of sleep.
C Yes, because there is a clear pattern of improvement in the quality of sleep experienced by participants.
D No, because despite an initial improvement in quality of sleep, participants' levels of sleepiness began to return to baseline levels.

Question 30 ©VCAA 2019 Q39 ADAPTED HARD

Which one of the following is not a limitation of Coral's study?

A Individual participant differences such as other sleep hygiene factors were not controlled for.
B There was no control to work as a comparison.
C There was a small sample size.
D Self-reports are subjective, and often individuals may expect a particular result so report that rather than the true quality of sleep.

Section B: Short answer

Question 1 ©VCAA 2021 SECTION B Q4 ADAPTED

For her honours project at university, Selin was interested in understanding how Advanced Sleep Phase Disorder (ASPD) affected individuals who work a 9 a.m.–5 p.m. job.

a Describe what is meant by the term 'Advanced Sleep Phase Disorder'.

[2 marks]

b Why is ASPD described as a circadian rhythm sleep disorder? [2 marks]

c How can bright light therapy be used to help treat ASPD in Selin's study?

[2 marks]

[Total = 6 marks]

Question 2 ©VCAA 2020 SECTION B Q5 ADAPTED

Lui is a businessman who travels quite frequently for work, often being across several time zones each week.
He has found he is getting increasingly tired and finding it harder and harder to adapt to the different time shifts.

a Name the circadian rhythm sleep disorder Lui is most likely to be suffering from, justifying your response.
[2 marks]

b Describe the role of melatonin in Lui's sleep–wake cycle. Why is this causing sleep deprivation?
[3 marks]

c Describe how Lui's affective and cognitive functioning may be disrupted due to his sleep deprivation, including examples.
[4 marks]

d Explain how bright light therapy could be used to help Lui regulate his sleep-wake cycle.
[3 marks]

e What might be a potential barrier for bright light therapy being an effective treatment for Lui?
[2 marks]
[Total = 14 marks]

Question 3 ©VCAA 2019 SECTION B Q4 ADAPTED

Jessica is a 20-year-old student who is relatively unhealthy and is struggling with falling and staying asleep.
She often goes out partying on the weekends, staying up until 5 a.m., but tries to be asleep by 11 p.m. on weeknights. She also binge eats quite frequently late at night. Jayne, her mother, is a 65-year-old who is very active and healthy for her age and has frequent good-quality sleeps.

a Outline a possible explanation for Jessica's poor-quality sleep.
[2 marks]

b Differentiate between the typical sleeping patterns of someone Jessica's age versus someone of Jayne's age.
[2 marks]

c What is one thing Jessica can implement in her sleep routine to improve her sleep hygiene? How will this help improve the quality of her sleep?
[2 marks]

d Why would video recording, rather than electromyograph (EMG), be a more appropriate method for monitoring Jessica's sleep?
[2 marks]
[Total = 8 marks]

Question 4

Anna is an elderly woman in her 70s. She engages in healthy eating, takes care of her mental wellbeing and is overall quite active for her age.

a Create a hypnogram of a typical night's sleep for Anna based on her age group.
[2 marks]

b Recently, Anna's granddaughter bought her an iPad to read eBooks on before bed. Since getting the iPad Anna noticed she is falling asleep later and getting poorer quality sleep. Give one reason for this.
[1 mark]
[Total = 3 marks]

Question 5

Differentiate between the sleep of an adolescent and that of an elderly person.
[2 marks]

Defining mental wellbeing

Key knowledge

- ways of considering mental wellbeing, including levels of functioning; resilience, as the ability to cope with and manage change and uncertainty; and social and emotional wellbeing (SEWB), as a multidimensional and holistic framework for wellbeing that encapsulates all elements of being (body, mind and emotions, family and kinship, community, culture, country, spirituality and ancestors) for Aboriginal and Torres Strait Islander people
- mental wellbeing as a continuum, with an individual's mental wellbeing influenced by the interaction of internal and external factors and fluctuating over time, as illustrated by variations for individuals experiencing stress, anxiety and phobia

Key science skills

Analyse, evaluate and communicate scientific ideas

- use appropriate psychological terminology, representations and conventions, including standard abbreviations, graphing conventions and units of measurement
- discuss relevant psychological information, ideas, concepts, theories and models and the connections between them
- analyse and explain how models and theories are used to organise and understand observed phenomena and concepts related to psychology, identifying limitations of selected models/theories
- critically evaluate and interpret a range of scientific and media texts (including journal articles, mass media communications, opinions, policy documents and reports in the public domain), processes, claims and conclusions related to psychology by considering the quality of available evidence
- analyse and evaluate psychological issues using relevant ethical concepts and guidelines, including the influence of social, economic, legal and political factors relevant to the selected issue
- use clear, coherent and concise expression to communicate to specific audiences and for specific purposes in appropriate scientific genres, including scientific reports and posters
- acknowledge sources of information and assistance, and use standard scientific referencing conventions

Source: VCE Psychology Study Design (2023–2027) page 40, 13

8 Defining mental wellbeing

R U OK? It's OK if you are not. We all feel a bit not OK at times. During VCE, a time which can be particularly stressful, it is important that you take care of your mental wellbeing and be on the lookout for when you need to ask for help.

Adobe Stock/ rashmisingh

p. 325

8.1 Ways of considering mental wellbeing

Your mental wellbeing is often associated with your ability to function well in your everyday life: work, study and relationships. Social and emotional wellbeing is a much more holistic concept in First Nations' cultures that encapsulates body, mind and emotions, family and kinship, community, culture, Country, spirituality and ancestors.

8.2 Mental wellbeing as a continuum

p. 343

Mental wellbeing can be considered as being on a continuum where optimal mental health is at one extreme and poor mental health at the other. Where you sit on the continuum changes over time and depends on both internal and external factors. You may have a Psychology SAC due tomorrow so you feel stressed today. But tomorrow, once you have completed the SAC, your stress may disappear.

Adobe Stock/ Antonioguillem

Everyone feels stress, anxiety and fear. It is what makes us human. It is when these feelings become overwhelming, or when they don't seem to go away, that we need to seek help from others.

Flashcards
Chapter 8 flashcards

Slideshow
Chapter 8 slideshow

Test
Chapter 8 pre-test

Assessment
- Pre-test
- End-of-chapter exam

Revision
- Chapter map
- Key term flashcards
- Key concept summary
- Slideshow

Investigation
- Investigation: What has been the influence of the COVID-19 pandemic on mental wellbeing?
- Logbook template: Literature review

Worksheet
- Prevalence of anxiety disorders

To access these resources, visit
cengage.com.au/nelsonmindtap

Nelson MindTap

Know your key terms

Aboriginal and Torres Strait Islander peoples
Anxiety
Continuum
External factors
Holistic framework
Internal factors
Levels of functioning
Mental wellbeing
Multidimensional framework
Phobia
Resilience
Social and emotional wellbeing (SEWB)
Specific phobia

Weblink
R U OK?

Are you OK? This is a question we have all heard many times. However, since the establishment of the R U OK? charity (see weblink) and the annual R U OK?Day, we now associate this question more with our mental health than with our physical health (Figure 8.1). The term **mental health** has often been misunderstood as being about whether or not someone has a diagnosable **mental disorder**, such as an anxiety disorder or depressive disorder. As you will learn in this chapter, though, whether we are living with a mental disorder or not is only one part of what determines our overall mental health.

Contemporary understandings of mental health recognise that it is something positive that we can experience more or less of at different times in our lives, and that many different personal (internal) and situational (external) factors affect our mental health. This positive dimension of mental health is emphasised in the World Health Organization's (WHO's) definition of mental health as not just the absence of mental ill-health, but as 'a state of wellbeing in which the individual realises his or her own abilities, can cope with the normal stresses of life, can work productively and fruitfully, and is able to make a contribution to his or her community'. WHO has a separate definition of mental disorder, to recognise that people who live with a mental disorder may still experience a level of mental wellbeing.

In this chapter, we define mental wellbeing as the positive dimension of our overall mental health and explore its relationship to mental ill-health and disorder. In the first part of the chapter, we see how mental wellbeing and **mental ill-health** each exist along a 'continuum' of experiences. That is, we can experience more or less of each at different times in our lives. We explore how the two *continua* (plural of *continuum*) of mental wellbeing and mental ill-health interact with each other. We consider Western understandings of the factors that influence mental wellbeing and mental disorder and how these affect our level of functioning in daily life and our resilience to change and uncertainty. We then expand our thinking through considering **Aboriginal and Torres Strait Islander peoples'** understandings of mental wellbeing within a holistic framework of social and emotional wellbeing. We learn how cultural, political, and historical factors affect the wellbeing of individuals and of whole communities.

In the second part of this chapter, we take a deeper look at the internal and external factors that influence our mental wellbeing and how changes in these factors at different times in our lives can cause us to move between locations on the continua of mental wellbeing and mental ill-health. We illustrate how mental wellbeing and mental ill-health can fluctuate (change) over time by understanding how experiences of stress and anxiety in response to daily hassles and life events differ from the experience of phobia as a mental disorder.

Figure 8.1 Casey Donovan, pictured, addresses guests during R U OK? Day celebrations at Barangaroo Reserve. She is an ambassador for R U OK?, a harm prevention charity that promotes mental wellbeing by encouraging people to stay connected and have conversations that can help others through difficult times in their lives.

8.1 Ways of considering mental wellbeing

If you analyse WHO's definition of mental health as a state of **mental wellbeing**, you will see that mental wellbeing has three parts. The first part is about how we function within ourselves to set and meet our goals and adapt to challenges. The second part is about the quality of relationships we have with other people. The third part is about how we relate and contribute to the broader community and society. This makes mental wellbeing a **multidimensional framework**.

WHO's multidimensional definition of mental health is based on models of 'positive psychology' that seek to shift psychological theories of mental health away from a purely medical approach that is focused on the presence or absence of mental disorders, towards one that places equal value on understanding the factors that support and promote mental wellbeing. Taking this approach allows us to consider mental wellbeing and mental ill-health as separate but interacting factors that together determine our overall level of mental health (e.g. Diener et al., 1999; Keyes, 2002; 2007; Ryff, 1989; Ryan & Deci, 2001; Seligman & Csikszentmihalyi, 2000).

A dual-continuum model of mental health and wellbeing

One of the most influential models of mental health to come from positive psychology is the **dual-continuum model** (Keyes, 2002; 2007; Westerhof & Keyes, 2010). As the name suggests, the dual-continuum model proposes that our overall mental health is determined by a combination of two factors: (1) our subjective sense of mental wellbeing, and (2) whether or not we are living with a mental disorder. The mental wellbeing and mental disorder factors are represented as two separate continua. The idea is that mental wellbeing and mental disorder are each experienced along a **continuum** – a range of experiences from low to high levels – and that we may be in different places on each continuum at different times of our lives. Where we sit along each continuum affects our level of functioning in the various aspects of our lives. Our **level of functioning** is the degree to which we are able to engage effectively in the different domains of our lives, such as in our personal care, relationships and work.

There are distinct descriptions of levels of functioning for mental wellbeing and mental disorder. Importantly, although the two continua of mental wellbeing and disorder are separate, they can interact with each other.

The following sections outline each continuum in turn and how the different levels of functioning in each are described. We then put the two pieces together into the dual-continuum model and consider how they interact to determine our overall state of mental health.

Table 8.1 The Mental Health Continuum Scale (short form) (adapted from Keyes et al., 2008). Each item is scored according to respondents' experiences over the last month on a 6-point Likert scale (0 = 'never', 1 = 'once or twice', 2 = 'about once a week', 3 = '2 or 3 times a week', 4 = 'almost every day', or 5 = 'every day').

Component of mental wellbeing	Item During the past month, how often did you feel
Emotional wellbeing	
Positive affect	happy
Positive affect	interested in life
Life satisfaction	satisfied
Social wellbeing	
Social contribution	that you had something important to contribute to society
Social integration	that you belonged to a community
Social actualisation	that our society is becoming a better place for people like you
Social acceptance	that people are basically good
Social coherence	that the way our society works makes sense to you
Psychological wellbeing	
Self-acceptance	that you liked most parts of your personality
Environmental mastery	good at managing the responsibilities of your daily life
Positive relations with others	that you had warm and trusting relationships with others
Personal growth	that you had experiences that challenged you to grow and become a better person
Autonomy	confident to think or express your own ideas and opinions
Purpose in life	that your life has a sense of direction or meaning to it

Source: Adapted from Hone, L. C., Jarden, A., Schofield, G. M., & Duncan, S. (2014). Measuring flourishing: The impact of operational definitions on the prevalence of high levels of wellbeing. *International Journal of Wellbeing*, 4(1), 66.

The continuum of mental wellbeing and levels of functioning

The dual-continuum model describes mental wellbeing as having three core components, which are similar to those in the WHO definition we discussed previously. These are as follows.

1 **Emotional wellbeing**, which relates to how a person feels. For example, people who experience a high level of mental wellbeing regularly experience positive emotions of happiness, calmness and satisfaction with life.
2 **Psychological wellbeing**, which relates to how a person thinks about (perceives or appraises) themselves and their goals, their sense of purpose and meaning, the way they respond to challenges (e.g. coping style), and the quality of their relationships.
3 **Social wellbeing**, which relates to a person's sense of belonging to a community and their sense of being able to contribute meaningfully to society.

Psychologists measure these three components of mental wellbeing using self-report questionnaires. Table 8.1 shows a commonly used measure called the Mental Health Continuum Scale (Keyes et al., 2008). Notice that the three components of emotional, psychological and social wellbeing are each measured with a separate set of questions (three questions about emotional wellbeing, six about psychological wellbeing, and five related to social wellbeing). People are asked to rate how often in the past month they have experienced each item on a scale from zero to five, where zero equals 'never' and five equals 'every day'. A person's responses across the three components gives them a mental wellbeing score between zero and 70.

Using people's responses to scales like the Mental Health Continuum Scale, psychologists have described three levels of functioning that relate to people's self-reported mental wellbeing. These are flourishing, languishing and moderate mental wellbeing.

Flourishing describes people who respond '4 = almost every day' or '5 = every day' to at least one of the emotional wellbeing questions and six of the questions across psychological and social wellbeing. Flourishing describes a state of optimal mental wellbeing in which a person both feels good (has emotional wellbeing) and functions effectively (has psychological and social wellbeing). People who are flourishing are able to set goals and work effectively towards them; they make their own decisions and are independent in managing tasks like personal hygiene, going to work and maintaining their physical heath; they nurture healthy relationships; and they actively contribute to and feel connected with the wider community.

In contrast, people are said to be **languishing** if they respond '0 = never' or '1 = once or twice' to at least one of the emotional wellbeing questions and to at least six of the questions across the psychological and social wellbeing scales. People who are languishing experience both low levels of positive emotions and low levels of psychological and social functioning. For example, they may struggle with things like functioning independently, developing and maintaining close relationships, holding onto a job, and connecting with the broader community.

People who score in between languishing and flourishing are classified as having **moderate mental wellbeing**. Studies show that the majority of people fall into this category (Keyes et al., 2008; Wissing et al., 2021). Figure 8.2 shows the three levels of functioning that are associated with emotional, psychological and social wellbeing.

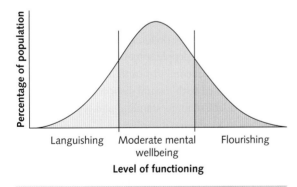

Figure 8.2 Three levels of functioning based on people's responses to questions assessing their mental wellbeing in terms of emotional, psychological and social wellbeing

The continuum of mental disorder and levels of functioning

WHO defines mental disorder separately from mental health. According to WHO, mental disorders are clinically significant conditions of mental ill-health that have a range of different presentations 'characterised by a combination of abnormal thoughts, perceptions, emotions, behaviour and relationships with others' (WHO, 2022). A **mental disorder** usually causes distress to the person experiencing the disorder and impairs their ability to function effectively in personal, family, social, educational and occupational domains of life. Mental disorders can also be called 'mental health disorders', 'mental illnesses', or 'psychological disorders'. We use the term 'mental disorder' consistently in this chapter as it is the term used in the Study Design.

Mental health professionals diagnose mental disorders by comparing a person's symptoms to a set of criteria that are developed by teams of mental health professionals and published in diagnostic manuals. For example, the American Psychiatric Association publishes the *Diagnostic and statistical manual of mental disorders* (DSM). The DSM is currently in its fifth edition (published in 2013), known as the DSM-V *(DSM-5-TR™)*. WHO publishes a diagnostic manual of all health disorders, including mental disorders, that is commonly referred to as the *International classification of disease* (or ICD), now in its 11th edition (ICD-11; WHO, 2019). Both the DSM and ICD classify mental disorders into broad categories, such as depressive disorders and anxiety disorders, and then specify particular mental disorders within these categories. For example, the DSM-V lists specific **phobia** under the category of 'anxiety disorders' and the ICD lists it under 'anxiety and fear disorders'.

When diagnosing a mental disorder, mental health professionals consider the severity of symptoms and how these affect the person's **level of functioning**. Clinicians assess the level of dysfunction experienced by someone with a mental disorder using self-report scales that measure the person's functioning in six domains over the past month. These domains are:

1 **cognition** – understanding and communicating
2 **mobility** – moving and getting around
3 **self-care** – hygiene, dressing, eating and staying alone
4 **getting along with people** – interacting with other people
5 **life activities** – domestic responsibilities, leisure, work and school
6 **participation** – joining in community activities.

The clinician assesses a person's level of dysfunction on a continuum with five points: (1) none, (2) mild, (3) moderate, (4) severe, and (5) extreme. People who live with a mental disorder may experience symptoms across a range of severity, but must have met specific criteria for diagnosis. The range of experiences of mental disorder represent one end of the broader continuum of experiences of mental ill-health.

A person who is experiencing some symptoms of mental disorder, but does not meet all of the criteria for a specific diagnosis, is said to be experiencing a **mental health problem**. Mental health problems are often triggered by stressful life events and resolve once the impacts of the event have passed. For example, we may experience a very low mood, 'feeling down' or 'feeling blue', for a period of time after a relationship break-up or loss of a job. Although distressing at the time, these are normal emotional and behavioural responses to challenging life events, and we can work through them when we have effective coping strategies and support networks. However, a mental health problem may develop into a mental disorder if the symptoms increase and/or persist so that they significantly impact a person's ability to function.

Figure 8.3 shows the continuum of mental ill-health and levels of functioning.

Seeking advice about mental health problems and mental disorder

When learning about mental disorders it is natural to apply what we are learning to ourselves and others and to wonder whether we or someone we know may be experiencing a mental disorder. If the content in this chapter or other chapters raises questions for you, then you could begin by seeking the help of a trusted teacher or school counsellor who can refer you to appropriate further resources and sources of help (Figure 8.4).

There are also many excellent online mental health resources for adolescents and young adults where you can seek further information, advice and support. Beyond Blue provides a list of resources specific for young people that is updated regularly (see weblink). Aboriginal and Torres Strait Islander students can contact 13YARN for culturally responsive mental health assistance (see weblink).

Mental disorders affect many people across all life stages. In fact, most people will experience some form of mental disorder during their lives. For example, the Australian National Health Survey 2017–18 estimated that one in five (20 per cent, or 4.8 million) Australians reported experiencing a mental disorder during the past 12 months (Australian Institute of Health and Welfare, 2020).

Mental disorders are serious health conditions that require professional diagnosis and treatment, just like physical illnesses. For this reason, it is important that we learn to recognise symptoms and know when to seek help, just as we do for symptoms of a physical disorder. However, people sometimes fail to seek help for symptoms of mental ill-health due to **stigma** that can be associated with mental disorders. Stigma is caused by a lack of understanding and is perpetuated by myths about mental disorders.

Mental disorder	Mental health problem	No mental ill-health
Severe emotional and/or behavioural impairment	Disruption to an individual's usual level of social and emotional wellbeing	Function at an effective level and meet the demands of everyday life
• Psychological dysfunction • Unable to function independently • Marked distress • Ongoing impairment • Excessive anxiety • Significant changes in sleep patterns and appetite • Withdrawal and avoidance of social functions	• Mild to moderate distress • Temporary impairment • Difficulties in coping • Some changes in sleep patterns • Some changes in appetite • Experience a loss of energy • Difficulty concentrating	• Able to cope with stress • Able to form positive relationships • Able to manage feelings and emotions • Independent • Displays resilience • Few sleep difficulties • Physically and socially active

FIGURE 8.3 The continuum of mental ill-health and levels of functioning

Figure 8.4 Talking about your worries with a school counsellor is a positive step towards maintaining good mental health.

disorders, including initiatives such as R U OK?Day. The more we can talk openly about mental health, and the factors that contribute to mental wellbeing and mental disorders, the healthier our society will become.

How you can help fight stigma

- Educate yourself about mental disorders. Having the facts can help you challenge the misinformation that leads to stigma.
- Be aware of words. Do not reduce people to a diagnosis. Instead of 'a schizophrenic', say 'a person with schizophrenia'. Correct people who use hurtful language such as 'psycho' or 'crazy' to describe people with a mental disorder.
- Challenge media stereotypes. Write letters to any newspapers, TV or radio stations that negatively portray people with a mental disorder.

Table 8.2 shows some of the common myths about mental disorders and corrects these with the facts. There is much good work being done to reduce the stigma associated with mental

Table 8.2 Mental disorder: myth versus fact

Myth	Fact
Mental disorder is fairly rare and does not affect average people.	Mental disorder is quite common. According to WHO, one in four people in the world will be affected by mental disorder at some point in their lives. Around 450 million people currently live with such conditions, placing mental health issues among the leading causes of ill-health and disability worldwide. Mental disorder can strike people of any age, race, religion or income status.
If you have a mental disorder, you can will it away. Being treated for a psychiatric disorder means an individual has in some way 'failed' or is weak.	A serious mental disorder cannot be willed away. Ignoring the problem does not make it go away, either. It takes courage to seek professional help.
Most people with a mental disorder are receiving treatment.	Only one in five people affected with a mental disorder seeks treatment.
People with mental disorders can never live productive lives.	Science has made great strides in the treatment of mental disorders in recent decades. With proper treatment, many people with mental disorders live happy, productive lives.
People with mental disorders are dangerous.	This powerful myth has been fed by the media. In fact, the vast majority of people living with mental disorders are not dangerous. They are much more likely to be the victims of violence and crime than perpetrators.
Depression and other mental disorders, such as anxiety disorders, do not affect children or adolescents. Any problems they have are just a part of growing up.	Children and adolescents can develop mental disorders. One in ten children or adolescents has a disorder severe enough to cause impairment.
Mental disorder is more like a weakness than a real illness.	Mental disorders are as real as other diseases like diabetes or cancer. Some mental disorders are inherited, just as some physical illnesses are. They are not the result of a weak will or a character flaw.

Source: Adapted from Mental Health Foundation Australia (2022). *Fight stigma*. https://www.mhfa.org.au/CMS/FightStigma

- » Support those with mental disorders. Treat them with respect. Help them find jobs or housing. Encourage them to get or stick with treatment.
- » Share your story. If you or someone in your family has had a mental disorder, speak up about it. Your example could help someone else.

'Mental illness is nothing to be ashamed of, but stigma and bias shame us all.' – Bill Clinton

Source: Adapted from Mental Health Foundation Australia (2022). *Fight stigma*. https://www.mhfa.org.au/CMS/FightStigma

The interaction between mental wellbeing and mental disorder

Treating mental wellbeing and mental ill-health as separable continua is important because it recognises that it is possible for someone who is living with a mental disorder to experience a sense of mental wellbeing, and that it is possible for someone who does not have a mental disorder to still be languishing. It also recognises that the two continua can influence each other.

Figure 8.5 shows the structure of the dual-continuum model. The model describes a person's overall level of mental health as the relationship between their experience of mental wellbeing and mental ill-health. The mental wellbeing continuum forms the vertical dimension (top to bottom) of the model and the mental ill-health continuum forms the horizontal dimension (left to right).

The model forms four quadrants of experiences of mental health. In the top right quadrant is the state of complete mental health. People in this category are flourishing in terms of their subjective mental wellbeing and level of functioning, and they are not experiencing mental ill-health (i.e. they are not experiencing a mental health problem or mental disorder). In the lower right quadrant are people who are languishing, with low levels of subjective mental wellbeing, but not experiencing a mental disorder (although they may still experience some mental health problems depending on how far to the left they are located within the quadrant). In the top left quadrant are people who are living with a mental disorder but experience a high level of mental wellbeing and functioning. In the lower left quadrant are people who have a complete lack of mental health because they are both languishing and living with a mental disorder.

In the dual-continuum model, although mental wellbeing and mental disorder can be separated, they can also influence each other.

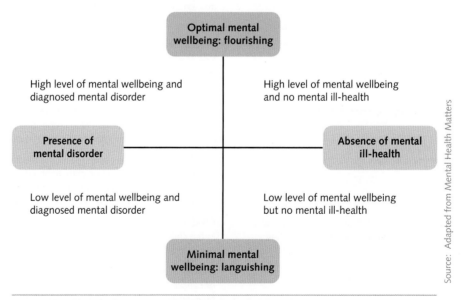

Figure 8.5 The dual-continuum model of mental health. In this model, a person's overall mental health is a combination of their location on the mental wellbeing continuum (from languishing at the bottom, to moderate mental wellbeing in the middle, to flourishing at the top) and their location on the mental ill-health continuum (from mental disorder on the left, to mental health problem in the middle, to no mental ill-health on the right).

For example, someone who is languishing may be experiencing some mental health problems. This may place them at higher risk of developing more serious mental health problems or a mental disorder than someone who is flourishing.

Or, when someone who is flourishing experiences a mental disorder they may improve more quickly, or may experience less severe symptoms, than someone who is languishing and experiences a mental disorder (Westerhof & Keyes, 2010).

EXAM TIP
Make sure you know the differences between mental health, mental wellbeing, mental ill-health and mental disorder.

ACTIVITY 8.1 THE DUAL-CONTINUUM MODEL

Try it

Figure 8.5, depicting the dual-continuum model, shows how mental health reflects a combination of subjective mental wellbeing and the presence or absence of mental ill-health. People can fit into different quadrants of the continuum at different stages in their lives. An individual's mental health and wellbeing can vary over time. For example, Carlton footballer Jake Edwards suffers from depression. He began to experience mental health problems early in his football career, sometimes locking himself in his room, crying inconsolably. He no longer plays football but he continues to contribute to the sport with a program he has developed called Outside the Locker Room. This program helps AFL clubs understand and support their employees and players with mental health issues.

1 Where would you place Jake Edwards in the continuum at the different stages in his life – before his AFL career, during and after?

Apply it

Place the people described below on the continuum.

2 a Jo has just ended a difficult relationship. The difficult break-up has left Jo emotionally and physically exhausted.

   b Sam has just been made redundant but is using this as an opportunity to retrain for a new job.

   c Lee has been diagnosed with an anxiety disorder. Lee uses medication to help with symptoms, therapy to develop coping strategies and has a strong and supportive group of friends.

   d Sia has been diagnosed with the anxiety disorder specific phobia. Sia is having trouble coping and has not found the best medication to manage a disorder like a phobia.

Exam ready

3 Which of the following statements about mental wellbeing is incorrect?

A A person's mental wellbeing can fluctuate over time.

B People will fall into one discrete category of either mentally healthy or mental illness.

C Mental wellbeing can be influenced by both internal and external factors.

D Mental wellbeing can be influenced by life events like a relationship break-up.

Resilience

One way of thinking about the relationship between mental wellbeing and our overall mental health is that more positive states of mental wellbeing and functioning are protective against developing mental health problems and mental disorders. Another way to put this is that positive levels of mental health help to make us **resilient** to the risk factors that contribute to mental disorder.

Resilience is a term used by psychologists to describe a person's ability to respond adaptively to stressful life events and to cope with uncertainty.

A person is described as resilient when they 'bounce back' from challenging circumstances, such as family or relationship problems, health problems, financial stress or more severe forms of adversity including homelessness, war, famine and physical and/or psychological trauma (Davydov et al., 2010; Stainton et al., 2019).

Studying the factors that make people resilient is one way that psychologists can develop approaches that are protective against mental ill-health. This work has shown that although resilience is influenced by internal factors, such as our personality traits and other heritable traits, it is not fixed.

8.1.2 RESILIENCE

Instead, resilience is a process that involves behaviours, thoughts and actions that can be learned. So, what factors support resilience?

Research has identified a range of **internal factors** and **external factors** that support resilience. Internal factors are personal characteristics that include having effective coping strategies for managing stress and regulating emotional responses, and having a sense of meaning and purpose in life. Internal factors can also be biological and genetic influences that affect how resilient we are, such as our physical health and our family history of health problems. External factors are social and environmental influences, such as having access to supportive relationships, whether these are found in the family and/or in social networks outside of the family such as schools, as well as having access to health and social services and to safe, stable and clean housing.

Resilience is not the same as mental wellbeing. Instead, resilience and mental wellbeing have a reciprocal relationship – that is, they influence each other. Resilience is both the outcome of factors that promote and support mental wellbeing and a resource that influences our current state of mental wellbeing. Like our mental wellbeing, our level of resilience changes over time and it is put to the test when we come under stress.

Building resilience does not mean we will not ever experience difficult and challenging life circumstances. In fact, we often build resilience by working through difficult experiences. Resilience helps us to manage stressful situations more effectively so that we may emerge stronger on the other side.

We can help develop our resilience by practising skills and taking part in activities that support our wellbeing. The following suggestions are adapted from the Victorian Government's Better Health Channel (2020) and the American Psychological Association (APA) – see the weblinks.

» **Connect with others**. Develop and maintain strong relationships with people around you who will support and enrich your life.
» **Take time to enjoy yourself**. Make time for activities, hobbies and projects you enjoy.
» **Participate and share interests**. Join a club or group of people who share your interests.
» **Contribute to your community**. Volunteer your time for a cause or issue that you care about. Helping others builds our sense of purpose and meaning in life.
» **Take care of yourself**. Be active and eat well – these help maintain a healthy body. Combine physical activity with a balanced diet to nourish your body and mind and keep you feeling good, inside and out.
» **Challenge yourself**. Learn a new skill or take on a challenge to meet a goal. Learning improves your mental fitness, while striving to meet your own goals builds skills and confidence and gives you a sense of progress and achievement.
» **Develop coping skills to help you manage stress and regulate your emotions**. Be aware of what triggers your stress and how you react. You may be able to avoid some of the triggers and learn to prepare for or manage others.
» **Rest and refresh**. Get plenty of sleep. Go to bed at a regular time each night and practise good habits to get better sleep. Sleep restores both your mind and body.
» **Notice the here and now**. Take a moment to notice each of your senses each day. Practising mindfulness, by focusing your attention on being in the moment, is a good way to do this. Practise gratitude for the positive things in your life, no matter how small they may be.
» **Ask for and accept help**. Dealing with challenges can lead us to isolate ourselves, but it is important to reach out and accept help and support from those who care about you (Figure 8.6).

FIGURE 8.6 Developing resilience sometimes means reaching out for support from others. We can also build our own resilience, and the resilience of the community, by supporting others in need.

WB
8.1.3 FACTORS THAT CONTRIBUTE TO POSITIVE SOCIAL AND EMOTIONAL HEALTH AND WELLBEING

Weblink
The Better Health Channel – *Wellbeing*

The APA – *Building your resilience*

EXAM TIP
Resilience is a person's ability to successfully adapt to stress and cope with uncertainty. It is influenced by internal and external factors. Do not confuse it with mental wellbeing.

ANALYSING RESEARCH 8.1

Three hours is enough to help prevent mental health problems in Australian teens

Research led by Dr Patricia Conrod from the University of Montreal in Canada has found that just two 90-minute sessions of an innovative mental health education program delivered in schools can reduce by up to 50 per cent the likelihood of harmful substance use and related mental health problems in at-risk adolescents. The research group has reproduced their findings in studies conducted in the US, UK, Europe and Canada, and has recently found similar results in a large sample of Australian secondary school students.

Fourteen schools from New South Wales participated in the Australian study between 2013 and 2015. Seven of the schools were randomly allocated to the control condition and seven to the intervention condition. Students were first evaluated for their risk of developing mental health and substance-use problems using a self-report scale. The scale measured four personality factors that are known to predict the development of substance use and mental health problems: impulsivity, anxiety sensitivity, sensation seeking and hopelessness. Only students who were identified as high risk were included in the intervention; 195 students met the criteria in the schools allocated to the intervention and 202 students in the control schools.

Staff from the schools in the intervention condition were trained to deliver the two intervention sessions to their high-risk students. The control schools followed their standard educational programs. The intervention involved two 90-minute sessions in which students learned about the risk factors associated with their personality profile and how to use cognitive-behavioural strategies to manage these.

The students then applied this information to real-life scenarios that were developed from experiences described by young people with similar personality profiles. The groups discussed thoughts, emotions and behaviours described in the scenarios, including identifying situational triggers. Then, with the guidance of the teacher, they explored ways to manage the issues.

Over the next two years, students completed questionnaires every six months that enabled the researchers to track their substance use, and the symptoms of depression, anxiety, panic attacks and conduct problems. The findings of the Australian study were similar to those from other countries. The intervention was successful in reducing alcohol and substance-use harms by up to 50 per cent in high-risk adolescents compared to the control group. The intervention group also showed a 25 per cent reduction in symptoms related to mental health problems and disorders, such as anxiety, depression and conduct problems when compared to the control group. The researchers noted that embedding the program within schools is a key factor in improving student outcomes because it provides intervention to high-risk adolescents who might not otherwise have access to effective family support and social support services.

Sources: adapted from Edalati, H., & Conrod, P. J. (2019). A review of personality-targeted interventions for prevention of substance misuse and related harm in community samples of adolescents. *Frontiers in Psychiatry*, 770; Newton, N. C., Conrod, P. J., Slade, T., Carragher, N., Champion, K. E., Barrett, E. L., ... & Teesson, M. (2016). The long-term effectiveness of a selective, personality-targeted prevention program in reducing alcohol use and related harms: A cluster randomized controlled trial. *Journal of Child Psychology and Psychiatry*, 57(9), 1056–1065; and Université de Montréal. (2013, October 3). Three hours is enough to help prevent mental health issues in teens. *ScienceDaily*.

Questions

1. State the aim of this study and one possible hypothesis.
2. Identify the independent and dependent variables.
3. Describe how the results support the hypothesis you identified in question 1.
4. Identify and explain one ethical guideline relevant to this study.
5. Analyse the research scenario to identify the internal and external factors in the intervention that were designed to build resilience to substance abuse and mental disorder.

INVESTIGATION 8.1 WHAT HAS BEEN THE INFLUENCE OF THE COVID-19 PANDEMIC ON MENTAL WELLBEING?

Scientific investigation methodology
Literature review

Aim
To conduct a literature review of contemporary research into the influence of the COVID-19 pandemic on mental wellbeing.

Logbook template
Literature review

Introduction
A literature review involves collating and analysing secondary data related to other people's scientific findings and/or viewpoints. The aim of a literature review is to answer a question or provide background information to help explain observed events, or as preparation for an investigation to generate primary data.

Research and select two to four articles relevant to the effect of the COVID-19 pandemic on mental wellbeing. Remember, it is important to be able to distinguish between different sources of information. Try to include at least one peer-reviewed journal article in your review. To find journal articles you can start with *The Conversation* (see weblink). *The Conversation* is written by researchers for the general public and includes links and references to original journal articles. To find journal articles directly you can search using Google Scholar (weblink). You might also find the student-led publication *Journal of Emerging Investigators* (weblink) a useful source of peer-reviewed articles written at a level that is easier to understand. For all of these sites you just need to enter the relevant search terms into the search window. You could try entering terms like 'COVID-19 AND mental wellbeing' or 'mental wellbeing AND pandemic'. The use of upper-case AND ensures that all search words are included in searches.

Structure of literature review
Use the following structure to complete a literature review of your selected articles.

Introduction
1. Discuss why you are writing a review and why it is an important topic.
2. Describe the scope of the review, including what aspects of the topic will be discussed.

Summary of articles
3. Provide a summary and analysis of each article, including scientific findings, viewpoints, results/data and conclusions.

Conclusions
4. Discuss the common findings from your articles.
5. Do the findings from all your articles agree with each other? If not, what are the contentious areas?
6. Describe any gaps in the current research.
7. Suggest areas for further research.
8. Discuss your overall perspective on the topic.

References
9. List all sources of information using scientific referencing conventions (see weblink).

Weblink
The Conversation

Google Scholar

Journal of Emerging Investigators (JEI)

Citation and referencing guides

KEY CONCEPTS 8.1a

» Mental wellbeing is a multidimensional construct that includes emotional wellbeing, psychological wellbeing and social wellbeing. It can be measured using self-report scales such as the Mental Health Continuum Scale.

» Emotional wellbeing relates to the experience of positive emotions such as happiness, joy or love, and feeling generally satisfied with life. Psychological wellbeing relates to how a person thinks about (appraises) themselves and their goals, the way they respond to challenges (e.g. coping style) and the quality of their relationships. Social wellbeing relates to a person's sense of belonging to a

- community, and their sense of being able to contribute meaningfully to society.
» Levels of functioning describes the extent to which someone can operate effectively in the different domains of everyday life such as self-care, personal relationships and work.
» People who report a high degree of emotional, psychological and social wellbeing have a level of functioning described as flourishing. People who report low levels of wellbeing have a level of functioning described as languishing. People in between (most people) experience moderate mental wellbeing.
» Mental disorders are clinically defined conditions of mental ill-health that have a range of distinct presentations. This may include a combination of abnormal thoughts, perceptions, emotions, behaviour and relationships with others.
» Mental disorders affect a person's level of functioning, ranging from no dysfunction to extreme dysfunction.
» Mental health problems are relatively mild experiences of mental ill-health that do not meet the criteria for a mental disorder.
» Mental ill-health can be represented as a continuum from no mental ill-health, to mental health problem, to mental disorder.
» According to the dual-continuum model, complete mental health includes both a high state of mental wellbeing (flourishing) and the absence of mental ill-health. Other states of mental health are represented by the intersection between a person's position on the mental wellbeing continuum and their position on the mental ill-health continuum (Figure 8.5).
» Resilience is a person's ability to successfully adapt to stress and cope effectively with uncertainty. Resilient people possess and are able to use effective coping strategies and adaptive ways of thinking, and can maintain healthy personal relationships and social connectedness. Resilience is protective against developing mental health problems or a mental disorder.

Concept questions 8.1a

Remembering
1. Using an example of each, distinguish between a mental disorder and a mental health problem. **r**
2. Describe the dual-continuum model of mental health and its relationship to the terms 'mental wellbeing' and 'mental disorder'. **r**

Understanding
3. Explain the difference between the WHO definition of mental health, and the definition of mental health from the dual-continuum model. **e**
4. Explain the relationship between resilience and mental wellbeing. **e**
5. Describe flourishing and languishing in terms of levels of functioning. What is their relationship to mental disorder? **e**

Applying
6. Describe three strategies that you could utilise to improve your resilience to cope with and manage change and uncertainty as a VCE student. **e**

HOT Challenge
7. Create a table or diagram that identifies the similarities and differences between mental health, mental wellbeing, mental ill-health and mental disorder. **d**

Aboriginal and Torres Strait Islander peoples' understandings of social and emotional wellbeing

The dual-continuum model of mental health that we have explored so far comes from Western understandings of health and wellbeing. It is a mistake to assume that the model provides an appropriate approach for people from different cultural backgrounds, who bring different perspectives, experiences and understandings of mental health and social and emotional wellbeing. Here, we explore a framework of **social and emotional wellbeing (SEWB)** that reflects **Aboriginal and Torres Strait Islander peoples' understandings of mental wellbeing**.

8.1.4 SOCIAL AND EMOTIONAL WELLBEING (SEWB)

I am Dr Graham Gee, an Aboriginal-Chinese man, also with Celtic heritage, originally from Darwin. My Aboriginal-Chinese grandfather was born near Belyuen on Larrakia Country, a small community located not far from Darwin on the eastern side of the Cox Peninsula. As a clinical psychologist, I worked at the Victorian Aboriginal Health Service for 11 years before taking up a senior research fellow position at the Murdoch Children's Research Institute. My early research focused on Aboriginal and Torres Strait Islander mental health, social and emotional wellbeing, and healing and recovery from complex trauma. In 2022, I received a fellowship from the Eisen Family Private Fund that has supported my team to commence work with Aboriginal services dedicated to healing child sexual abuse. I recently joined the National Clinical Reference Group for the Prime Minister and Cabinet National Office for Child Safety, and the Research Advisory Committee for the National Centre for Action on Child Sexual Abuse. Through my work with Aboriginal and Torres Strait Islander families and communities, and in collaboration with other Aboriginal and Torres Strait Islander psychologists, I co-developed the framework of social and emotional wellbeing you are learning about and I have co-authored this content for your textbook.

Figure 8.7 Dr Graham Gee, descendent of the Larrakia people from the lands surrounding what is now known as Darwin, N.T. Graham is a clinical psychologist and he co-developed the Aboriginal and Torres Strait Islander framework of social and emotional wellbeing.

Social and emotional wellbeing (SEWB)

The SEWB framework situates mental wellbeing within a broader, **holistic framework** that recognises the importance of culture and history as factors that influence wellbeing and reflects the understandings and specific needs of Aboriginal and Torres Strait Islander peoples. The framework of SEWB that we explore in this chapter was developed by Aboriginal psychologists in consultation with Aboriginal and Torres Strait Islander communities (Gee et al., 2014). The framework reflects the strength and wisdom of Aboriginal and Torres Strait Islander peoples' ways of knowing, being and doing, which have supported the wellbeing of individuals, communities and *Country* for over 60 000 years (see Chapter 4). It also recognises that the current health and wellbeing challenges faced by many Aboriginal and Torres Strait Islander children, families and communities are the outcome of the ongoing impacts of government policies, racist attitudes and discrimination that began when Australia was colonised by the British in 1788.

The SEWB framework is a general framework to help health professionals gain a broad understanding of concepts of mental health and wellbeing that are often core components of Aboriginal and Torres Strait Islander world views. However, it is important to recognise that there is great diversity of experiences and understandings within different Aboriginal and/or Torres Strait Islander cultural group(s) and communities. For example, people living in major cities will have different experiences and needs to those in rural communities, and from those in remote communities. For this reason, the SEWB framework needs to be adapted to each local context. As noted by Gee and colleagues, 'for health practitioners to gain an in-depth understanding of mental health and SEWB as it relates to their local Aboriginal and Torres Strait Islander community, they must engage with Elders, families and leaders of the community over time. There are no alternatives or short cuts' (Gee et al., 2014, p. 65).

Table 8.3 The nine guiding principles of Aboriginal and Torres Strait Islander social and emotional wellbeing (SEWB)

1	Aboriginal and Torres Strait Islander health is viewed in a *holistic* context, that encompasses mental health and physical, cultural and spiritual health. Land is central to wellbeing. Crucially, it must be understood that when the harmony of these interrelations is disrupted, Aboriginal and Torres Strait Islander ill-health will persist.
2	*Self-determination* is central to the provision of Aboriginal and Torres Strait Islander health services.
3	*Culturally valid understandings* must shape the provision of services and must guide assessment, care and management of Aboriginal and Torres Strait Islander peoples' health in general, and mental health in particular.
4	It must be recognised that the *experiences of trauma and loss*, present since European invasion, are a direct outcome of the disruption to cultural wellbeing. Trauma and loss of this magnitude continues to have *intergenerational effects*.
5	*The human rights* of Aboriginal and Torres Strait Islander people must be recognised and respected. Failure to respect these human rights constitutes continuous disruption to mental health. Human rights relevant to mental illness must be specifically addressed.
6	*Racism*, *stigma*, environmental adversity and social disadvantage constitute ongoing stressors and have negative impacts on Aboriginal and Torres Strait Islander peoples' mental health and wellbeing.
7	The centrality of Aboriginal and Torres Strait Islander *family and kinship* must be recognised as well as the broader concepts of family and the bonds of reciprocal affection, responsibility and sharing.
8	There is *no single* Aboriginal or Torres Strait Islander culture or group, but numerous groups, languages, kinships and clans, as well as ways of living. Furthermore, Aboriginal and Torres Strait Islander people may currently live in urban, rural or remote settings, in traditional or other lifestyles, and frequently move between these ways of living.
9	It must be recognised that Aboriginal and Torres Strait Islander people have great *strengths*, *creativity and endurance* and a deep *understanding* of the relationships between human beings and their environment.

Source: Fact Sheet, Social and Emotional Wellbeing. With permission Professor Pat Dudgeon, (https://timhwb.org.au/wp-content/uploads/2021/04/SEWB-fact-sheet.pdf)

A key aim of the SEWB framework is to help build the knowledge and skills of non-Indigenous health professionals and social service providers to be **culturally responsive**. Cultural responsiveness involves health professionals learning about local Aboriginal and Torres Strait Islander culture and history, reflecting on their power and privilege, demonstrating cultural humility and addressing racism and discrimination. Cultural responsiveness ensures that services provided to Aboriginal and Torres Strait Islander peoples and communities are culturally safe and respectful.

The SEWB framework provides a foundation for the Gayaa Dhuwi (Proud Spirit) Declaration and is reflected in the approach of Aboriginal and Torres Strait Islander-led organisations and projects like Transforming Indigenous Mental Health and Wellbeing, the Aboriginal and Torres Strait Islander Healing Foundation and the Australian Indigenous Psychology Education Project (see weblinks). Nine guiding principles provide the foundation of the model, shown in Table 8.3.

The visual representation of the SEWB framework in Figure 8.8 immediately signals a very different cultural understanding of mental wellbeing compared to the Western dual-continuum model. Note how the circular shape and embedding of the individual 'self' within layers of relationships and connections contrasts with the dual-continuum model's location of the individual along two lines that intersect horizontally and vertically. The placement of the self in the centre of the SEWB framework communicates the key concept of a **collectivist** understanding of human development, rather than individualistic. In collectivist cultures, individuals are embedded within, defined and shaped by interconnected networks of relationships. The framework defines seven domains of relationships that surround the self, which we will explore in detail shortly.

The ring that encircles the seven domains represents **the historical, political, cultural and social contexts** that influence the strength of connections between an individual and the seven domains. These broader contexts are very important because they significantly shape the environments and circumstances into which Aboriginal and Torres Strait Islander children are born. For this reason, these contextual factors are often referred to as **determinants** of social and emotional wellbeing. For example, social, political and historical factors strongly influence the capacity of a child's family, cultural group and community to retain and/or revitalise language, traditions, cultural values and practices.

Weblink
Gayaa Dhuwi (Proud Spirit) Declaration

Transforming Indigenous Mental Health and Wellbeing

The Healing Foundation

Australian Indigenous Psychology Education Project

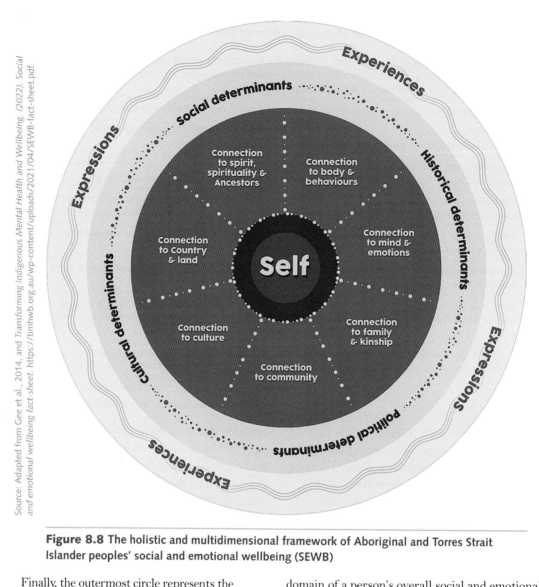

Figure 8.8 The holistic and multidimensional framework of Aboriginal and Torres Strait Islander peoples' social and emotional wellbeing (SEWB)

Finally, the outermost circle represents the **experiences** and **expressions** of individuals and communities that change over time in response to social, historical, political and cultural factors that influence connection to the seven domains. The rippling wave symbol represents 'the ebb and flow of change as risk factors disrupt connections and protective factors restore and strengthen connections' (Transforming Indigenous Mental Health and Wellbeing, 2022, p.3). This framework is **holistic** because it recognises the interconnectedness of the seven domains on SEWB. It is **multidimensional** because there are multiple domains and broader contextual factors that influence SEWB.

Seven domains of connections

Rather than being specifically about mental health as a separate construct, with social and emotional wellbeing as parts, the SEWB framework sees mental wellbeing as one (important) interconnected domain of a person's overall social and emotional wellbeing. Notice that the closest match to the concept of mental wellbeing is the domain called 'connection to mind and emotions'.

At any one time, a person may experience different levels of connection to each of the seven domains, some being healthy while others are in need of support and healing. Importantly, a person will experience changes in their connections to the domains at different times in their lives. Broadly, a person's SEWB (and often by extension that of their family and extended kin) will decrease when connections to one or more of the domains are disrupted, and will increase when connections are restored (Figure 8.8).

Table 8.4 defines the seven domains. Next to the description of each domain are examples of the risk factors that can disrupt connections and the protective factors that can restore connections. Note that although it is helpful to define each of

Table 8.4 Seven domains of connection within the SEWB framework

Domain of connection	Description	Risk factors that disrupt connection	Protective factors that restore connection
Body and behaviours	Biological elements of physical health, including diet and exercise	Smoking, alcohol, exposure to disease, exclusion from health systems, poor quality housing and overcrowding	Playing sports, living on *Country*, hunting, fishing and gathering traditional foods and medicines, access to health services
Mind and emotions	Mental health and wellbeing, including culturally specific disorders or expressions of distress, experience of positive emotions, not just absence of disorder	Threats to safety from expressions of racism, denial of human rights, *over-incarceration*, effects of *intergenerational trauma* on families and communities	Access to social supports, healthcare, education, recognition of human rights, *truth-telling* about colonisation
Family and kinship	Importance of family and wider kinship groups, reciprocal caring through gender and age roles, including respect for and learning from Elders	History of *Frontier Wars, killings and massacres*; policies of *forced removal of children* (historical *Stolen Generations* and ongoing child-protection policies)	Connecting to family history, reconnecting families, parenting and family support programs, spending time with Elders
Community	Cohesion of community, enacting community responsibilities (e.g. caring for *Country*); cultural identity of community	Forced removal to *missions* and *reserves*; disruption of communities; fragmented relationships between different family groups, causing feuding and violence	Restoration of *self-determination* and community control; community as a source of support, strength and connection
Culture	Cultural practices, including yarning, ceremony, traditional fire management of land, art, dance, song, storytelling; cultural knowledge of languages, law, ethics; pride in and sense of belonging to cultural identity	Loss of culture through *removal from lands*, splitting up of communities and removal of children; policies of *assimilation*	Cultural and language revitalisation; participation in cultural practices
Country	Deep feeling of belonging to *Country*; spiritual connection to *Country* through kinship, culture and caring for *Country*	Removal from and stealing of land and waterways; impacts of mining of land; pollution of land and water	Returning to *Country*, *land rights and sovereignty* of *Country*; being able to care for and heal *Country* through traditional land management, reconnecting with community and culture tied to *Country*
Ancestors	Knowledge of belief systems, Dreaming, Songlines, healing practices, wisdom and hope	Loss of knowledge through Mission life (imposition of Christianity) and assimilation	Evolving expressions of cultural knowledge and identity and expressions of spirituality coexisting with Christianity; mindfulness practices, such as Dadirri (see Chapter 4).

Source: Fact Sheet, Social and Emotional Wellbeing. With permission Professor Pat Dudgeon, (https://timhwb.org.au/wp-content/uploads/2021/04/SEWB-fact-sheet.pdf)

the seven domains separately, there is overlap and interconnection between the domains within a **holistic framework** such as the SEWB framework (Gee et al., 2014).

Social, cultural, historical and political determinants of SEWB

To understand the risk and protective factors that affect the domains of connection outlined in Table 8.4, all Australians need to develop their knowledge about the history of colonisation and of the ongoing traumatic impacts of racist policies across generations of families and communities.

Table 8.4 refers to a number of these policies and their impacts. These include killings and massacres during the period of the Frontier Wars (1788–1938); the introduction of infectious diseases, including smallpox, that decimated entire populations of Aboriginal and Torres Strait Islander people; intentional poisoning of water sources and food; forced removal of children from families (Stolen Generations, from the 1800s to the 1970s); removal from lands and the breaking up of families and communities onto Missions and Reserves; forced labour; and the denial of language and culture through assimilation and the White Australia policy.

In 1997, the impact of the forced removal of children from families and communities under policies of 'protection' was described by the Royal Commission into the Stolen Generations as systematic racial discrimination and genocide (Wilkie, 1997).

The impact of these policies on families and communities is described as **intergenerational trauma**. This means that the effects of trauma are passed down, and can be amplified, over generations (Atkinson et al., 2014). The effects of intergenerational trauma are evident in the social and health challenges that face many Aboriginal and Torres Strait Islander peoples today. Truth telling, healing and reconciliation requires recognising these facts and responding by advocating for genuine social justice outcomes.

It is equally important to recognise the great strengths and the rich and incredible diversity of the oldest continuing cultures in the world. Aboriginal and Torres Strait Islander peoples draw on the strength of their enduring cultures, their Elders and ancestors, and their history of resistance and activism in seeking justice. The phrase 'Always was, always will be', the theme of NAIDOC week in 2020, is a proud statement that asserts the fact that sovereignty was never ceded and celebrates the resilience and deep connection of Aboriginal and Torres Strait Islander peoples to the lands we now call Australia. In Chapter 10, we explore Aboriginal and Torres Strait Islander peoples' ongoing efforts to achieve **self-determination** and the importance of this for maintaining social and emotional wellbeing.

ACTIVITY 8.2 SEWB MODEL CONTEXTS

These activities will help you to develop your understanding of the cultural, social, political and historical contexts that are described in the SEWB model.

Try it

1. The Aboriginal and Torres Strait Islander Healing Foundation provides an explanation of intergenerational trauma in a short video and accompanying text (see weblink). Watch the video, make notes and use this to explain the concept of intergenerational trauma to a classmate. Identify the relevant contextual influences on domains of connection over generations.

Apply it

2 a Explore the Victorian Aboriginal Health Service (VAHS) guide to the history of colonisation and Deadly Story's timeline of Aboriginal and Torres Strait Islander activism and self-determination in the weblinks. Use this to develop your understanding of some of the policies and practices mentioned in Table 8.4, and the ongoing efforts of Aboriginal and Torres Strait Islander peoples in seeking justice.

Apply the SEWB model to identify examples of social, cultural, historical and political influences on domains of connection from Table 8.4.

b Listen to personal accounts of the Stolen Generations from Faye Clayton and Ian Hamm at the Aboriginal and Torres Strait Islander Healing Foundation website (see weblinks). Apply the SEWB model to identify examples of social, cultural, historical and political influences on domains of connection from Table 8.4.

c Celebrated Aboriginal activist and singer-songwriter Archie Roach (1956–2022) released the song *Took the children away* in 1990, describing the experiences of the Stolen Generations. Archie and his brothers and sisters were taken from their parents when Archie was aged three. They were split up and sent to different institutions and foster homes in New South Wales and Victoria. Archie was told that his parents had died in a house fire. *Took the children away* is the only song to ever have won an International Human Rights Award. Listen to and read the lyrics (see weblinks). Then research the answers to these questions.

  i To what experience and policy does 'fenced us in like sheep' refer?

  ii To what does 'mission land' and 'Framlingham' refer?

Weblink
The Healing Foundation – intergenerational trauma

VAHS – *History fact-sheet*

Deadly Story timeline

Faye Clayton

Ian Hamm

Took the children away

Took the children away – lyrics

d Watch the first 50 minutes of Episode 3 of *The First Australians*, which features the First Nations of Victoria – see weblink. Learn of Simon Wonga's and William Barak's achievements for their people. Explain how knowledge of this history serves as a strength base for current generations. (Figure 8.9).

Exam ready

3. Which of the seven domains of connection in the model of social and emotional wellbeing (SEWB) is most relevant to language revitalisation as a protective factor?

   A Family and kinship
   B Mind and emotions
   C Culture
   D *Country*

Weblink
The First Australians, Episode 3

Figure 8.9 Restoring family connections and connections to Elders for healing family, kin and culture

Experiences and expressions

There is great diversity in the way that Aboriginal and Torres Strait Islander people, families and communities experience and express SEWB. Aboriginal and Torres Strait Islander peoples are not defined by the historical impacts of colonisation and denial of human rights. Aboriginal and Torres Strait Islander history is a proud history of survival and change over millennia. This strength, resilience and dignity is evident in the often unsung and unrecognised contributions of many Aboriginal and Torres Strait Islander Elders and leaders in shaping Australia's multicultural society; the ongoing efforts of Aboriginal and Torres Strait Islander leaders and organisations to resist injustice and oppression; in the revitalisation of culture and languages; and in continued calls for land - rights, self-determination and sovereignty. This proud history of survival is most recently and powerfully articulated in the Uluru Statement from the Heart (First Nations National Constitutional Convention, 2017).

ACTIVITY 8.3 ULURU STATEMENT FROM THE HEART

Try it

1 Listen to and read the Uluru Statement from the Heart – see weblink. Analyse the content to identify the strengths of First Nations Australians and references to impacts of government policies on domains of connection.

Apply it

2 Identify the restorative actions in the Uluru Statement from the Heart that could help heal our shared history and be protective of Aboriginal and Torres Strait Islander peoples' connections to the seven domains of SEWB.

Exam ready

3 Using Table 8.4 as a reference, to which of the seven domains of connection in the SEWB framework is the 'Voice' most relevant?

   A Community
   B Body and behaviours
   C Culture and *Country*
   D Spirituality and Ancestors

Weblink
Uluru Statement from the Heart

One of the most powerful ways in which Aboriginal and Torres Strait Islander peoples express and support social and emotional wellbeing is through the revitalisation of cultural practices (e.g., Andrews, 2020). In Case study 8.1, Dr Vicki Couzens tells her story of revitalising traditional possum-skin cloak making within contemporary contexts as a powerful way to restore connections.

CASE STUDY 8.1

Cloaked in strength: possum-skin cloaks in social and emotional wellbeing

WARNING: Aboriginal and Torres Strait Islander students are advised that the following case-study contains the names and images of people who have died.

I am Dr Vicki Couzens, a Gunditjmara, Keerray Woorroong citizen whose lands are located in what is now known as the western districts of Victoria (Figure 8.10). I am a Senior Knowledge Holder of Gunditjmara Language and the Possum Cloak Story. I am also an artist, and I hold a PhD from RMIT University where I am a Research Fellow. Our great-grandfather was one of six men who made the historic Lake Condah cloak shown in Figure 8.11. You can see men wearing traditional possum-skin cloaks in a photograph from 1858 shown in Figure 8.12.

Figure 8.11 The historic Lake Condah cloak, collected in 1872, fur side and skin side up

In 1999 I attended a workshop with other Aboriginal artists from across Victoria that was hosted by the Melbourne Museum. The staff showed us the magnificent nineteenth-century Lake Condah possum-skin cloak that had been made by my Ancestors.

Figure 8.12 A group of First Nations men in possum-skin cloaks and blankets in 1858 at Penshurst, in Victoria's western districts

Figure 8.10 Dr Vicki Couzens, Senior Knowledge Holder of Gunditjmara Language and the Possum Cloak Story

It was during this encounter that I had a profound and transformative experience. I experienced a vision, which was gifted by the makers of the cloak. It was a call from the Ancestors to re-awaken the Songlines of the Possum Cloak Story, to return the cloaks to our People, to reclaim, regenerate, revitalise and remember the Old Peoples' story – that it our journey to carry this vision forward.

From that encounter I undertook a journey of cultural revitalisation sharing the learnings and teachings of possum-skin cloaks across south-eastern Australia. This journey included my younger sister Debra (now deceased), Treahna Hamm, Lee Darroch and Maree Clarke, along with many others. We work with communities, family clan groups, and individuals in reconnecting, revitalising and regenerating possum-skin cloak cultural practice. The workshops we share create the opportunities for deep cultural connection, learning of cultural knowledges, skills, story, song, dance and ceremony. No one is unaffected. Experiences from the workshop contribute to a deepening and strengthening of identity, connection and belonging.

Explore the video weblinks showing the application of Dr Couzens' work. Then answer the questions that follow.

Weblink
Culture is healing

Possum skin cloaks then and now

Questions

1 Apply the domains of the SEWB to the video. Identify which of the seven domains of connections are involved in using possum-skin cloaking for healing.

2 Identify the cultural, historical, political and/or social influences that are relevant to this case study of Dr Couzens' work.

KEY CONCEPTS 8.1b

» Social and emotional wellbeing (SEWB) is a holistic and culturally responsive framework of the factors that influence health outcomes for Aboriginal and Torres Strait Islander peoples and communities, including mental health.

The framework is holistic and multidimensional because a person's SEWB is influenced by the seven domains of connection and their interactions with broader social, cultural, historical and political determinants.

Concept questions 8.1b

Understanding

1 Describe the benefits for the wellbeing of Aboriginal and Torres Strait Islander peoples in using the social and emotional wellbeing (SEWB) framework. **e**

HOT Challenge

2 Create a table or diagram that identifies the similarities and differences between Western definitions of mental wellbeing and the Aboriginal and Torres Strait Islander framework of social and emotional wellbeing. **d**

8.2 Mental wellbeing as a continuum

In section 8.1, we learned about the dual-continuum model of mental health, which reflects the combined effects of mental wellbeing and mental ill-health. Wellbeing can vary from flourishing to languishing, and we can experience an absence of mental ill-health, a mental health problem or a mental disorder.

In this section, we take a deeper look at how our location within the dual continuum can change over time due to the interaction between different internal (personal) and external (social and environmental) factors. This is illustrated by considering the range of experiences involved in feeling stressed, experiencing anxiety and having a diagnosis of phobia.

The influence of internal and external factors

Our experience of mental health and wellbeing fluctuates (changes) over time as we experience the ups and downs of life. At a given point in our lives, we may be flourishing – feeling happy, hopeful and purposeful, achieving our goals, sharing rewarding experiences with a partner, friends and/or family, being engaged with our community. However, this state can be disrupted by stressful life events that challenge our ability to cope and may put us at risk of developing a mental health problem or mental disorder. Our location along the two continua of mental wellbeing and mental ill-health at any particular time in our lives is affected by internal and external factors, which can be either risk factors (stressors) or protective factors (resources).

Internal factors

8.2.1 MENTAL WELLBEING

Internal factors are influences on our wellbeing that come from sources within ourselves. Internal factors include genetically influenced factors, such as a family history of mental ill-health, that may make us more susceptible to developing a mental health problem, and physical and psychological characteristics, such as our personality. They also include internal characteristics that we have acquired or learned through experience, such as our state of physical fitness, our habits, dietary choices, and learned coping and problem-solving strategies.

Some internal factors are more stable than others, such as personality traits and inherited risk of mental or physical disorders. For example, people who are high on trait anxiety have a tendency (a cognitive bias) to focus on negative emotions such as fear and worry across many situations. Although this is a stable personality trait, people can learn techniques to understand the strengths and weaknesses associated with their personality traits to manage the way they respond to situations (as we learned in Analysing research 8.1). Similarly, knowing whether we have a family history of mental or physical disorders can help us develop practices that reduce our risk of ill-health, such as practising yoga or meditation, exercising, making healthy food choices, reducing alcohol intake and stopping smoking. Internal factors

8.2.2 BIOPSYCHOSOCIAL APPROACH TO MENTAL WELLBEING

include personal characteristics that are protective (e.g. high self-esteem) and those that are risk factors (e.g. family history of anxiety disorders). Internal factors are generally biological and psychological factors.

External factors

External factors are those influences on our wellbeing that come from sources outside of ourselves, including the physical and social environments. They include biological or social factors.

The physical environment affects our mental wellbeing through things like toxins in the environment that can affect the brain and our levels of stress (e.g. exposure to diesel fumes and noise for people living on a busy road), or infectious diseases or unsanitary conditions (e.g. due to poor-quality housing). The biological environment in the womb impacts our early development and ongoing physical and mental wellbeing through the effects of maternal stress and intake of drugs and alcohol.

The social environment can include stressors ranging from strained or unhealthy interpersonal relationships, to peer pressure to engage in risky behaviours, to experiences of stigma or prejudice. The social environment can also provide protective external factors, like a supportive family or social network. Other social factors include our financial stability, which in turn affects our access to resources such as healthcare and educational opportunities, and the wider cultural and political context that affects the stability and peacefulness of society.

Interactions between factors

Internal and external factors often interact with each other to affect our mental wellbeing. For example, those of us with higher trait anxiety may find it more difficult to manage our responses to stressors from the external physical and social environments than people with low trait anxiety, which can negatively impact our mental wellbeing. However, through even more complex sets of interactions, the same individual can reduce the negative impact of their anxious personality style by developing internal resources to manage their anxiety.

Social and environmental external factors often interact with each other and with internal factors. For example, loss of income (an external social factor) reduces a person's ability to buy healthy foods and keep the heater on in winter (external environmental factors), which affects their physical health (an internal factor). In this way, the interaction between and within internal and external factors can produce cumulative (growing) or compounding (multiplying) effects on our wellbeing.

Changes in internal and external risk and protective factors over time cause us to move between regions of the mental health continuum at different stages of our lives. In Chapter 10 you will explore in detail the ways we can optimise and maintain our mental wellbeing to protect us against moving towards mental health problems and mental disorder. For now, read the story of Finn in Case study 8.2 to identify how internal and external factors interact to influence his experience of mental wellbeing.

CASE STUDY 8.2

Finding yourself

As you get older you may start to think about what you want to do when you "grow up". For a lot of people, this type of soul searching happens while travelling overseas. After all, people always say 'you find yourself while travelling'.

This was just the situation 25-year-old Ted* found himself in. Unfortunately for him, travelling didn't give him the answers he was looking for – in fact it raised more questions than answers. Believing that his identity was tied to what he was going to do, Ted began to panic.

Anxiety was something that he had grappled with throughout his life and for the most part he was able to manage it well. However, something about this felt different.

One day, he received a call from a friend inviting him to join him at the local gym. Realising that his home environment wasn't make him feel any better, he agreed to meet up with his friend.

"Sure, the first few sessions were tough. I wasn't used to the intensity. But as time went on I began to get the hang of things. The gym also had a rock climbing wall so I decided to try that too. Not only did I get fitter, I began to have more energy and my mind was less cluttered. That decision was probably the best decision I've ever made."

Many people think about the physical improvements of exercise but, as Ted experienced, exercise can also improve mental wellbeing. The endorphins and serotonin released in the body during physical exercise help to improve mood, increase energy levels and also improve sleep quality.

"Whether it was lifting weights or climbing the wall, there's a particular way of doing things. You need to focus on adjusting your body and problem solve when things aren't working. It's both mental and physical, and when you combine the two the dopamine really kicks in."

Having to focus and continually problem solve boosts brain activity – which was really important for Ted and his mental wellbeing journey. Having a consistent schedule and tackling these challenges every week helped him better manage his stress and anxiety.

"Exercise has definitely played an important role in managing my mental wellbeing. Not just the physical and mental aspect, but the social aspect of the gym is also really great."

Ted now works in the health industry while he works out what he wants to do next. He still goes to the gym regularly with his friends.

Source: Adapted from 'Your Mental Wellbeing', Queensland Health 2020.

Questions

1. List the internal and external factors that were influencing this young man's mental wellbeing.
2. Explain what the author is referring to in stating 'The body and mind are intrinsically linked'.
3. List the benefits of exercise as presented in the case study.
4. Describe the influence support from friends has on mental wellbeing.
5. What have you learned from this case study that could benefit your mental wellbeing?

EXAM TIP
In questions that assess the application of psychological knowledge to a scenario, it is important to make clear reference to the scenario in your responses. You can do this by referring by name to the person in the scenario.

Stress, anxiety and phobia

In this section, we apply the dual-continuum model to distinguish between the experiences of stress, anxiety and phobia. Each of these experiences exist on a continuum from mild to more severe forms, which affects our position within the continua of mental wellbeing and mental ill-health.

Stress

In Chapter 3, we learned that stress is the combination of physiological and psychological responses we experience when confronted with a situation that is threatening or challenging. We experience stress when the demands on us exceed our perceived ability to cope. Stress is a subjective experience, which means our personal interpretation (appraisal) of a situation or event determines how stressed we feel. Indeed, the very same situation can be experienced differently depending on the person. For example, a shy person may find a loud party stressful, but a more outgoing person may experience very little stress in this situation. This example highlights the influence of internal personal factors on the experience of stress and its impact on subjective mental wellbeing.

In terms of the dual-continuum model, a stressful situation or event may lower our sense of mental wellbeing and reduce our level of functioning for a period of time, but once the stressor has passed we can return to an optimal level of feeling and functioning. As you learned in Chapter 3, stress can be positive, such as when it helps us to avoid danger or perform at an optimal level (this is called eustress). However, when stress is severe, such as when we experience a traumatic event, and/or if it becomes chronic (long-lasting), it can cause us to move towards regions of the dual-continuum that put us at risk of mental health problems or mental disorder. For example, the catastrophic floods that occurred in Queensland and New South Wales during 2022 caused widespread distress and suffering. The degree to which different people and families will continue to experience mental health problems or mental disorder depends on the range of internal and external resources they are able to access and make use of to cope with the ongoing impacts of their experience.

Anxiety

We have all experienced the feelings of worry and nervousness that come with situations like waiting to take an important test, or wondering why we are being followed by a police car while driving. **Anxiety** is the term used to describe the emotional state we feel when we anticipate threat or danger. Anxiety causes feelings of fear, worry, dread and uneasiness, and is associated with heightened arousal of the sympathetic nervous system, causing faster breathing and heart rate, and increased muscle tension and sweating.

Although stress and anxiety both cause a heightened sympathetic response, they are different. Stress occurs as a response to an experienced event (i.e. the stress response), whereas anxiety occurs when we anticipate or sense the potential for threat or danger in the future, even if we are unsure of the nature of the threat. Stress and anxiety interact with each other. For example, someone who has experienced a stressful event may experience anxiety afterwards, worrying that a similar event might occur in future.

Feeling mildly or moderately anxious is a normal emotional response when we anticipate an upcoming stressful event. In fact, we know that moderate levels of anxiety can actually improve performance through increasing our focus (attention). Even a very high level of anxiety is appropriate when it is consistent with the level of potential threat, as long as we can return to a state of mental wellbeing after the threat has passed (see Figure 8.13).

Different people experience anxiety in different ways, even when facing a similar level of threat. This is due to the specific interactions between internal and external factors that influence their personal experience of wellbeing at the time. For example, while one person may feel relaxed while enjoying a bushwalk in a particular area, another may feel highly anxious due to the memory of being confronted by a snake in the same area (an internal factor). However, this is not evidence that the anxious person is suffering from an anxiety disorder.

Figure 8.13 Short periods of higher-than-normal levels of anxiety can be beneficial. For example, snowboarders who challenge themselves to go faster may experience anxiety while also improving their performance.

So, if fear and anxiety are normal emotions, when do they signify a problem? Just as we can experience varying levels of stress, the level of anxiety that we experience at any particular time sits along a continuum. Anxiety can range from mild and moderate levels that do not impair our daily functioning and that resolve once the potential threat is gone, to more severe anxiety that may be a sign of a mental health problem or a mental disorder, such as a phobia (see Figure 8.14).

When someone experiences anxiety that is out of proportion to the situation and that interferes with aspects of their daily functioning, the person is experiencing an **anxiety disorder**. This is a clinical term used to describe a class of mental disorders that feature intrusive feelings of fear, panic and the anticipation of danger. Anxiety disorders cause high levels of sympathetic nervous system arousal and avoidance behaviours. A person living with an anxiety disorder can have maladaptive levels of arousal and avoidance that can interfere with their capacity to learn new information, attend social events, plan an appropriate response to an issue or carry out the routine activities of daily life. Further, they recognise that their fears are irrational, unrealistic and intrusive, but they feel unable to control them.

Worksheet Prevalence of anxiety disorders

8.2.3 STRESS, ANXIETY AND PHOBIA

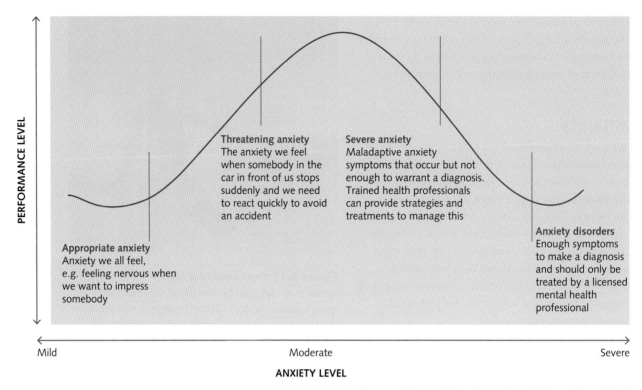

Source: Adapted from Mental Health Foundation (2014)

Figure 8.14 Feeling anxious or fearful is a normal emotional response to some situations – there is a difference between feeling anxious in some situations and suffering from an anxiety disorder.

The DSM-V category of anxiety disorders includes separation anxiety disorder, selective mutism, specific phobia, social phobia, panic disorder, agoraphobia and generalised anxiety disorder. These anxiety disorders are separate from the category of stress and traumatic disorders (which includes post-traumatic stress disorder (PTSD)), and from the category of obsessive compulsive disorders. Similar distinctions are made in the ICD-11.

Anxiety disorders are the most commonly diagnosed mental disorders in Australia. On average, one in four people – one in three women and one in five men – will experience anxiety during their lifetime. In any one year, approximately two million Australian adults have a diagnosed anxiety disorder (Australian Demographic Statistics, 2016). Table 8.5 summarises the continuum of anxiety experiences we have discussed in this section. In the next section, we focus on phobia as an example of an anxiety disorder.

Table 8.5 Levels of anxiety

Anxiety type	Description
Mild anxiety	Mild anxiety is commonly experienced when a person is presented with challenging situations and circumstances. You may feel nervous about an upcoming test or oral presentation but these feelings are not overwhelming. Mild anxiety is anxiety that is usually manageable and you can cope and participate in everyday activities. Mild anxiety can have positive effects; for example, it can help you to solve problems, and motivate you to succeed and engage in goal-directed activity.
Moderate anxiety	Moderate anxiety is the disturbing feeling that something is wrong. It is considered normal when experienced in short periods in response to difficult life experiences. You may feel nervous, agitated and find it difficult to concentrate on the task at hand. Moderate anxiety and its consequences interfere with a normal lifestyle.
Severe anxiety	A person experiencing severe anxiety will feel restless, irritable and angry. The person will experience the flight-or-fight-or-freeze response (increased heart rate, breathing and blood pressure). Their cognitive skills decrease significantly, and they have feelings of being totally overwhelmed. Severe anxiety interferes significantly with a person's ability to cope with everyday life.

Source: Adapted from Videbeck (2017)

Phobia

Phobia comes from the Greek word 'phóbos', meaning fear or panic. Almost everyone experiences some mild fears. For example, fears of heights, enclosed spaces or insects are common. Phobias differ from common fears. They are a form of anxiety disorder in which a person experiences an intense and irrational fear of a specific object, activity or situation. Phobias cause a person to experience an overwhelming sense of anxiety, often accompanied by symptoms like nausea, vomiting, shaking, fainting, uncontrollable sweating, increased heart rate or hot flushes.

Phobias are most likely to begin in childhood and continue into adulthood if left untreated. There is often a family history of phobia, but it is unclear whether it is transmitted genetically or through observation (or both). People who experience phobia can identify what triggers their phobia and they understand that their reaction is irrational. However, they feel powerless to control their fear and they may not know how or why it originated (Figure 8.15).

People with specific phobia will avoid the feared stimulus in any way possible.

Specific phobias are usually grouped into the following sub-categories:
» animals (e.g. spiders, insects, dogs)
» natural environment type (e.g. heights, storms, water)
» blood injection or injury type (e.g. needles, invasive medical procedures)
» situational type (e.g. aeroplanes, lifts, enclosed spaces)
» other (rare and unusual kinds).

Specific phobia is a form of mental disorder because the feelings of anxiety are overwhelming and out of proportion to the actual threat, and because the level of anxiety and avoidance interferes with a person's ability to function normally in everyday situations. For example, someone who fears needles (trypanophobia) may avoid seeking medical treatment (Figure 8.16). Table 8.6 summarises the DSM-V criteria for specific phobia.

Figure 8.15 Many people are naturally fearful of heights, but people with a phobia have an intense and irrational fear that is out of proportion to the danger faced.

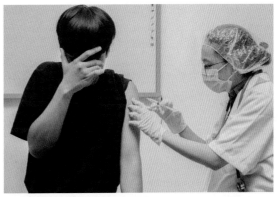

Figure 8.16 An individual's specific phobia of injection (trypanophobia) can be so intense and frightening that they avoid medical treatment.

A **specific phobia** is one of three kinds of phobia listed under anxiety disorders in the DSM-V and ICD-11. The other two are agoraphobia and social phobia (social anxiety disorder), which both relate to more complex situations than a specific phobia. People diagnosed with specific phobia experience severe anxiety and excessive and irrational fear of a particular stimulus, which could be an object, entity or situation. Specific phobia occurs in the presence of the feared stimulus, in anticipation of exposure, and can even occur when hearing of, speaking about, or looking at a picture of the feared stimulus.

In this section, we have seen how variations in peoples' experiences of stress, anxiety and phobia illustrate fluctuations in mental wellbeing and mental disorder over time and in response to interactions between internal and external factors. When stress and anxiety become severe and begin to impact everyday functioning, they shift from being normal responses to stressful situations, towards becoming mental health problems and mental disorder. Phobia is a specific form of anxiety disorder that is positioned squarely in the dual-continuum region of mental disorder.

Table 8.6 DSM-V criteria for specific phobia

Criterion	Description
Unreasonable, excessive fear	The person exhibits excessive or unreasonable, persistent and intense fear triggered by a specific object or situation.
Immediate anxiety response	The fear reaction must be out of proportion to the actual danger and appears almost instantaneously when the person is presented with the object or situation.
Avoidance or extreme distress	The individual goes out of their way to avoid the object or situation, or endures it with extreme distress.
Life-limiting	The phobia significantly impacts the individual's school, work or personal life.
Six months duration	In children and adults, the duration of symptoms must last for at least six months.
Not caused by another disorder	Many anxiety disorders have similar symptoms. A doctor or therapist would first have to rule out similar conditions such as agoraphobia, obsessional-compulsive disorder (OCD), and separation anxiety disorder before diagnosing a specific phobia.

Source: Adapted text from DSM-5, American Psychiatric Association.

8.2.4 ANALYSIS OF A RESEARCH INVESTIGATION

However, if a person has an effective treatment plan and uses coping strategies successfully, they may still experience otherwise high levels of mental wellbeing. This is a good example of how a person's position within the dual-continuum model can change over time, depending on their progress during treatment and on the interaction of internal and external factors that may cause their symptoms to improve or deteriorate at different times.

The psychological and physiological responses that accompany stress, anxiety and phobia are very similar. All involve the activation of the sympathetic nervous system (flight-or-fight-or-freeze response) to help prepare our body to deal with a threat. Learning to recognise the difference between stress, anxiety and a phobia is important because management strategies and treatments vary. If left untreated, mental disorders may affect relationships with family and friends, job performance and overall quality of life. Table 8.7 summarises the key similarities and differences between stress, anxiety and phobia. Figure 8.17 shows the relationship between stress, anxiety and phobia.

Many factors lead to the development of a phobic disorder, and no single factor can determine whether someone will experience phobic anxiety. The experience of a specific phobia can be explained through the biopsychosocial approach to understanding and treating mental disorders; this is explored in Chapter 9.

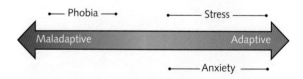

Figure 8.17 The continuum of stress, anxiety and phobia. Stress, anxiety and phobia can be experienced across a range of severity. Phobia is associated with more severe experiences of stress and/or anxiety and is maladaptive. Stress and anxiety can range from mild to severe and can be adaptive responses

8.2.5 DEFINING MENTAL WELLBEING CONCEPTS

Table 8.7 The key similarities and differences between stress, anxiety and phobia

Stress	Anxiety	Phobia
Activation of the sympathetic nervous system	Activation of the sympathetic nervous system	Activation of the sympathetic nervous system
Response to an actual situation that is experienced	Response to a potential threat that may be unclear or ambiguous	Response to a known stimulus
Positive (eustress) or negative (distress) response	Negative (distress) response	Negative (distress) response
Stress can be adaptive	Anxiety can be adaptive	Phobia is maladaptive

ACTIVITY 8.4 INTERNAL AND EXTERNAL FACTORS INFLUENCING WELLBEING AND PHOBIA

Try it

1. The forced lockdowns in Victoria during the COVID-19 pandemic affected the mental wellbeing of many people. Identify one internal and one external factor that may have been protective for individuals during this time, and discuss how these factors may interact with each other.

Apply it

2. Go to the Australian Psychological Society (APS) website topic on phobia (see weblink). Read the section on 'Causes' and identify the internal and external factors that contribute to the development of phobias.

Exam ready

3. Which of the following would be considered external factors that influence mental wellbeing?
   A. Physical fitness and diet
   B. Problem-solving strategies
   C. Peer pressure
   D. All of the above

Weblink
APS – causes of phobia

ANALYSING RESEARCH 8.2

Writing about exam worries boosts exam performance

Research by Professor Sian Beilock from the University of Chicago has found that a simple intervention can help students who suffer from exam anxiety to boost their performance on high-stakes exams. It is well known that high levels of test anxiety impair exam performance. The researchers tested whether having students write down their worries about an upcoming exam could improve their performance on the exam.

The first two studies involved creating a high-stakes testing environment in the laboratory. In Study 1, 20 college students took two short maths tests. Before taking the first maths test (the pre-test), students were simply told to perform their best. After completing the pre-test, students were placed in a high-pressure scenario designed to mimic the high-stakes of admissions to the US college system. Students were informed that they would receive a monetary reward if both they and a partner they had been paired with performed well. Then they were told that their partner had already completed the test and scored highly, so it was now entirely up to the student if they won or lost the money. They were also told that their performance would be videotaped and that teachers would watch the tapes.

Students were then randomly allocated to either spend 10 minutes sitting quietly (this was the control group) or to the expressive writing task. The writing group spent 10 minutes writing about their thoughts and feelings about the upcoming test. Everyone then took the maths post-test. The results showed that pre-test maths performance did not differ between the groups. However, for the post-test, the control participants 'choked' under pressure, showing a 12 per cent drop in math accuracy from the pre-test to post-test, whereas the writing group showed a significant 5 per cent improvement. Figure 8.18 shows the results.

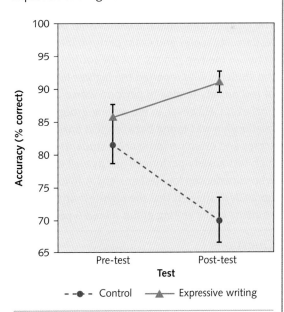

Figure 8.18 Results of Study 1 showing the beneficial effect of expressive writing about exam worries on subsequent exam performance

Source: Ramirez, G., & Beilock, S. L. (2011). Writing about testing worries boosts exam performance in the classroom. *Science, 331*(6014), 211.

The Study 1 results were promising, but could not reveal whether writing about test-related worries caused the improvement in student performance or whether any writing task might be beneficial. Study 2 tested this with a new sample; it also added a third group that spent 10 minutes writing about a mundane life event before taking their test. Again, only the group that wrote about their test worries improved at the post-test. Both the control group and the other writing group performed worse on the post-test than on the pre-test.

Studies 3 and 4 were conducted in a real high-stakes exam situation. Two cohorts of Grade 9 students sitting their final Biology exam took part, one year apart (51 students in Study 3 and 55 students in Study 4). In the US, the Grade 9 final exam is high stakes because the results count towards college admission scores. All students first completed a questionnaire to measure their tendency to feel anxiety in test situations. Students rated items on a scale from one to four to statements like 'During tests, I find myself thinking about the consequences of failing' (where one equals 'not typical of me' and four equals 'very typical of me').

The exam was held six weeks later. Half the students in each cohort were randomly allocated to either the control or expressive writing group. The analysis compared the relationship between test anxiety scores and exam performance between the two groups. For the control group, the students who reported the highest levels of test anxiety tended to perform worse on the final exam. However, in the writing group, no relationship was found between test anxiety scores and exam performance – students with high test anxiety performed just as well as those who experienced low test anxiety. A follow-up analysis showed that the high-test anxiety students in the writing group scored 6 per cent higher on their final exams than the high-test anxiety students in the control group, and that this could not be attributed to their being stronger students overall.

When interviewed about her research, Professor Beilock explained that the idea for the writing exercise came from the use of writing to treat mental disorders like anxiety and depression. Expressive writing about a traumatic or emotional experience over several weeks has been shown to decrease feelings of anxiety in people with mental disorders. Professor Beilock believes that the technique works by freeing cognitive resources from worry, leaving more 'brain power' for the exam (Figure 8.19). The next stage of the research will involve looking inside the anxious brain to see how it changes during stressful situations.

Sources: Adapted from Ramirez, G., & Beilock, S. L. (2011). Writing about testing worries boosts exam performance in the classroom. *Science*, *331*(6014), 211–213 and AP. (2011, January 14). Exam anxiety relief found – research. *The Australian*. https://www.news.com.au/breaking-news/exam-anxiety-relief-found--research/news-story/b23131a4bd1261b3bf9dc3547e6881e9

Figure 8.19 Students who spend 10 minutes before an exam writing about their exam worries can free up cognitive resources previously occupied by test worries and do their best work.

Questions

1. Explain how reproducibility applies in this series of studies.
2. Focus on the description of results for Studies 3 and 4 to answer the following questions.
    a. In terms of correlation, what kind of relationship is described between test anxiety scores and final exam performance for the control group and expressive writing groups in Studies 3 and 4?
    b. Explain how the difference in the relationship between the two groups supports the hypothesis that the intervention would be effective.
    c. Draw the relationship between anxiety scores and test scores for the two groups using a scatterplot.

3 Explain the reason for conducting Study 2 and why this was important.
4 Discuss the pros and cons of the laboratory-based approach used in Studies 1 and 2 and those of the real exam context used in Studies 3 and 4.
5 How could you use the findings of this study to benefit your assessment preparation?

KEY CONCEPTS 8.2

» Our location on the dual-continuum of mental wellbeing and mental disorder is influenced by the interaction between internal and external factors and fluctuates over time. We can move between states of languishing and flourishing, and between states of no mental ill-health, mental health problems and mental disorder.
» Internal factors are related to our personal characteristics and internal resources, both inherited and learned. External factors relate to the effects of our physical and social environments. Both internal and external factors can be either risk or protective factors.
» Stress is the physiological and psychological response that occurs when a person is confronted with a situation that is threatening or challenging and that they perceive as exceeding their ability to cope.
» Anxiety is an emotional state characterised by the anticipation of danger, as well as dread or uneasiness as a response to a potential or ambiguous threat.
» People can experience varying levels of stress and anxiety, ranging from mild to extreme. These responses are normal and can be adaptive when they occur in response to an appropriate threat.
» When anxiety is out of proportion to a situation, and is intrusive and uncontrollable, it becomes detrimental to wellbeing, and may result in an anxiety disorder such as a specific phobia.
» A specific phobia is an anxiety disorder characterised by a persistent, intense and irrational fear and avoidance of a particular object, activity or situation.

Concept questions 8.2

Remembering
1 List three internal and three external factors that can influence an individual's mental wellbeing. For each, indicate whether it is a risk factor or a protective factor. r

Understanding
2 Using an example, explain how internal and external factors can interact with each other to influence an individual's experience of mental wellbeing. c

Applying
3 Without reference to your textbook or notes, sketch the dual-continuum model of mental health. Then check your sketch against Figure 8.5 and correct it. Then, complete the following two activities. c
   a Annotate your drawing to indicate the location of three separate individuals within the model: one experiencing stress related to daily hassles, one experiencing anxiety about the outcome of an exam, and one with a diagnosis of specific phobia. Assume that all these individuals have access to good social support and coping strategies and report feeling positive emotions most of the time.
   b Now, indicate the location of the same individuals when their circumstances change so that they are alone in a new country with little access to social support and mental health services.

HOT Challenge
4 Construct a list of questions you could ask an individual in order to identify where they may be located on the dual-continuum model. c

8 Chapter summary

KEY CONCEPTS 8.1a

» Mental wellbeing is a multidimensional construct that includes emotional wellbeing, psychological wellbeing and social wellbeing. It can be measured using self-report scales such as the Mental Health Continuum Scale.

» Emotional wellbeing relates to the experience of positive emotions such as happiness, joy or love, and feeling generally satisfied with life. Psychological wellbeing relates to how a person thinks about (appraises) themselves and their goals, the way they respond to challenges (e.g. coping style) and the quality of their relationships. Social wellbeing relates to a person's sense of belonging to a community, and their sense of being able to contribute meaningfully to society.

» Levels of functioning describes the extent to which someone can operate effectively in the different domains of everyday life such as self-care, personal relationships and work.

» People who report a high degree of emotional, psychological and social wellbeing have a level of functioning described as flourishing. People who report low levels of wellbeing have a level of functioning described as languishing. People in between (most people) experience moderate mental wellbeing.

» Mental disorders are clinically defined conditions of mental ill-health that have a range of distinct presentations. This may include a combination of abnormal thoughts, perceptions, emotions, behaviour and relationships with others.

» Mental disorders affect a person's level of functioning, ranging from no dysfunction to extreme dysfunction.

» Mental health problems are relatively mild experiences of mental ill-health that do not meet the criteria for a mental disorder.

» Mental ill-health can be represented as a continuum from no mental ill-health, to mental health problem, to mental disorder.

» According to the dual-continuum model, complete mental health includes both a high state of mental wellbeing (flourishing) and the absence of mental ill-health. Other states of mental health are represented by the intersection between a person's position on the mental wellbeing continuum and their position on the mental ill-health continuum (Figure 8.5).

» Resilience is a person's ability to successfully adapt to stress and cope effectively with uncertainty. Resilient people possess and are able to use effective coping strategies and adaptive ways of thinking, and can maintain healthy personal relationships and social connectedness. Resilience is protective against developing mental health problems or a mental disorder.

KEY CONCEPTS 8.1b

» Social and emotional wellbeing (SEWB) is a holistic and culturally responsive framework of the factors that influence health outcomes for Aboriginal and Torres Strait Islander peoples and communities, including mental health. The framework is holistic and multidimensional because a person's SEWB is influenced by the seven domains of connection and their interactions with broader social, cultural, historical and political determinants.

KEY CONCEPTS 8.2

- Our location on the dual-continuum of mental wellbeing and mental disorder is influenced by the interaction between internal and external factors and fluctuates over time. We can move between states of languishing and flourishing, and between states of no mental ill-health, mental health problems and mental disorder.
- Internal factors are related to our personal characteristics and internal resources, both inherited and learned. External factors relate to the effects of our physical and social environments. Both internal and external factors can be either risk or protective factors.
- Stress is the physiological and psychological response that occurs when a person is confronted with a situation that is threatening or challenging and that they perceive as exceeding their ability to cope.
- Anxiety is an emotional state characterised by the anticipation of danger, as well as dread or uneasiness as a response to a potential or ambiguous threat.
- People can experience varying levels of stress and anxiety, ranging from mild to extreme. These responses are normal and can be adaptive when they occur in response to an appropriate threat.
- When anxiety is out of proportion to a situation, and is intrusive and uncontrollable, it becomes detrimental to wellbeing, and may result in an **a**nxiety disorder such as a specific phobia.
- A specific phobia is an anxiety disorder characterised by a persistent, intense and irrational fear and avoidance of a particular object, activity or situation.

8 End-of-chapter exam

Section A: Multiple-choice

1. The dual-continuum model describes mental wellbeing as having three core components. Which of the following is not one of these core components?
   A cognitive wellbeing
   B emotional wellbeing
   C psychological wellbeing
   D social wellbeing

2. Which of the following is not a characteristic of mental wellbeing?
   A coping with the normal stresses of life
   B having feelings of apprehension
   C realising own abilities
   D working productively and fruitfully

3. Which of the following is not a characteristic of a mentally healthy person?
   A self-confidence
   B optimism
   C lack of resilience
   D positive thinking

4. The ability for a person to successfully adapt to stress and cope effectively with uncertainty is known as:
   A mental health.
   B a mental disorder.
   C mental toughness.
   D resilience.

5. The physiological and psychological responses that a person experiences when confronted with a situation that is threatening or challenging are known as:
   A a phobia.
   B a mental disorder.
   C anxiety.
   D stress.

6. A persistent, irrational and intense fear of a specific object, activity or situation is known as:
   A a specific phobia.
   B a mental disorder.
   C anxiety.
   D stress.

7. Phobia can be distinguished from anxiety because only phobia:
   A involves distress.
   B triggers the flight-or-fright-or-freeze response.
   C involves an irrational fear.
   D is influenced by biological, psychological and social factors.

8. Which of the following may activate the flight-or-fright-or-freeze response?
   A stress
   B anxiety
   C phobia
   D all of the above

9. Biological factors that commonly enhance (or reduce) an individual's risk of, or vulnerability to, a mental health problem are known as:
   A social factors.
   B cognitive factors.
   C external factors.
   D genetic factors.

10. A disruption to how a person thinks, feels and behaves is known as:
    A a mental health problem.
    B stress.
    C a mental illness.
    D a mental disorder.

11. Which of the following identifies an external factor that may influence an individual's mental wellbeing?
    A genetic predisposition to anxiety
    B family relationships
    C emotional state
    D low self-esteem

12 A fundamental difference between mental health and a mental disorder (illness) is that everyone has some level of _____ all of the time, whereas it is possible to be without a _____.
   A mental disorder; mental health
   B mental health; mental disorder
   C physical illness; mental disorder
   D mental disorder; physical illness
13 Which of the following highlights one difference between anxiety and phobia?
   A Anxiety activates the parasympathetic nervous system, whereas phobia activates the sympathetic nervous system.
   B Anxiety activates the sympathetic nervous system, whereas phobia activates the parasympathetic nervous system.
   C Anxiety is maladaptive, whereas phobia can be adaptive.
   D Anxiety can be adaptive, whereas phobia is maladaptive.
14 Yarning, ceremony, traditional fire management of land, art, dance, song and storytelling are all examples of Aboriginal and Torres Strait Islander practices that fall within which domain of the SEWB model?
   A community
   B culture
   C Country
   D family and kinship
15 Which of the seven domains of connection in the Aboriginal and Torres Strait Islander model of social and emotional wellbeing is most clearly related to Western concepts of mental wellbeing?
   A mind and emotions
   B body and behaviours
   C spirituality and ancestors
   D Country
16 The current health and wellbeing challenges faced by many Aboriginal and Torres Strait Islander children, families and communities result from the ongoing impacts of which of the following?
   A government policies
   B racist attitudes
   C discrimination
   D all of the above
17 Which of the following identifies internal factors that could influence an individual's wellbeing?
   A immune system functioning and our emotions
   B income and a lack of emotional resilience
   C a physiological response to stress and stigma
   D substance use and friendship groups
18 Which of the following accurately describes the impact of external or internal factors on mental wellbeing?
   A Biological factors are likely to have a greater impact than social factors.
   B External factors are likely to have a bigger impact than internal factors.
   C External factors will not have an impact on mental wellbeing.
   D The impact on wellbeing depends on the number and the nature of internal and external factors.
19 An individual's interpretation of the object, situation or event that determines the extent of stress experienced is known as:
   A an objective experience.
   B a subjective experience.
   C an emotional experience.
   D a coping strategy.
20 Which level of anxiety is characterised by normal feelings of short-term nervousness, agitation and difficulties in concentrating on a task?
   A low anxiety
   B mild anxiety
   C moderate anxiety
   D severe anxiety

Section B: Short-answer

1 Distinguish between mental wellbeing and mental disorder. [2 marks]
2 Identify and describe each of the three levels of functioning related to people's self-reported mental wellbeing [3 marks]
3 Explain why the Social and Emotional Wellbeing (SEWB) framework is considered a holistic approach to wellbeing. [3 marks]

Application of a biopsychosocial approach to explain specific phobia

9

Key knowledge

» the relative influences of factors that contribute to the development of specific phobia, with reference to gamma-amino butyric acid (GABA) dysfunction and long-term potentiation (biological); behavioural models involving precipitation by classical conditioning and perpetuation by operant conditioning, and cognitive biases including memory bias and catastrophic thinking (psychological); and specific environmental triggers and stigma around seeking treatment (social)

» evidence-based interventions and their use for specific phobia, with reference to the use of short-acting anti-anxiety benzodiazepine agents (GABA agonists) in the management of phobic anxiety and breathing retraining (biological); the use of cognitive behavioural therapy (CBT) and systematic desensitisation as psychotherapeutic treatments of phobia (psychological); and psychoeducation for families/supporters with reference to challenging unrealistic or anxious thoughts and not encouraging avoidance behaviours (social)

Key science skills

Analyse, evaluate and communicate scientific ideas

» discuss relevant psychological information, ideas, concepts, theories and models and the connections between them

» critically evaluate and interpret a range of scientific and media texts (including journal articles, mass media communications and opinions, policy documents and reports in the public domain), processes, claims and conclusions related to psychology by considering the quality of available evidence

» analyse and evaluate psychological issues using relevant ethical concepts and guidelines, including the influence of social, economic, legal and political factors relevant to the selected issue

Source: VCAA VCE Psychology Study Design (2023–2027) + page 40, 13

9 Application of a biopsychosocial approach to explain specific phobia

Do you have a fear of heights? The dark? Clowns? Spiders? If you do, then you have a specific phobia. A specific phobia is an anxiety disorder in which a person experiences intense fear or anxiety towards a particular object or situation.

Adobe Stock/Pixel-Shot

p. 363

9.1 Development of specific phobia

You weren't born with a fear of clowns or spiders. You developed it over your lifetime from a combination of biological, psychological and social factors (biopsychosocial model). Reduced GABA (a neurotransmitter) means that a person cannot regulate their flight-or-fight response, which increases anxiety (biological). Classical conditioning can contribute to associating objects with an unpleasant situation (psychological) and your mum or dad might have been scared of heights so they modelled this fear to you all your life (social).

Adobe Stock/ Valerii Honcharuk

9.2 Evidence-based interventions

p. 373

The good news is that specific phobias can be treated. There are a number of evidence-based treatments or interventions that can help reduce your phobia. This includes medications and anxiety-management treatments such as relaxation techniques (biological interventions), cognitive behavioural therapy and systematic desensitisation (psychological interventions) and psychoeducation (social intervention).

Adobe Stock/Monkey Business

Getting your anxiety and phobias under control are all part of maintaining your mental wellbeing. Even in a busy year like Year 12 you need to take the time to put your health and wellbeing first. Try some mindfulness meditation – it might help.

Flashcards
Chapter 9 flashcards

Slideshow
Chapter 9 slideshow

Test
Chapter 9 pre-test

Assessment
- Pre-test
- End-of-chapter exam

Revision
- Chapter map
- Key term flashcards
- Key concept summary
- Slideshow

Investigation
- Investigation: Investigation of a specific phobia
- Logbook template: Case study

Worksheet
- Animal phobias

To access these resources, visit
cengage.com.au/nelsonmindtap

Nelson MindTap

Know your key terms

Avoidance behaviour
Behavioural models
Benzodiazepine
Beta blocker
Biopsychosocial model
Breathing retraining
Catastrophic thinking
Cognitive behavioural therapy (CBT)
Cognitive bias
Cognitive model
Evidence-based interventions
Fear hierarchy
Maladaptive behaviour
Memory bias
Modelling
Psychoeducation
Psychotherapy
Reciprocal inhibition
Stigma
Systematic desensitisation

Could there be a stranger phobia than one who fears tiny holes? The condition is known as trypophobia – a person who fears irregular patterns or clusters of small holes (Figure 9.1). While the phobia is not listed in scientific literature (yet), little holes, however, are a deep-rooted fear Kendall Jenner suffers from. In particular, she fears pancakes. (How can she tell that the pancake is ready to be flipped over in the fry-pan if she can't look at the bubbling holes on top of the mixture?) 'Anyone who knows me knows that I have really bad trypophobia', the then 20-year-old posted on her app. 'Trypophobics are afraid of tiny little holes that are in weird patterns. Things that could set me off are pancakes, honeycomb or lotus heads (the worst!). It sounds ridiculous but so many people actually have it! I can't even look at little holes – it gives me the worst anxiety. Who knows what's in there???'

Research found that the symptoms of trypophobia were long term and persistent and it was common for sufferers to also be diagnosed with other mental wellbeing issues such as depression and generalised anxiety.

Jenner is not the only celebrity who has a strange phobia. She's beaten by Orlando Bloom who is afraid of little pigs (zoophobia), Cameron Diaz cannot stand doorknobs (ostiumtractophobia), Oprah Winfrey isn't at all a fan of chewing gum (chiclephobia), Nicole Kidman fears butterflies (lepidopterophobia) and perhaps the strangest of all comes from film director Woody Allen, who reportedly has a fear of peanut butter sticking to the roof of his mouth (arachibutyrophobia).

Adapted from: https://graziamagazine.com/articles/kendall-jenner-phobia/

Figure 9.1 Trypophobia – the fear of repetitive patterns of closely packed holes or protrusions, such as with a pancake or in honeycomb – is not recognised as an official phobia, but it has been widely discussed in social media.

Reading stories of other people's fears of everyday items or situations highlights how different, unique and individual we all are. One person's love can be another person's fear.

Who doesn't love the sight of a colourful butterfly? … Nicole Kidman for one!

The previous chapter investigated how anxiety is our body's natural and normal response to stress. It is commonly experienced in situations where we are presented with a threat such as completing an exam, going to a job interview or giving a speech. It is understandable that in these situations we feel fearful. However, if the fear of a specific object, activity or situation is intense, persistent and irrational it is considered a phobia.

In this chapter, we will explore how a phobia may develop and the treatment options available to those who suffer from a specific phobia.

Weblink
An introduction to phobias

9.1 Development of specific phobia

Phobic disorders are a perfect example of how the interaction of a range of factors can result in the development of an anxiety disorder. The **biopsychosocial model** proposes that health and illness outcomes are determined by the interaction and contribution of biological, psychological and social factors. Biological factors such as the autonomic nervous system's (ANS) response to a perceived threat, the impact of neurotransmitters and a person's physical health state contribute to the development of anxiety disorders such as a phobia. In addition, genetic factors, or inherited biological factors, predispose some individuals to develop and sustain an anxiety disorder (Figure 9.2). Medications targeted to increase the release of specific inhibitory neurotransmitters may help a person regulate their ANS activity and manage their disorder more effectively.

The biopsychosocial approach suggests that psychological factors (for example, **memory bias**) also contribute to anxiety disorders such as phobias. Management involves therapy that focuses on identifying the cause of the conflict and learning to think in a more rational way. This approach also considers the role that learning plays in anxiety disorders. It suggests that if we learn a faulty association between two or more stimuli, or we learn a phobic reaction by observing the reaction of others to specific stimuli, we may develop an anxiety disorder. The biopsychosocial model focuses on the nature of communication between parents or caregivers and children, which, in turn, is dependent on the societal norms that are learned and how they are learned. Social factors – such as the culture a person is raised in, the social experiences they have and the style of parenting they experience – are also considered to contribute to a learned response that results in an incorrect appraisal of a situation as threatening. Figure 9.3 lists the top 10 phobias of Australians.

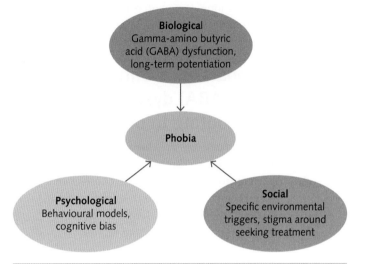

Figure 9.2 Phobias are the result of the interaction between biological, psychological and social factors.

The biopsychosocial model has enabled more personalised treatment plans; however, it does have limitations. These include that not all disorders have equally relevant biological, psychological and social aspects, the model is vague regarding the boundaries of each category and making a diagnosis and treatment plan becomes very difficult when considering the complexity of so many factors.

THE NATION'S TOP TEN PHOBIAS
1 Arachnophobia – spiders
2 Social phobia – social or public situations
3 Aerophobia – flying
4 Agoraphobia – open or public spaces
5 Claustrophobia – enclosed spaces
6 Emetophobia – vomiting
7 Acrophobia (vertigo) – heights
8 Cancerphobia – developing cancer
9 Brontophobia – thunderstorms
10 Necrophobia – death (your own and others)

Figure 9.3 The nation's top 10 phobias. Most are specific phobias, with the exception of 2 and 4, which are complex phobias. Complex phobias are beyond the scope of VCE Psychology.

Biological factors contributing to phobia

A range of biological factors contribute to the development of specific phobia. These include the impact of heredity, neurotransmitters, automatic physiological responses and the biological basis of learning. Emerging evidence about the role of the neurotransmitter gamma-amino butyric acid (GABA) and the process of long-term potentiation (LTP) has contributed greatly to our understanding and management of phobic anxiety.

Gamma-amino butyric acid (GABA) dysfunction (contributing to phobia)

Neurotransmitters are chemicals released from the axon terminal buttons of a presynaptic neuron into the synaptic cleft. These chemicals will either excite or inhibit the postsynaptic neuron. They act as messengers that carry signals and information about stimuli (originating in the external world or in internal organs) from the presynaptic neuron to the postsynaptic neuron and impact on the activity level of the receiving neuron. There are several neurotransmitters known to be influential in anxiety disorders, including gamma-amino butyric acid (GABA), noradrenaline (also called norepinephrine) and serotonin. GABA is one of the neurotransmitters most strongly implicated.

As we learned in Chapter 2, when released, some neurotransmitters (such as adrenaline) have an excitatory effect, preparing the body to flee fight, or evade (the flight-or-fight-or-freeze response); other neurotransmitters (such as GABA) have an inhibiting effect, and therefore calm or slow neural transmission by making receiving neurons less likely to fire.

Studies have shown that when GABA activates its receptors, the cells that have GABA receptors are inhibited, and the system becomes calm or slows down to counteract the excitability of the neurons (through the glutamate or noradrenergic system) (Figure 9.4). This means that the specific features of the body's response to stress such as the increased heart rate, respiration or blood pressure, are reduced. When there is reduced GABA in the brain, circuits that regulate emotional responses to threatening stimuli become dysfunctional. Therefore, an individual with low levels of GABA may not be able to regulate their flight-or-fight-or-freeze response and therefore may be more vulnerable to anxiety and have a greater chance of developing a specific phobia. A deficiency in GABA may also lead to insomnia, depression, mood disorders, excessive stress and hypertension (Nuss, 2015).

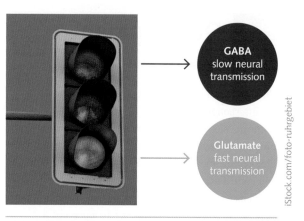

Figure 9.4 GABA is an inhibitory neurotransmitter. It slows neural transmission by making receiving neurons less likely to fire. Glutamate is an excitatory neurotransmitter that makes receiving neurons more likely to fire.

Role of long-term potentiation (LTP) in contributing to phobia

Fear plays a vital role in our survival; however, we are not born with a fear of spiders and snakes – they are learned responses. When we associate fear with a specific stimulus, such as a spider, a new memory circuit with connections within the amygdala (the brain area responsible for adding the emotional content to a memory) is thought to be made.

As you learned in Chapters 2 and 5, long-term potentiation (LTP) is the neural mechanism that allows us to learn associations between stimuli. LTP is a form of synaptic plasticity that results in a long-lasting strengthening of neural connections at the synapse as a result of repeated stimulations from a presynaptic to postsynaptic neuron. This has been demonstrated to occur in cells of the hippocampus, producing the neural changes that underlie the formation of memory. Because LTP strengthens the connections between neurons, it enables them to transmit information more efficiently and more quickly next time a similar experience is encountered.

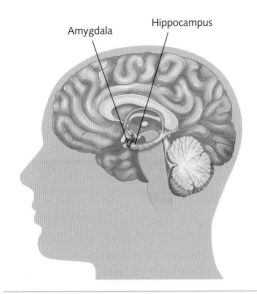

Figure 9.5 The amygdala is an almond-shaped brain structure located deep within each temporal lobe and is responsible for the physiological experience of emotions, especially the emotional response of fear.

The hippocampus is a seahorse-shaped structure located deep within each temporal lobe that is crucial for consolidating declarative (explicit) memories, especially episodic memories.

As outlined in Chapters 4 and 5, LTP in the hippocampus underlies normal fear learning through binding neural information from the amygdala related to fear responses with information from the neocortex related to the perception and comprehension of experiences (Figure 9.5). For example, the first time a young child sees a spider they may not be afraid, just curious. However, if on another occasion the neurons that fire in response to the child seeing a spider do so at the same time as the neurons in the amygdala that cause the flight-or-fight-or-freeze response, then the association between these two groups of neurons will be strengthened through their mutual connections with the hippocampus. Now, each time the child encounters a spider, the connections in this memory circuit will be strengthened, causing physiological symptoms such as increased heart rate, blood pressure and sweating.

As discussed in Chapter 4, fear responses can be learned through direct experience, through observational learning of someone else's fear response, or even through hearing a very scary story about spiders (Figure 9.6). This kind of fear learning through LTP is normally an adaptive process, allowing us to learn protective fear responses. It seems that the extreme and maladaptive fear responses that occur in specific phobia may develop due to an extremely frightening early experience with the feared stimulus, and/or because the person has a predisposition to develop a phobia due to the dysregulation of GABA.

Figure 9.6 A fear of spiders is a learned response. Exposure to the fear stimulus will strengthen the memory circuit (LTP) and, via communication with the amygdala, activate multiple brain regions that then produce a variety of symptoms (increased blood pressure, heart rate and sweating).

KEY CONCEPTS 9.1a

- A specific phobia is an anxiety disorder in which a person experiences intense fear or anxiety towards a particular object or situation.
- Phobias are the result of the interaction between biological, psychological and social factors.
- Reduced GABA is related to dysfunction in the brain circuits that regulate emotional responses to threatening stimuli.
- An individual with low levels of GABA may not be able to regulate their flight-or-fight-or-freeze response and therefore may be more vulnerable to anxiety and have a greater chance of developing a specific phobia.
- Long-term potentiation (LTP) contributes to the development of specific phobia by strengthening connections between neurons in the amygdala that fire during the flight-or-fight-or-freeze response and neurons involved in perceiving the feared stimulus, so that the connections become consolidated in the hippocampus.

> ## Concept questions 9.1a
>
> ### Remembering
> 1. Without reference to your textbook or notes, sketch a diagram of the biopsychosocial model. **d**
> 2. Identify two factors that may influence the development of a specific phobia for each category of the biopsychosocial model. **r**
>
> ### Understanding
> 3. Define 'specific phobia'. **r**
> 4. Compare the influence on neural transmission of two different types of neurotransmitters. **e**
>
> ### Applying
> 5. Construct a flowchart explaining how a dysfunction in natural GABA production may influence the development of a specific phobia. **d**
> 6. Using an example not presented in the text, explain the role of long-term potentiation in the development of a specific phobia. **e**
>
> ### HOT Challenge
> 7. Research a case study of an individual who suffers from a specific phobia. Classify the factors that have influenced the specific phobia for each category of the biopsychosocial model. **c**

Psychological factors contributing to phobia

In this section we explore the psychological factors that influence the development and maintenance of specific phobia. Behavioural approaches focus on the roles of classical and operant conditioning in precipitating (starting) and perpetuating (maintaining) specific phobias. Cognitive approaches focus on the ways in which our attention, memory and thought processes are involved. We explore each in turn.

Behavioural models

Behavioural models propose that phobic anxiety could be the result of learning. Just as we have learned to associate feelings of joy when we hear a song that has been associated with happy memories, it is possible to explain the precipitation of a phobia as a learned association between two stimuli (classical conditioning) and its perpetuation due to rewards and punishment (operant conditioning).

Classical conditioning and the precipitation of a specific phobia

As we noted in Chapter 4, classical conditioning is a form of learning. It occurs when a previously neutral stimulus becomes associated with a reflex response, so that it becomes a conditioned stimulus that causes a conditioned reflex response. For example, walking past a playground may elicit a fear response if the individual was once attacked by a dog at a playground. Because phobias are learned fears, the consistent pairing of a neutral stimulus with an unpleasant stimulus can cause phobic reactions.

The role of classical conditioning in the precipitation of a specific phobia can also be seen in the case of Little Albert (see weblink). Little Albert's phobia of white fluffy objects was presumably the result of repeatedly pairing the white rat (neutral stimulus) with an unpleasant unconditioned stimulus (the loud noise). Once learned, Little Albert's fear was then generalised to other similar stimuli such as a rabbit, a seal-skin coat and cotton balls (Figure 9.7).

To explain how classical conditioning contributes to the precipitation of a phobia, let's look at an everyday example of a young girl,

Figure 9.7 Little Albert acquired a specific phobia of rats through the process of classical conditioning. By avoiding the feared stimulus of the rat his level of fear is reduced (negative reinforcement). However, this also perpetuates his phobia through the process of operant conditioning, generalising his fear to other stimuli such as rabbits.

Weblink
The Little Albert experiment

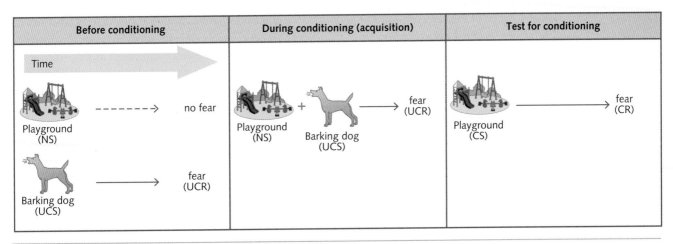

Figure 9.8 The behavioural model proposes that phobic anxiety is precipitated (starts) by classical conditioning.

Jazmin, becoming phobic about dogs (Figure 9.8). When she was growing up, Jazmin's neighbour's large dog chased her through a public playground a number of times, barking loudly and jumping at her each time. At the time, Jazmin experienced extreme fear including a rapid pulse, sweating palms and an increase in breathing. Ten years later, whenever Jazmin is approached by a dog, she experiences the same reactions (sweating palms and increases in heart rate and breathing). Jazmin will now avoid areas where dogs may be present and in doing so, she feels more comfortable and relaxed.

In the example above, the development of Jazmin's phobia can be explained through the process of classical conditioning. The dog was an initially neutral stimulus that became a conditioned stimulus through its association with the unconditioned stimuli of loud barking and being chased. This terrifying early experience has left Jazmin with a persistent phobia of all dogs, so that even 10 years later the mere thought of a dog, an image, or being in the presence of a dog is sufficient to cause a conditioned extreme fear response, including powerful physiological symptoms of a racing heart and sweating. Linking this to the previous section, the associations learned in classical conditioning are consolidated through the process of LTP.

Operant conditioning and the perpetuation of a specific phobia

In Chapter 4, we saw that operant conditioning is a learning process in which the likelihood of a behaviour being repeated is determined by the consequence of that behaviour. The perpetuation (or continuation) of the learned fear response can be explained through the process of operant conditioning. If we take the example of Jazmin from the previous section, it was noted that Jazmin now avoids visiting playgrounds because of their association with being menaced by a dog. This avoidance behaviour is a voluntary behaviour that is maintained through providing Jazmin with relief from the fear response. In the language of operant conditioning, Jazmin's avoidance of playgrounds is negatively reinforced by the consequence of removing fear.

If we apply the three-phase model of operant conditioning we can say that the playground is an antecedent stimulus that has become associated with the fear of being chased by a dog. The playground triggers Jazmin's avoidance behaviours, which are reinforced through reduction of fear. Figure 9.9 maps the components of this example to the three-phase model.

Operant conditioning therefore provides an explanation of the crippling avoidance behaviours that can occur in specific phobia.

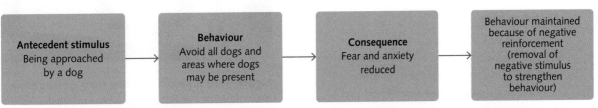

Figure 9.9 The behavioural model proposes that phobic anxiety is perpetuated by operant conditioning.

Avoidance is one of the key symptoms of phobia and it impacts people's lives by restricting their activities and enjoyment of experiences. The avoidance behaviours are maintained by the reinforcing consequence of the reduction of fear.

Cognitive model

The **cognitive model** of psychology is an approach based on understanding how people's thought patterns, memories and beliefs affect their emotions, attitudes and behaviours. This model examines the influence that inaccurate mental processes have in the development and maintenance of phobias.

Our cognitions (thoughts) are the psychological result of perception, learning and reasoning. We attempt to correctly perceive objects, situations and events so that our understanding of a particular object, situation or event is helpful in our everyday functioning. However, we may sometimes pair faulty reasoning and rationale with fearful stimuli from the environment; hence, a new, faulty cognition is formed. This is known as a cognitive bias.

A **cognitive bias** is an automatic tendency or preference for processing or interpreting information in a particular way, producing systematic errors in thinking when making judgements or decisions. Cognitive biases may sometimes lead to an inaccurate judgement or illogical interpretation of a situation. There are many cognitive biases that are associated with the development and maintenance of specific phobias. These include memory bias and catastrophic thinking.

For example, if you faint when you are having a blood sample taken, you may store this experience wrongly in your memory as, 'Oh, no! Needles are dangerous and drawing blood is frightening – I must keep away from needles and blood'. This can develop into a phobia where you experience a panic attack at the sight of blood or a needle, or simply at the thought of a blood sample being drawn from you. Part of the treatment of the phobic illness therefore revolves around 'correcting' these cognitive biases.

Memory bias

A memory bias is a tendency to remember information of one kind at the expense of another kind; including the bias towards remembering negative and threat-related experiences that is associated with specific phobias. This type of error in thinking may either enhance or impair the recall of memory, or it may alter the content of what we report remembering. Research suggests that memory biases for threatening information may contribute to the development of a phobia because they cause a person to recall negative information more readily than positive information about a specific object, situation or event they have experienced in the past (LeMoult & Joormann, 2012). Memory bias can also alter recalled memories so that they are different from what actually happened. For example, a person who has developed a phobia of horses because of one negative experience with an aggressive horse will tend to recall this experience of the horse being overly aggressive more readily than all other positive experiences with horses.

Catastrophic thinking

Catastrophic thinking occurs when an individual repeatedly overestimates the potential dangers and assumes the worst of an object or event. The catastrophic thinker predicts an outcome of a future event that others would consider unrealistic and irrational. When this occurs, the person will typically experience heightened levels of distress and anxiety, and underestimate their ability to cope with the situation. For example, a person with a phobia of public speaking (glossophobia) may think that every time they speak publicly they are going to be embarrassed or rejected (Figure 9.10). The physical symptoms they experience with catastrophic thinking, such as increased heart rate and sweating, act to reinforce their irrational thoughts. An individual's catastrophic thoughts may even extend to their family members, such as not wanting a sibling to learn to drive because they will definitely be involved in a serious car accident.

Figure 9.10 A person suffering from glossophobia (fear of public speaking) may have irrational catastrophic thoughts that the audience will dislike them so much that they will walk out mid-speech.

Social factors contributing to phobia

Suffering a traumatic experience in response to a specific stimulus is a common contributing factor in the development of a specific phobia. This specific event in the environment has triggered a fearful response, such as a dog attack, being bitten by a spider or nearly drowning. We classify these specific environmental triggers as social factors. Social factors are the aspects of our interactions with other people, groups, society and culture that influence how we think and behave.

A significant reason for the maintenance (perpetuation) of anxiety disorders in society is the social factor of stigma. **Stigma** is a negative social attitude about a characteristic of a person or social group that implies some form of deficiency, often leading to unfair discrimination against or exclusion of the person or social group. With the lack of knowledge and understanding surrounding mental illness in general by a large proportion of society, stigma associated with having a mental illness is one of the biggest barriers that prevents people from getting treatment.

Specific environmental triggers

Several factors in the environment can predispose an individual to the development of specific phobias. Traumatic events – for example, being attacked by a dog or being trapped in a closet – can result in the development of a phobic reaction. A phobic reaction can also result from having unexpected anxiety attacks in situations that are perceived to be threatening, but in reality present no threat.

Learning occurs primarily through interactions with others, especially through observing the behaviours of others and the consequences that follow, including active cognitive processes that determine whether the behaviour will be imitated. Consider the example of a child who pats a neighbour's dog that appears docile, but that dog unexpectedly attacks (Figure 9.11). The trauma of this unexpected incident may become paired with the fear response and the child may develop an irrational belief that all dogs are dangerous and unworthy of affection.

The social and cognitive processes of learning by observing another person's behaviour and the consequences of the behaviour are known as **modelling**. Models are people in our lives who we admire and whose behaviour we consciously or unconsciously tend to replicate. For example, if we observe a parent or other model responding to a stimulus with fear, then we may also learn to fear this stimulus.

Learning by modelling can maintain an association between an object and the emotions that a person experiences. This may occur by observing a reaction in other people (for example, when children start crying when they see another child cry after receiving an injection) or by being taught or warned about particular objects (for example, being warned about the serious consequences of speeding by way of television advertisements) (Tyng et al., 2017). Negative events and information that we are exposed to via the media could result in learning that these things are potentially threatening (even though we haven't experienced them directly) and this could lead to the development of a phobia.

Figure 9.11 Being frightened by a dog as a child may lead to a phobia of all dogs later in life.

Stigma around seeking treatment contributing to phobia

Individuals who experience mental illness such as a specific phobia are often faced with stigma that results from a lack of understanding about their illness. Because phobias involve irrational and unrealistic emotions and behaviours, it is often difficult for other people to empathise with the sufferer.

9.1.2 THE BIOPSYCHOSOCIAL FRAMEWORK: APPLICATION TO UNDERSTANDING SPECIFIC PHOBIAS

9.1.3 DEVELOPMENT OF SPECIFIC PHOBIA

Weblink
Reducing stigma

9.1.4 BEHAVIOURAL FACTORS INFLUENCING THE DEVELOPMENT OF SPECIFIC PHOBIAS

People who do not have a phobia often apply a negative stereotype to the phobic person. As a result, a stigma about the illness develops and the sufferer may even encounter discrimination (Figure 9.12).

Stigma can lead to feelings of embarrassment, shame, hopelessness and distress that cause sufferers to hide the symptoms of their illness. As a result, many people with a mental illness fail to seek readily available support from friends, family and professional services for fear that they will be viewed in a negative way.

Some of the harmful effects of stigma associated with anxiety disorders can include:
» reluctance to seek help or treatment
» a lack of understanding by family, friends, co-workers or others you know
» fewer opportunities for work, school or social activities or trouble finding housing
» bullying, physical violence or harassment
» the belief that you'll never be able to succeed at certain challenges or that you can't improve your situation (Mayo Clinic, 2014).

Figure 9.12 A lack of understanding about mental illnesses can lead to stigma, which can compound symptoms for sufferers.

ACTIVITY 9.1 BIOLOGICAL, PSYCHOLOGICAL AND SOCIAL CONTRIBUTING FACTORS TO SPECIFIC PHOBIA

Try it

1. For each of the statements regarding specific phobia below, tick one of the boxes to illustrate the origin of the statement according to the biopsychosocial framework.

	Biological	Psychological	Social
'Even thinking about the object of my fear makes me feel anxious'			
'Phobias are fuelled by avoidance behaviour'			
'Lower levels of GABA increase a person's chance of developing a phobia'			
'Phobias develop when a person creates an association between a stimulus and fear'			
'Phobic responses occur in the presence of the phobic stimulus'			
'I try to manage my phobia because I don't want to be seen as different'			
'My friends and family say I'm exaggerating, but the bird nearly killed me!'			

Apply it

Jamie has a fear of dogs. Jamie walks to and from school each day, careful to ensure that he walks down streets where there are no dogs.

2. With reference to the scenario above, explain the influence of operant conditioning in shaping Jamie's behaviour.

Exam ready

Dr Ben works as a research psychologist. He conducted a study to test the effectiveness of a new medication, Melazepam, as a treatment method for the specific phobia of birds. Twenty otherwise healthy adults with a phobia of birds were selected for the study.

Half of the participants received 25 mg of Melazepam twice daily for four weeks; the other half took a placebo treatment twice daily for four weeks.

3 When constructing a hypothesis, Dr Ben would need to know that, respectively, participants in the experimental group and control group were exposed to taking:

A the placebo treatment; Melazepam.

B Melazepam; the placebo treatment.

C the placebo treatment twice daily for four weeks; 25 mg of Melazepam twice daily for four weeks.

D 25 mg of Melazepam twice daily for four weeks; the placebo treatment twice daily for four weeks.

ANALYSING RESEARCH 9.1

Explainer: why are we afraid of spiders?

I have personal interest in arachnophobia – the fear of spiders – because I am a spider expert, but also because my daughter has it. She is not alone. According to the American Psychiatric Association, phobias affect more than one in ten people in the US, and of those individuals, up to 40% of phobias are related to bugs (including spiders), mice, snakes and bats.

There are clearly a lot of arachnophobes. But do they know why they fear spiders? Can they do something to control those fears?

Once bitten, twice shy?

Psychologists believe that one reason why people fear spiders is because of some direct experience with the arachnids that instilled that fear in them. This is known as the 'conditioning' view of arachnophobia.

In 1991, Graham Davey at City University London ran a study to understand more about this view. He interviewed 118 undergraduate students about their fear of spiders. About 75% of the people sampled were either mildly or severely afraid of spiders. Of those, most were female. (This gender bias in arachnophobia has been supported in subsequent research.)

There was also an effect from family. Those people fearful of spiders reported having a family member with similar fears, but the study was unable to separate genetic factors from environmental ones. What is surprising is that Davey found that arachnophobia wasn't the result of specific 'spider trauma', which means there was no support for the conditioning view.

So what makes spiders so terrifying? Surely it must be the threat of being bitten? Davey looked at that issue too. It turns out that it is not so much a fear of being bitten, but rather the seemingly erratic movements of spiders, and their 'legginess'. Davey said: 'Animal fears may represent a functionally distinct set of adaptive responses which have been selected for during the evolutionary history of the human species.'

Genes or environment?

In 2003, John Hettema at the Virginia Institute for Psychiatric and Behavioural Genetics and his colleagues conducted twin studies to tease apart genetic factors.

Identical twins have identical DNA but tend to live in different environments in adult life, which allows researchers to find out how genes affect behaviour. Hettema recorded the responses of twins to 'fear-relevant' images (spiders, snakes) compared to 'fear-irrelevant' images (circles, triangles). Statistical analysis of the results revealed that genetic influences were 'substantial', which means that arachnophobia is inheritable.

You need not necessarily experience spiders to be fearful of them.

Scare tactics

So, to my dissatisfaction, arachnophobia is here to stay. But there may be a simple technique to reduce the fear these bugs cause. In 2013, Paul Siegel at the State University of New York and his colleague published a study that helped volunteers lessen their arachnophobia.

They first split the volunteers into phobic and non-phobic groups, based on simple spider-fear tests. After a week of doing these tests, both the groups were then exposed to images of flowers or spiders, but the exposure was for a very short time. The idea was that people can't recognise the images consciously, but it has an effect on their subconscious. When the spider-fear tests were carried out on both these groups again, those who feared spiders had become less afraid.

While other general conclusions are hard to draw from the literature on arachnophobia, arachnologists like me should rejoice at the results of Hettema's study. If nothing else, at least sharing images of spiders may help reduce arachnophobia.

Source: Chris Buddle, Assoc. Professor, McGill University, 'Explainer: why are we afraid of spiders?' May 8, 2014. Originally published on The Conversation. CC BY-NC-SA (https://creativecommons.org/licenses/by-nd/4.0/)

Questions

1. Construct a hypothesis for the 2013 arachnophobia study done by Paul Siegel.
2. Identify the independent variable(s) and dependent variable(s) for this study.
3. State the results of the experiment.
4. Identify and explain two ethical concepts the researcher would need to follow in this study.

KEY CONCEPTS 9.1b

» Behavioural models propose that phobic anxiety could be the result of learning. It is possible to explain the precipitation (acquisition) of a phobia as a learnt association between two stimuli (classical conditioning) and its perpetuation (maintenance) due to negative reinforcement of avoidance behaviours (operant conditioning).

» The cognitive model claims we may sometimes couple faulty reasoning and rationale with fearful stimuli from the environment and hence a new cognition is formed into a phobia.

» Specific environmental triggers such as a traumatic event can predispose an individual to the development of a specific phobia.

» According to social learning theory and parental modelling, children learn from their parents and often mirror what their parents do. Children learn how to respond to stress and difficulties by observing how their parents respond to these.

» Stigma is a negative social attitude about a characteristic of a person or social group that implies some form of deficiency, often leading to unfair discrimination against or exclusion of the person or social group.

Concept questions 9.1b

Remembering

1. Define 'classical conditioning'. **r**
2. What do the behavioural models propose in regard to the precipitation and perpetuation of a specific phobia? **r**

Understanding

3. Outline the process of how a specific phobia may be perpetuated by operant conditioning. **e**
4. Explain how memory bias may influence the development of an anxiety disorder. **e**

Applying

5. Using an example, explain how catastrophic thinking may result in the perpetuation of a specific phobia. **e**
6. Identify **three** harmful effects of stigma associated with anxiety disorders. **r**

HOT Challenge

7. By referring to Figure 9.8 and the associated text, construct a similar set of diagrams that demonstrates how a child may develop a phobia of cats. **d**

9.2 Evidence-based interventions

Phobias can have a significant impact on sufferers' lives, from limiting their daily activities to triggering severe anxiety and depression. Phobias can be treated with a variety of different interventions. **Evidence-based interventions** are psychological treatments whose effectiveness has been supported by the integration of clinical research findings and clinical expertise. These treatments try to maximise the benefit and minimise the risk of harm to the patient. Empirical methods have been applied to gather the evidence – the treatments are based on clinical research that demonstrates their positive outcomes. Intervention refers to all direct services given to the patient by healthcare professionals. This may include assessment, diagnosis, prevention, treatment, **psychotherapy** and consultation (Figure 9.13).

Figure 9.13 Psychotherapists help clients to understand their maladaptive behaviours and learn new ways to cope with their disorder.

Every individual brings with them a unique set of experiences, personality attributes and genetic make-up. It is this individual set of characteristics that makes the experience, and treatment, of mental illness different for every sufferer. Just as it is important to consider how the biological, psychological and social factors combine and interact to influence a person's physical and mental health (biopsychosocial model), it is also important to consider each of these three domains when devising a treatment plan for sufferers of mental illnesses such as a specific phobia.

The management and treatment of specific phobias focuses on the phobic person learning to manage their anxiety as well as learning how to change unrealistic thoughts and behaviours associated with the phobic stimulus.

Biological interventions

Biological interventions used to manage the body's response associated with a specific phobia include medications such as the use of short-acting anti-anxiety benzodiazepine agents and relaxation techniques such as **breathing retraining**.

Short-acting anti-anxiety benzodiazepine agents (GABA agonists): managing anxiety

Medications can relieve symptoms of anxiety; however, their use is not a long-term solution because they do not treat the underlying cause of the anxiety disorder. Anti-anxiety medications work by slowing down the central nervous system. These medications are known as **beta blockers** and work by blocking the stimulating effects of adrenaline on the body, such as an increased breathing rate, elevated blood pressure and a pounding heart, that are caused by anxiety.

Other medications used in the treatment of phobic anxiety are closely related to the role of the neurotransmitter gamma-amino butyric acid (GABA) and its effects on the body. GABA functions as an inhibitory neurotransmitter – meaning that it blocks nerve impulses, calming nervous activity.

Biological treatments often target the enhancement of GABA neurotransmission, in order to inhibit the hyper and overexcited bodily responses seen in anxiety. A group of medications known as **benzodiazepines** can be used in the short-term treatment of anxiety (also anxiolytics); as GABA agonists, they enhance the GABA-induced inhibition of overexcited neurotransmitters, calming nervous activity. Benzodiazepines are commonly known as tranquilisers and sleeping pills. By enhancing GABA's function, benzodiazepines bring about a state of calm as they reduce any over-activity in the brain. The calming effect of benzodiazepines makes them an effective treatment for the management of specific phobias. For example, a person who has a phobia of boats (naviphobia) may take a benzodiazepine medication such as Valium before getting on a boat.

9.2.1 TERMS USED IN EVIDENCE-BASED INTERVENTIONS

9.2.2 UNDERSTANDING EVIDENCE-BASED INTERVENTIONS

EXAM TIP
A common misunderstanding is the role of benzodiazepines in the treatment of anxiety and specific phobias. Benzodiazepine medications act to promote GABA (the primary inhibitory neurotransmitter), thus calming physiological arousal and reducing extreme anxiety associated with a specific phobia. Benzodiazepines do not 'replace' GABA.

This will reduce the person's physiological arousal, promote relaxation and make the boat trip more manageable.

However, because of their high potential for addiction, benzodiazepines are typically prescribed by doctors for short-term use only (usually no longer than 4 weeks). People who take benzodiazepines for an extended time may report increased anxiety. This is because the body becomes accustomed to the drug and, unless the dose is increased, withdrawal symptoms are experienced (Beyond Blue, 2015).

Benzodiazepines are the most common class of anti-anxiety drugs. They include Valium (diazepam), Ativan (lorazepam), Klonopin (clonazepam) and Xanax (alprazolam) (Figure 9.14). Benzodiazepines are very effective when taken during an anxiety episode; however, they also come with side effects such as the risk of addiction; drowsiness, sleepiness and fatigue; slow reflexes; slurred speech; confusion and disorientation; impaired thinking and judgement; memory loss; nausea and loss of appetite; double or blurred vision; and mood swings and aggression (Alcohol and Drug Foundation, 2021).

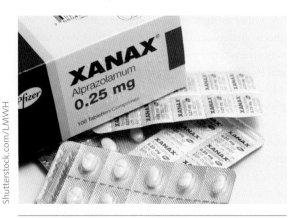

Figure 9.14 Short-term medications relax some of the physical symptoms of anxiety. They are not intended to be taken for long periods of time as they may be habit-forming.

Breathing retraining

Relaxation techniques are methods that can be learned to reduce physiological and psychological arousal associated with stress-related anxiety. Relaxation techniques such as breathing retraining can help individuals cope effectively with the stresses related to their specific phobia. Breathing and relaxation strategies are taught to minimise physical symptoms of anxiety and manage stress in general.

People suffering from anxiety disorders, such as specific phobias, may experience abnormal breathing patterns. A normal relaxed breathing rate for a healthy adult is 12–20 breaths per minute, using the diaphragm rather than the shoulders to move air in and out of the lungs. When a person is under stress, their breathing patterns may change. An anxious person's breathing may consist of small, shallow breaths (hyperventilation), which may raise oxygen levels and reduce the amount of carbon dioxide in the blood. This can prolong feelings of anxiety by making the physical symptoms of stress worse.

Breathing retraining is the process of identifying incorrect breathing habits and replacing them with correct ones. Symptoms of incorrect breathing include loud, noisy, rapid or shallow breathing.

A person who suffers from anxiety should learn how to breathe from their diaphragm, rather than their chest, through their nose in a slow, even and gentle way. Correct breathing means the abdomen moves, rather than the chest. Holding your breath for a few seconds will help to boost carbon dioxide levels in the blood. Carbon dioxide assists in the regulation of the body's reaction to anxiety and panic.

Deliberately copying a relaxed breathing pattern seems to calm the nervous system that controls the body's involuntary functions. Controlled breathing can cause physiological changes that counteract the sympathetic response, which include:
» lowered blood pressure and heart rate
» reduced levels of stress hormones in the blood
» balanced levels of oxygen and carbon dioxide in the blood
» improved immune system functioning
» increased physical energy
» increased feelings of calm and wellbeing.

Breathing retraining can help to correct breathing habits and help individuals have more control of their anxiety (Figure 9.15). The psychological benefits of reduced anxiety include:
» improved sleep, which in turn reduces stress
» an increased sense of control over shortness of breath
» increased self-confidence
» increased feelings of calm.

Learned breathing techniques should be implemented at the first signs of anxiety. By changing breathing patterns, breathing retraining can help to correct breathing habits and help individuals have more control of their anxiety (Better Health Channel, 2021).

Figure 9.15 Breathing retraining improves oxygen supply to your body and helps you relax when you are anxious or stressed.

Weblink
Breathing retraining protocol

KEY CONCEPTS 9.2a

» Medications can relieve symptoms of anxiety; however, their use is not a long-term solution because they do not treat the underlying cause of the anxiety disorder.
» Benzodiazepine medications act to promote GABA (the primary inhibitory neurotransmitter), thus calming physiological arousal and reducing extreme anxiety associated with a specific phobia.
» Relaxation techniques such as breathing retraining can help individuals cope effectively with the stresses related to their specific phobia.
» Breathing retraining is the process of identifying incorrect breathing habits and replacing them with correct ones.
» By changing breathing patterns, breathing retraining can help to correct breathing habits and help individuals have more control of their anxiety.

Concept questions 9.2a

Remembering
1 Define 'evidence-based interventions'. r
2 Without reference to your textbook or notes, list **three** possible side effects of the use of benzodiazepines in the treatment of anxiety. r

Understanding
3 Explain why medications are not considered long-term treatments for anxiety disorders. e
4 Outline the role of GABA agonists in the treatment of anxiety disorders such as specific phobia. e

Applying
5 Construct a flowchart of the steps involved in using breathing retraining in the management of phobic anxiety. d
6 Describe **one** limitation of the use of benzodiazepines for an extended period of time in the treatment of anxiety. r

HOT Challenge
7 Develop a hypothesis for an experiment investigating the effectiveness of breathing retraining in the management of anxiety. i

Psychological interventions

In most cases, the answer to treating people with mental illness is some form of psychotherapy. Psychotherapy is any psychological technique used for treating mental health disorders, with the goal of producing positive changes in thinking, emotions, personality, behaviour or adjustment. Psychotherapy most often refers to verbal interaction between trained mental health professionals and their clients. Many therapists also use learning principles to directly alter troublesome behaviours.

Psychotherapy is a vast field and there are several schools of thought based on learning theories and behavioural theories. A psychologist is educated extensively in the theories behind most of these therapies, but would then specialise in a few modes of treatment, particularly cognitive behavioural therapy. Mental health practitioners do not attempt to fit the patient to the therapies they are trained in; rather, they attempt to use treatments that fit the patient's needs.

Evidence now clearly shows that psychotherapies are as effective, if not more effective, than biological interventions such as medications alone in managing anxiety and phobias. Medications have several biological side effects, whereas psychotherapies do not. Psychotherapy, or 'talk therapy', is a way to treat people with a mental disorder by helping them understand their illness.

Education is often the first step in the management of any disorder, and access to quality information is often very important in helping to reduce our fears about what we are dealing with. As we saw previously, cognition is the psychological result of perception, learning and reasoning. Therefore, education about the disorder and how to deal with it is an important part of the treatment. Psychological interventions used in the treatment of a specific phobia include psychotherapies such as cognitive behavioural therapy and systematic desensitisation.

Cognitive behavioural therapy

Cognitive behavioural therapy (CBT) is an evidence-based psychological treatment approach that teaches clients to apply cognitive behavioural strategies to recognise and change negative and unproductive patterns of thinking and behaving. It is intended as a brief treatment program that can be effective within weeks to treat conditions such as anxiety, depression and other mental health problems (Figure 9.16). It is based on the premise that the way a person thinks about something determines how they feel about it and respond to it. Therefore, if they change the way they think about it, they can then change their behaviour.

Figure 9.16 Destructive, maladaptive behaviours – such as drinking, smoking and avoiding work – can be changed by using cognitive behavioural therapy, which teaches a person new, healthier ways to cope with their stress.

Using knowledge and information to overcome irrational thinking and replace it with reasonable, realistic thinking forms the 'cognitive' part of CBT. The behavioural component of the treatment involves modifying the unhelpful behaviours (such as avoidance) that have developed as a result of the faulty cognitions.

Research suggests that CBT is the most effective method of treating specific phobia. Wolitzky-Taylor et al. (2008) carried out a review of a large number of clinical trials in the treatment of specific phobias and found that exposure-based treatments were the most successful and seen as the treatment of choice for specific phobias.

Some common CBT techniques for the treatment of a specific phobia include:
» *relaxation training*. This method helps to calm the mind and body in the presence of the specific phobic stimulus. Techniques include meditation, guided imagery, progressive muscle relaxation and breathing exercises.

» *flooding*. This involves exposing a phobic person to the real feared stimulus all at once. This exposure is continued until the anxiety response disappears. If you had a phobia of snakes (ophidiophobia), for example, your therapist may take you to the zoo on a therapy session, where the snake handler 'helps' you hold and touch a snake.

» *imaginal flooding*. Exposure to the feared stimulus is attempted in the imagination of the client rather than in a real-life situation. In this case, the therapist will describe the fearful situation in graphic detail, perhaps even with the use of pictures, while the client attempts to gain awareness of the components of their body's response such as heart rate, sweating and respiration. This technique can be used if there is an element of actual, not just perceived, danger involved in exposure to the feared stimulus.

The purpose of CBT in treating phobia is to challenge the irrational and negative thoughts and replace them with realistic thoughts, and to examine related behaviours. To achieve this, psychotherapists often use systematic desensitisation.

Systematic desensitisation

Systematic desensitisation is a type of behaviour therapy that uses counterconditioning to reduce the anxiety a person experiences when in the presence of, or thinking about, a feared stimulus. It involves first learning specific muscle relaxation techniques and then practising these while the psychologist exposes the client to experiences with the feared stimulus by systematically increasing the intensity of the experience, beginning in the imagination and ending in reality. Counterconditioning refers to learning the relaxation response that is incompatible with the fear response. The target response is for the sufferer of the phobia to be relaxed and anxiety free in the presence of the feared stimulus. In this procedure, a relaxation technique is used to reduce the anxiety. Systematic desensitisation is based on **reciprocal inhibition,** a term coined by Joseph Wolpe (Wolpe & Plaud, 1997). In reciprocal inhibition, one emotional state is used to block another; that is, it is impossible to be anxious and relaxed at the same time.

Many phobias result from experience – they are learned associations between an object, activity or situation and an unpleasant experience. Systematic desensitisation uses classical conditioning principles to unlearn (or extinguish) the association the person has made and to learn a new relaxed response.

With repeated pairing of relaxation and the phobic stimulus, the stimulus loses its power to provoke anxiety. Systematic desensitisation can be used in a real setting (where the phobic stimulus is actually presented) or in an imagined setting (where the phobic stimulus is recalled using imagination), or in a combination of both.

This process takes a long time and relies on exposing sufferers gradually (or systematically) to the feared stimulus until they can remain relaxed in its presence (or desensitised) using a methodical step-by-step approach.

For example, a person with a phobia of spiders would work with the professional to develop a hierarchy of a number of things to do with the phobia they are fearful of, going from most fearful to least. The professional would then teach the patient relaxation techniques to lower their anxiety level (e.g. breathing, imagery, progressive muscle relaxation). The patient would then be exposed to stimuli that are progressively more threatening – initially, perhaps a picture of a spider, then a toy spider, then a dead spider, and finally a live spider. This is known as a **fear hierarchy** (Figure 9.17). During each step, the person tries to keep calm and relaxed, perhaps by trying to regulate their heart rate or breathing rate. They do not move to the next step until they have achieved this relaxed state. When this is achieved, the feared stimuli is said to have become extinct.

Note, however, that all phobias are different, so individual treatment plans need to be designed by qualified psychologists. For many fears, systematic desensitisation works best when people are directly exposed to the stimuli and situations they fear (McLeod, 2015) (Figure 9.18).

EXAM TIP
A common omission in examination responses by students is that when explaining the process of cognitive behavioural therapy, they fail to mention that the basic premise of CBT is that by changing a person's thoughts, this leads to changes in the way they behave.

Weblink
Puppy scares man to tears

My extreme animal phobia

9.2.3 SYSTEMATIC DESENSITISATION

Figure 9.17 A sample fear hierarchy for a dog phobia. Dogs are the fear stimulus and petting a dog off a leash is the target response.

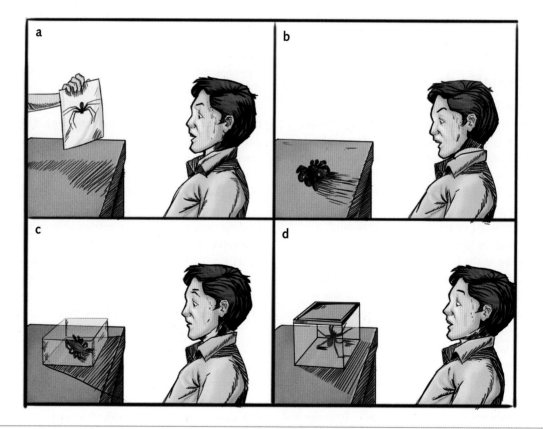

Figure 9.18 A person with a phobia of spiders is exposed to stimuli that are progressively more threatening. **a** The person is shown a picture of a spider. **b** The person is presented with a toy spider. **c** The person is presented with a dead spider in a glass case. **d** The person is presented with a live spider in a glass case. These progressions occur over a long period of time.

CASE STUDY 9.1

How virtual reality spiders are helping people face their arachnophobia

Gradually exposing people to the thing they fear, say a spider, in a controlled environment has long been the mainstay of treating phobias.

But with exposure therapy you don't have to have a spider physically present in the room for you to feel the benefits. Psychologists and researchers are using virtual reality to help people face their fears.

What is exposure therapy?

Psychologists originally proposed systematic desensitisation in the 1950s as a way of treating specific phobias.

The idea is that if you are presented with the phobic stimulus (for example, spiders or heights) repeatedly, but safely, then your fear reduces over time. In the case of a spider phobia (arachnophobia), systematic desensitisation may start with the spider in a cage or container so it cannot physically harm you.

What happens during systematic desensitisation?

Systematic desensitisation has three main elements.

First, you identify the situations or objects that make you feel afraid or anxious, or situations you avoid because of the phobia. You then rank them according to which ones provoke the most fear. This part of the therapy is known as constructing a fear hierarchy.

Then a psychologist teaches you how to progressively relax your muscles. This involves focusing on how you feel when you tense and relax different muscles. You use this relaxation technique when facing items on your fear hierarchy (or list).

Try following this progressive muscle relaxation exercise.

Finally, the therapist gradually shows you items on your fear list, while you practise your relaxation techniques. You start with the least fear-provoking item until you feel you can manage your fear, before being ready to move sequentially on to more fear-provoking ones.

In the case of a spider phobia, this could mean progressing from looking at an image, to a spider in a container far away, to having one sit in your hand.

Systematic desensitisation with virtual reality is the same, except instead of being directly exposed to the items on your fear hierarchy, you experience them through a headset.

Why use virtual reality systematic desensitisation?

Real-life systematic desensitisation has long been considered the most effective form of treatment for phobias. Yet, despite strong evidence systematic desensitisation works, about one third of people with a specific phobia don't seek treatment, or if they do, they avoid systematic desensitisation. The idea of facing their phobias is just too distressing.

Because they avoid the things they fear, there is no experience of safe exposure and, in turn, no decrease in their fear, so unfortunately phobias often persist.

But people tend to be more willing to take part in virtual reality systematic desensitisation than in the real-life kind. Researchers found that people prefer it mainly because confronting the phobia in real life is too fearsome.

So, virtual reality systematic desensitisation is a promising alternative, especially for people who find exposing themselves to their fear in real life is too difficult or stressful.

Does it work?

Since its introduction in the 1990s, virtual reality systematic desensitisation has been effective in treating a variety of phobias. These include acrophobia (fear of heights), aviophobia (fear of flying), claustrophobia (fear of confined spaces) and arachnophobia (fear of spiders).

Virtual reality exposure is just as effective as traditional forms of systematic desensitisation when assessed immediately after treatment, a year later and even up to three years after treatment.

For example, people who seek treatment for a spider phobia are less likely to avoid spiders, and less likely to feel anxious when they see spiders after virtual reality systematic desensitisation.

Another advantage is that psychologists can provide their clients with a range of experiences (phobic stimuli), which can be difficult to achieve in the real world. Consider how time-consuming it can be to provide real-life systematic desensitisation for someone with a fear of flying.

Additionally, virtual reality allows the psychologist to control the types of experiences clients have as they face items on their fear hierarchy.

Clients can also be assured of their safety and confidentiality, as the therapy is conducted in their psychologist's office.

The future of virtual reality systematic desensitisation

Although research shows that virtual reality systematic desensitisation is effective, there are concerns about its cost, accessibility and quality.

However, its quality continues to improve, as does its cost. There are now clinics that specialise in treating specific phobias this way.

Source: Boynton, R., Swinbourne, A. James Cook University. (2017) How virtual reality spiders are helping people face their arachnophobia. This article is republished from The Conversation.com under a Creative Commons license.

Questions

1 Identify and describe the three main elements of systematic desensitisation.
2 Summarise the advantages and disadvantages of virtual reality systematic desensitisation.
3 What ethical concepts and guidelines could the use of virtual reality systematic desensitisation raise?
4 What is your reaction to this case study? Do you think it would be a worthwhile treatment for phobias?
5 Would you use virtual reality systematic desensitisation to treat your phobia? Explain your reasons.

EXAM TIP
Ensure you provide a response to every examination multiple-choice question. You do not lose marks for an incorrect response. A guess (preferably an educated guess) has a chance of being correct, whereas no response has no chance of a mark and could disrupt the recording of your responses on the multiple-choice question answer sheet. An effective strategy is to eliminate incorrect options before selecting the most correct answer.

ACTIVITY 9.2 DEVELOPING A FEAR HIERARCHY

Try it

1 Create a fear hierarchy for an individual with a phobia of birds. Add more rows as required.

Situation	Score (0–100)

Apply it

2 Explain how systematic desensitisation can be used to help a person overcome their phobia of birds.

Exam ready

3 Systematic desensitisation utilises classical conditioning as a means of treating phobias. Which one of the statements below correctly explains how classical conditioning is applied via the use of systematic desensitisation?

A Classical conditioning is the form of learning that occurs when a person acquires a phobia.

B Systematic desensitisation attempts to create an association between the phobic stimulus and a relaxation response.

C Systematic desensitisation attempts to replace negative reinforcement with positive reinforcement.

D Classical conditioning is the form of learning that occurs when a person develops a phobia, and operant conditioning perpetuates the phobia.

Social interventions

A supportive family and social environment is an important factor in the treatment of mental illnesses. A person with a specific phobia who is surrounded by a social network of people who understand the illness will more readily receive support to help manage their fears. The social interventions used in the treatment for a specific phobia include psychoeducation for families and supporters in challenging sufferers' unrealistic or anxious thoughts and not encouraging **avoidance behaviours**.

Psychoeducation

A psychosocial approach in which a person experiencing a mental health problem or disorder and their family are provided with information to help them understand the condition and how they can contribute to managing it is known as **psychoeducation**. Psychoeducation is offered to the phobic person as well as their families and supporters, and can be delivered individually or in groups, tailored to the needs of the specific individuals involved.

The goal of psychoeducation for the phobic person is to empower them to understand their illness and how to develop strategies to cope with it and recover from it. This is one of the first steps in cognitive behavioural therapy. For example, people tend to be less fearful of symptoms if they understand why and how their bodies respond to a threatening situation. Education of the naturally occurring symptoms of the flight-or-fight-or-freeze response in preparing our body to deal with a threat may assist the person in becoming less fearful of the symptoms themselves.

The goal of psychoeducation for the families and supporters of a phobic person is to help them understand the illness (Figure 9.19). This helps to reduce the stigma associated with the illness. Additional goals are to teach them the skills needed to help support the phobic person to cope and recover, and to help themselves adapt to living with a person suffering from a phobia.

Challenging unrealistic or anxious thoughts

Families and those supporting a person with a specific phobia can play an important role

Figure 9.19 The goal of psychoeducation for the families and supporters of a phobic person is to help them understand the illness.

in assisting the sufferer to challenge their unrealistic or anxious thoughts towards the phobic stimulus. Sufferers of specific phobias overestimate the dangers involved in a particular object, activity or situation and treat every negative thought as if it were fact. The specific phobia sufferer usually knows that their fear is excessive or unreasonable but feels powerless to control it (Figure 9.20). These irrational and pessimistic attitudes are known as cognitive distortions.

Cognitive distortions fall into the following categories:
» *fortune-telling.* For example, 'It's spring. All the bees are out. If I go outside, I will get stung and have a really bad allergic reaction.'
» *overgeneralisation.* For example, 'I saw a person get bitten by a dog once. All dogs are dangerous.'
» *catastrophising.* For example, 'If this plane goes through turbulence, it is going to crash.'

Families and supporters can help a sufferer to challenge these unhelpful thoughts and develop a more balanced perspective. This can be done by family members and supporters encouraging the phobic person to consider the following questions.
» What's the probability of that actually happening?
» Is there any evidence to support that thought?
» What's happened in the past when you were presented with this object, activity or situation?
» Is there a more realistic way of looking at the situation? (ReachOut Australia 2022)

9.2.4 BIOPSYCHOSOCIAL APPROACH STUDY CARDS

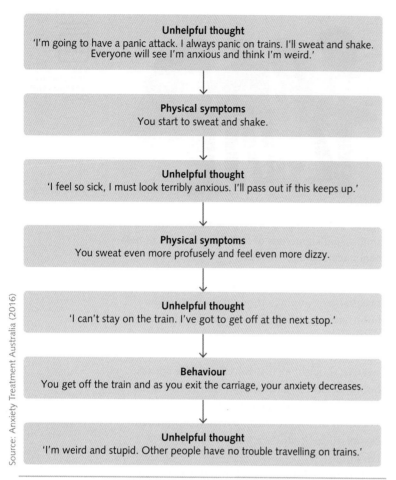

Figure 9.20 This example of someone who fears having a panic attack on a train highlights the interaction between thoughts, physical symptoms and behaviour.

attempt to remove the fear response to the phobic stimulus but it does assist in maintaining the anxiety disorder. Avoiding the phobic stimulus provides the person with negative reinforcement because they avoid the unpleasant fear symptoms associated with it. The absence of fear that results from their avoidance becomes a reward and reinforces the likelihood of the avoidance behaviour being repeated in the future. This is an example of operant conditioning in practice. For example, an individual with a phobia of insects (entomophobia) (Figure 9.21) will feel a significant decrease in anxiety once they decide to avoid a location where insects are present. This avoidance removes the unpleasant anxiety symptoms, thereby reinforcing their avoidance behaviour.

A sufferer's family and supporters should not criticise or encourage the use of avoidance behaviours. Instead, they can assist by providing more adaptive alternatives in a supportive environment. The principles of operant conditioning and systematic desensitisation may be used in the treatment of an anxiety disorder. By providing gradual exposure to the feared object, activity or situation in a safe and reassuring manner, the maladaptive behaviours may become extinct. For example, the family and supporters of an individual with a phobia of insects may assist by encouraging the person to visit locations where they are likely to encounter insects, such as natural environments.

Not encouraging avoidance behaviours

Psychoeducation for the families and supporters of a person with a specific phobia may also focus on not encouraging avoidance behaviours. Avoidance behaviours are behaviours that attempt to prevent exposure to the fear-provoking object, activity or situation. For a person with a specific phobia this may involve completely avoiding or escaping from the feared object, activity or situation.

Avoidance behaviour can significantly restrict a sufferer's life. As a strategy to cope with the specific phobia, avoidance behaviour is considered maladaptive. **Maladaptive behaviour** is potentially harmful and prevents a person from meeting and adapting to the demands of everyday living. This is because it does not

EXAM TIP
In questions that assess the application of psychological knowledge to a scenario, it is important to make clear reference to the scenario in your responses. This can be done by referring by name to the person in the scenario in your response.

Figure 9.21 Avoidance behaviours are behaviours that attempt to prevent exposure to the fear-provoking object, activity or situation, such as not going to natural environments in order to avoid insects.

ANALYSING RESEARCH 9.2

If you're afraid of spiders, they seem bigger: phobia's effect on perception of feared objects allows fear to persist

A better understanding of how a phobia affects the perception of feared objects can help clinicians design more effective treatments for people who seek to overcome their fears, according to researchers from Ohio State University.

In the study, participants who feared spiders were asked to undergo five encounters with live spiders – tarantulas, in fact – and then provide size estimates of the spiders after those encounters ended. The researchers recruited 57 people who self-identified as having a spider phobia. Each participant then interacted at specific time points over a period of 8 weeks with five different varieties of tarantulas varying in size from about 2.5 to 16 cm long.

The spiders were contained in an uncovered glass tank. Participants began their encounters 3.6 m from the tank and were asked to approach the spider. Once they were standing next to the tank, they were asked to guide the spider around the tank by touching it with a 20-cm probe, and later with a shorter probe.

Throughout these encounters, researchers asked participants to report how afraid they were feeling on a scale of 0–100 according to an index of subjective units of distress. After the encounters, participants completed additional self-report measures of their specific fear of spiders, any panic symptoms they experienced during the encounters with the spiders, and thoughts about fear reduction and future spider encounters.

Finally, the research participants estimated the size of the spiders – while no longer being able to see them – by drawing a single line on an index card indicating the length of the spider from the tips of its front legs to the tips of its back legs.

An analysis of the results showed that higher average peak ratings of distress during the spider encounters were associated with estimates that the spiders were larger than they really were.

Source: Adapted from Ohio State University. (2012, February 22). If you're afraid of spiders, they seem bigger: Phobias effect on perception of feared objects allows fear to persist. *ScienceDaily*. www.sciencedaily.com/releases/2012/02/120222204241.htm

9.2.5 ANALYSIS OF A RESEARCH INVESTIGATION

Questions
1. State the aim of the study.
2. Write a hypothesis for the study.
3. What are the results of the study?
4. Explain whether the results support or refute the hypothesis.
5. Identify and explain an extraneous variable that may have affected the results and suggest how this variable could be controlled in future research.
6. Identify and explain the ethical concepts or guidelines relevant to this study.

INVESTIGATION 9.1 INVESTIGATION OF A SPECIFIC PHOBIA

Scientific investigation methodology
Case study

Aim
To apply a biopsychosocial approach to explain the development and management of specific phobia.

Introduction
A specific phobia is a kind of anxiety disorder in which a person experiences intense fear or anxiety towards a particular object or situation. The response is persistent, overwhelming and out of proportion to the actual threat. It produces avoidance and/or extreme distress that interferes with normal functioning. Typical examples of a specific phobia include fear of flying, heights, spiders, receiving an injection and seeing blood (adapted from the *Diagnostic and Statistical Manual of Mental Disorders*, 5th ed.).

Logbook template
Case study

Specific phobias are thought to affect around 11 per cent of the Australian population. The first symptoms of specific phobias usually arise in childhood or early adolescence (Beyond Blue, 2022).

Your task

Research a well-known personality or sportsperson who has been diagnosed with a specific phobia.

Use the questions below to help you develop your understanding of the application of a biopsychosocial approach to explain their specific phobia.

Questions

1. Name the person and their specific phobia.
2. Explain how this specific phobia is different to a fear.
3. Describe the factors involved in the development of the specific phobia (if identified). Categorise these factors using the biopsychosocial model.
4. Discuss the impact of the specific phobia on the person's wellbeing and everyday life.
5. Has the person experienced any stigma associated with the phobia? If so, discuss.
6. For each category of the biopsychosocial model, discuss one strategy that could be utilised in the management of the specific phobia.
7. Discuss the quality of the sources of information you have used for this case study.
8. Provide a list of all sources of information you have used for this case study using scientific referencing conventions.

KEY CONCEPTS 9.2b

- Cognitive behavioural therapy (CBT) is an evidence-based psychological treatment approach that teaches clients to apply cognitive behavioural strategies to recognise and change negative and unproductive patterns of thinking and behaving; intended as a brief treatment program that can be effective within weeks to treat conditions such as anxiety, depression and other mental health problems.
- Systematic desensitisation is a type of behaviour therapy that uses counterconditioning to reduce the anxiety a person experiences when in the presence of, or thinking about, a feared stimulus.
- Psychoeducation is a psycho-social approach in which a person experiencing a mental health problem or disorder and their family are provided with information to help them understand the condition and how they can contribute to managing it.
- Avoidance behaviours are behaviours that attempt to prevent exposure to the fear-provoking object, activity or situation.

Concept questions 9.2b

Remembering

1. Define and provide an example of a maladaptive behaviour. **e**
2. List the steps involved in the use of cognitive behavioural therapy to treat specific phobia. **r**

Understanding

3. Compare the use of psychotherapies with the use of medications in the treatment of specific phobia. **e**
4. Using an example, explain how avoidance behaviours may influence the maintenance of a specific phobia. **e**

Applying

5. By referring to Figure 9.17 and the associated text, construct a similar 10-step fear hierarchy for a cat-specific phobia. **d**
6. Create a list of **three** questions that families of a person suffering from a specific phobia could ask to challenge unrealistic or anxious thoughts. **c**

HOT Challenge

7. With reference to the phobia of dogs in the 'My extreme animal phobia' video (page 377), explain how systematic desensitisation could be used to treat the sufferer of this specific phobia. **c**

9 Chapter summary

KEY CONCEPTS 9.1a

» A specific phobia is an anxiety disorder in which a person experiences intense fear or anxiety towards a particular object or situation.
» Phobias are the result of the interaction between biological, psychological and social factors.
» Reduced GABA is related to dysfunction in the brain circuits that regulate emotional responses to threatening stimuli.
» An individual with low levels of GABA may not be able to regulate their flight-or-fight-or-freeze response and therefore may be more vulnerable to anxiety and have a greater chance of developing a specific phobia.
» Long-term potentiation (LTP) contributes to the development of a specific phobia by strengthening connections between neurons in the amygdala that fire during the flight-or-fight-or-freeze response and neurons involved in perceiving the feared stimulus, so that the connections become consolidated in the hippocampus.

KEY CONCEPTS 9.1b

» Behavioural models propose that phobic anxiety could be the result of learning. It is possible to explain the precipitation (acquisition) of a phobia as a learnt association between two stimuli (classical conditioning) and its perpetuation (maintenance) due to negative reinforcement of avoidance behaviours (operant conditioning).
» The cognitive model claims we may sometimes couple faulty reasoning and rationale with fearful stimuli from the environment and hence a new cognition is formed into a phobia.
» Specific environmental triggers such as a traumatic event can predispose an individual to the development of a specific phobia.
» According to social learning theory and parental modelling, children learn from their parents and often mirror what their parents do. Children learn how to respond to stress and difficulties by observing how their parents respond to these.
» Stigma is a negative social attitude about a characteristic of a person or social group that implies some form of deficiency, often leading to unfair discrimination against or exclusion of the person or social group.

KEY CONCEPTS 9.2a

» Medications can relieve symptoms of anxiety; however, their use is not a long-term solution because they do not treat the underlying cause of the anxiety disorder.
» Benzodiazepine medications act to promote GABA (the primary inhibitory neurotransmitter), thus calming physiological arousal and reducing extreme anxiety associated with a specific phobia.
» Relaxation techniques, such as breathing retraining, can help individuals cope effectively with the stresses related to their specific phobia.
» Breathing retraining is the process of identifying incorrect breathing habits and replacing them with correct ones.
» By changing breathing patterns, breathing retraining can help to correct breathing habits and help individuals have more control of their anxiety.

KEY CONCEPTS 9.2b

» Cognitive behavioural therapy (CBT) is an evidence-based psychological treatment approach that teaches clients to apply cognitive behavioural strategies to recognise and change negative and unproductive patterns of thinking and behaving; intended as a brief treatment program that can be effective within weeks to treat conditions such as anxiety, depression and other mental health problems.
» Systematic desensitisation is a type of behaviour therapy that uses counterconditioning to reduce the anxiety a person experiences when in the presence of, or thinking about, a feared stimulus.
» Psychoeducation is a psycho-social approach in which a person experiencing a mental health problem or disorder and their family are provided with information to help them understand the condition and how they can contribute to managing it.
» Avoidance behaviours are behaviours that attempt to prevent exposure to the fear-provoking object, activity or situation.

9 End-of-chapter exam

Section A: Multiple choice

1. According to the biopsychosocial model, the influence of emotions, memories and learning on an individual's wellbeing would be considered _____ factors.
   A social
   B psychological
   C biological
   D social, psychological and biological

2. Which one of the following characteristics is not accurate in describing the fear associated with a phobia?
   A It is persistent.
   B It is irrational.
   C It is reasonable.
   D It is intense.

3. A chemical in the brain that is released between the synapses of two neurons is known as a:
   A neuron.
   B neurotransmitter.
   C neurochemical.
   D hormone.

4. The key role of gamma-amino butyric acid in the experience of anxiety is that it should have an:
   A inhibiting response, calming the individual.
   B excitatory response, calming the individual.
   C inhibiting response, preparing the individual for action.
   D excitatory response, preparing the individual for action.

5. The experience of stress results in activation of the _____, which is known as the flight-or-fight-or-freeze response.
   A central nervous system
   B somatic nervous system
   C sympathetic nervous system
   D parasympathetic nervous system

6. Exposure to the fear stimulus will strengthen the memory circuit in which part of the brain?
   A the amygdala
   B the hypothalamus
   C the hippocampus
   D the thalamus

7. A type of error in thinking that occurs when people are interpreting information is known as a _____ bias.
   A cognitive
   B irrational
   C thinking
   D filtering

   ©VCAA 2016 Q8 ADAPTED EASY

8. In terms of the development and management of a specific phobia, the biopsychosocial model considers:
   A psychological factors more important than biological and social factors.
   B biological, psychological and social factors to be equally important.
   C that biological, psychological and social factors do not contribute to the development and management of a specific phobia.
   D biological factors to be more important than psychological and social factors.

9. Anti-anxiety medications work by slowing down the _____ nervous system.
   A central
   B sympathetic
   C parasympathetic
   D somatic

10. Which of the following is a physical symptom of anxiety that may be triggered by hyperventilation?
    A feeling overwhelmed
    B an increase in oxygen levels
    C an increase in carbon dioxide in blood
    D memory loss

©VCAA 2020 Q49 ADAPTED MEDIUM

11 Which of the following is not a characteristic of systematic desensitisation?
   A It requires the development of a hierarchy.
   B It involves the use of classical conditioning.
   C It is based on the way a person thinks about something to determine how they feel about it and respond to it.
   D It uses counterconditioning to reduce anxiety.

12 The _____ model proposes that phobic anxiety could be the result of learning.
   A cognitive
   B behavioural
   C emotional
   D social

13 A negative social attitude about a characteristic of a person or social group that implies some form of deficiency is known as:
   A prejudice.
   B discrimination.
   C a stigma.
   D an attitude.

14 _____ can relieve symptoms of anxiety; however, their use is not a long-term solution as they do not treat the underlying cause of the anxiety disorder.
   A Medications
   B Psychologists
   C Psychiatrists
   D Psychoeducation

15 A person who suffers from anxiety should learn how to breathe from their _____, rather than their chest.
   A shoulders
   B chest
   C diaphragm
   D nose

16 Reciprocal inhibition is:
   A the concept that one emotional state is used to block another.
   B a type of behavioural therapy whereby an individual with a phobia or fear is exposed to the fear-producing object very slowly.
   C a type of behavioural therapy whereby an individual with a phobia or fear is exposed to the fear-producing object very quickly.
   D the education about a mental illness such as the nature of the illness, its treatment and management strategies.

17 CBT involves the application of what type of principles to change thought processes and human behaviour?
   A The way a person thinks about something determines how they feel about it and respond to it.
   B Psychological problems are based on faulty or unhelpful ways of thinking.
   C Psychological problems are based on learned patterns of unhelpful behaviour.
   D All of the above

©VCAA 2020 Q42 ADAPTED MEDIUM

18 Which one of the following accurately categorises a contributing factor in the development of a specific phobia?
   A classical conditioning (biological)
   B catastrophic thinking (social)
   C stigma around seeking treatment (psychological)
   D long-term potentiation (biological)

19 The education about a mental illness such as the nature of the illness, its treatment and management strategies is known as:
   A psychoeducation.
   B systematic desensitisation.
   C reciprocal inhibition.
   D avoidance behaviours.

20 Behaviours that attempt to prevent exposure to the fear-provoking object, activity or situation are known as:
   A avoidance behaviours.
   B maladaptive behaviours.
   C unhelpful behaviours.
   D covert behaviours.

Section B: Short answer

1 Outline how breathing techniques may help individuals cope effectively with the stresses related to their specific phobia. [4 marks]

2 Why is the psychoeducation of family members and supporters of sufferers an important factor in the treatment of a mental illness? [4 marks]

3 How could a benzodiazepine agent help to manage a phobia? [4 marks]

Maintenance of mental wellbeing

10

Key knowledge

- the application of a biopsychosocial approach to maintaining mental wellbeing, with reference to protective factors including adequate nutritional intake and hydration and sleep (biological), cognitive behavioural strategies and mindfulness meditation (psychological) and support from family, friends and community that is authentic and energising (social)
- cultural determinants, including cultural continuity and self-determination, as integral for the maintenance of wellbeing in Aboriginal and Torres Strait Islander peoples

Key science skills

Develop aims and questions, formulate hypotheses and make predictions
- identify independent, dependent and controlled variables in controlled experiments
- formulate hypotheses to focus investigation

Comply with safety and ethical guidelines
- demonstrate ethical conduct and apply ethical guidelines when undertaking and reporting investigations

Generate, collate and record data
- systematically generate and record primary data, and collate secondary data, appropriate to the investigation
- record and summarise both qualitative and quantitative data, including use of a logbook as an authentication of generated or collated data
- organise and present data in useful and meaningful ways, including tables, bar charts and line graphs

Analyse and evaluate data and investigation methods
- process quantitative data using appropriate mathematical relationships and units, including calculations of percentages, percentage change and measures of central tendencies (mean, median, mode), and demonstrate an understanding of standard deviation as a measure of variability
- evaluate investigation methods and possible sources of error or uncertainty, and suggest improvements to increase validity and to reduce uncertainty

Construct evidence-based arguments and draw conclusions
- evaluate data to determine the degree to which the evidence supports the aim of the investigation, and make recommendations, as appropriate, for modifying or extending the investigation
- use reasoning to construct scientific arguments, and to draw and justify conclusions consistent with the evidence and relevant to the question under investigation
- discuss the implications of research findings and proposals

Source: VCE Psychology Study Design (2023–2027) + pages 40, 12–13

10 Maintenance of mental wellbeing

You learnt in Chapter 8 that mental wellbeing is a state of wellbeing in which an individual realises their own abilities, can cope with the normal stresses of life, can work productively and fruitfully, and is able to make a contribution to their community (World Health Organization). This chapter is going to look at ways that you can maintain your own mental wellbeing and contribute to the mental wellbeing of others.

Adobe Stock/Okeas

Adobe Stock/fizkes

p. 393

10.1
The application of a biopsychosocial approach to maintaining mental wellbeing

Again, we look to the biopsychosocial model to maintain our mental wellbeing. It is important that we get adequate sleep and that our diet includes all the nutrition that our body needs for the stage of life that we are at (biological). It is also important that we make time for ourselves to reflect on our thoughts, feelings and actions (psychological), community, family and friends (social).

p. 406

**10.2
Cultural determinants of social and emotional wellbeing**

We explore cultural continuity and self-determination as two key cultural determinants that maintain social and emotional wellbeing within the holistic and multidimensional framework of Aboriginal and Torres Strait Islander peoples' SEWB.

Adobe Stock/Iurii Sokolov

The world is changing quickly and there is lots of pressure on people to be happy. This can sometimes be hard. Take the time to care for yourself, and remember, this is not a one-size-fits-all solution. Find out what you enjoy doing and make sure that you pamper yourself from time to time.

Slideshow
Chapter 10 slideshow

Flashcards
Chapter 10 flashcards

Test
Chapter 10 pre-test

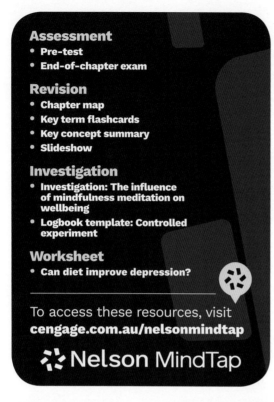

Know your key terms

Adequate diet
Adequate sleep
Anxiety
Authentic social support
Cognitive behavioural strategies
Cultural continuity
Cultural determinants
Diet
Energising social support

Hydration
Meditation
Mindfulness
Mindfulness meditation
Nutritional intake
Protective factor
Risk factor
Self-determination
Social support

Weblink
Daily Calm

In July 2018, 12 Thai boys and their soccer coach were trapped for more than two weeks in a flooded cave during a trip intended to last just an afternoon. The complexity of the rescue attracted global attention, and an international team of experts gathered to tackle the predicament. When the first group of British divers finally discovered the victims, they were amazed to find that the boys and their coach were sitting in the dark, meditating. All of them came out alive. According to CNBC, the coach had spent a decade as a Buddhist monk (Wang, 2022).

This chapter explores the application of a biopsychosocial approach to maintaining mental wellbeing, including the influence of **meditation**. As an introduction to this chapter the following activity provides a practical application of the Unit 4 Psychology key knowledge and an opportunity to utilise this 5000-year-old method to improve your mental wellbeing (Figure 10.1).

Alternatively, watch and participate in the Daily Calm YouTube video in the weblink for a guided 10-minute **mindfulness meditation** session.

A simple meditation practice (as detailed on this page) focuses on the breath, not because there is anything special about it, but because the physical sensation of breathing is always there and you can use it as an anchor to the present moment. Throughout the practice you may find yourself caught up in thoughts, emotions, sounds – wherever your mind goes, simply come back again to the next breath. Even if you only come back once, that's okay.

A simple meditation practice

1. Sit comfortably. Find a spot that gives you a stable, solid, comfortable seat.
2. Notice what your legs are doing. If on a cushion, cross your legs comfortably in front of you. If on a chair, rest the bottoms of your feet on the floor.
3. Straighten your upper body—but don't stiffen. Your spine has natural curvature. Let it be there.
4. Notice what your arms are doing. Situate your upper arms parallel to your upper body. Rest the palms of your hands on your legs wherever it feels most natural.

5 Soften your gaze. Drop your chin a little and let your gaze fall gently downward. It's not necessary to close your eyes. You can simply let what appears before your eyes be there without focusing on it.

6 Feel your breath. Bring your attention to the physical sensation of breathing: the air moving through your nose or mouth, the rising and falling of your belly, or your chest.

7 Notice when your mind wanders from your breath. Inevitably, your attention will leave the breath and wander to other places. Don't worry. There's no need to block or eliminate thinking. When you notice your mind wandering, gently return your attention to the breath.

8 Be kind about your wandering mind. You may find your mind wandering constantly–that's normal, too. Instead of wrestling with your thoughts, practise observing them without reacting. Just sit and pay attention. As hard as it is to maintain, that's all there is. Come back to your breath over and over again, without judgement or expectation.

9 When you're ready, gently lift your gaze (if your eyes are closed, open them). Take a moment and notice any sounds in the environment. Notice how your body feels right now. Notice your thoughts and emotions.

Source: mindful.org 2022

Figure 10.1 Meditation has been practised for over 5000 years. Its benefits include: reduction in blood pressure, increasing calmness and physical relaxation, coping with illness and enhancing overall health and wellbeing.

10.1 The application of a biopsychosocial approach to maintaining mental wellbeing

As discussed in Chapter 3, there are many sources of stress and adversity, including daily pressures such as working long hours, and life events such as loss of a job, illness and moving to a new home. Stress is a normal part of life and it can motivate us to achieve our goals. However, if stress becomes chronic, or when we feel we're no longer able to cope, it can negatively affect our wellbeing.

As discussed in Chapter 8, a number of factors, known as **risk factors**, increase the likelihood that an individual will experience mental health problems or a mental disorder. Risk factors can have an internal (biological or psychological) or external (biological or social) source.

There are also many internal and external factors that influence an individual's **resilience** and ability to recover from a negative experience. Any behavioural, biological, psychological or environmental characteristic that decreases the likelihood of a person developing a particular mental health problem or disorder is known as a **protective factor**.

In this section we explore the protective factors that help us maintain mental wellbeing through the lens of the **biopsychosocial model**. Biological protective factors include making sure that we have a nutritious diet (adequate nutritional intake), drink sufficient fluids (hydration), and get enough sleep. Psychological factors that help us maintain mental wellbeing include cognitive-behavioural strategies and mindfulness meditation. Social factors include support from our family, friends and community that is authentic and energising.

Biological protective factors that maintain mental wellbeing

Nutritional intake, hydration and sleep are everyday biological protective factors that influence our physiological functioning and can help to maintain our mental wellbeing.

10.1.1 THE APPLICATION OF A BIOPSYCHOSOCIAL APPROACH TO MAINTAINING MENTAL WELLBEING

10.1.2
AUSTRALIAN GUIDE TO HEALTHY EATING

Worksheet
Can diet improve depression?

Adequate nutritional intake and hydration

Brain functioning is very sensitive to what we eat and drink, therefore **diet** can play a vital role in maintaining mental wellbeing as well as physical health. There is growing evidence that nutritional intake and hydration play an important contributory role in specific mental health problems, including attention deficit hyperactivity disorder (ADHD), depression, schizophrenia and Alzheimer's disease (Mental Health Foundation of New Zealand).

An **adequate diet** is one where the **nutritional intake** and **hydration** includes sufficient energy for a person's needs, including protein for growth and maintenance of body cells; minerals, vitamins and water for growth and regulation of body processes; and fats and carbohydrates for energy. A person's nutritional intake will not cause or prevent the development of a mental illness; however, it can help to promote good mental wellbeing and contribute to a person's level of resilience when presented with adversity (Figure 10.2). A balanced diet consists of adequate amounts of complex carbohydrates, essential fats, amino acids, vitamins, minerals and water in the correct proportions.

Figure 10.2 Following a healthy pattern of eating is linked with better stress management, improved sleep quality, increased concentration and better mental wellbeing in general.

Nutritional intake can affect our mental wellbeing both directly and indirectly: directly, through the nutrients that keep our body and brain physically healthy so they can function at an optimal level, and indirectly, through the emotional impact of having a physical condition caused by an unhealthy diet. Eating a balanced diet can help improve mood, maintain healthy brain functioning and help people cope with a range of adversities, including mental disorders such as anxiety disorders (Figure 10.3).

Nutritional intake and hydration contributes to resilience by protecting against diet-related diseases that affect physical and cognitive functions and by reducing vulnerability to stress and depression (Flórez et al., 2014).

For good mental wellbeing, it is important to keep adequately hydrated. The human body can last weeks without food, but only days without water. The body is made up of 50–75 per cent water. Water forms the basis of blood, digestive juices, urine and perspiration, and is contained in lean muscle, fat and bones. Water is needed to make neurotransmitters, the chemicals that transmit signals between brain cells and hormones that control the processes of the body and brain. Even mild dehydration can make you irritable and affect mental performance (Health Direct, 2018). The best source of fluids is fresh tap water. The Australian Dietary Guidelines recommends that women should have about 2 litres (8 cups) and men about 2.6 litres (10 cups) of water per day to stay adequately hydrated (Better Health Channel, 2022).

The following nutritional intake and hydration guidelines from the National Health and Medical Research Council provide advice on the impact on mental wellbeing of specific foods.

» Drinking plenty of water helps prevent dehydration. Even mild dehydration can affect mood, causing irritability and restlessness.
» Wholegrain cereals (those with intact kernels) and many fruits, vegetables and legumes have a low 'glycaemic index', which means that the sugar in these foods is absorbed slowly into the bloodstream. This helps to stabilise blood sugars and optimise mental as well as physical performance.
» For people who experience **anxiety**, avoiding caffeine is wise. Caffeine, especially for those who are particularly sensitive to it, increases anxiety and contributes to insomnia. Coffee, tea, energy drinks (such as Red Bull) and cola drinks all contain caffeine, as do cocoa and chocolate in lesser amounts.

Adequate sleep

As discussed in Chapters 6 and 7, sleep is vitally important to us for replenishing and revitalising the physiological processes that keep the mind and body functioning at an optimal level. NREM sleep

Weblink
How to manage your mood with food 8 tips

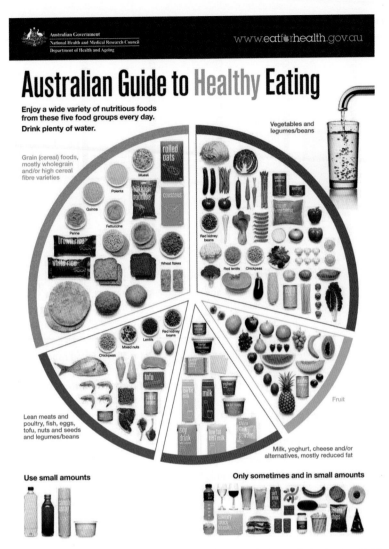

Source: National Health and Medical Research Council. Australian Guide to Healthy Eating. www.eatforhealth.gov.au CC BY 4.0 licence (https://creativecommons.org/licenses/by/4.0/)

Figure 10.3 Adequate nutritional intake and hydration can directly contribute to your mental wellbeing.

plays a role in restoring the body and REM sleep in the restoration of mental processes. Ensuring we get adequate sleep is essential in maintaining good physical health and mental wellbeing.

Adequate sleep is the necessary amount for individuals to function effectively during the daytime and cope with normal daily stress. The amount of adequate sleep will differ between individuals depending on a range of factors such as age, physical activity levels and general health.

The US National Sleep Foundation has published sleep recommendations for age groups across the life span (Figure 10.4). These recommendations highlight that as people get older, they generally need less sleep. In line with this, research shows that among Australian children aged 9 to 18 years, school-night sleep duration decreases at a rate of 12 minutes per year. Data from Victoria and South Australia shows adolescents are only getting an average of between 6.5 and 7.5 hours of sleep on school nights, rather than the recommended 8–10 hours.

One of the many effects that anxiety has on the body is that it can disrupt sleep patterns. Negative and anxiety-provoking thoughts can disrupt our sleep pattern by disrupting the onset and maintenance of sleep.

Scientists have found that a lack of sleep, common in anxiety disorders, may play a key role in activating brain regions that contribute to excessive worrying. The psychological repercussions of inadequate sleep include changes in emotions (being irritable, anxious, depressed and moody), difficulty concentrating, poor behaviour control and difficulty in social settings.

10.1.3 ADEQUATE SLEEP

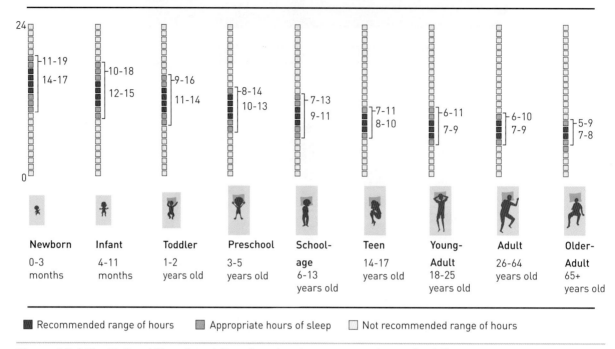

Figure 10.4 The National Sleep Foundation sleep duration recommendations by different age groupings.

How to get better sleep

Despite knowing why you need sleep and how much you need, the act of sleeping is easier said than done for most people. Here are a few tips on how to sleep more soundly tonight (Figure 10.5):

» Stick to a sleep schedule: Going to bed at the same time every night and waking up at the same time every morning helps regulate your internal body clock so your body knows when to prep for sleep.

» Exercise: No matter what the health issue is, exercise is always good for you. Cardio-heavy workouts not only help you sleep longer, but also help you sleep better. With that said though, try to avoid vigorous exercise 4 hours before bedtime so that your body has time to properly cool down.

» Cut that afternoon caffeine jolt: Even though the afternoon lull might be hitting you at your desk, try to resist the temptation for caffeine after 2 p.m. Caffeine typically stays in your body for about 8 hours and will prevent you from entering deep sleep.

» Take time to wind down: Sleep isn't an instantaneous thing; it's usually a process to fall asleep. Though it may be second nature by now, try to avoid those pesky smartphones, tablets, laptop screens – all of those screens emit blue fluorescent light that prevents the brain from producing melatonin, the sleep-regulating hormone. Plus, all of those games, emails and online shows only stimulate our brains rather than relax them. Instead, relax in bed with your comfy pjs and a book – that way your brain is primed and ready for sleep by the time your head hits the pillow.

Source: adapted from: https://humanhealthproject.org/

Watch the video at the weblink to learn how sleep and mental disorders are mentally linked in the brain.

Figure 10.5 Strategies to promote sleep will contribute to increased energy levels, clearer and more rational thinking, and boost our resilience to cope with everyday stress more effectively.

Weblink
Why do we sleep?

EXAM TIP
The VCE Unit 3 & 4 Psychology key knowledge requires students to learn many definitions. To help in your knowledge and understanding of the key terms, link as many as possible to real life examples, e.g. think about how you feel when you are sleep deprived.

ANALYSING RESEARCH 10.1

Reduced sleep quality can aggravate pre-existing psychological conditions

Disturbed sleep is a commonly reported symptom among individuals diagnosed with anxiety disorders. However, the direct cause of disrupted sleep is poorly understood. Proper sleep is critical for cognitive and daily functioning, and reduced quality of sleep has the potential to exacerbate pre-existing psychological conditions, according to new research.

To effectively evaluate differences in sleep architecture* after induced stress, Robert Ross MacLean, of Boston University, used an objective measure of anxiety and recorded subsequent sleep-wake behavior in rats. In the rodent model, many previous studies had observed differences in sleep-wake behavior after shock exposure, but the level of anxiety was merely assumed or absent.

MacLean's study exposed naïve rats** to one of three paradigms: escapable shock, inescapable shock or fear conditioning. Immediately after experimental manipulation, individual level of anxiety was assessed using the elevated-plus maze apparatus, and polygraphic signs of sleep-wake behavior were recorded for 6 hours.

By measuring individual anxiety level before recording sleep, MacLean was able to make comparisons between sleep architecture and level of anxiety. In doing so, MacLean intended to establish a direct link between variation in sleep architecture and heightened anxiety in the rodent model.

"These changes could elucidate sleep-wake behavior associated with the subjective complaint of disrupted sleep, thus creating the potential for new diagnostic and assessment criteria for anxiety disorders," said MacLean. "This information is relevant given the recent influx of psychological disorders in Iraq war veterans, particularly generalized anxiety and post-traumatic stress disorder."

The amount of sleep a person gets affects his or her physical health, emotional wellbeing, mental abilities, productivity and performance. Recent studies associate lack of sleep with serious health problems such as an increased risk of depression, obesity, cardiovascular disease and diabetes.

* *The structure and pattern of sleep*
** *Not previously subjected to experimentation*

Source: American Academy of Sleep Medicine. (2007, June 13). Reduced sleep quality can aggravate pre-existing psychological conditions. https://aasm.org/reduced-sleep-quality-can-aggravate-pre-existing-psychological-conditions

Questions
1. Write a possible hypothesis for this study.
2. Identify the independent variable(s).
3. Identify the dependent variable(s).
4. How did MacLean assess the levels of anxiety displayed by the rats?
5. MacLean's results are not cited in this abstract. Given that disturbed sleep is commonly associated with anxiety disorders, what do you predict would be the results of this experiment?
6. What model was used in this experiment? Explain why this model was used.

Psychological protective factors that maintain mental wellbeing

As discussed in Chapter 8, the characteristics of a mentally healthy person include being able to realise their own potential, cope with the normal stresses of life, work productively, and be able to make a contribution to their community. However, we cannot function at optimal levels at all times, as our behavioural and emotional adjustment can be influenced by the world around us. At times, we need to draw upon a range of strategies to help us maintain our mental wellbeing. Cognitive behavioural strategies and mindfulness meditation are psychological protective factors that contribute to mental wellbeing.

Cognitive behavioural strategies

Cognitive behavioural strategies are structured psychological techniques that can be applied to help people recognise how negative or unproductive patterns of thinking (cognition) and behaviour affect their emotions.

Weblink
The science of mindfulness

10.1.4 ANALYSIS OF A RESEARCH INVESTIGATION

These collective strategies are commonly known as cognitive behavioural therapy (CBT) (Chapter 9) and are used by healthcare professionals in treating a person's physical and mental health and promoting resilience.

In cognitive behavioural therapy, a person works with a professional to evaluate and change unhelpful patterns of thinking that cause high levels of anxiety and lead to negative behaviours. It is based on the premise that the way a person thinks about something determines how they feel about it and respond to it. Therefore, if they change the way they think about it, they can then change their behaviour. Cognitive behavioural strategies aim to help a patient recognise these unhelpful patterns of thinking and use techniques to replace these patterns and improve their ability to cope (Figure 10.6).

Cognitive behavioural strategies used by health professionals will be tailored to a patient's individual needs. These strategies may include:

» educating patients about the body's natural reactions to threatening objects and situations
» helping patients recognise the difference between unhelpful and productive thoughts
» identifying situations that are often avoided and confronting these through systematic desensitisation (Chapter 9)
» teaching relaxation and breathing techniques to manage stress and anxiety
» helping patients establish effective daily routines that promote engaging in enjoyable activities and exercise.

One of the goals of cognitive behavioural strategies is to help patients develop or strengthen their resilience so that they have an increased capacity to cope with adversity when they encounter it.

Figure 10.6 Cognitive behavioural strategies used by health professionals will be tailored to a patient's individual needs.

Mindfulness meditation

Meditation has been a spiritual and healing practice in some parts of the world for more than 5000 years. It has also become an increasingly common practice in Western cultures within the last 50 years. In the recent past, meditation and mindfulness-based approaches have been used for increasing calmness and physical relaxation, improving psychological balance, coping with illness, and enhancing overall health and wellbeing. **Mindfulness** is at the heart of Buddhist practices. The Sanskrit word for mindfulness, *smriti*, means 'that which is remembered'. Therefore, mindfulness can be defined as 'remembering to come back to the present moment' (Hanh, 1998).

Mindfulness meditation is a meditation practice in which a person focuses attention on their breathing, with thoughts, feelings and sensations being experienced freely as they arise and without judgement. It is intended to enable people to become highly attentive to sensory information and to focus on each moment as it occurs.

Mindfulness meditation has two main parts: attention and acceptance. Attention is about tuning into your experiences to focus on what's happening in the present moment. Acceptance involves observing those feelings and sensations without judgement. Instead of responding or reacting to those thoughts or feelings, you aim to note them and let them go.

Mindfulness meditation has been shown to have beneficial effects on health, wellbeing and a number of clinical disorders, including anxiety, depression, addiction, stress-related symptoms and chronic pain (Tang & Tang, 2015).

Studies using magnetic resonance imaging to monitor the brain during and after sessions of mindfulness meditation have highlighted changes in specific areas of the brain. Results indicate a decrease in the activation of connections in the amygdala, which is responsible for the activation of the stress response. This may benefit sufferers of anxiety disorders. The prefrontal cortex, associated with higher order brain functions such as awareness, concentration and decision-making, showed an increase in activation during and after mindfulness meditation (www.scientificamerican.com).

Researchers believe the benefits of mindfulness are related to its ability to dial down the body's response to stress. Chronic stress can impair the body's immune system and make many other health problems worse. By lowering the stress response, mindfulness may have downstream effects throughout the body (www.apa.org).

Spending too much time planning, problem-solving, daydreaming or thinking negative or random thoughts can be draining. It can also make you more likely to experience stress, anxiety and symptoms of depression. Practising mindfulness exercises can help you direct your attention away from this kind of thinking and engage with the world around you.

There are many simple ways to practise mindfulness. Some examples include:

- » Pay attention. It's hard to slow down and notice things in a busy world. Try to take the time to experience your environment with all your senses – touch, sound, sight, smell and taste. For example, when you eat a favourite food, take the time to smell, taste and truly enjoy it.
- » Live in the moment. Try to intentionally bring an open, accepting and discerning attention to everything you do. Find joy in simple pleasures.
- » Accept yourself. Treat yourself the way you would treat a good friend.
- » Focus on your breathing. When you have negative thoughts, try to sit down, take a deep breath and close your eyes. Focus on your breath as it moves in and out of your body. Sitting and breathing for even just a minute can help.

You can also try more structured mindfulness exercises (see weblinks), such as:

- » Body scan meditation. Lie on your back with your legs extended and arms at your sides, palms facing up. Focus your attention slowly and deliberately on each part of your body, in order, from toe to head or head to toe. Be aware of any sensations, emotions or thoughts associated with each part of your body.
- » Sitting meditation. Sit comfortably with your back straight, feet flat on the floor and hands in your lap. Breathing through your nose, focus on your breath moving in and out of your body. If physical sensations or thoughts interrupt your meditation, note the experience and then return your focus to your breath.
- » Walking meditation. Find a quiet place 3 to 6 metres in length and begin to walk slowly. Focus on the experience of walking, being aware of the sensations of standing and the subtle movements that keep your balance. When you reach the end of your path, turn and continue walking, maintaining awareness of your sensations.

(Source: Mayo Clinic, 2022)

Weblink
Everyday mindfulness explained

Weblink
Smiling Mind

ANALYSING RESEARCH 10.2

Focusing awareness on the present moment can enhance academic performance and lower stress levels

by Anne Trafton, MIT News Office. August 26, 2019

Figure 10.7 An MIT study suggests that mindfulness can improve mental health and academic performance in middle school students.

Two new studies from MIT suggest that mindfulness – the practice of focusing one's awareness on the present moment – can enhance academic performance and mental health in middle schoolers. The researchers found that more mindfulness correlates with better academic performance, fewer suspensions from school and less stress.

"By definition, mindfulness is the ability to focus attention on the present moment, as opposed to being distracted by external things or internal thoughts. If you're focused on the teacher in front of you, or the homework in front of you, that should be good for learning," says John Gabrieli, the Grover M. Hermann Professor in Health Sciences and Technology, a professor of brain and cognitive sciences, and a member of MIT's McGovern Institute for Brain Research.

The researchers also showed, for the first time, that mindfulness training can alter brain activity in students. Sixth-graders who received mindfulness training not only reported feeling less stressed, but their brain scans revealed reduced activation of the amygdala, a brain region that processes fear and other emotions, when they viewed images of fearful faces.

Together, the findings suggest that offering mindfulness training in schools could benefit many students, says Gabrieli, who is the senior author of both studies.

"We think there is a reasonable possibility that mindfulness training would be beneficial for children as part of the daily curriculum in their classroom," he says. "What's also appealing about mindfulness is that there are pretty well-established ways of teaching it."

In the moment

Both studies were performed at charter schools in Boston. In one of the papers, which appears today in the journal *Behavioral Neuroscience*, the MIT team studied about 100 sixth-graders. Half of the students received mindfulness training every day for eight weeks, while the other half took a coding class. The mindfulness curriculum, created by the nonprofit program 'Calmer Choice', was designed to encourage students to pay attention to their breathing, and to focus on the present moment rather than thoughts of the past or the future.

Students who received the mindfulness training reported that their stress levels went down after the training, while the students in the control group did not. Students in the mindfulness training group also reported fewer negative feelings, such as sadness or anger, after the training.

About 40 of the students also participated in brain imaging studies before and after the training. The researchers measured activity in the amygdala as the students looked at pictures of faces expressing different emotions.

At the beginning of the study, before any training, students who reported higher stress levels showed more amygdala activity when they saw fearful faces. This is consistent with previous research showing that the amygdala can be overactive in people who experience more stress, leading them to have stronger negative reactions to adverse events.

"There's a lot of evidence that an overly strong amygdala response to negative things is associated with high stress in early childhood and risk for depression," Gabrieli says.

After the mindfulness training, students showed a smaller amygdala response when they saw the fearful faces, consistent with their reports that they felt less stressed. This suggests that mindfulness training could potentially help prevent or mitigate mood disorders linked with higher stress levels, the researchers say.

Richard Davidson, a professor of psychology and psychiatry at the University of Wisconsin, says that the findings suggest there could be great benefit to implementing mindfulness training in middle schools.

"This is really one of the very first rigorous studies with children of that age to demonstrate behavioral and neural benefits of a simple mindfulness training," says Davidson, who was not involved in the study.

Source: https://news.mit.edu/2019/mindfulness-mental-health-benefits-students-0826

Questions

1. What was the aim of this research?
2. Why was a control group used in the research?
3. State the results of the experiments.
4. According to the article, what is the link between stress levels and the amygdala?
5. Identify two ethical guidelines that participants must be made aware of and are able to exercise in this study.

INVESTIGATION 10.1 THE INFLUENCE OF MINDFULNESS MEDITATION ON WELLBEING

Scientific investigation methodology

Controlled experiment

Aim

To conduct an investigation into the influence of mindfulness meditation on heart rate and perceived level of stress.

Logbook template
Controlled experiment

Introduction

This investigation gives you the opportunity to design and conduct your own investigation relating to mental processes and psychological functioning. Remember that your design must include the generation of primary quantitative data.

Data calculator

Make sure your investigation has an aim, method, results, discussion and conclusion while complying with safety and ethical guidelines. Remember to use your logbook throughout this investigation. Present your findings as a scientific poster. The poster may be produced electronically or in hard-copy format and should not exceed 600 words. Please see the VCAA Psychology Study Design pages 15 and 16 for further specifications of the poster format.

Pre-activity preparation

Conduct a literature review of at least three articles on the influence of mindfulness meditation on heart rate and perceived level of stress. Record your literature review in your logbook. Don't forget to record the full references that you use. From your literature review, decide what aspect of the topic you want to investigate. Develop an aim, research question and hypothesis for your investigation.

Method

Work out what you need to do in order to answer your research question and test your hypothesis. This will include:
- Population and sample – how big does your sample need to be?
- Variables – dependent, independent, controlled. Are there any extraneous variables you have not accounted for?
- What quantitative data are you going to record and how are you going to record it?
- Are there any ethical or safety issues you have to take into account?

Results

Collate the results for all participants and process the quantitative data obtained, using appropriate mathematical calculations and units (for example, descriptive statistics such as a mean). Organise, present and interpret data using an appropriate table and/or graph.

Discussion

1. State whether the hypothesis was supported or refuted.
2. Describe the influence of the mindfulness meditation activity on heart rate and perceived level of stress.
3. Discuss any implications of the results.
4. State if it is possible to generalise any of the results to the wider population.
5. Discuss any confounding variables and how they have affected the data.
6. Evaluate the investigation methods and possible sources of error or uncertainty.
7. Suggest future improvements to address any confounding variables if the investigation was to be repeated.

Conclusion

Write a conclusion to respond to the research question.

Social protective factors that maintain mental wellbeing

Social protective factors involve our interactions with other people, groups, society and culture that influence how we think and behave. The support from family, friends and community is critical in building resilience and enhancing the ability of individuals to cope with difficulties.

Support from family, friends and community

Social support is the assistance and comfort we receive from people in our social network when we are facing a stressful or challenging situation. This could be from family members and friends through to support groups and social institutions. One reason this support can be beneficial is that support from family, friends and community groups serves as a buffer to cushion the impact of stressful events. Talking about problems and expressing tensions can be extremely helpful. Research has shown that isolation can increase stress-related bodily changes, including hormones and anxiety symptoms. Seeking social support to avoid isolation is an important method of coping with stress. During the COVID pandemic many people experienced a great deal of disconnection from their family, friends and regular routines.

It may be difficult for family and friends to understand why specific objects and situations cause extreme levels of anxiety. However, these close relationships can help sufferers develop their inner resilience through engagement in energising daily activities with others (Figure 10.8). Just as we get nourishment from food and sleep that provides energy, we can gain nourishment from our relationships with others. Support can be considered energising when it provides the enthusiasm and determination to complete a task, try something new, or reach a goal.

This **energising social support** is also found through participating in physical activities; for example, playing a sport, going for an organised group walk, or joining a gym. These activities provide both physical exercise and opportunities to strengthen support networks.

A network of good friends is also a key to feeling supported. Friends often support us with unconditional positive regard and this enhances our trust that we can rely on them in times of crisis, thereby making the relaxation response accessible. It is important that family and friends are authentic in their support. Authenticity is a personal trait that reflects the extent to which a person is genuine and true to themselves and in their interactions with others. It involves self-understanding in terms of recognising and owning our own thoughts and emotional responses, including personal weaknesses, strengths, goals and values. Authentic people behave with truth and integrity towards others and provide **authentic social support** when they truly listen to and connect with the experiences of another person.

Research has identified four distinct types of social support: emotional support (physical comfort, such as a hug, listening to problems and empathising), esteem support (expressions of confidence or encouragement), informational support (advice-giving, gathering and sharing information) and tangible support (taking on responsibilities for someone else, such as providing meals or financial support) (Verywell Mind, 2020).

A sense of belonging and self-confidence is often most effectively achieved through connection with a community of people with common conditions and experiences. Partnerships between professionals and consumers, and support and clinical services, will ensure that the experience and knowledge of people with mental health issues will be valued and heeded, and contribute to the development of appropriate and effective services (Anxiety Recovery Centre Victoria). Community support organisations such as Beyond Blue, the Black Dog Institute, Headspace and SANE Australia offer support, training and education for the families of individuals with mental health problems and illnesses.

Figure 10.8 Families are often the main support for people affected by mental illness.

CASE STUDY 10.1

How online mindfulness training can help students thrive during the pandemic

By Adam Austen Kay, Lecturer, School of Business, The University of Queensland
27 August 2021

Figure 10.9 Online mindfulness training has been shown to have a distinct benefit. It improved psychological wellbeing by helping students cultivate authenticity.

COVID-19 is reasserting itself, with the Delta variant posing a serious threat to young people. The pandemic has made physical distancing an inescapable new reality of post-secondary education as universities continue to deliver courses online. Our research shows online mindfulness meditation can also be effective when delivered online, bringing benefits previously unknown to science.

One year into the pandemic, students are showing signs of wear. The 2020 Student Experience Survey shows post-secondary students' engagement with learning has dropped. Responses indicated they were 4% more likely to drop out due to stress or health concerns.

Universities thus face a pressing need to help their students cope. Fortunately, a promising new resource is available: online mindfulness meditation.

Mindfulness is the process of focusing attention and awareness on present moment experience with an open, curious and accepting attitude. It's usually taught in person. However, given the advantages of online delivery in a pandemic, the popularity of online mindfulness meditation has boomed.

Online mindfulness meditation: What did the study find?

In recent decades, a mountain of research has shown mindfulness is broadly effective for relieving symptoms of psychological suffering like anxiety, depression and stress. However, our study, published in the journal *Academy of Management Learning & Education*, shows online mindfulness training can do more than alleviate such symptoms. It can help students flourish.

We examined the effects of online mindfulness training on the psychological well-being of 227 graduate students. Half of them took part in a free, evidence-based online program. It involved 30 minutes a day of mindfulness meditation, five days a week, for eight weeks.

As a placebo control, the other half took part in an equal amount of training also known to promote health and well-being: physical exercise.

The psychological well-being of students in both groups improved. These gains were indicated by criteria like self-acceptance, personal growth, meaning and purpose in life, and positive relationships with others.

However, online mindfulness training had a distinct benefit. It improved psychological wellbeing by helping students cultivate authenticity.

What is authenticity?

Authenticity is one of the most powerful indicators of psychological health. Authentic individuals are self-aware, meaning they are in touch with their thoughts and emotions. They act in accordance with their values and beliefs. This study revealed that online mindfulness training helped students develop authenticity.

Some students benefit more than others

While these findings are promising, the benefits were not the same for all students. Online mindfulness training improved the authenticity of nearly 60% of students but not others (although they still gained other well-being benefits).

What was the difference? The answer lies in personality.

Every educator knows personality has important implications for student performance. Similarly, every psychologist knows the single most important dimension of personality for student performance is 'conscientiousness'. Highly conscientious students perform better because they show self-discipline, attention to detail, reliability, thoughtfulness and persistent hard work.

We reasoned that conscientiousness would be even more important in an online learning environment where students don't have access to the dedicated learning space and shelter from distractions that classrooms can provide.

Results supported our reasoning: only highly conscientious students benefited from online mindfulness training in terms of authenticity and the psychological well-being that flows from it. Even though students who were low in conscientiousness undertook a similar amount of training, they did not develop in authenticity. In other words, conscientiousness appears to have improved the quality of online mindfulness training.

It's not a cure-all for student well-being

This study is the first to show that, despite the advantages of online mindfulness training for helping students cultivate authenticity and thereby flourish in a remote learning environment, it's not a one-size-fits-all solution.

As the impacts of the pandemic stretch universities financially and educators struggle to respond with innovative content for engaging and effective online delivery, this research offers timely evidence for incorporating online mindfulness training into higher education.

However, these findings also serve to caution educators not to view online mindfulness training as a panacea for student well-being. Instead, it should be seen as one part – albeit a promising one – of a broader strategy for helping students cope with the psychological consequences of physically distanced education in a time of COVID-19.

Source: Adam Austen Kay, University of Queensland, 'How onliine mindfulness training can help students thrive during the pandemic' August 27, 2021. Originally published on The Conversation. CC BY 4.0 Licence https://creativecommons.org/licenses/by-nd/4.0/

Questions

1. What were the findings of the 2020 Student Experience Survey?
2. According to the article, what is mindfulness and what are some of its benefits?
3. What were the effects of online mindfulness training on the psychological wellbeing of students?
4. Suggest how mindfulness meditation could be implemented to improve student mental wellbeing in a secondary school.
5. Do you think you would like to try a mindfulness meditation session? Why or why not?

ACTIVITY 10.1 PROTECTIVE FACTORS

Try it

1. Protective factors are assets to help an individual build their mental wellbeing. Make your own plan to access biological, psychological and social factors to help you maintain your mental wellbeing.

   Biological: Write your own menu for a day that will support your mental wellbeing.
   Breakfast:
   Lunch:
   Dinner:
   Write down the time you need to go to bed tonight to give you adequate sleep: _____

   Psychological: Find a mindfulness meditation app to use like Smiling Mind or Calm. You can get some good advice from the 'What is mindfulness meditation?' video (see weblink).

   Social: Write down the name of someone you can rely on.

Weblink
What is mindfulness meditation?

Apply it

Dylan feels like his lifestyle is taking a toll on his mental wellbeing. He is working long hours at his new job. He often skips breakfast to get into the office early and most days he does not stop to eat lunch, so he grabs a muesli bar and a coffee to keep him going. Because he is working such long hours, he generally comes home exhausted, but he is so stressed that he can't unwind to get to sleep before midnight. Dylan tries to catch up with his friends over the weekend, but he finds he is just too tired to go out in the evenings when his friends like to catch up.

2 Using the biopsychosocial approach, develop a plan to help Dylan identify changes he could make to promote his mental wellbeing.

Exam ready

Mitchell has just finished school and is about to start university. He has just got his driver's licence. He works long hours during the week, often until very late at night, at his two part-time jobs. On the weekend he catches up with friends for a game of golf.

3 What sort of protective factor has Mitchell employed to support his mental wellbeing?

A A biological factor because he works long hours.

B A psychological factor as his time is being taken up with two part-time jobs.

C A psychological factor as he is about to start university.

D A social factor because he spends time with friends who share his interests.

KEY CONCEPTS 10.1

» Protective factors are any behavioural, biological, psychological or environmental characteristic that decreases the likelihood of a person developing a particular mental health problem or disorder.
» Brain functioning is very sensitive to what we eat and drink, therefore diet can play a vital role in maintaining mental wellbeing.
» Nutritional intake and hydration contribute to resilience by protecting against diet-related diseases that affect physical and cognitive functions and by reducing vulnerability to stress and depression.
» Adequate sleep is the necessary amount for individuals to function effectively during the daytime and cope with normal daily stress.

» Cognitive behavioural strategies are structured psychological treatments that recognise that a person's way of thinking (cognition) and acting (behaviour) affect the way they feel.
» Mindfulness meditation is a meditation practice in which a person focuses attention on their breathing, with thoughts, feelings and sensations being experienced freely as they arise and without judgement.
» Social support is the authentic and energising assistance and comfort we receive from people in our social network when we are facing a stressful or challenging situation, from family members and friends through to support groups and social institutions.

Concept questions 10.1

Remembering

1 Using an example of each to demonstrate your understanding, describe how risk and protective factors may influence mental wellbeing. **e**

2 What is meant by authentic social support? **r**

3 What is meant by energising social support? **r**

Understanding

4 Explain how nutritional intake may affect our mental wellbeing both directly and indirectly. **e**

EXAM TIP
The number of lines provided after each question, together with the number of marks allocated, will indicate the appropriate length of the response. If you need more space, then clearly indicate where you have written the rest of your response. Do not go outside the border lines of the page.

5 Using an example, outline what is meant by a cognitive behavioural strategy.

Applying

6 Create a brochure that is designed to inform secondary students about the importance of nutritional intake, hydration and sleep in maintaining mental wellbeing.

7 Using an example not presented in the text, explain how stress can motivate us to achieve our goals.

HOT Challenge

8 Research a case study of an individual who has used mindfulness meditation to benefit their mental wellbeing.

10.2 Cultural determinants of social and emotional wellbeing

10.2.1 THE MAINTENANCE OF WELLBEING IN ABORIGINAL AND TORRES STRAIT ISLANDER PEOPLES

As we learned in Chapter 8, mental wellbeing is only one part of Aboriginal and Torres Strait Islander peoples' holistic and multidimensional framework of social and emotional wellbeing (SEWB). In the SEWB framework the maintenance of mental wellbeing is supported through the interweaving of connections between the seven domains of body and behaviours, mind and emotions, family and kinship, community, culture, Country and spirituality and ancestors.

We also saw that the capacity of people and communities to build and strengthen these connections is influenced by a broader set of contextual factors, including cultural, social, historical and political determinants. These contextual factors are called **determinants** of SEWB because they are over-arching factors that can strongly influence (i.e. contribute to or determine) how people and communities can restore and strengthen connections across the seven domains.

Culture can be defined broadly as a body of collectively shared values, principles, practices, customs and traditions (Hovane et al., 2014). **Cultural determinants** can be viewed, in part, as **protective factors** that are integral to Aboriginal and Torres Strait Islander peoples' right and capacity to learn, practise and pass on their traditional and contemporary ways of knowing, being and doing. Cultural determinants of SEWB support Aboriginal and Torres Strait Islander peoples to maintain a strong and secure sense of cultural identity and cultural values, and to participate in cultural practices that allow them to exercise their cultural rights and responsibilities, and their fundamental human right to self-determination. Two important cultural determinants of SEWB are **cultural continuity** and **self-determination** (Dudgeon et al., 2020; Liddle et al., 2022).

Cultural continuity

Cultural continuity can be defined as the process that enables cultural knowledge, values and practices to be transmitted over generations, and that integrates and connects individuals and communities with the past and future of their culture.

The continuity of Aboriginal and Torres Strait Islander cultures has been deeply disrupted by the ongoing effects of colonisation. Restoring cultural continuity is crucial because, without it, people and communities cannot build and strengthen connections across the domains of family and kinship, culture, community, Country, spirituality and ancestors.

Cultural continuity is not only about reviving and preserving traditional knowledges. It is also about the continuous adaptation of a culture to ensure its ongoing survival. Living, thriving cultures are continuously transforming.

When people feel connected to the past and engaged in the future of a continuous culture, their sense of personal identity and continuity is strengthened (Dudgeon et al., 2020; Liddle et al., 2022; Zubrick et al., 2014). In this way, the intergenerational transmission of culture provides powerful protection against the impact of intergenerational trauma and racism (Macedo et al., 2019).

The revival and adaptation of possum-skin cloak making as a contemporary practice for healing (described in Chapter 8) is one example of how cultural continuity can improve current SEWB and help build resilience and strengthen identity for the future. Other examples include the revival of Indigenous Australian languages, art, songs and stories, as well as practising traditional methods of caring for Country (Sivak et al., 2019). The passing down of these knowledges from Elders strengthens connections across all seven of the domains that support and maintain SEWB.

Cultural continuity is achieved through Aboriginal and Torres Strait Islander peoples and communities leading initiatives that enable intergenerational learning and cultural practices to be shared. It is about Aboriginal and Torres Strait Islander peoples being able to connect with their traditional lands and seas, and caring for Country. Physical and mental health is strengthened through both the revival and renewal of ancient traditional healing practices and the creation of contemporary healing practices. This is particularly important because experiences and expressions of ill-health are influenced by culture in ways that are not recognised in Western diagnostic and treatment methods (Gee, 2016; Westerman, 2021).

An example of a powerful community-led initiative to restore cultural continuity is the annual Tanderrum gathering of the five clans of the Kulin Nation of south-central Victoria. The event is held each October in Federation Square to open the Melbourne International Arts Festival. Explore Case study 10.2 to learn more.

10.2.2 ABORIGINAL AND TORRES STRAIT ISLANDER PEOPLES' CULTURAL DETERMINANTS OF HEALTH

CASE STUDY 10.2

Tanderrum

Tanderrum is a word from the Wurundjeri Woi-wurrung language. It is the name of the traditional ceremony used by Wurundjeri people for welcoming visitors to their Country.

The Wurundjeri are one of the five clans of the Kulin Nation of Victoria, which also includes the Boon Wurrung, Wathaurung, Taungurung and Dja Dja Wurrung clans. Figure 10.10 shows the geographical locations of the five Kulin tribal groups. The Tanderrum involves months of preparation by the different tribal groups and is a powerful way for people to connect with culture.

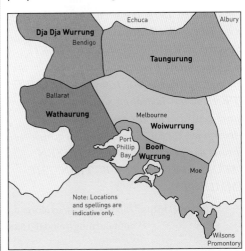

Figure 10.10 The five clans of the Kulin Nation of south-central Victoria

Figure 10.11 The annual Tanderrum ceremony of Kulin Nation clans in Naarm (Melbourne).

Question

1. Watch the 9-minute video to learn about Tanderrum. Analyse the video and explain why Tanderrum is an expression of cultural continuity. Support your explanation with at least two examples of how the event strengthens cultural identity, belonging and social and emotional wellbeing.

Weblink
Tanderrum

Self-determination

Continuity of culture is closely connected to self-determination. **Self-determination** is defined by the Australian Human Rights Commission (AHRC) as '… the fundamental right of people to shape their own lives, so that they determine what it means to live well according to their own values and beliefs' (AHRC, 2022).

The right to self-determination is based on the recognition that Aboriginal and Torres Strait Islander peoples are Australia's First Peoples. It recognises that the loss of self-determination is at the heart of the current disadvantage experienced by many Aboriginal and Torres Strait Islander peoples. Cultural continuity cannot be supported and nourished without Aboriginal and Torres Strait Islander peoples and communities being able to exercise self-determination.

Self-determination has been fought for by Aboriginal and Torres Strait Islander peoples continuously since colonisation. It is evident in the resistance to the invasion of lands during the Frontier Wars, through to the establishment of the Indigenous-led political organisations that have called for equal rights for Aboriginal and Torres Strait Islander peoples. This proud history of resistance and activism provides a source of strength that builds pride in a strong and resilient cultural identity.

For example, the establishment of the Australian Aborigines League and the Aborigines Progressive Association in the early 20th century led to the first 'Day of Mourning' on 26 January 1938 (the 150th anniversary of the landing of the First Fleet in Botany Bay) (Figure 10.11). This paved the way for the successful campaign for constitutional change to recognise First Nations peoples as part of the population in 1967, and for the Mabo case against the Commonwealth that overturned the myth of 'terra nullius' (land belonging to no one) in 1992. Self-determination is at the core of the powerful Uluru Statement of the Heart: 'In 1967 we were counted, in 2017 we seek to be heard' (First Nations National Constitutional Convention, 2017).

Mitchell Library, from the collections of the State Library of New South Wales.

Figure 10.12 The first 'Day of Mourning' protest, 26 January 1938

On 26 January 1938, on the 150th anniversary of the landing of the First Fleet, a group of Aboriginal men and women gathered at Australia Hall in Sydney. They met to move the following resolution:

> *WE, representing THE ABORIGINES OF AUSTRALIA, assembled in conference at the Australian Hall, Sydney, on the 26th day of January, 1938, this being the 150th Anniversary of the Whiteman's seizure of our country, HEREBY MAKE PROTEST against the callous treatment of our people by the white men during the past 150 years, AND WE APPEAL to the Australian nation of today to make new laws for the education and care of Aborigines, we ask for a new policy which will raise our people TO FULL CITIZEN STATUS and EQUALITY WITHIN THE COMMUNITY.*
>
> (Source: https://aiatsis.gov.au/explore/day-ofmourning#modal-15389)

Self-determination is the driving force behind the development of Aboriginal and Torres Strait Islander community-controlled health organisations. These grass-roots organisations began in the 1970s and were developed by Aboriginal and Torres Strait Islander peoples to provide Aboriginal and Torres Strait Islander-designed services that better met the health needs of communities. Today there are over 140 community-controlled health services. The National Aboriginal Community Controlled Health Organisation (NACCHO) represents these organisations at a national level, providing policies and advice to government.

Within psychology, the Australian Indigenous Psychologists Association (AIPA) has led reforms to the training of psychologists to ensure that the profession is culturally responsive, and to increase the representation of Indigenous psychologists.

In 2016, Australia's first Indigenous psychologist, Professor Pat Dudgeon (Figure 10.13), led the Australian Psychological Society (APS) in a formal apology to Aboriginal and Torres Strait Islander peoples for its role in the Stolen Generations, perpetuating racist stereotypes, and for denying the value of traditional knowledge and practice for maintaining social and emotional wellbeing.

Figure 10.13 Professor Pat Dudgeon, a Bardi woman from the Kimberley, Australia's first Indigenous psychologist and a national leader in developing policies and practices to improve the social and emotional wellbeing of Indigenous Australians.

Source: https://cbpatsisp.com.au/the-manual-of-resources/about-the-manual-of-resources/

10.2.3 MAINTENANCE OF MENTAL WELLBEING CONCEPTS STUDY CARDS

KEY CONCEPTS 10.2

- Cultural determinants of social and emotional wellbeing are protective factors integral to Aboriginal and Torres Strait Islander peoples' right and capacity to learn, practise and pass on their traditional and contemporary ways of knowing, being and doing. They include factors and processes that support Aboriginal and Torres Strait Islander peoples to maintain a strong and secure sense of cultural identity and cultural values, and to participate in cultural practices that allow them to exercise their cultural rights and responsibilities, and their fundamental human right to self-determination.
- Two key cultural determinants of social and emotional wellbeing are cultural continuity and self-determination.
- Cultural continuity refers to the transmission of Aboriginal and Torres Strait Islander peoples' ways of knowing, being and doing over generations, connecting the present with the past and future. When people and communities connect with a continuous culture, their own sense of personal continuity and cultural identity is strengthened, which protects against the effects of intergenerational trauma and racism.
- Cultural continuity can be supported through revival, renewal and preservation of traditional and contemporary cultural knowledge and practices.
- Self-determination is inherent within cultural continuity. It involves recognition of the rights of Aboriginal and Torres Strait Islander peoples to practise their culture and shape their future based on their unique cultural worldview. This includes recognising Aboriginal and Torres Strait Islander sovereignty, and issues of land rights and control of resources.

> » Self-determination has been central to the survival of Aboriginal and Torres Strait Islander peoples' cultures. It is expressed through a proud legacy of resistance, activism and grassroots organisation and provides a source of strength and resilience for maintaining social and emotional wellbeing.
>
> » Self-determination in the context of psychological practice and research is evident in the work of Aboriginal and Torres Strait Islander psychologists to transform the discipline to be culturally responsive, recognising its biases and past contributions to the disempowerment of Aboriginal and Torres Strait Islander peoples.

EXAM TIP
If you are stuck on a question, place a star next to it, move on to the next question and return to these questions once all other questions are complete. Don't leave the exam room early; proofread, add detail and check your responses.

Weblink
The APS Apology to Aboriginal and Torres Strait Islander peoples

Concept questions 10.2

Remembering
1 What are two of the most powerful cultural determinants of Aboriginal and Torres Strait Islander peoples' social and emotional wellbeing? **r**

Understanding
2 Explain how cultural continuity is protective against the impacts of intergenerational trauma and racism? **e**

3 Explain the relationship between self-determination and cultural continuity. **e**

Applying
4 Read the APS apology to Aboriginal and Torres Strait Islander peoples (see weblink) and identify statements that recognise cultural continuity and self-determination as cultural determinants of Aboriginal and Torres Strait Islander peoples' social and emotional wellbeing. **c**

HOT Challenge
5 Practise drawing the SEWB framework from Chapter 8 for yourself. Then:
   a annotate your drawing to include cultural continuity and self-determination as cultural determinants of SEWB
   b give a specific example of an experience or expression of each (from the text or from your own knowledge or research) and explain how each example strengthens one or more of the domains of SEWB. **d**

10 Chapter summary

KEY CONCEPTS 10.1

» Protective factors are any behavioural, biological, psychological or environmental characteristic that decreases the likelihood of a person developing a particular mental health problem or disorder.
» Brain functioning is very sensitive to what we eat and drink, therefore diet can play a vital role in maintaining mental wellbeing.
» Nutritional intake and hydration contribute to resilience by protecting against diet-related diseases that affect physical and cognitive functions and by reducing vulnerability to stress and depression.
» Adequate sleep is the necessary amount for individuals to function effectively during the daytime and cope with normal daily stress.
» Cognitive behavioural strategies are structured psychological treatments that recognise that a person's way of thinking (cognition) and acting (behaviour) affect the way they feel.
» Mindfulness meditation is a meditation practice in which a person focuses attention on their breathing, with thoughts, feelings and sensations being experienced freely as they arise and without judgement.
» Social support is the authentic and energising assistance and comfort we receive from people in our social network when we are facing a stressful or challenging situation, from family members and friends through to support groups and social institutions.

KEY CONCEPTS 10.2

» Cultural determinants of social and emotional wellbeing are protective factors integral to Aboriginal and Torres Strait Islander peoples' right and capacity to learn, practise and pass on their traditional and contemporary ways of knowing, being and doing. They include factors and processes that support Aboriginal and Torres Strait Islander peoples to maintain a strong and secure sense of cultural identity and cultural values, and to participate in cultural practices that allow them to exercise their cultural rights and responsibilities, and fundamental human right to self-determination.
» Two key cultural determinants of social and emotional wellbeing are cultural continuity and self-determination.
» Cultural continuity refers to the transmission of Aboriginal and Torres Strait Islander peoples' ways of knowing, being and doing over generations, connecting the present with the past and future. When people and communities connect with a continuous culture, their own sense of personal continuity and cultural identity is strengthened, which protects against the effects of intergenerational trauma and racism.
» Cultural continuity can be supported through revival, renewal and preservation of traditional and contemporary cultural knowledge and practices.
» Self-determination is inherent within cultural continuity. It involves recognition of the rights of Aboriginal and Torres Strait Islander peoples to practise their culture and shape their future based on their unique cultural worldview This includes recognising Aboriginal and Torres Strait Islander sovereignty, and issues of land rights and control of resources.
» Self-determination has been central to the survival of Aboriginal and Torres Strait Islander peoples' cultures. It is expressed through a proud legacy of resistance, activism and grassroots organisation and provides a source of strength and resilience for maintaining social and emotional wellbeing.
» Self-determination in the context of psychological practice and research is evident in the work of Aboriginal and Torres Strait Islander psychologists to transform the discipline to be culturally responsive, recognising its biases and past contributions to the disempowerment of Aboriginal and Torres Strait Islander peoples.

10 End-of-chapter exam

Section A: Multiple choice

1. Resilience is
   A. an individual's inability to adapt to stressful situations.
   B. an individual's ability to properly adapt to stress and cope with adversity.
   C. the object, entity or event that causes a feeling of stress.
   D. the physiological and psychological responses that a person experiences when confronted with a situation that is threatening or challenging.

2. The influences that decrease the likelihood that an individual will develop mental health problems are known as
   A. hazard factors.
   B. risk factors.
   C. protective factors.
   D. defensive factors.

3. Which one of the following is correct in relation to adequate nutritional intake?
   A. A person's nutritional intake will not cause or prevent the development of a mental illness.
   B. A person's nutritional intake does not contribute to a person's level of resilience.
   C. A person's nutritional intake will prevent the development of a mental illness.
   D. All of the above.

4. The body is made up of _____ per cent water.
   A. 10–25
   B. 25–50
   C. 50–75
   D. 75–90

5. The Australian Dietary Guidelines recommend that women should have about _____ litres and men about _____ litres of water per day to stay adequately hydrated.
   A. 1; 1.6
   B. 2; 2.6
   C. 3; 3.6
   D. 4; 4.6

6. Which one of the following is a psychological factor that decreases the likelihood of a person developing a particular mental health problem or disorder?
   A. adequate nutritional intake
   B. mindfulness meditation
   C. sleep
   D. hydration

7. Which one of the following is an incorrect statement regarding nutritional intake and hydration guidelines?
   A. Drinking plenty of water helps prevent dehydration.
   B. Vegetables and legumes help to stabilise blood sugars and optimise mental as well as physical performance.
   C. Mild dehydration can affect mood, causing irritability and restlessness.
   D. Caffeine calms the central nervous system and brings about a sense of calm.

8. The National Sleep Foundation recommends that teenagers should get _____ hours of sleep on school nights.
   A. 7–9
   B. 8–9
   C. 8–10
   D. 9–10

9. Which of the following may be a psychological repercussion of inadequate sleep?
   A. changes in emotions
   B. difficulty concentrating
   C. poor behaviour control
   D. all of the above

10. Which of the following is a cognitive behavioural strategy used by health professionals in maintaining a person's mental health?
    A. identifying situations that are often avoided and confronting these through systematic desensitisation.
    B. teaching relaxation and breathing techniques to manage stress and anxiety.
    C. helping patients establish effective daily routines that promote engaging and enjoyable activities and exercise.
    D. all of the above

11. Which type of therapy aims to help a patient recognise unhelpful patterns of thinking and use techniques to replace these patterns and improve their ability to cope?
    A cognitive behavioural strategies
    B systematic desensitisation
    C psychotherapy
    D rehabilitation

12. Mindfulness meditation is a meditation practice in which a person focuses attention on their breathing, with thoughts, feelings and sensations being experienced freely as they arise and without judgement. The two main parts of mindfulness meditation are
    A attention and acceptance.
    B focus and breath.
    C sensation and perception.
    D calmness and stillness.

13. Which one of the following is an example of a social support protective factor in maintaining mental wellbeing?
    A participating in a mindfulness meditation session
    B drinking the recommended amount of water per day
    C playing football at a local park with friends
    D recognising the difference between unhelpful and productive thoughts

14. Which of the following is a benefit of a network of good friends in maintaining mental wellbeing?
    A Friends can increase a person's engagement in energising daily activities.
    B Friends often support us with unconditional positive regard.
    C We can rely on friends in times of crisis, thereby making the relaxation response accessible.
    D All of the above.

15. Brain functioning is very sensitive to what we eat and drink, therefore diet can play a vital role in maintaining mental wellbeing. Diet is an example of a _____ protective factor.
    A psychological
    B social
    C biological
    D environmental

16. Which of the following statements is true of the cultural determinants of Aboriginal and Torres Strait Islander peoples' social and emotional wellbeing (SEWB)?
    A Cultural determinants are protective factors that support and maintain SEWB.
    B Cultural continuity and self-determination are two key cultural determinants of SEWB.
    C Cultural determinants are contextual factors that affect the strength of connections in the seven domains of SEWB.
    D all of the above

17. Which of the following is an example of cultural continuity?
    A activism to seek the right to access traditional lands
    B the establishment of Indigenous-controlled community health services
    C revival of Indigenous Australian languages to be taught in schools
    D all of the above

18. An Aboriginal-controlled health service in Melbourne establishes a yarning circle for local Indigenous people to attend and share knowledge of traditional healing practices. This initiative is an example of:
    A cultural determination.
    B cultural continuity.
    C self-determination and cultural continuity.
    D social and emotional wellbeing.

19. For which of the following reasons is knowledge of the history of self-determination of Aboriginal and Torres Strait Islander peoples in response to colonisation protective of social and emotional wellbeing?
    A Because this knowledge empowers Indigenous Australians and communities through building a shared sense of pride and resilience.
    B Because this knowledge contributes to a sense of cultural continuity and builds a strong sense of cultural identity.
    C Because these achievements challenge negative stereotypes about Indigenous Australians.
    D all of the above

20. Children listening to Elders tell traditional stories is an example of
    A self-determination.
    B cultural continuity.
    C self-determination and cultural continuity.
    D social and emotional wellbeing.

Section B: Short answer

1. Identify three community support organisations that offer support, training and education in maintaining a person's mental health. [3 marks]

2. How are an adequate diet and enough sleep beneficial to a person's mental health? [3 marks]

3. Describe two ways in which cultural continuity relates to the domain of spirituality and ancestors in the framework of social and emotional wellbeing, and give one reason why this is a protective factor. [3 marks]

Unit 4, Area of Study 2 review

Section A: Multiple choice

Question 1 ©VCAA 2019 Q41 ADAPTED MEDIUM
Which one of the following statements is true of a person with high resilience?
A They do not experience stress.
B They experience low levels of functioning.
C They have the ability to overcome life's stressors.
D They cannot be taught; individuals are born resilient.

Use the following information to answer Questions 2–5.
Sue was an active individual who was highly successful in her job as a lawyer. She was able to juggle work, gym, her hobbies of theatre and reading as well as spend time with friends and family. Recently her mum was diagnosed with Alzheimer's disease and Sue has not taken the news well. Since being told, Sue has called in sick to work every day and hasn't left the house. She hasn't been working out and has been living on fast food.

Question 2 ©VCAA 2019 Q41 ADAPTED MEDIUM
Contrast Sue's levels of functioning before and after her mother's diagnosis.
A Sue's level of functioning remained relatively consistent.
B Sue's level of functioning was high before the news of her mother's health but declined significantly after, due to being unable to function with her normal everyday life.
C Sue's level of functioning was low to begin with, which is why she was unable to deal with any form of stressor.
D Sue's level of functioning was lower before the stressor, as it is teaching her to adapt and overcome life's issues.

Question 3 ©VCAA 2019 Q43 ADAPTED MEDIUM
Which one of the following identifies the internal and external factors contributing to Sue's mental health problem?

	Internal	External
A	Poor diet	Her mother's diagnosis
B	Genetic predisposition to depression	Lack of solutions
C	Emotional state	Workload
D	Low self-esteem	Conflict resolution skills

Question 4 ©VCAA 2019 Q44 ADAPTED MEDIUM
An appropriate emotion-focused coping strategy that Sue could use could be
A exercise to help her cope.
B seeing a psychologist to discuss her concerns and help her with coping strategies.
C changing the subject when her partner begins to discuss the issue.
D applying a coping strategy that she used successfully to deal with a difficult colleague in the past.

Question 5 ©VCAA 2019 Q45 ADAPTED EASY
Sue's friend Lisa encourages her to go to the gym with her to make her feel better. This is an example of
A a biological protective factor.
B a psychological protective factor.
C a social protective factor.
D an economic protective factor.

Question 6 ©VCAA 2017 Q15 ADAPTED EASY
Cheryl has developed a specific phobia of birds and covers her eyes whenever she sees one. With reference to gamma-amino butyric acid (GABA), it is most likely that Cheryl has
A a deficiency of this excitatory neurotransmitter.
B a deficiency of this inhibitory neurotransmitter.
C too much of this excitatory neurotransmitter.
D too much of this inhibitory neurotransmitter.

Question 7
Which one of the following is true of specific phobia?
A It is something everyone experiences when they have a fear of a specific event or object.
B It is intense and short-lived anxiety that is a one-off experience with no known cause.
C It is a mental health problem leading to fear of objects.
D It is a mental disorder in which individuals experience anxiety in response to a particular object or situation.

Question 8 ©VCAA 2021 Q41 ADAPTED

Which one of the following describes both an internal and external influencer the development of specific phobia?

	Internal factor influencing wellbeing	External factor influencing wellbeing
A	Stigma around seeking treatment	GABA dysfunction
B	Catastrophic thinking	Stigma around seeking treatment
C	GABA dysfunction	Catastrophic thinking
D	GABA agonists	Not encouraging avoidance behaviours

Use the following information to answer Questions 9 and 10.

Elaine was experiencing stress in response to her upcoming assessment task due in her Psychology class. She was finding it hard to keep up with her everyday tasks such as part-time work and netball in the lead-up to the assessment.

Question 9 ©VCAA 2021 Q45 ADAPTED MEDIUM

Based on the mental health continuum, which of the following statements is true regarding Elaine's mental health?

A Elaine is mentally healthy as she is not experiencing any distress.

B Elaine is experiencing a mental health problem as the stress and interruption to her daily life is short term in response to the SAC.

C Elaine has developed a mental disorder as she is experiencing impacts on her ability to function in her daily life.

D Elaine is experiencing a mental disorder as she is experiencing short term but intense stress as a result of the assessment task.

Question 10 ©VCAA 2021 Q46 ADAPTED MEDIUM

What is an external factor impacting on Elaine's mental health?

A the pressure to do well in her psychology assessment from family and friends

B the fear of failing the assessment

C catastrophic thinking about the worst possible outcome on the assessment

D Her mother also suffers from anxiety so the stress has a genetic basis.

Use the following information to answer Questions 11 and 12.

For the past 2 years Ava has been developing stronger obsessions with cleanliness. Ava washes her hands four times after touching any surface. This is becoming increasingly difficult at school; she struggles to get her work done as she feels the need to wash her hands. Ava has been avoiding leaving the house as she finds it easier to avoid touching surfaces at home, but she is now starting to fail all of her subjects.

Question 11 ©VCAA 2020 Q43 ADAPTED MEDIUM

Ava might be showing signs of a mental disorder because

A washing her hands a lot indicates a mental disorder.

B both internal and external factors are contributing to her behaviour.

C she is experiencing long-term behaviour that is impeding her ability to function in everyday life.

D her behaviour is uncharacteristic and has had a negative impact on her wellbeing.

Question 12 ©VCAA 2020 Q44 ADAPTED HARD

Which one of the following protective factors may improve Ava's mental wellbeing according to the biopsychosocial framework?

	Biological	Psychological	Social
A	Ensuring Ava has adequate sleep	Establishing a mindfulness routine for when Ava is overwhelmed	Reminding Ava of her support systems such as family
B	Ensuring Ava has an adequate diet	Getting Ava to join a support group	Getting Ava an exercise program
C	Organising genetic testing	Getting Ava to expose herself to dirty surfaces	Ensuring Ava gets enough sleep
D	Organising mindfulness sessions	Challenging Ava's thoughts around the constant need to wash her hands	Getting Ava's friends to visit her regularly

Use the following information to answer Questions 13–15.

Moira was always a smart student in high school, getting straight As and always completing all her work. However, once making it to university Moira found she was struggling significantly with her work. Instead of studying what she didn't know, Moira started constantly going out with friends to parties and ignoring her work. Eventually, Moira hadn't completed enough work and failed her first year of university. This made Moira upset, and she found she was struggling to cope with the failure, leading to her dropping out of university and becoming unemployed.

Question 13 ©VCAA 2017 Q39 ADAPTED MEDIUM
What type of coping is Moira demonstrating?
A avoidance coping
B problem-focused coping
C emotion-focused coping
D approach coping

Question 14 ©VCAA 2017 Q40 ADAPTED MEDIUM
Which of the following is true regarding Moira's situation?
A External factors had little impact on her mental health, it was mainly influenced by internal factors.
B Internal factors had little impact on her mental health, it was mainly influenced by external factors.
C There were only external factors and no internal factors impacting on her mental health.
D Internal and external factors equally impacted on her mental health.

Question 15 ©VCAA 2017 Q41 ADAPTED MEDIUM
Which one of the following would best help Moira better cope with university workloads if she were to return to university?
A ensuring she gets adequate sleep.
B engaging in mindfulness.
C setting up strong support networks with fellow university students.
D venting to family and friends about how hard university is.

Question 16 ©VCAA 2021 Q43 ADAPTED
Long-term potentiation can contribute to the development of a mental disorder because it is a
A biological risk factor that results in decreased stimulation of neural pathways related to the stressor.
B psychological risk factor that encourages the individual to engage in catastrophic thinking.
C social risk factor that leads to the inability to seek help and support.
D biological risk factor resulting in overstimulation of neural pathways related to the stressor.

Use the following information to answer Questions 17–19.

Carmel is working in the garden when suddenly a snake comes out and bites her. Ever since that day Carmel has a major fear of her backyard. Whenever Carmel attempts to go into her backyard she feels anxiety, with intense heart palpitations and sweats, and decides to turn back to her house, which makes her feel relieved.

Question 17 ©VCAA 2021 Q48 ADAPTED MEDIUM
Explain how in terms of classical conditioning Carmel is being precipitated to have a phobia of the backyard.
A The snake acted as a conditioned stimulus to create the conditioned response of fear.
B The backyard is the unconditioned stimulus; when paired with the conditioned stimulus of the snake this led to a fear response.
C The snake is the unconditioned stimulus; when paired with the neutral stimulus of the backyard this elicited a conditioned fear response of the backyard.
D The conditioned response of fear was only exhibited when the snake was actively in the backyard.

Question 18 ©VCAA 2021 Q49 ADAPTED HARD
In terms of operant conditioning, explain how Carmel's fear of going into the backyard was perpetuated.
A Carmel was positively reinforced by rewarding herself with relief when she avoids the backyard.
B Carmel was negatively punished by being forced out of the backyard, which she once enjoyed.
C Carmel was negatively reinforced as when she avoided the backyard the negative feelings of anxiety disappeared.
D Carmel was positively punished by the negative effect of being anxious being introduced when she entered the backyard.

Question 19 ©VCAA 2021 Q50 ADAPTED HARD
Which one of the following would be the most appropriate psychological intervention to help Carmel overcome her phobia?
A breathing retraining to help Carmel feel less anxious through controlling her breath when trying to go into the backyard
B taking GABA agonists to help combat her anxiety when entering the backyard
C cognitive behavioural therapy by slowly exposing Carmel to increasingly stressful phobic stimuli
D cognitive behavioural therapy to help Carmel address the underlying psychological basis of her phobia and help with strategies to improve her behaviour towards entering the backyard

Use the following information to answer Questions 20–23.

Mai was five when she was playing with her dolls and a spider crawled across her hand. Mai was so frightened by this event that she developed a phobia of spiders. Anytime Mai sees a spider, even if it is small, she reacts with a strong fear reaction and must remove herself from the room.

Question 20 ©VCAA 2019 Q46 ADAPTED EASY
Which one of the following identifies what was released by Mai's neurons during the encoding of the fear?

	Released into neuron	Function
A	Glutamate	Help Mai form a fearful memory
B	GABA	To trigger the response of the sympathetic nervous system
C	Cortisol	To energise Mai's body to be able to deal with the sudden threat
D	Adrenaline	To enable neural pathways to fire, causing anxiety

Question 21 ©VCAA 2019 Q47 ADAPTED MEDIUM
Which one of the following best describes the effect of GABA dysfunction in the development of Mai's phobia?
A The memories of her experiences have the potential to affect her in the long term.
B The memories of the spider have been encoded into her episodic long-term memory.
C The neural signals representing the connection between the doll and her fear of it have been strengthened.
D There is excessive firing of neural impulses when experiencing anxiety from encountering spiders.

Question 22 ©VCAA 2019 Q48 ADAPTED EASY
Which one of the following describes the likely role of catastrophic thinking in the development of Mai's phobia?
A Mai is unable to recall the events accurately due to a fallible memory system.
B Each time Mai thinks of the events, they seem more threatening than they actually were.
C Mai's encoding of the events when playing with her dolls has been distorted to appear less threatening over time.
D Each time Mai talks about the events that happened when playing with her dolls, she incorporates new information.

Question 23 ©VCAA 2019 Q49 ADAPTED HARD
Which one of the following identifies the most appropriate strategies that Mai's family could use to help reduce Mai's phobia of spiders?

	Strategy 1	Strategy 2
A	Teach Mai relaxation techniques to use when she feels anxious	Remind Mai of a funny event of playing with dolls
B	Encourage Mai to think positively	Discourage Mai from avoidance behaviour
C	Use systematic desensitisation to slowly expose Mai to spiders	Assist Mai in challenging unrealistic thoughts
D	Help Mai learn about mental illness	Use CBT to help overcome her phobia of spiders

Question 24 ©VCAA 2020 Q42 ADAPTED MEDIUM
Which one of the following does not describe both a contributing factor in the development of a specific phobia and an evidence-based intervention used to treat a specific phobia?

	Contributing factor	Evidence-based intervention
A	GABA dysfunction (biological)	Support from family and friends (social)
B	Precipitation by classical conditioning (biological)	Challenging unrealistic thoughts (psychological)
C	Catastrophic thinking (psychological)	Adequate diet (biological)
D	Stigma around seeking treatment (social)	Breathing retraining (biological)

Question 25 ©VCAA 2019 Q50 ADAPTED HARD
Which one of the following evidence-based interventions is the most useful in the treatment of specific phobias in the long term?
A GABA agonists, as they treat the biological basis of phobias
B Psychoeducation, because it helps inform loved ones on how to best support sufferers with their phobias
C Seeking counselling, as it helps sufferers express themselves freely
D Systematic desensitisation, as it progressively reconditions an individual until no fear response is present and a relaxation response takes its place

Question 26 ©VCAA 2021 Q44 ADAPTED HARD
Which one of the following is true regarding risk factors and protective factors in the progression of mental disorders?
A A catastrophic event can be a risk factor and a protective factor.
B Coping abilities and strategies can be risk factors and protective factors.
C Biological factors, such as genetics, can present as risk factors but are not likely to be protective factors.
D Risk factors and protective factors increase a person's chances of developing a mental disorder.

Question 27 ©VCAA 2018 Q34 ADAPTED HARD
Which one of the following is the most accurate in terms of protective factors?
A If an individual lacks protective factors, they will develop a mental disorder.
B Mindfulness is a highly effective biological protective factor.
C Individuals with highly effective protective factors are likely to maintain a more positive wellbeing than those who lack strong protective factors.
D Social support is not important in maintaining positive mental wellbeing.

Question 28
In the SEWB framework wellbeing is supported through the interweaving of connections between the _____ domains.
A four
B five
C six
D seven

Question 29
Two of the strongest cultural determinants of the SEWB framework are
A minds and emotions.
B cultural continuity and self-determination.
C family and kinship.
D family and spirituality.

Question 30
Cultural continuity can be supported through
A revival of traditional cultural knowledge.
B activism through grassroots organisations.
C the rights of Aboriginal and Torres Strait Islander peoples to shape their own lives.
D increasing the number of Aboriginal-controlled health services.

Section B: Short answer

Question 1 ©VCAA 2018 SECTION B Q7 ADAPTED
Lauren is a 17-year-old student who has an intense fear of balloons. This is due to a balloon being popped when she accidentally sat on it as a child, causing a loud bang which caused her a lot of distress. Each time she sees or is near a balloon she gets extremely anxious and cries.

a Identify the biological, social and psychological aspects contributing to Lauren's specific phobia. Justify your response. [3 marks]

b Describe how one relevant external factor may help intervene with Lauren's specific phobia. [2 marks]

c Explain how one relevant biological protective factor could influence Lauren's ability to maintain positive mental health. [2 marks]
[Total = 7 marks]

Question 2 ©VCAA 2017 SECTION B Q6 ADAPTED
Dushani is a first-year university student who is engaged in multiple extracurricular activities such as sports and debating club. Despite a busy workload, Dushani can manage her stress quite well. However, Dushani then breaks up with her boyfriend and finds she is now unable to attend any of the activities she usually would enjoy and is starting to drop her grades at university.

a According to Lazarus and Folkman's Transactional Model of Stress and Coping, describe Dushani's appraisals in the scenario regarding her relationship break-up. [4 marks]

b Identify a different type of coping strategy that Dushani could have used to help her in the break-up with her boyfriend. [1 mark]

c Describe one protective factor that could have helped Dushani in not developing more serious mental health issues from the break-up. [2 marks]

d Describe resilience and explain how this can assist Dushani in coping with the stressor of the break-up. [2 marks]
[Total = 9 marks]

Question 3 ©VCAA 2020 SECTION B Q6 ADAPTED
Louis is constantly feeling anxious and can't work out the specific trigger for his stress. It is getting to a point where he is struggling to sleep or interact with friends because of this anxiety.

a According to the scenario, what are Louis' levels of functioning with the stressor? Justify your response. [2 marks]

b According to the mental health continuum, describe whether it is likely Louis is experiencing a mental disorder or mental health problem. [2 marks]

c Describe one internal and one external factor contributing to Louis' anxiety. [2 marks]
[Total = 6 marks]

Question 4
Differentiate between internal and external factors with reference to specific phobias. [4 marks]

Question 5
Explain what is meant by cultural determinants in the SEWB framework. Discuss the two strongest cultural determinants and provide examples of each. [6 marks]

11 Using scientific inquiry

Key knowledge

Investigation design
- psychological concepts specific to the selected scientific investigation and their significance, including definitions of key terms
- characteristics of the selected scientific methodology and method, and appropriateness of the use of independent, dependent and controlled variables in the selected scientific investigation
- techniques of primary quantitative data generation relevant to the selected scientific investigation
- the accuracy, precision, repeatability, reproducibility and validity of measurements
- the health, safety and ethical guidelines relevant to the selected scientific investigation

Scientific evidence
- the nature of evidence that supports or refutes a hypothesis, model or theory
- ways of organising, analysing and evaluating primary data to identify patterns and relationships, including sources of error and uncertainty
- authentication of generated primary data using a logbook
- assumptions and limitations of investigation methodology and/or data generation and/or analysis methods
- criteria used to evaluate the validity of measurements and psychological research

Science communication
- conventions of science communication: scientific terminology and representations, symbols, formulas, standard abbreviations and units of measurement
- conventions of scientific poster presentation, including succinct communication of the selected scientific investigation and acknowledgements and references
- the key findings and implications of the selected scientific investigation

Source: VCE Psychology Study Design (2023–2027) page 41

11 Using scientific inquiry

Chapter 11 provides you with guidance and advice on how to complete Unit 4 Outcome 3. This assessment gives you the opportunity to select an area from Unit 3 or 4, or across Units 3 & 4, to undertake a student-designed scientific investigation. Choose an area that interests you or one that you want to find out more about. When you are thinking about what to investigate, you could consider testing an existing model or theory, or building on research that has been carried out by someone else.

11.2 Conducting an investigation
p. 431

Now you are ready to conduct your investigation. You will be generating primary quantitative data, which you will go on to display, analyse and evaluate. Finally, you will determine whether your findings support or refute your hypothesis. These are all skills that you have built up and practised during your studies of Psychology.

11.1 Designing an investigation
p. 425

Make sure you have your logbook ready to go. This will form part of your assessment so be diligent in maintaining it. Whatever you choose to do, you need to start off your research with the question you want to answer. From this, develop an aim and hypothesis. Once you have done this you can organise your population and sample, methodology and method. Make sure you respect the mental and physical wellbeing of your sample by considering ethical concepts and guidelines and safety (refer back to Chapter 1 if you need a refresher on this).

Adobe Stock/Minerva Studio

p. 434

11.3 Science communication

It is now time to show others what you have discovered. Unit 4 Outcome 3 dictates that you will do this using a poster format. Make sure you understand the format of the poster before you start creating it. Ask your teacher if you are not sure.

Science investigation gives you the opportunity to add to the total sum of psychological knowledge. This is what scientific investigation is all about.

Slideshow
Chapter 11 slideshow

Flashcards
Chapter 11 flashcards

Test
Chapter 11 pre-test

Assessment
- Pre-test

Revision
- Chapter map
- Key term flashcards
- Key concept summary
- Slideshow

Investigation
- Investigation: The process of learning based on performance on a puzzle maze
- Logbook template: Controlled experiment
- Data calculator

To access these resources, visit
cengage.com.au/nelsonmindtap

Nelson MindTap

Know your key term

Communication statement

Research Australia is an organisation that supports the health and medical research sector to be a global force when it comes to medical research. The goals of Research Australia are to:
» connect researchers, funders and the public to maximise health funding
» engage Australians in a discussion to make citizens aware of the health benefits and economic value of research in Australia
» influence government policies to support research in Australia while showing how this produces better health outcomes.

One event run by Research Australia since 2002 is the Health and Medical Research Awards. The awards celebrate the researchers and doctors who are striving to improve health and, as a result, change lives. One of the most prestigious Australian awards is the GlaxoSmithKline (GSK)-organised award presented at the event, the GSK Award for Research Excellence. In 2021, it was awarded to researchers at Monash University, professors Jamie Cooper and Rinaldo Bellomo, who worked on changing approaches to the treatment of critically ill patients (Figure 11.1). Their work has significantly improved outcomes of patients with traumatic brain injury, acute kidney failure and acute respiratory failure. The award recognises their contributions to medicine over a period of 20 years.

Research Australia is an example of an organisation that promotes and recognises

Figure 11.1 Professor Jamie Cooper and Professor Rinaldo Bellomo, co-directors of the Australian and New Zealand Intensive Care Research Centre (ANZIC-RC), have been acknowledged for their global leadership and innovative research in critical care medicine, with the prestigious GSK Award for Research Excellence in 2021.

Source: https://au.gsk.com/en-au/research-and-development/supporting-research/gsk-australia-award-for-research-excellence

research of a high quality and value. Scientific research that is conducted in Australia is often at the forefront of research. The credibility of the research – and also of Australia – is influenced by how that research is conducted and reported. Research must always be conducted with integrity.

You have already had experience at undertaking a literature review to analyse and evaluate evidence (Unit 1 Outcome 3) and adapting or designing a research investigation (Unit 2 Outcome 3). You are now going to bring the skills gained in completing these Outcomes to design and conduct your own scientific investigation (Unit 4 Outcome 3). Within this chapter, you will be provided with guidance on how to design and conduct your own scientific investigation relating to the psychological concepts covered in Unit 3 and/or Unit 4. This chapter, used in association with Chapter 1, outlines the conventions psychologists use when conducting scientific investigations. It is then up to you to conduct an investigation in a professional manner by following these conventions. Upon completing your investigation, you will be required to present key elements of your research and findings as a scientific poster.

11.1 Designing an investigation

This chapter will explain the steps you will need to undertake to complete a research investigation and poster and will model an example. This will be an example only. The methodology and design of your research investigation will be unique to you and may involve slightly different steps to those shown in our example.

A good research design ensures that you have appropriate background understanding of the relevant psychological concepts, key terms and other research in the field or fields that have explored the relationship between these concepts. This background will assist you in developing a research question. Go back to the previous chapters in this resource and re-read sections if you need to revise any psychological concepts. Go back to Chapter 1 in this resource to refresh your understanding of psychological research methods.

Step 1: Get your logbook ready

Prior to this Outcome, you will have become familiar with the use of a logbook. For Unit 4 Outcome 3, it is a requirement that the logbook is submitted along with your scientific poster for assessment. This is so your teacher can authenticate your work. That is, your teacher must be able to verify that the work submitted by you was completed by you. If you record your work as you go in a logbook, it becomes a record of what was completed and when it was completed. Take some time now to set up your logbook. Setting up a well-constructed logbook now will save you time later. Your logbook should contain a table of contents page, page numbers and a place to put a date and heading on each page as you use it. Anything that you print out from a website can be glued into the logbook. You can set up a digital document, or a page in your logbook to record all references that you use. Recording references as you go is much easier than having to go back later and try to find them again.

Step 2: Determine your research question

Before you can write a research question (see Chapter 1) you will need to identify the area (or topic) in which to conduct a research study. Your topic can come from the content covered in Unit 3 or Unit 4 or can be relevant to both Units 3 and 4. You may wish to choose a topic that you found particularly interesting. Alternatively, you may wish to use this outcome to consolidate a topic that you did not understand well. Start off by looking at all the topics covered in Units 3 and 4 (Table 11.1). Decide on which topics you may like to choose for this outcome.

Table 11.1 Topics covered in VCE Psychology Units 3 and 4

	Topic
Unit 3	Nervous system functioning
	Stress as an example of psychobiological process
	Approaches to understand learning
	The psychobiological process of memory
Unit 4	The demand for sleep
	Importance of sleep in mental health
	Defining mental wellbeing
	Application of a biopsychosocial approach to explain specific phobia
	Maintenance of mental wellbeing

Once you have decided which topic you want to focus on for your research investigation then look at the dot points in the VCAA VCE Psychology Study Design related to that topic. You can find these dot points in the Study Design or by looking at the start of the relevant chapter in this resource. Write out two dot points that you could possibly investigate. Developing two means that if one becomes too difficult to complete later on, you have a second option or Plan B. Discuss your ideas with teachers and peers; this may help you gain insight into whether your research investigation is viable.

For example, you could choose the following two dot points:

(Unit 3 Area of Study 2, The psychobiological process of memory)

» the use of mnemonics (acronyms, acrostics and the method of loci) by written cultures to increase the encoding, storage and retrieval of information as compared with the use of mnemonics such as sung narrative used by oral cultures, including Aboriginal peoples' use of Songlines

(Unit 4 Area of Study 2, Maintenance of mental wellbeing)

» mental wellbeing as a continuum, with an individual's mental wellbeing influenced by the interaction of internal and external factors and fluctuating over time, as illustrated by variations for individuals experiencing stress, anxiety and phobia

Source: VCAA VCE Psychology Study Design 2023–2027

In your logbook, write a possible research question for each of the dot points you have chosen. Remember, a research question defines the question that the research study attempts to answer. Sentence starters for a research question include: How, Do, Does, Are, Is.

Examples:
Does the use of mnemonics increase the retrieval of information over time?
Is our mood affected by the season (an external factor)?

11.1.1 FORMULATING HYPOTHESES

Step 3: Write an aim

To make a prediction about the direction of your results, you first need to complete a literature review. Remember, a literature review investigates what is already known on your chosen topic. The findings from several studies can be combined to provide an overview of a particular topic. This will provide secondary data which you can compare to your primary data. Find at least two articles from a website or journal and summarise (3–4 sentences) each one in your logbook. Make sure you include the complete reference for all articles that you use.

Example:
The use of mnemonics reorganises the brain's functional network organisation to enable superior memory performance. Improvements were sustained for up to 4 months after training.

Dresler, M., Shirer, W. R., Konrad, B. N., Müller, N., Wagner, I. C., Fernández, G., Czisch, M., & Greicius, M. D. (2017). Mnemonic Training Reshapes Brain Networks to Support Superior Memory. *Neuron, 93*(5), 1227–1235.e6. https://doi.org/10.1016/j.neuron.2017.02.003

Example:
Seasonal affective disorder (SAD) begins around the start of winter and ends in spring. Symptoms are lacking energy and feeling moody. A possible cause is a change in levels of melatonin. Treatments include exposure to light and medications.

Mayo Foundation for Medical Education and Research. (2021, December 14). *Seasonal affective disorder (SAD)*. Mayo Clinic. Retrieved April 20, 2022, from https://www.mayoclinic.org/diseases-conditions/seasonal-affective-disorder/symptoms-causes/syc-20364651

Remember to check your resources using the CRAP test. The CRAP acronym stands for how current and reliable the source is, and its level of authority and purpose (Table 11.2).

You can now write an aim for your two topics. Remember, an aim is a broad statement about the goal of the research investigation, and usually begins with 'To ...'.

Example:
To investigate whether the use of mnemonics increases our ability to retrieve information from memory over time.
To investigate whether our mood is affected by the season.

Step 4: Write a hypothesis

Before you can write a hypothesis, you must decide on the population of research interest. Who do you want to apply your result to? For example, everyone in your suburb, your school or your class? Or will you focus on people from a specific demographic, region or background? The more

Table 11.2 C.R.A.P. test to check the credibility of your secondary sources

Name of source:		
Source type:		
URL:		
Date reviewed:		
Currency	Yes/No	Details
Was the information published or updated recently?		
If using a website, do the links within the article work?		
Reliability		
Is the information supported by evidence like data or quotes? Are there references for the evidence?		
Does the source make reasonable claims about what the evidence shows?		
Has the information (or its references) been reviewed?		
Can you confirm the information using another source?		
Does the language or tone seem unbiased and professional?		
Authority		
Is the author, publisher, or sponsor of the information a trustworthy source, such as an educational or government institution?		
Is the author qualified to write on the topic?		
Is the author likely to be unbiased about the topic?		
Is there any contact information?		
Purpose		
Is the purpose of the information to teach or inform rather than to sell, entertain or persuade?		
Is the information fact, rather than opinion or anecdote?		
Overall judgement		

precisely you define your population, the more likely it will be that you gather a representative sample.

Example: a possible population is VCE students at Rooftop Secondary College. This population is limited in age and also location. VCE students is probably too general; you cannot sample such a general population.

Residents of the suburb of Banksia Park. This population is not age limited. Your sample could therefore include school students, but also family members or neighbours.

To write your hypothesis you will need to do the following.
» Identify the independent variable (IV) that will be manipulated.
» Identify the dependent variable (DV) that will be measured.
» Nominate the population of research interest.
» Predict the direction of the relationship between the IV and the DV.

Now is a good time to look at your three statements (research question, aim and hypothesis) and highlight any psychological concepts and terms. Look up a definition of the terms and write them in your logbook. The glossary in the back of this book or the APA Dictionary of Psychology are good references to assist you here (see weblink). You will need to be able to correctly use these concepts and terms when you write your introduction and discussion for your report.

Example: Psychological concepts
1 Mnemonics, acronyms, acrostics, method of loci, encoding, storage, retrieval
2 Mental wellbeing, seasonal affective disorder, internal and external factors, stress, anxiety, phobia

Weblink
APA Dictionary of Psychology

11.1.2 EXTRANEOUS VARIABLES OF ALL SHAPES AND SIZES

Example: Hypothesis

Year 12 Maths students who use mnemonics will be able to retrieve more words after 2 weeks than those people who use repetition to learn a list of words.

    IV: strategy used to learn a list of words
    DV: number of words retrieved after 2 weeks
    Population of interest: Year 12 Maths students
    Predict direction: use of mnemonics (IV) will increase DV

VCE students at Rooftop Secondary College will report more symptoms of seasonal affective disorder in winter (July) but not in spring (October).

    IV: Time of the year: short days (July) vs. long days (October)
    DV: Symptoms of SAD as measured by (amb) a self-report
    Population of interest: VCE students at Rooftop Secondary College
    Predict direction: more symptoms in winter than spring

Step 5: Choose a methodology

11.1.3 WHICH EXPERIMENTAL DESIGN IS WHICH?

Read the methodologies, as outlined on pages 14–22 of Chapter 1. Decide which methodology would be best to test your research question. As both examples compare two groups and involve an IV and a DV, a controlled experiment methodology would suit best. Record your choice of methodology in your logbook.

Example:
Memory study: Controlled experiment: within subjects
Seasonal affective disorder: Controlled experiment: within subjects

    This is the time to choose between your two options and develop one into a full research investigation. Example 2, seasonal affective disorder, will be further developed in this chapter.

Step 6: Write a method

The method will need to cover the following:
- Participants: how many do you need to create a representative sample? Are there any important variables such as age or gender that you need to consider? A loose rule is to include 10 per cent of a population in a sample.
- Sampling technique: how will you create a sample from your population? Do you need to stratify your sample? How will participants be allocated to groups if you have an experimental and a control group?
- Data: what sort of data are you going to collect? One of the requirements of this outcome is that you generate primary quantitative data. Refer to Chapter 1 to find out what this means.
- Materials: create a list of materials, including things you may need to purchase. Do you need several copies of anything? Remember to create a consent form.
- Procedure: write the steps that you will undertake using dot points. Include where you will conduct your study and any potential extraneous variables that you will need to control (for example, distracting noises).
- It is worthwhile writing a script which you will read (or send) to each participant when obtaining their consent to participate. This increases internal validity as it standardises your instructions.

Example:
Method:
Seasonal affective disorder
Participants: 50 VCE students from Rooftop Secondary College. The names of all members of VCE will be placed against a number. A random number generator will be used to choose participants.
Materials:
- Script to use when introducing the experiment (Figure 11.2), self-report questionnaire (Figure 11.3), consent form.

Procedure:
- Participants will complete a self-report questionnaire in July and October of the same year.
- Select 60 participants using a random number generator. (Ten more participants will be selected than needed in case some participants do not wish to participate, or they withdraw during the year.) Random allocation to groups is not relevant to this example.
- Contact each participant by email with a consent form attached. Request that the consent form is returned within a week.
- Collate and store the consent forms.
- On 25 June, one copy of the self-report questionnaire is delivered by email along with instructions on what to do.
- This is repeated in October (20th) when each participant completes another self-report.
- Answers to questionnaires will be collated. Two totals will be calculated for each participant by

11.1.4 SAMPLING PROCEDURES

adding up the numbers on the form completed in July and the form completed in October. Each form will be given a score between 0 and 24.

An extraneous variable that could affect results is that the participants are self-reporting, and some individuals may not be honest in their self-reporting. Some individuals may prefer to providing the answer they think is more acceptable, or that the researcher wants, rather than providing a truthful answer (this is called self-reporting bias).

> As part of VCE Psychology, I am studying whether mood changes during the year in VCE students. I invite you to participate in this study. You will be asked to fill out a questionnaire describing your mood in July and then again in October.
>
> After today, if you change your mind about participating, please let me know, in writing.
>
> Also, please let me know if you have already participated in a study on this topic.

Figure 11.2 A script is used when introducing your experiment to your sample.

Step 7: Consider safety and ethics

In your logbook, you need to record how you will ensure the ethical concepts and guidelines are managed (Table 11.3). You also need to apply the ethical concepts when conducting your study (Chapter 1). This means you will behave in a professional manner (integrity) and treat each participant in the same way (justice) and with respect for their customs and beliefs (respect). You also need to consider ethical guidelines to ensure you do not cause any harm to your participants.

Instructions: Seasonal Affective Disorder Questionnaire

Thank you for participating in this study! Please rate yourself on the following scale by ticking the option that best relates to you. Complete this form on 1 July, or as close to this date as possible.

Name: .... Date completed: ....

Participant number: ....

	0. Very low	1. Low	2. High	3. Very high
How happy I feel overall	☐	☐	☐	☐
How motivated I feel	☐	☐	☐	☐
My sleep quality	☐	☐	☐	☐
How restless I feel	☐	☐	☐	☐
My energy levels	☐	☐	☐	☐
My ability to concentrate	☐	☐	☐	☐
My ability to complete tasks	☐	☐	☐	☐

Figure 11.3 A questionnaire for the seasonal affective disorder study is completed for July. The same questionnaire will be sent out again for completion in October.

11.1.5 ETHICS IN RESEARCH

Table 11.3 Ethical concepts and guidelines that need to be considered before you start your study

Ethical concepts/guidelines	Questions to consider
Beneficence and non-maleficence	How will you ensure 'no harm' is done to your participants?
Confidentiality	Will you be taking photographs? How will you modify them? How will you ensure that each participant cannot be identified?
Voluntary participation	Does your script include an invitation to participate? Are you using any form of coercion?
Withdrawal rights	Are withdrawal rights included in your script?
Informed consent	Have you created a consent form? Does the study include participants who are under the age of 18? If so, have you included a line for a signature from a parent/guardian?
Deception	Are you using deception? If so, describe it.
Debriefing	What do you need to inform your participants of once they have finished the experiment? Do you need to offer your participants time to talk about the experiment at a later date?

Example: Ethics, Seasonal affective disorder

Confidentiality: I will remove students' names from their self-reports and allocate a number. I will store the consent forms in a locked filing cabinet at home.

Voluntary participation: I will not just ask my friends to participate. I will make sure that my initial approach to participants informs them that participation is voluntary.

Withdrawal rights: When releasing the self-report, I will remind participants of their right to leave the study at any time.

Informed consent: I will include an area for a parent/guardian signature. I will ensure that participants are clear on what they have to do.

Debriefing: I will offer participants a time to view my poster and ask questions about the study and its results.

Table 11.4 Research issues that need to be addressed before you start collecting data

Research issue	Questions to consider
Accuracy	What aspects of your investigation will need to be accurate? How will you ensure your measurements will be as accurate as possible?
Precision	What aspects of your investigation will need to be repeated, to demonstrate precision? (e.g. a task is completed 5 times, then the average is recorded)
Repeatability	How many times does each participant need to complete the tasks set? (e.g. you need to record each participant catching a ruler 8 times, not just once)
Reproducibility	How will you ensure that other researchers can reproduce your study?
Validity	What aspects of your investigation might decrease internal validity?

Before you begin: things to consider

Research must be considered in terms of how carefully the experiment was run. Was there good preparation and procedures? If not, the results cannot be applied to the population of interest. You should be aiming for high internal validity by addressing the information shown in Table 11.4 in your logbook. You may not end up using this table in your scientific poster, however, it will stand as evidence of your thought process.

KEY CONCEPTS 11.1

- A logbook must be used and submitted along with your poster, to authenticate your work.
- You must write a research question, based on a dot point from the study design.
- Write an aim which is a broad statement about the goal of the research investigation.
- Conduct a literature review, recording a summary of at least two relevant articles.
- If you choose a controlled experiment methodology you will write a hypothesis that includes the IV, DV, prediction about how the IV influences the DV and population of research interest.
- Write a method. Method includes participants, sampling technique, materials and procedure.
- Consider potential extraneous variables, ethics and safety.

Concept questions 11.1

Remembering
1. What is a logbook? r
2. Why do you need to maintain a logbook throughout your research investigation? r
3. What are the three general headings in a method? r

Understanding
4. Describe an ethical guideline relevant to your chosen topic. e
5. Why do you need to think of potential extraneous variables before you begin collecting data? e
6. Describe two safety issues that would need to be considered in the study on seasonal affective disorder. e

HOT Challenge
7. A student decided to investigate how often VCE students woke up during the night. Write a research question, aim and hypothesis, and choose a methodology for this research. Write a method that could be used to investigate the research question. c

11.2 Conducting an investigation

When a researcher conducts an investigation, they collect data – information gained from direct observation and measurement. Researchers can use any of a variety of data collection techniques including surveys, direct observation, psychological tests or examination of files and documents. The data collected can be qualitative or quantitative data. For Unit 4 Outcome 3, you need to collect primary quantitative data. This means that the data needs to be collected by yourself (primary) and it is to be in numerical form (quantitative).

Step 8: Generate primary quantitative data

Before you start testing your participants, you will need to create a data table to record your results into. The type of data table you create will depend on the type of data you want to record. For the seasonal affective disorder example, the researcher will need to record two sets of scores out of 24 (July and October) for each of 50 participants (Table 11.5).

When creating your data table make sure that:
» your table has a descriptive title
» you create the number of columns and rows that you will need
» your columns are headed and include any units (for example, sec) you may be using. Do not put units into each cell.

You might like to practise collecting data for your investigation using the data table you have created to make sure it works properly. If not, it may need a bit of adjustment before you start collecting data for real.

Step 9: Organise and summarise your data

Decide what numerical calculations are relevant for your data. You can include descriptive statistics in your data table. The results section is intended to show *what* happened but must not explain *why*. Possible calculations or summaries include:
» a summary or frequency table. A frequency table shows the frequency of various outcomes in a sample such as the number of words remembered correctly from a 10-word list (Table 11.6).

Table 11.5 Seasonal Affective Disorder data table

Participant number	July score /24	October score /24
1		
2		
3		
↓		
50		

Table 11.6 A frequency table for a study where a particular number of words was retrieved from a 10-word list

Number of words correctly retrieved	Frequency	Total
10	I	1
9	II	2
8	III	3
7	IIII	5
6	IIII I	6
5	IIII	4
4	III	3
3	III	3
2	II	2
1	I	1
0	0	0

» a column for percentages, or percentage change
» measures of central tendency: mean, median, mode
» measures of variability: standard deviation (if you know how) or the range
» a graph (bar graph or line graph), labelled, with outliers identified. Make sure your graph is large and clear. The axes should be labelled with the names of the variables and their units (if relevant). Choose a scale so that your data takes up most of the plot area (Figure 11.4).

11.2.1 EXPERIMENTAL DATA

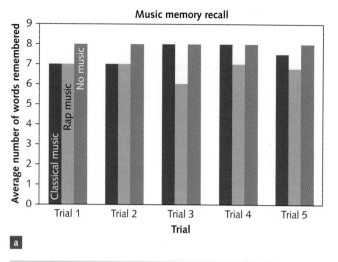
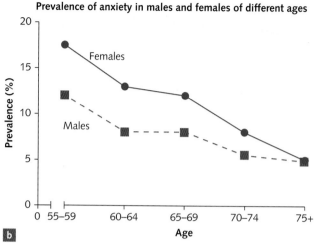

Figure 11.4 These are well-drawn graphs showing: **a** number of words remembered over 5 trials; **b** the prevalence of anxiety in males and females of different ages. Note that the axes are labelled, the graph is titled and each graph is labelled.

11.2.2
APPLYING YOUR KNOWLEDGE

Step 10: Analyse and evaluate your data

Before you can analyse your data you have to decide if it is good quality data. Questions to ask about your data are:

» Is the data accurate?
» Is the data precise?
» Is the data repeatable?
» Does the data contain any random, systematic or personal errors? (Table 11.7)
» Am I confident that I used any measuring devices correctly and that they measured correctly?
» Was my sample size large enough?
» Do I have to repeat my experiment to collect more data or make sure the data I did collect is correct?
» Are there any confounding variables?

Note: Errors do not have to appear on your poster unless they are particularly significant in their effect. But errors should appear in your logbook. List the error and the potential effect it may have had on the data.

As well as evaluating your data, you also need to evaluate your investigation methods for internal and external validity (see Chapter 1).

Step 11: Determine uncertainty and limitations of your data

As you saw in Chapter 1, when we carry out a psychological investigation, we do not know at the beginning of the investigation what the data is going to be. Therefore, there is a degree of uncertainty within each measurement (because you do not know the value of the quantity you are measuring). In VCE Psychology you only need to deal with uncertainty in a qualitative way. You need to ask the following questions about your data.

» Is there any contradictory data?
» Is there any missing or incomplete data?
» Is there any bias within the data?

Uncertainty does not appear on your poster but it does need to appear in your logbook. Make a subjective evaluation about the uncertainty in your data. Try to create a measure of the level of uncertainty, such as: low, medium or high.

Table 11.7 Types of errors that may occur during the experiment

Type of error	Description of error	Example
Random	Any random events that occurred in your experiment	Interruptions from other people, announcements or sudden noises
Systematic	Issues that cause readings to differ from the true value by a consistent amount or by the same proportion each time a measurement is made	Measuring devices that are slightly inaccurate or reading a device (e.g. a ruler) incorrectly
Personal	Mistakes made by you, the researcher	Incorrect calculations, recording errors, observation errors

Ask yourself: if I repeated the experiment, how similar would the results be to the experiment I just did?

Outliers (results that lie a long way from other results) in your data need to be dealt with and not just ignored. They could be due to recording errors such as a personal error, where the researcher copied the wrong answer. Outliers may affect the validity of the research. Repeating measurements could reduce outliers.

You also need to identify any limitations within your method. You need to be honest about any limitations within your findings. There is no place for opinion, anecdotes or non-scientific ideas. Was your sample large enough? Was the survey sensitive enough to show small differences? Was the measuring tool the correct one to use? These sorts of limitations within the method may affect the type of evidence-based conclusions that you can draw. Identifying limitations is part of the discussion section of your poster. In your logbook, list any limitations, plus ways to avoid them should the experiment be repeated.

Step 12: Interpret your data

Once you have analysed your data, you need to interpret them. This means being able to either answer your research question or state whether your results support or refute your hypothesis. If your hypothesis was refuted, it is not enough to simply say 'the hypothesis is wrong'. You need to suggest why the data did not support the hypothesis. It may be that the underlying psychological concepts that you based your hypothesis on were incorrect or outdated. Go back to your literature review and check this. If this is the case, how would you change or extend this investigation if you had to conduct it again?

Generalisability is the extent to which your findings can be applied to other settings. How generalisable are your results? A generalisation can only be made if the sample used was representative of that population. Remember, any decision you make must be based on evidence provided by your data, not on your subjective opinion.

Step 13: Draw evidence-based conclusions

Your conclusion must be firmly grounded in the evidence provided by the data in your investigation. In your conclusion you answer the question posed at the start. The conclusion is word limited so you need to be succinct (try to keep it below 50 words). An example of a well-written conclusion is:

This study investigated whether season affects mood. The hypothesis was supported as VCE students reported greater levels of SAD during winter than in spring. This could have a significant effect on students' abilities to perform well and gain their best score at VCE.

KEY CONCEPTS 11.2

- » Conducting an investigation means the generation of primary quantitative data.
- » Primary data is easier to interpret if it is organised. Summary tables, frequency tables, measures of central tendency and graphs are all ways of organising data.
- » Analysing data involves mathematical processes including mean, percentage or percentage change and understanding the variability within the data.
- » Evaluating data includes reviewing it for accuracy, precision, validity and errors.
- » Evaluating data is an assessment of error, uncertainty and confounding variables.
- » Interpreting data is deciding whether the data supports or refutes the hypothesis and deciding how generalisable the results are.
- » Any conclusions that you draw must be based on evidence provided by your data.

Concept questions 11.2

Remembering
1. What features do you need to include on a well-drawn graph? **r**
2. List the three types of errors relevant to research. **r**
3. Define 'generalisability'. **r**
4. What is included in a conclusion? **r**
5. What is primary quantitative data? **r**

Understanding
6. How do random and systematic errors affect data? **e**
7. Under what circumstances would you mention errors on your poster? **e**

HOT Challenge
8. The table below shows the results of a memory investigation. Thirty participants were asked to reorder 10 objects so they were in the same order as a picture they were shown for varying amounts of time. **c**
   a. Construct a graph to display these results.
   b. What mathematical analysis has been done on the raw data?
   c. Write a hypothesis for this investigation.

Amount of time shown original picture (sec)	Mean number of mistakes in reordering objects
20	3
15	5
10	6
5	9

11.3 Science communication

11.3.1 ANALYSING RESEARCH

Unit 4 Outcome 3 must be presented as a scientific poster similar to one that is used by researchers to display their findings at conferences or universities. This is an opportunity for you to demonstrate your science communication skills, so think carefully about each component of the poster. The scientific poster is intended to present the key features of an investigation. With any presentation, you must consider your audience. In this case, your audience includes readers with no scientific background. The poster can be produced either electronically or in hard-copy.

Step 14: Create your poster

Refer to the VCAA VCE Psychology Study Design 2023–2027 for all the requirements of a scientific poster. An effective scientific poster will include:
» not more than 600 words. The parts of your poster that are NOT included in the word count are the title, your name, tables, figure captions, references and acknowledgements. This is why your logbook entries are important: so your teacher can assess the detail of your investigation in your logbook.
» use of graphics to reduce the word count and to convey information. Graphics include tables, photos, diagrams and graphs. All graphics must be labelled. Write the text to support the graphics, not the other way round.
» Graphs should not have a coloured background or grid lines. Axes should not be labelled in ALL CAPS.
» Text must be checked for spelling, and written in a way that a reader with no background in

Figure 11.5 The presentation of your scientific poster is important. It must be readable from a distance and the number of words must be minimised.

the area would be able to understand. It helps to use plain language with short sentences or bullet points.

» Use colour conservatively. Stick to a theme of two or three colours. Bright colours are not recommended. Dark letters on a light background are easier to read.
» White space. The inclusion of white space makes the poster look less cluttered.
» A font type and font size that will be easily read (Figure 11.6). Fonts that are curly are more difficult to read.

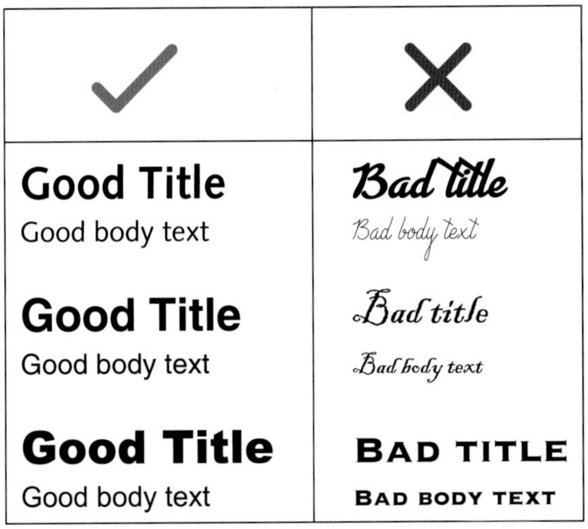

✓	✗
Good Title Good body text	*Bad title* *Bad body text*
Good Title Good body text	*Bad title* *Bad body text*
Good Title Good body text	**BAD TITLE** **BAD BODY TEXT**

Figure 11.6 Straight (sans serif) fonts are easier to read than curly (serif) fonts.

Poster sections

If a viewer remembers one thing from your poster, what should it be? Most importantly, they need to know the main finding of the research. The key point needs to be described in one sentence. The sentence is then placed in the middle of the poster as a **communication statement** (Figure 11.7). For example, the communication statement for the SAD study could be: 'VCE students showed signs of seasonal affective disorder, including lack of energy and motivation'.

Around the communication statement, the other sections of your poster must follow the given format under seven titles. This content is outlined below.

1 Title

The title of your poster is written as a research question. For example: Is our mood affected by the season?

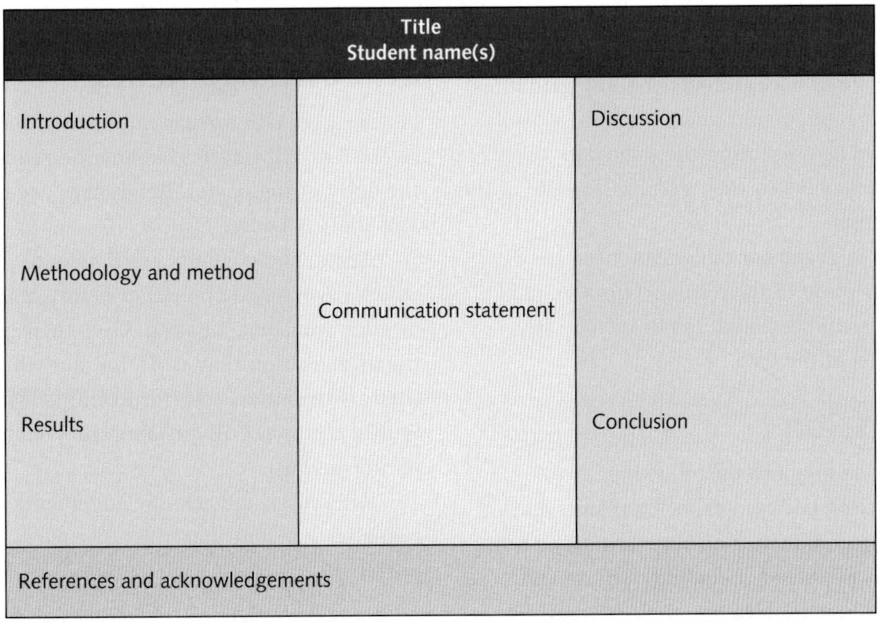

Title Student name(s)		
Introduction		Discussion
Methodology and method	Communication statement	
Results		Conclusion
References and acknowledgements		

Figure 11.7 VCAA scientific poster format showing the required headings

Source: Adapted from VCAA Study Design 2023–2027

2 Introduction

The introduction must include:
- the aim
- the hypothesis
- background information providing the setting, or context, for your experiment. Include relevant key terms and psychological theories, concepts, and secondary sources outlining the research that has been completed previously on this topic
- a photograph or graphic can be included to explain a theory/model/concept, or other background information.

3 Methodology and method

The methodology includes the investigation type (for example, controlled experiment) and why this was chosen. Procedure is what you did do, and so it is written in the past tense. It is summarised ending with the data generation method. A flowchart, labelled drawing or a photograph (for example, you doing something) can all be used for the procedure. Only include the essential details.

4 Results

Select the data trends you wish to highlight. For example, you may wish to highlight the mean and the range. Decide how you can present this data. An easy-to-read figure, a table of results or a labelled graph could be used. Graphs must be clear with relevant scales, labels and annotations. Do not have coloured backgrounds or show grid lines. If you use a graph, do not include a table as well (as you would be presenting the same data twice). Write a sentence below each graphic describing the key information.

As well as graphs or tables, you need to write a description of the results. This should be done in one sentence and does not include any interpretation of the data.

5 Discussion

The discussion includes the following:
- a comparison between expected results and the results obtained. Note any deviations from what was predicted, including outliers and confounding variables.
- a statement about what the results indicate: what is the significance of the results?
- a statement about whether the data supports, partly supports or refutes the hypothesis. Avoid terms like 'proved', 'disproved', 'correct'. Use terms like 'supported', 'indicated', 'suggested'.
- comparing findings to earlier work referenced in the introduction. If you were testing a particular theory, how well did your data support the theory?
- identifying the limitations of the experimental design, including errors and flaws in the procedure. Discuss how these errors may have affected the data.
- suggested explanations for problems in the data.
- identifying further questions or areas for future research.

6 Conclusion

This is where you answer your research question. You also need to decide on the significance of your results. Think beyond your experiment and ask: how do your findings relate to the relevant psychological concept and to everyday applications? Make sure that any specific details that you use are supported by evidence from your investigation.

7 References and acknowledgements

Harvard or APA format can be used for references. All references must be referred to in the body of the poster. References are listed in alphabetical order.

Acknowledgements are thanks to people who have assisted you either before, during or after your investigation. Give their name, the organisation they work for and what their contribution was (for example: Mrs Smith, Uptown College laboratory technician: photocopying).

References and acknowledgements can be in a smaller size font than the other parts of your poster.

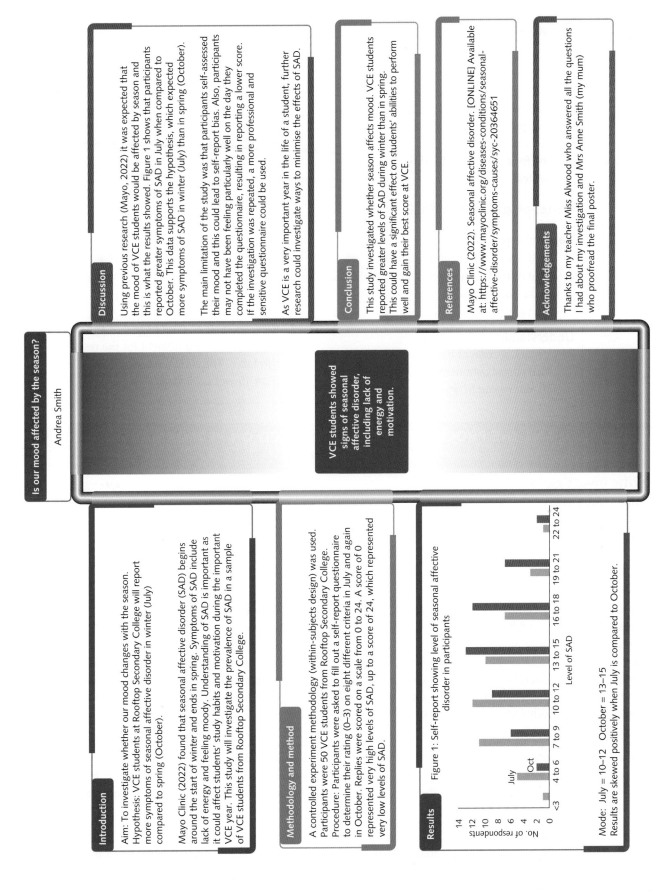

Figure 11.8 An example VCE Psychology poster

KEY CONCEPTS 11.3

» A scientific poster is like an illustrated summary of your research investigation.
» Poster conventions include simple colour and font themes and the inclusion of white space.
» Your poster should contain no more than 600 words.
» The focus of the poster should be graphics, not text.
» The centre of the poster is a single sentence known as a communication statement.
» There are seven sections or headings that must be included in order: title, introduction, methodology, results, discussion, conclusion, references and acknowledgements.

Concept questions 11.3

Remembering
1 What is a communication statement? r
2 What is the recommended number of colours for the poster? r
3 What is meant by white space? r

Understanding
4 In which section of the poster would you place ethical considerations that are significant for your research? r

5 If your data does not support your hypothesis, what extra sentence is required? r
6 Why is there a word limit for the poster? r

HOT Challenge
7 Refer to Table 11.6 on page 423 and write a communication statement for the results in this table. The statement should be no more than 15 words long. c

---

INVESTIGATION 11.1 THE PROCESS OF LEARNING BASED ON PERFORMANCE ON A PUZZLE MAZE

Scientific investigation methodology
Controlled experiment

Logbook template
Controlled experiment

Aim
To design and conduct a scientific investigation related to the process of learning based on performance on a puzzle maze.

Introduction
This investigation gives you the opportunity to design and conduct your own investigation relating to the process of learning based on performance on a puzzle maze. Remember, your design must include the generation of primary quantitative data.

Data calculator

   Make sure your investigation has an aim, methodology and method, results, discussion and conclusion while complying with safety and ethical guidelines. Remember to use your logbook throughout this investigation. Present your findings as a scientific poster. The poster may be produced electronically or in hard-copy format and should not exceed 600 words. Please see the VCAA Psychology Study Design pages 15 and 16 for further specifications of the poster format.

Pre-activity preparation
Conduct a literature review of at least three articles on the process of learning based on performance on a puzzle maze. Record your literature review in your logbook. Do not forget to record the full references that you use. From your literature review, decide what aspect of the topic you want to investigate. Develop an aim, research question and hypothesis for your investigation.

Method
This scientific investigation requires students to use a puzzle maze, similar to the one on page 431, to obtain data on the effect of practice on performance.

The task requires students to access several participants, who are required to complete the same puzzle maze a number of times. The students will be required to collect data, use evidence and transfer their findings to explain the relationship between practice and performance on a puzzle maze.

Work out what you need to do in order to answer your research question and test your hypothesis. This will include:
- Population and sample – how big does your sample need to be?
- Variables – dependent, independent, controlled. Are there any extraneous variables you have not accounted for?
- What quantitative data are you going to record and how are you going to record it?
- Are there any ethical or safety issues you have to take into account?

Results

Collate the results for all participants and process the quantitative data obtained, using appropriate mathematical calculations and units (for example, descriptive statistics such as a mean). Organise, present and interpret data using an appropriate table and/or graph.

Discussion

1. State whether the hypothesis was supported or refuted.
2. Describe the relationship between practice and performance on the puzzle maze.
3. Discuss any implications of the results.
4. State if it is possible to generalise any of the results to the wider population.
5. Discuss any confounding variables and how they have affected the data.
6. Suggest future improvements to address the confounding variables if the investigation was to be repeated.

Conclusion

Write a conclusion to respond to the research question.

11 Chapter summary

KEY CONCEPTS 11.1

- » A logbook must be used and submitted along with your poster, to authenticate your work.
- » You must write a research question, based on a dot point from the study design.
- » Write an aim which is a broad statement about the goal of the research investigation.
- » Conduct a literature review, recording a summary of at least two relevant articles.
- » If you choose a controlled experiment methodology you will write a hypothesis that includes the IV, DV, prediction about how the IV influences the DV and population of research interest.
- » Write a method. Method includes participants, sampling technique, materials and procedure.
- » Consider potential extraneous variables, ethics and safety.

KEY CONCEPTS 11.2

- » Conducting an investigation means the generation of primary quantitative raw data.
- » Primary data is easier to interpret if it is organised. Summary tables, frequency tables, measures of central tendency and graphs are all ways of organising data.
- » Analysing data involves mathematical processes including mean, percentage or percentage change and understanding the variability within the data.
- » Evaluating data includes reviewing it for accuracy, precision, validity and errors.
- » Evaluating data is an assessment of error, uncertainty and confounding variables.
- » Interpreting data is deciding whether the data supports or refutes the hypothesis and deciding how generalisable the results are.
- » Any conclusions that you draw must be based on evidence provided by your data.

KEY CONCEPTS 11.3

- » A scientific poster is like an illustrated summary of your research investigation.
- » Poster conventions include simple colour and font themes and the inclusion of white space.
- » Your poster should contain no more than 600 words.
- » The focus of the poster should be graphics, not text.
- » The centre of the poster is a single sentence known as a communication statement.
- » There are seven sections or headings that must be included in order: title, introduction, methodology, results, discussion, conclusion, references and acknowledgements.

Glossary

Aboriginal and/or Torres Strait Islander peoples One way to refer collectively to First Nations Australians (Indigenous Australians), inclusive of the wide range of nations, cultures and languages across mainland Australia and throughout the Torres Strait

Accuracy How close a measurement is to the 'true value'; it cannot be a fixed number, but may be described as more or less accurate

Acronym An abbreviation formed from the first letters of each word in a title or procedure that can be pronounced as a word (e.g; ANZAC); can include initialisms that cannot be pronounced as words (FBI)

Acrostic A poem, word puzzle or other composition in which initial letters in each line form a word or words that the composition is about; or, a memorable phrase in which the first letter of each word in the phrase matches the first letter of a term in a sequence of terms to be remembered

Action potential The electrical charge that travels down the axon of a neuron during transmission of a neural impulse, occurring as the result of the rapid depolarisation of the neuron's membrane and prompting the release of neurotransmitters

Acute stress A state of brief but intense physiological arousal in response to an immediate perceived psychological stressor that normally has no long-term negative effects on health and wellbeing

Adequate diet Food and water intake that includes sufficient energy and nutrients to meet basic requirements for healthy living

Adequate sleep The necessary amount of sleep for individuals to function effectively during the daytime and cope with normal daily stress

Adrenaline A neurohormone that is released by the adrenal glands during the stress response, acting on the heart, lungs and muscles to optimise the body's flight-or-fight-or-freeze response to the stressor by increasing heart rate, oxygenation of blood and blood sugar levels, and relaxing smooth muscles to open airways; also called referred to as epinephrine

Advanced Sleep Phase Disorder (ASPD) A circadian rhythm sleep disorder characterised by a sleep pattern that is significantly earlier (by at least two hours) than a conventional or socially desirable sleep pattern, resulting in evening sleepiness and early-morning insomnia (inability to sleep); also known as advanced sleep phase syndrome; may lead to impairments in social and/or occupational functioning; more common in the elderly

Affective functioning The aspect of a person's behavioural responses that relate to the display and control of emotions and feelings

Agonist In neural communication, a substance that binds to a neuroreceptor to produce a similar effect to that of a neurotransmitter in either exciting or inhibiting a postsynaptic neuron; for example, benzodiazepine drugs are GABA (gamma-aminobutyric acid) receptor agonists that have sedative effects similar to GABA

Aim A question or statement about what the researcher intends to investigate; it describes the purpose of the study

Alarm-reaction stage The first stage of Selye's general adaptation syndrome response to stress, during which an initial physiological shock response changes to countershock, releasing cortisol and adrenaline to adapt to the stressor

Alpha waves The brainwave pattern seen in an electroencephalogram (EEG) that is typical of normal resting wakefulness, usually with eyes closed, or when a person is practising a meditative state of awareness; characterised by moderately high amplitude and moderate frequency

Altered states of consciousness (ASC) A psychological state that is characteristically different from normal waking consciousness, including altered levels of self-awareness, perceptions, emotions, sense of reality, orientation in time or space, responsiveness to stimuli, and memorability; including sleep; may be drug induced

Alzheimer's disease (AD) A progressive and fatal neurodegenerative brain disease in which amyloid plaques and neurofibrillary tangles disrupt neural functions, causing cell death and atrophy of the brain; early hippocampal damage disrupts consolidation of explicit memory first followed by progressive loss of existing episodic and semantic memories due to neocortical damage

Amygdala An almond-shaped brain structure located within each temporal lobe in front of the hippocampus; associates emotional information with explicit memories; plural amygdalae

Anecdote A factual claim based only on personal observation, collected in a casual or non-systematic way

Antagonist In neural communication, a substance that suppresses the release of a neurotransmitter or blocks the receptor sites, making the postsynaptic neuron less likely to fire

Antecedent In the three-phase model of operant conditioning, a stimulus that occurs before a voluntary behaviour and its consequence that serves to cue the behaviour

Anxiety The emotional state we feel when we anticipate threat or danger, causing feelings of fear, worry, dread or uneasiness; the threat may be specific or ambiguous

Aphantasia The term used to describe people who experience reduced or absent voluntary mental imagery

Approach strategy Any response to managing stress (coping) that involves direct engagement with the stressor; consisting of behavioural or psychological responses designed to change (remove or diminish) the nature of the stressor and/or how one thinks about it

Attention In observational learning, the cognitive process used to focus awareness on a model

Authentic social support Authenticity is a personal trait that reflects the extent to which a person is genuine and true to themselves and in their interactions with others; involves self-understanding in terms of recognising and owning own our own thoughts and emotional responses, including personal weaknesses, strengths, goals and values; authentic people behave with truth and integrity towards others and provide authentic social support when they truly listen to and connect with the experiences of another person

Autobiographical memory (ABM) The component of explicit memory that represents our episodic memories of personally experienced events and semantic self-knowledge

Autonomic nervous system (ANS) The component of the peripheral nervous system (PNS) that innervates involuntary (smooth) muscles and glands, including the organs of the circulatory, digestive, respiratory and reproductive systems; transmits information from the brain to organs and glands, and from these systems back to the brain

Avoidance behaviour Any behaviour that attempts to prevent exposure to a fear-provoking object, activity or situation

Avoidance strategy Any response to managing stress (coping) in which the person does not address the problem directly; diverting attention away from a threat or disengaging from a problem to escape painful or threatening thoughts, feelings, memories or sensations associated with the stressor

Axon The long, thin fibre of the neuron through which action potentials are transmitted

Axon terminal The swollen tip of a presynaptic neuron's axon, containing the synaptic vesicles from which neurotransmitters are released, that forms a synapse with the dendrite of the postsynaptic neuron

Basal ganglia A group of brain structures located at the base of the forebrain and in the midbrain that play important roles in controlling voluntary movement

Behaviour The responses of an individual to externally or internally generated stimuli, including voluntary and involuntary responses

Behavioural models Models proposing that phobic anxiety could be the result of learning

Behaviourism An approach in psychology based on the study of objective, observable behaviours rather than subjective, qualitative processes, such as feelings, motives and consciousness

Behaviourist approach An approach that emphasises the role of the environment in shaping observable behaviours based on learning associations between stimuli (stimulus–stimulus associations) and between stimuli and behavioural responses (stimulus–response associations), including the three-phase processes of classical and operant conditioning

Beneficence The commitment to maximising benefits while minimising risk and harm when taking a particular position or course of action

Benzodiazepine Any one of a group of medications used in the short-term treatment of anxiety (also anxiolytics); as GABA agonists, they enhance the GABA-induced inhibition of overexcited neurotransmitters, calming nervous activity

Beta blocker Medication used to treat high blood pressure, congestive heart failure, abnormal heart rhythms and anxiety by reducing heart rate and blood pressure through inhibiting the binding of adrenaline (epinephrine) to neuroreceptors

Beta waves Brainwaves characteristic of normal waking consciousness, with a low amplitude and high frequency

Between subjects design A research design where participants are tested once only, usually within an experimental group or a control group; also called independent groups

Biological rhythm A cyclical natural rhythm our body follows to perform a function

Biopsychosocial model An approach that proposes that health and illness outcomes are determined by the interaction and contribution of biological, psychological and social factors

Blood alcohol concentration (BAC) The amount of alcohol present in the bloodstream

Brain The master organ of the central nervous system, responsible for receiving and processing information from the rest of the body and generating responses to it

Brain imaging studies Studies that use imaging techniques, including magnetic resonance imaging (MRI) and functional magnetic resonance imaging (fMRI) to study brain structure and function

Brain lesion studies A brain research technique in which patients with damage (lesions) to a specific region of the brain are studied to determine the effects of the lesion on behaviour and cognition; can include post-mortem dissection and structural and functional brain imaging of living patients

Breathing retraining The process of identifying incorrect breathing habits and replacing them with correct ones

Bright light therapy A treatment for circadian rhythm phase disorders that exposes people to intense but safe amounts of artificial light for a specific and regular length of time to help synchronise their sleep–wake cycle with a normal external day–night cycle

Case study A type of research investigation that focuses on a particular person or event, which is studied in-depth; case studies usually involve direct observation and the gathering of qualitative data and provide insight into a particular psychological phenomenon

Catastrophic thinking Occurs when a person repeatedly overestimates the potential dangers and assumes the worst of an object or event

Central nervous system (CNS) A major division of the nervous system consisting of all the nerves in the brain and spinal cord

Cerebellum The major hindbrain structure; it controls movement, balance and coordination, and affects cognitive function by regulating the speed, consistency and appropriateness of mental processes

Cerebral cortex The layers of grey matter that cover the outside of the cerebral hemispheres, consisting mostly of the neocortex; it includes multiple distinct functional regions associated with the higher cognitive processes of attention, thought, perception, memory and language as well as sensory-motor processing; also called the telencephalon

Chronic stress A state of prolonged physiological arousal in response to a persistent stressor that negatively affects health and wellbeing

Circadian rhythm sleep disorders Any sleep disorder caused by a mismatch between a person's internal circadian rhythm and their actual or required sleep schedule; includes jet lag, shift work sleep disorder, Advanced Sleep Phase Disorder and delayed sleep phase disorder; referred to in *DSM-TR-5*™ as circadian rhythm sleep–wake disorder

Circadian rhythm Any regular, automatic variation in physiological or behavioural activity that repeats at approximately 24-hour intervals, including the sleep–wake cycle, body temperature

Classical conditioning A fundamental form of associative learning shared across species in which an organism learns to associate an originally neutral stimulus, such as an environmental sound, with the occurrence of a naturally rewarding or threatening event (unconditioned stimulus) that causes a reflex response (unconditioned response), such as the presence of a predator causing fear; through repeated experiences of the neutral stimulus preceding or co-occurring with the unconditioned stimulus, the neutral stimulus becomes a conditioned stimulus that on its own causes an automatic conditioned response similar to the unconditioned response, as if the unconditioned stimulus had occurred

Classification A system for organising similar information

Cognitive behavioural strategies Structured psychological techniques that can be applied to help people recognise how negative or unproductive patterns of thinking and behaviour affect their emotions; aimed at helping people think in new ways to promote more positive feelings and behaviours

Cognitive behavioural therapy (CBT) An evidence-based psychological treatment approach that teaches clients to apply cognitive behavioural strategies to recognise and change negative and unproductive patterns of thinking and behaving; intended as a brief treatment program that can be effective within weeks to treat conditions such as anxiety, depression and other mental health problems

Cognitive bias An automatic tendency or preference for processing or interpreting information in a particular way, producing systematic errors in thinking when making judgements or decisions

Cognitive model A psychological approach based on understanding how people's thought patterns, memories and beliefs affect their emotions, attitudes and behaviours

Communication statement A short one-sentence summary of the key findings of your research investigation

Conclusion Part of a research report; a statement that makes a judgement about the meaningfulness of the findings of the investigation; it should answer the question posed in the aim

Conditioned response (CR) A reflex response to a conditioned stimulus in the absence of the unconditioned stimulus that would usually cause it

Conditioned stimulus (CS) A previously neutral stimulus that acquires the ability to cause a reflex response through its association with an unconditioned stimulus

Confidentiality A participant's right to privacy and security of their personal information, including not being identifiable in the results

Confounding variable A type of extraneous variable that ends up changing the dependent variable (DV) in an unwanted way; this confounds the results as it is impossible to determine the cause of the change in the DV; confounding variables interfere with the internal validity of the study

Consciousness An organism's awareness of internal or external events, including awareness of sensations, perceptions, emotions and thoughts

Consequence In the three-phase model of operant conditioning, the feedback a learner receives from the environment as an outcome of a voluntary behaviour; can be reinforcing, punishing or neutral (no consequence)

Context-specific effectiveness In relation to coping strategies, the effectiveness of a strategy is influenced by the degree to which it provides a good match to the situation (context)

Control group In a controlled experiment, the group of participants that is not exposed to the independent variable (or treatment); it provides a comparison for the group that is exposed to the treatment, so ideally, its members are matched to the members of the experimental group on other relevant variables

Controlled experiment A type of research investigation where the researcher manipulates one or more independent variables and then measures the effect on a dependent variable (DV); the researcher attempts to control the influence of other variables that could also affect the DV; usually involves the comparison of outcomes for a control group and an experimental group

Controlled variable An extraneous variable whose influence has been eliminated from an experiment so that it cannot affect results; it has been controlled using a particular strategy (for example, matching groups)

Convenience sampling A method for selecting participants; they are selected because they are readily available to the researcher

Coping The process of adapting one's thinking and/or behaviour to manage the demands of a stressful or unpleasant situation

Coping flexibility The ability to stop an ineffective coping strategy (or evaluate the coping process) and implement an alternative effective coping strategy (or adapt the coping process)

Coping skills Learned behaviours or techniques that help us solve problems or meet the demands of a stressor

Coping strategy A deliberate action or thought process used to manage a stressful or unpleasant situation and/or to regulate one's response to such a situation

Correlational study A scientific investigation that involves measuring variables in an uncontrolled (natural) setting to identify and understand any relationships that may exist between them

Cortisol The primary stress hormone secreted into the bloodstream; it enables adaptive changes to energise the body in response to a stressor; however, prolonged abnormal levels in the bloodstream can cause health problems

Counterbalancing A technique used to control order effects in experiments that use participants more than once; participants are exposed to the different parts of the experiment in a different order (e.g; part A first or part B first)

Countershock The stage of alarm-reaction in Selye's general adaptation syndrome (GAS), during which the sympathetic nervous system is aroused and adrenaline increases, triggering a defensive fight, flight or freeze response

Country An Indigenous Australian understanding of place as a system of interrelated living entities, including the learner, their family, communities and interrelationships with land, sky, waterways, geographical features, climate, animals and plants

Cultural continuity The transmission and transformation of the traditional knowledges, values and practices of a cultural group over generations, connecting the past with the present and future.

Cultural determinants The protective factors that support the social and emotional wellbeing of Aboriginal and Torres Strait Islander peoples that come from connecting to culture, *Country* and spirituality (embedded within family/kinship and community connections)

Data The observed facts that constitute the results of an experiment

Debriefing At the end of a research study, participants are informed of the study's true purpose, essential in studies where deception has been necessary; mistaken beliefs are corrected and information is provided about services to help with distress resulting from participation

Deception Withholding information from participants about the true nature of and procedures used in a study; used in cases where giving participants the information beforehand might influence their responses and affect the internal validity of the study

Delayed Sleep Phase Syndrome (DSPS) A circadian rhythm sleep disorder characterised by a sleep pattern that is significantly later (by at least two hours) than conventional sleep patterns, resulting in later sleep onset and wake times; causing impaired alertness and performance during the day; common in adolescence; also referred to as delayed sleep phase disorder (DSPD) or delayed sleep–wake phase sleep disorder

Delta waves The lowest frequency brainwaves with high regular amplitude, characteristic of the deepest stages of non-REM sleep (stages 3 and 4)

Dendrites The threadlike branches that extend from the cell body (soma) of the neuron and receive signals from other neurons

Dependent variable (DV) The variable that is measured in an experiment; it is expected to change when exposed to the independent variable (IV); represented in graphs on the vertical (y) axis

Descriptive statistics Statistics used to summarise and organise data; includes measures of central tendency (mean, median, mode), the range and spread (standard deviation) of data, frequency tables and graphs

Diet The food a person regularly eats, considered in terms of its quality, composition and effects on health

Distress A negative psychological response to a stressor that results from being overwhelmed by the perceived demands of a situation, loss or threat

Dopamine A modulatory neurotransmitter produced in the midbrain and adrenal glands that plays a major role in the coordination of movement and in the regulation of reward; dopamine imbalance is associated with many neurological disorders (especially Parkinson's Disease) and with many mental health problems, including addictive behaviours

Double-blind procedure A procedure where neither the experimenter or the participants know which of the participants are in the experimental group and are therefore exposed to the independent variable

Dreams Rich, internally generated sensory, motor, emotional and other experiences that occur most often during periods of REM sleep

Encoding The processing of information in short-term memory to transfer it to long-term memory

Energising social support In interpersonal relationships, a person's capacity to promote feelings of motivation, engagement and action in others; social support is energising when there is a positive relational energy between people

Enteric nervous system (ENS) The largest component of the autonomic nervous system, which manages the functions of the digestive system (gastrointestinal tract); can function independently of the central nervous system and so is sometimes called the 'second brain'; the extensive two-way neural connections between the ENS and central nervous system (CNS), particularly via the vagus nerve, form the gut–brain axis

Episodic memory The component of explicit long-term memory used for storing and retrieving memories of personally experienced events and for imagining ourselves experiencing future events; accompanied by the feeling of mental time travel; also called episodic-autobiographical memory

Episodic-autobiographic memory See **Episodic memory**

Ethical concepts Considerations about how an investigation may affect human or non-human participants that must be taken into account before an investigation is carried out

Ethical guidelines Considerations about how an investigation may impact a human participant that must be taken into account before, during and after an investigation is carried out

Ethics Moral principles and values that provide guidance about making judgements and decisions about how to act; unethical behaviour is that which is morally wrong

Eustress A positive psychological response to a stressor that has been appraised as a challenge rather than a threat; characterised by positive psychological and physiological responses that allow the person to meet the challenge effectively

Evidence-based intervention Psychological treatment whose effectiveness has been supported by the integration of clinical research findings and clinical expertise

Excitatory neurotransmitter Neurotransmitters that increase the likelihood that a receiving neuron will fire an action potential

Exhaustion stage The last of the three stages of Selye's general adaptation syndrome (GAS), during which the adaptations made to resist the stressor break down, depleting the body of resources, causing symptoms such as increased risk of illness, sleep disturbance, fatigue and trembling

Experimental group In a controlled experiment, the group of participants exposed to the independent variable (IV) – the treatment

Experimental hypothesis A broad and general prediction about the direction of the relationship between variables in an experiment; that is, whether the variables increase or decrease in relation to one another

Experimenter effect Any expectations, beliefs or preferences of a researcher that may unintentionally influence their study, the recording of observations or any way that affects the outcome of the investigation

Explanatory power The ability of a model or theory to explain the phenomena of interest

Explicit memory (declarative memory) The kind of long-term memory we use when consciously remembering information about facts (semantic memory) or events (episodic memory)

External factors Influences on our mental wellbeing that come from biological and social sources outside of ourselves, in the physical and social environments

External stressor A stressor that comes from external factors, which are forces that you can't easily control; examples include major life events, discovering your pay has been cut, urgent deadlines or an upcoming exam

External validity The extent to which the results of an investigation can be applied (generalised) to people or situations beyond the sample

Extraneous variable Any variable, other than the independent variable, that may change the results (the dependent variable); researchers try to control extraneous variables before the research starts by thinking of what they could be and then taking steps to stop their effect

Fear hierarchy An assortment of stimuli related to a phobia, ordered from least fearful to most, used in desensitisation therapy to overcome the phobia

Fieldwork A data-collecting technique where an animal or person is observed in their natural environment; there is no experimental control of variables; also known as naturalistic observation

Flight-or-fight-or-freeze response (FFF) The body's automatic reaction to danger in which the autonomic nervous system mobilises energy and prepares the body for one of three responses: a) confront the stressor (fight); b) escape the stressor (flight); c) immobilise to evade detection and prepare (freeze)

Freeze response An automatic reaction to threat, controlled by the parasympathetic nervous system, that immobilises the body; adaptive to avoid detection from a predator and allowing time to respond; maladaptive when the response is so overwhelming that it prevents further adaptive action

Gamma-aminobutyric acid (GABA) The primary inhibitory neurotransmitter; its overall effects are to calm or slow neural transmission and therefore the body's response

General adaptation syndrome (GAS) The physiological model of the stress response proposed by Hans Selye which describes three stages of alarm-reaction (shock/countershock), resistance (or adaptation) and exhaustion

Generalisability The extent to which research findings can be applied to people and/or contexts outside of the research

Generalisation A decision or judgement about whether results obtained from a sample are representative of the population of interest

Glutamate The primary excitatory neurotransmitter in the brain responsible for the fast transmission of neural messages and involved in cognitive functions

Gut microbiota The system of microorganisms, including bacteria, that live in the gastrointestinal tract (digestive system), playing important roles in digestion and metabolism; also affecting brain health and functioning through extensive connections between the enteric nervous system and central nervous system (gut–brain axis)

Gut–brain axis (GBA) The network of bidirectional (two-way) communication pathways that allows communication between bacteria in the gastrointestinal tract (gut microbiota) and the brain; includes communication via chemical transmission through the bloodstream, neuronal and hormonal pathways, and via the immune system; causes disorders within the gut to affect the brain and vice versa

Hippocampus An organ deep in the temporal lobe that is involved in encoding, storing and retrieving explicit memories, and in particular the consolidation of episodic memories

Hydration Replenishing a lack of water in the body

Hypnogogic state A state when alpha waves begin to present on the EEG and a person is drifting from wakefulness to sleep

Hypothesis A testable prediction about the relationship between two variables; it is based on prior knowledge, so it is also considered to be an educated guess

Identification Placing a particular individual into one of the groups that was created

Implicit memory (nondeclarative memory) The kind of long-term memory that is demonstrated through changes in behaviour and adaptive responses as a result of repetition or practice, without conscious recollection of the knowledge that underlies the performance; can operate independently of the hippocampus

Independent variable (IV) The variable systematically manipulated by the experimenter to gauge its effect on the dependent variable; represented in graphs on the horizontal (x) axis

Informed consent Before participating, the researcher must explain the nature and purpose of the experiment, the potential risks and the participant's rights; participants then give their consent in writing should they wish to proceed

Inhibitory neurotransmitter Any neurotransmitter that decreases the likelihood that a receiving neuron will fire an action potential

Integrity When completing research, the commitment to searching for knowledge, the honest reporting of all sources of information and results (whether favourable or unfavourable) in ways that permit scrutiny and contribute to public knowledge and understanding

Internal factors Influences on our wellbeing that come from biological and psychological sources within ourselves, including genetic, acquired and learned factors

Internal stressor A stressor that is internal, such as stress-inducing thoughts from one's psychological mindset or expectations or behaviours; examples include putting pressure on yourself to be perfect or fear of public speaking

Internal validity Refers to whether the study was carried out following scientific procedures; in particular, variables other than the independent variable were properly controlled

Involuntary association A learned association between a conditioned stimulus and an unconditioned stimulus so that the conditioned stimulus produces an involuntary conditioned response (reflex) in the absence of the unconditioned stimulus

Justice The moral obligation to ensure that there is fair consideration of competing claims; that there is no unfair burden on a particular group and that there is fair distribution and access to the benefits of an action

K-complex A short burst of high-amplitude brainwaves, experienced in stage 2 non-rapid eye movement (NREM) sleep

Learning The biological, cognitive and social processes through which an individual makes meaning from their experiences, resulting in long-lasting changes in their behaviour, skills and knowledge

Levels of functioning The different degrees to which a person is able to engage effectively in the domains of life, such as in personal care, relationships and work

Literature review A method of scientific inquiry that involves collating, analysing and synthesising the existing scientific literature on a specific topic, with the goal of determining the 'state of the art' in the field, to provide the background to a new investigation, and/or to determine the extent of current consensus of views on the topic, articulate the other viewpoints and recommend directions for future research in the field

Logbook A complete, permanent record of how an experiment or research project was conducted; it shows what was done at every step along the way

Long-term depression (LTD) A form of neural plasticity that results in a long-lasting reduction in the strength of a neural response due to persistent weak stimulation

Long-term memory (LTM) The set of memory storage systems that enables us to store and retrieve knowledge and skills acquired over a lifetime with apparently unlimited capacity

Long-term potentiation (LTP) A form of neural plasticity that results in a long-lasting strengthening of neural connections at the synapse as a result of repeated stimulations from a presynaptic to postsynaptic neuron during learning; demonstrated to occur in cells of the hippocampus producing the neural changes that underlie the formation of memory

Maladaptive behaviour Behaviour that is potentially harmful and prevents a person from meeting and adapting to the demands of everyday living

Mean A measure of central tendency that gives the numerical average of all the scores in a data set; calculated by adding all the scores in a data set, then dividing the total by the number of scores in the set (see also **Median**, **Mode**)

Measure of central tendency A measure of the tendency for a majority of scores to fall in the mid-range of possible values

Median A measure of central tendency, the middle score in a data set; calculated by arranging scores in a data set from the highest to the lowest and selecting the middle score (see also **Mean**, **Mode**)

Meditation A set of techniques that is intended to encourage a heightened state of awareness and focused attention

Melatonin A hormone secreted by the pineal gland that causes drowsiness and helps to regulate the sleep–wake cycle

Memory bias A tendency to remember information of one kind at the expense of another kind; including the bias towards remembering negative and threat-related experiences associated with specific phobia

Mental imagery The conscious experience of perception-like representations without corresponding sensory input

Mental wellbeing A multidimensional construct that includes emotional, psychological and social wellbeing; it is experienced along a continuum from languishing to flourishing

Method of loci A mnemonic technique in which the items to be remembered are associated with specific locations on a familiar route or within a building or landscape, or even the night-sky

Microsleep Episodes of sleep lasting only a few seconds that are not detected by the brain, posing danger in situations like driving; characterised by momentary lack of awareness, sudden waking due to head falling forward or body jerks; occurs because of sleep deprivation

Mindfulness A state of awareness that arises when we pay non-judgemental attention to our thoughts and feelings in the present moment

Mindfulness meditation A meditation practice in which a person focuses attention on their breathing, with thoughts, feelings and sensations being experienced freely as they arise and without judgement; intended to enable people to become highly attentive to sensory information and to focus on each moment as it occurs

Mixed design A research design that includes both within- and between-subjects conditions as independent variables

Mnemonic Any device or technique used to assist encoding, storage and retrieval of memories; usually by creating an association between the information to be remembered and existing knowledge; also called memory-aid

Mode A measure of central tendency, it is the most frequently occurring score in a data set (see also **Mean**, **Median**)

Modelling The social and cognitive processes of learning by observing another person's behaviour and the consequences of the behaviour

Motivation and reinforcement In observational learning, the cognitive processes that influence whether the learner decides to reproduce an observed behaviour based on their understanding of the observed consequences

Motor neurons Specialised efferent neurons within the central nervous system that carry motor commands from the brain and spinal cord to muscles, organs and glands to control voluntary and involuntary movements; upper motor neurons carry information from the brain to the spinal cord; lower motor neurons form connections between the spinal cord and muscles, organs and glands

Multidimensional framework A conceptual structure (i.e. framework) that defines the roles of multiple intersecting factors that contribute to an experience. The Aboriginal and Torres Strait Islander framework of social and emotional wellbeing is an example. Western understandings of mental wellbeing also consider multiple dimensions

multimodal system (of knowledge) In Aboriginal and Torres Strait Islander cultures, the many formats in which knowledge is stored within *Country* and is expressed through language, stories, song, dance and art by human and more-than-human entities

Multi-store model of memory The model of human memory systems and processes described by Atkinson and Shiffrin (1968), including the three memory storage systems of sensory, short-term and long-term memory and the processes of encoding, storage and retrieval

Negative punishment The removal of a rewarding stimulus (reinforcer) as a consequence of a behaviour, making the behaviour less likely in the future

Negative reinforcement The removal of an aversive stimulus (punisher) as a consequence of a behaviour, making the behaviour less likely in the future

Neocortex The largest structure of the cerebral cortex, comprising 80 per cent of cortical grey matter other than the allocortex; modified through learning throughout life with specialised regions for higher cognitive functions involved in attention, thought, and language and sensory-motor processing

Nervous system The integrated network of neurons, nerves, nerve tracts and associated organs and tissues, including the brain, that together coordinate a person's functioning, behaviours and responses adaptively as they interact with and adapt to their external environment

Neuromodulator Any of a group of neurotransmitters that can affect a large number of neurons at the same time; they are slow acting but bring about long-lasting change to affected neurons and synapses; responsible for a range of human behaviour related to sleep, pain, motivation and voluntary movements

Neurotransmitter Any chemical released from the axon terminal buttons of a presynaptic neuron into the synaptic cleft following an action potential that either excites or inhibits the postsynaptic neuron

Neutral stimulus (NS) A stimulus (internal or external event) that does not naturally cause a reflex response

Non-maleficence A research ethic meaning to avoid causing harm; when it is not avoidable, the harm resulting from any course of action should not be disproportionate to the benefits from that course of action

Non-rapid eye movement (NREM) sleep The four stages of night-time sleep in which there is no rapid eye movement (REM); characterised by increasingly deep sleep as the stages progress, during which muscles become more relaxed and physiological functions slow; evident in electroencephalograph readings that show brainwaves of decreasing frequency and increasing amplitude, producing delta waves during slow-wave sleep typical of stage 4

Non-scientific ideas Knowledge that has not been obtained through the use of the scientific method

Noradrenaline (norepinephrine) A neurohormone that is produced in the brainstem and adrenal glands; as a hormone it works together with epinephrine to support the stress response, constricting arteries to increase blood pressure, heart rate and blood sugar levels; as a neurotransmitter it is released in response to emotional arousal and enhances the learning and memory for emotionally arousing events; also referred to as norepinephrine

Normal waking consciousness (NWC) The state of awareness we experience during wakefulness when we are aware of our surroundings and engage effectively in daily work, learning and social experiences; characterised by low-amplitude, high-frequency irregular activity in an electroencephalogram

Nutritional intake The daily eating patterns of an individual, including specific foods and calories consumed and relative quantities

Observational learning A form of social learning in which the learner attends to the behaviours of another person, encodes the behaviours in memory, and is motivated to rehearse and/or reproduce the behaviour based on their interpretation of the reinforcing consequences of the behaviour

Observer bias Bias in results of an observational study that occurs when an observer sees what they expect to see, or records only selected details of an observed behaviour

Observer effect Changes in the behaviour of a person being observed caused by their awareness of the presence of an observer; also called the Hawthorne effect

Operant conditioning A learning process in which the likelihood of a behaviour being repeated is determined by the consequences of that behaviour

Opinion A statement describing a personal belief or thought that cannot be tested (or has not been tested) and is unsupported by evidence (unless it is provided by an expert who uses evidence to support it)

Oral cultures Human cultural groups who use methods other than written language to store and pass down knowledge

Order effect Where prior knowledge of a task or situation influences a participant's performance, which in turn influences the results of the experiment (also known as the practice effect)

Outlier Data readings that lie a long way from other results; they may occur by chance or be the result of measurement and recording errors

Paradoxical sleep See **Rapid eye movement (REM) sleep**

Parasympathetic nervous system The branch of the autonomic nervous system that controls unconscious processes related to rest, repair and enjoyment, such as digestion, sleep, slowed heart rate, pupil constriction, sexual arousal; it calms the effects of the sympathetic nervous system

Parkinson's Disease (PD) A disease that causes tremors, muscular rigidity, slowness of movement and difficulty in initiating voluntary movement; believed to be caused by a lack of dopamine

Partial sleep deprivation Getting some sleep in a 24-hour period but less than normally required for optimal daytime functioning

Participant variables Individual differences in the personal characteristics of research participants that, if controlled, can confound the results of the experiment

Participants People or animals used in an experiment to study behaviour, characteristics or responses

(knowledge) Patterned on *Country* In Aboriginal and Torres Strait Islander approaches to learning, refers to the embedding of knowledge within the multimodal system of *Country*

Peripheral nervous system (PNS) The division of the nervous system that comprises all of the nerves outside the central nervous system (CNS), through which motor information is communicated from the CNS to muscles and organs of the body, and sensory information is communicated back to the CNS

Phobia A persistent, irrational and intense fear of a specific object, activity or situation

Physiological stress response The biological responses to stress that tend to follow the same patterns, regardless of the type and severity of the stressor

Pineal gland An endocrine organ (gland) located deep within the forebrain that secretes melatonin, which regulates body rhythms and the sleep–wake cycle

Placebo A medical treatment that looks real, but does not have any active ingredients, so it cannot have an effect on the condition being studied

Population The entire group of people that is of interest to a researcher, from which a sample will been drawn

Positive punishment The addition of an aversive stimulus (punisher) as a consequence of a behaviour, making the behaviour less likely in future

Positive reinforcement The addition of a rewarding stimulus (reinforcer) as a consequence of a behaviour, making the behaviour more likely in future

Postsynaptic neuron The neuron within a synapse that receives a neurotransmitter signal from a presynaptic neuron

Precision How close a set of measurements are to one another if conditions are not changed; precise measurements are repeatable and reproducible

Presynaptic neuron The neuron within a synapse that releases a neurotransmitter into the synaptic cleft to transmit a signal to postsynaptic neurons

Primary appraisal In the transactional model of stress and coping, the first judgement we make when evaluating whether a situation poses a threat, harm or challenge; followed by secondary appraisal

Primary data Data generated in a study by a researcher

Protective factor Any behavioural, biological, psychological or environmental characteristic that decreases the likelihood of a person developing a particular mental health problem or disorder

Psychoeducation A psychosocial approach in which a person experiencing a mental health problem or disorder and their family are provided with information to help them understand the condition and how they can contribute to managing it

Psychological construct A concept used in psychology to describe a mental process, psychological state or trait; they are used to describe something that is believed to exist, because we can measure its effects, but we cannot directly observe or measure it

Psychological model A construct built from current theoretical understandings to make theory more concrete and testable; models can have limitations, including incorrect assumptions or oversimplifications

Psychological stress response The cognitive, emotional and behavioural responses to stress; is subjective and unique to an individual

Psychological theory An organised set of interrelated psychological constructs, mechanisms and processes that describes and/or explains a psychological system, process or experience

Psychotherapy Any psychological technique used for treating mental health disorders, with the goal of producing positive changes in thinking, emotions, personality, behaviour or adjustment

Punishment A consequence of behaviour that weakens the likelihood of the behaviour being reproduced

Qualitative data Non-numerical data that describes the attitudes, behaviours or experiences of participants; often collected using questionnaires or interviews and includes descriptions of feelings and experiences

Quantitative data Data that is numerical; collected through systematic and controlled procedures and can be graphed and statistically analysed

Random allocation A procedure for assigning participants to either the experimental group or control group in an experiment, ensuring that all participants have an equal chance of being allocated to either group

Random error Unpredictable variations that are present in all measurements (except counting) and that result in a spread of readings; they affect the precision of a measurement

Random sampling A sampling technique that uses chance to ensure that every member of a population of interest has an equal chance of being selected to participate in the study

Rapid eye movement (REM) sleep The sleep stage that occurs between stages of non-REM sleep, in which most dreaming occurs, typically accounting for between one quarter to one fifth of total sleep time; characterised by high-frequency, low-amplitude electroencephalogram readings that resemble normal waking consciousness, accompanied by paralysis of skeletal muscles; also called paradoxical sleep because of the similarity in brainwave patterns with wakefulness

Raw data Any data collected in a study in its initial form; it has not been collated, sorted or otherwise processed

Receptor cells Cells located in the sense organs that are specialised to detect (receive) specific types of sensory information

Receptor sites Tiny areas on the membrane of a neuron that are sensitive to particular neurotransmitters, located mainly on dendrites but can be anywhere on the neuron

Reciprocal inhibition The concept that one emotional state is used to block another, as is the case in systematic desensitisation

Reinforcement A consequence of behaviour that strengthens the likelihood of the behaviour being reproduced

REM rebound A natural compensatory process that occurs after being deprived of REM sleep or after periods of stress in which a person experiences increased frequency, depth and intensity of the REM stage of sleep

Repeatability A procedure is considered repeatable if successive measurements that are carried out in the same way (using the same procedure, observer, equipment and location) produce the same or very similar results

Representative sample A randomly selected group that accurately reflects the characteristics of a larger population from which it is drawn; the group becomes the participants in the study

Reproducibility The closeness of the agreement between measurements of the same quantity that have been taken under different conditions (such as different method, observer, equipment, location or time); when findings are not reproducible, they may lack credibility

Reproduction In observational learning, the cognitive process used to re-enact an observed behaviour or to rehearse it mentally

Research hypothesis A hypothesis that operationalises the variables by precisely defining and describing how each variable is measured, and predicts the exact effect the independent variable is expected to have on the dependent variable

Resilience A person's ability to respond adaptively to stressful life events and cope with uncertainty, influenced by coping strategies, adaptive ways of thinking and social connectedness

Resistance stage The second of the three stages of Selye's general adaptation syndrome (GAS), during which the increases in cortisol and adrenaline, and other physiological changes, are stabilised to maintain the defensive response to the stressor; also called the adaptation stage

Respect Psychologists must conduct themselves showing respect for other people through their actions and language

Response A behaviour produced by an individual as an outcome of stimulus processing

Retention In observational learning, the cognitive process used to encode and store knowledge of observed behaviour

Retrieval The process of bringing to mind knowledge of events or facts stored in explicit memory, or of initiating and executing an implicit procedural memory

Reward system An area of the brain that responds to the release of dopamine to produce a state of motivation

Risk assessment Procedure to identify potential risks and hazards associated with an experiment

Risk factor The genetic and environmental conditions that influence the likelihood that a person will experience a mental health condition or another negative health outcome

Sample A group of participants selected to participate in a study, taken from a population of research interest

Sample size The number of participants in a study

Sampling The process of selecting participants from a population of interest; sampling techniques can be random, stratified or a sample of convenience

Secondary appraisal In the transactional model of stress and coping, the judgement we make after primary appraisal to evaluate our ability and resources to cope with the stressor

Secondary data Data that existed before a current research project was conducted and was collected by someone else; it may have been statistically summarised

Self-determination The fundamental right of Aboriginal and Torres Strait Islander peoples to shape their own lives, so that they determine what it means to live well according to their own values and beliefs

Self-report A data collection technique in which individuals are asked to freely express their attitudes (verbally or in writing) by answering questions

Semantic memory The component of explicit long-term memory that we use when we encode, store and retrieve factual and conceptual knowledge, and to recognise objects, people or places; accompanied by awareness of knowing without a feeling of reliving the past

Sensory memory The set of temporary memory stores with large capacity that enable sensory information to persist for a very brief duration so that goal-relevant information can be attended and encoded into short-term memory; includes iconic (visual) and echoic (auditory) sensory memory

Sensory neurons Specialised afferent neurons (sensory receptors) located in sense organs that detect and respond to information (physical energy) from the environment and transmit it to the central nervous system

Serotonin A major neuromodulator that is involved in pain, sleep and mood regulation

Shift work disorder A circadian rhythm sleep disorder caused by a person's work hours being scheduled during the normal sleep period (at night), causing their circadian rhythms to be out of step with their work patterns

Shift work Hours of paid employment that are outside the period of a normal working day and may follow a different pattern in consecutive periods of weeks

Shock The first of stage of alarm-reaction in Selye's general adaptation syndrome (GAS) during which the initial acute physiological response to the stressor occurs, causing a rapid drop in body temperature, blood pressure and muscle tone as though the body is injured

Short-term memory (STM) A temporary memory store that represents information that is the current focus of attention, with limited capacity (7 ± 2 items) and a duration of several seconds or for as long as information can be actively rehearsed

Simulation A way to imitate an environment in a realistic way to represent a large, complex system or a system that is difficult to access or dangerous

Sleep A naturally occurring altered state of consciousness governed by circadian rhythms during which awareness of ourselves and the environment is suspended; characterised by a series of typical changes in sleep electroencephalogram readings, muscle tension, eye-movements and other physiological changes that accompany the different stages of sleep

Sleep debt The amount of sleep loss accumulated from an inadequate amount of sleep, regardless of the cause

Sleep demand The regulation of sleep and wakefulness through the interaction between circadian rhythms and sleep–wake homeostasis

Sleep deprivation The condition of having had sufficient sleep to support optimal daytime functioning

Sleep diary A log of subjective behavioural and psychological experiences surrounding a person's sleep

Sleep disorder A persistent disturbance of typical sleep patterns (including the amount, quality and timing of sleep) or the chronic occurrence of abnormal events or behaviour during sleep

Sleep hygiene A set of behavioural treatment techniques to manage insomnia (inability to fall asleep) and improve sleep patterns, which may include: using the bed only for sleeping and sex, reduced daytime napping, decreasing caffeine and avoiding it after 4 p.m., a regular bedtime routine and keeping a sleep diary

Sleep laboratory A specialised research facility designed to study the physiological and behavioural activities that occur during sleep, equipped with specialised electronic monitors to record breathing, heart rate, brainwaves and muscle tone

Sleep spindles A type of brain activity characterised by a short burst of high-frequency brainwaves, experienced during stage 2 NREM sleep

Sleep-deprivation psychosis A disruption of mental and emotional functioning as a result of lack of sleep

Sleep–wake cycle The natural biological rhythm that produces the pattern of alternating sleep and wakefulness over a 24-hour period

Slow-wave sleep (SWS) A sleep state characterised by the emergence of high-amplitude, low-frequency delta waves, experienced during stages 3 and 4 NREM sleep

Social and emotional wellbeing (SEWB) A holistic model of the determinants of health, including mental health, for Aboriginal and Torres Strait Islander peoples and communities; it considers the influence of a person's connections to the seven domains of body and behaviours; mind and emotions; family and kin; community; culture; *Country*; and spirituality and ancestry

Social learning theory A theory in which human learning is proposed to occur primarily through interactions with others, especially through observing the behaviours of others and the consequences that follow, including active cognitive processes that determine whether the behaviour will be imitated

Social support The assistance and comfort we receive from people in our social network when we are facing a stressful or challenging situation, from family members and friends through to support groups and social institutions

Social-cognitive approach An approach that emphasises the active role of the learner in learning through observing and listening to others (social learning) and as actively making meaning of what they observe (cognitive learning)

Soma The main cell body of a neuron that contains the nucleus and other organelles

Somatic nervous system The part of the peripheral nervous system (PNS) that transmits sensory information received from sensory receptors to the central nervous system (CNS), and motor messages from the CNS to skeletal muscles

Somatic reflex An involuntary response involving skeletal muscles

Songlines The sung narratives encoded in physical routes across *Country* and in constellations in the night-sky that convey ancestral knowledge of *Country*; also known as Song-spirals or Dreaming

Specific phobia A kind of anxiety disorder in which a person experiences intense fear or anxiety towards a particular object or situation; the response is persistent, overwhelming and out of proportion to the actual threat

Spinal cord Part of the central nervous system consisting of a cable of sensory and motor nerve fibres that extend from the brainstem through a canal in the centre of the spine to the lumbar region of the spine; transmits sensory information from the peripheral nervous system (PNS) to the brain, and motor messages from the brain to the PNS

Spinal reflex The simplest kind of automatic, unlearned responses to stimuli (reflex) controlled by simple sensory-motor circuits in the spinal cord (bypassing the brain); includes reflexes comprising a sensory and motor neuron connected by an interneuron (e.g. withdrawal reflex) and reflexes in which there is a direct connection between a sensory and motor neuron (e.g. patellar knee-jerk response); also called simple reflex arcs

Spontaneous recovery The reappearance of a conditioned response after a period of apparent extinction

Sprouting The growth of new dendritic spines and/or axon terminals on a neuron, allowing stronger and more numerous connections with other neurons

Standard deviation (SD) A statistical measure representing how far each value in a data set is from the mean

Stigma A negative social attitude about a characteristic of a person or social group that implies some form of deficiency, often leading to unfair discrimination against or exclusion of the person or social group

Stimulus Any internal or external event that produces a response in an individual

Storage The retention of information in memory over time

Stratified sampling A sampling technique used to ensure that a sample contains the same proportions from each nominated strata that exist in the population

Stress The physiological and psychological responses that a person experiences when confronted with a situation that is threatening or challenging

Stress responses A set of physical and psychological responses that are automatically set in motion when the sympathetic nervous system is activated following the perception of a threat

Stressor An object, entity or event that causes a feeling of stress

Sung narratives See **Songlines**

Suprachiasmatic nucleus (SCN) A cluster of neurons in the hypothalamus located directly above the optic chiasm that regulates the body's circadian rhythms, particularly the sleep–wake cycle, using information about the intensity and duration of light received from the retina via the optic nerve

Sympathetic nervous system The branch of the autonomic nervous system that alters the activity level of internal muscles, organs and glands to physically prepare our body for increased activity during times of high emotional or physical arousal

Synapse The specialised junction between a presynaptic and postsynaptic neuron, separated by a small gap called the synaptic cleft; enables neural signals to be transmitted when an action potential causes the presynaptic neuron to release neurotransmitters into the synaptic cleft, where they can bind to receptors on the postsynaptic neuron

Synaptic cleft The 20–40 nanometre gap between the axon terminals of the presynaptic neuron and receptors on the dendrites of the postsynaptic neuron in the synapse

Synaptic plasticity The term used to describe the changes that occur to the synaptic connection between two or more neurons

Synaptic pruning A process of removing extra, weak or unused synaptic connections so that neural transmission is as efficient as possible

Synaptic vesicles Small spherical sacs within the axon terminals of presynaptic neurons that contain molecules of neurotransmitters; an action potential causes the vesicles to fuse with the cell membrane and release neurotransmitters into the synaptic cleft

Systematic desensitisation A type of behaviour therapy that uses counterconditioning to reduce the anxiety a person experiences when in the presence of, or thinking about, a feared stimulus; involves first learning specific muscle relaxation techniques and then practising these while being exposed to experiences with the feared stimulus that systematically increase

Systematic error An error that causes readings to differ in the same way each time a measurement is made, so that all the readings are shifted in one direction (up or down) from the true value; systematic errors affect the accuracy of a measurement

Theta waves A mix of brainwaves which indicate that the electrical activity in your brain is slowing and you are moving further away from consciousness

Three-phase process of classical conditioning A model of the three stages involved in learning a classically conditioned response, including three phases described as before conditioning, during conditioning and after conditioning

Three-phase process of operant conditioning A model of the three components involved in acquiring and maintaining an operantly conditioned voluntary behaviour, including antecedent, behaviour and consequence; also called the ABC model of operant conditioning

Total sleep deprivation The condition of going without any sleep in a 24-hour period

Transactional Model of Stress and Coping A model that proposes that stress is a subjective experience that varies between individuals depending on how they interpret the stressor and perceive their own ability to cope with it

True value The value (or range of values) that would be found if the quantity could be measured without error

Ultradian rhythm A biological rhythm that follow a cycle of less than 24 hours and determines the timing and duration of our sleep states

Uncertainty Potential sources of variation cause uncertainty in all measurements; the level of uncertainty in a measurement influences all inferences and conclusions that are based on that measurement

Unconditioned response (UCR) An involuntary reflex response to a biologically significant stimulus

Unconditioned stimulus (UCS) A biologically significant stimulus, such as food or a sudden loud sound, that causes a reflex response

Unconscious Describes the self-regulating, constant responses, controlled by the autonomic nervous system, that are not normally under our conscious control

Vagus nerve The tenth cranial nerve that is the major communication route in the gut–brain axis, extending from the brainstem to provide parasympathetic innervation to many organs, including the gut and digestive organs

Validity Whether a questionnaire or scale actually measures what it is supposed to measure; an investigation has internal validity when the study produces results that can be interpreted meaningfully in relation to the aims of the study; an investigation has external validity when the results can be meaningfully generalised from the sample to the population

Variability A single number that tells us the degree to which scores in a distribution are spread out or clustered together

Variable Any condition (trait, event or characteristic) that can have a range of values; can be manipulated or measured in an investigation

Voluntary participation When participants willingly agree to take part in an experiment free from pressure or fear of negative consequences, after understanding what is required of them

Ways of knowing Indigenous Australian approaches to learning and knowledge based on core understanding of the interrelationship between all entities, so that the learner and knowledge are embedded within a system of interrelationships; knowing takes place through being and doing

Withdrawal rights Participants are entitled to leave an experiment (withdraw) for any reason at any stage without any negative consequences; participants are informed of the right to withdraw before agreeing to participate

Within-subjects design A research design where the same group of participants makes up both the experimental and control groups; also known as repeated measures

Zeitgeber An external cue such as light, temperature, noise or food that influences the activation or timing of a biological rhythm, such as the circadian sleep–wake cycle; from German meaning 'time giver'

ANSWERS

CHAPTER 2

ACTIVITY 2.1
Apply it

2 Sensory receptors in your fingers detected a stimulus. The sensory messages were sent via the somatic division of the peripheral nervous system to the spinal cord.

↓

The message is now in the central nervous system.

↓

Interneurons in the spinal cord intercept the message and immediately connect with motor neurons.

↓

This is the motor role of the somatic division of the peripheral nervous system.

The motor message in the somatic nervous system would be to direct the muscles of the [hand/finger] to move quickly away from the sharp end of the compass, the pain stimulus. This is an example of an unconscious response known as a spinal reflex.

Exam ready

3 C

ACTIVITY 2.2
Apply it

2 Glutamate is the body's primary excitatory neurotransmitter. GABA is the body's primary inhibitory neurotransmitter. If you are low in glutamate, then you are likely to also be low in GABA. If you are low in GABA the body's ability to relax becomes impaired, resulting in anxiety.
Other possible symptoms include stress, difficulty concentrating, memory problems and sleep problems.

Exam ready

3 C

ACTIVITY 2.3
Apply it

2 Long-term potentiation is the long-lasting increase in the strength of neural pathways due to repeated stimulation. When Jorje rehearsed his monologue for his solo performance, the repeated rehearsal would have increased the strength of the relevant neural pathways, making it easier and easier for Jorje to remember the words of the monologue for his performance in front of the panel.

Exam ready

3 B

END-OF-CHAPTER EXAM

Section A: Multiple-choice

1 A	5 D	9 D	13 D	17 C
2 D	6 D	10 B	14 C	18 D
3 C	7 B	11 D	15 D	19 D
4 A	8 D	12 B	16 C	20 A

Section B: Short answer

1 a Somatic nervous system [1 mark]
  b A spinal reflex was responsible for Julie lifting her leg suddenly without any conscious awareness [1 mark]. Interneurons within the spinal cord will carry the message to the brain, where it is interpreted as pain, but this takes longer than the motor message to lift her foot [1 mark].

2 a LTP [1 mark]. When Trudy was learning to speak Spanish as a child, the repeated stimulation of neural pathways while she was learning strengthened these neural pathways [1 mark], so remembering the skill of speaking Spanish was strengthened [1 mark].
  b LTD [1 mark]. When Trudy stopped speaking Spanish as an adult, long-term depression caused the neural pathways for remembering Spanish words become weakened [1 mark] due to repeated low-level stimulation of these pathways [1 mark].
  c LTP [1 mark]. The original neural pathways that were weakened due to lack of stimulation over the years when Trudy was not speaking Spanish [1 mark] could become strengthened again by new stimulation in relearning Spanish and then speaking Spanish in her retirement [1 mark].

3 Increase in the activation of receptors [1 mark]; significant increase in the number of receptor sites on the postsynaptic neuron [1 mark]; increase in glutamate production in the presynaptic neuron [1 mark].

CHAPTER 3

ACTIVITY 3.1

Apply it

2 When we experience acute stress, the sympathetic nervous system is activated. The stress response is considered to be non-specific because the same physiological changes occur when we experience eustress and distress; however, our psychological interpretation of the event is different.

Exam ready

3 A

ACTIVITY 3.2

Apply it

2 a During the first half of the term, Oscar was most likely in the resistance stage of the GAS. Oscar is responding to the stressors in his environment and, despite feeling stressed at times, he is able to cope.

b In the last few weeks of the term, Oscar was most likely in the exhaustion stage of the GAS. Oscar's resources are depleted, he is struggling to keep up with his commitments and he is experiencing migraines.

Exam ready

3 A

ACTIVITY 3.3

Try it

1 1 B; 2 C; 3 F; 4 G; 5 D; 6 E; 7 A

Apply it

2 Possible appraisals that Jake might form are:
   » Threat: Jake might appraise the situation as a threat to his ambitions of playing football professionally.
   » Challenge: Jake might see this as a chance for personal growth because this is an obstacle to overcome.
   » Harm/loss: Jake might appraise the initial injury he suffered as harm/loss.

Exam ready

3 C

ACTIVITY 3.4

Apply it

2 The gut–brain axis is a bidirectional relationship that links our emotional and cognitive centres with our digestive system. An imbalance in our microbiome (dysbiosis) can lead to a variety of physiological conditions, which also has an influence on our psychological functioning. The use of a faecal transplant restores the diversity of the gut biome, which – according to recent studies – helps to reduce our experiences of depression and anxiety.

Exam ready

3 C

ACTIVITY 3.5

Apply it

2 Approach strategies – either of:
   » Mitch messages Toby for tips on how to study.
   » Mitch downloads a few past exams and completes the multiple-choice questions.

Avoidance strategies – either of:
   » Mitch watches hours of TV instead of studying.
   » Mitch spends time on his mobile phone scrolling through his social media feed instead of studying.

Exam ready

3 B

END-OF-CHAPTER EXAM

Section A: Multiple choice

1 D	5 C	9 D	13 D	17 B
2 D	6 B	10 D	14 B	18 B
3 A	7 B	11 B	15 A	19 B
4 D	8 B	12 C	16 D	20 A

Section B: Short answer

1 A VCE student may fall ill just before their exams because, according to Selye's general adaptation syndrome, the student will be dealing with the initial stressor (preparing for the exams) [1 mark]. Because physiological arousal remains at a higher-than-normal level to deal with the initial stressor, the student's immune system can become compromised and their ability to deal with additional stressors (such as infection) is lowered [1 mark]. Therefore, they become sick and their body will not have any resources left to fight the new stressor [1 mark].

2 Approach strategies are considered to be an adaptive response to stress because, through the use of practical strategies to deal with the nature of the stressor, and by adjusting to any changes and demands of the stressor, they help us to tolerate that stressor [1 mark]. Avoidance strategies are considered to be a maladaptive response to stress because the strategies do not attempt to deal with or defeat the stressor; rather, they are strategies that attempt to avoid or escape the stressor [1 mark].

3 Any two of the following limitations: [2 × 1 mark]
   » The model is too simplistic; it does not consider the automatic physical responses the body has to a stressor.
   » It is difficult to test through experimental research because of its subjective nature.
   » It is difficult to separate primary and secondary appraisals because they often occur simultaneously.

Unit 3 Area of Study 1 review

Multiple-choice questions

1 B	7 B	13 C	19 B	25 B
2 D	8 C	14 C	20 B	26 A
3 B	9 A	15 B	21 B	27 C
4 D	10 D	16 C	22 C	28 A
5 B	11 B	17 C	23 D	29 B
6 C	12 C	18 C	24 A	30 B

Short answer questions

1 Any two of: [2 × 2 marks]
   » Both consist of motor neurons.
   » Both stimulate muscles.
   » Both lack conscious awareness to be activated.

2 The subdivision involved is the somatic nervous system [1 mark].
The sensory receptors on Lucas' leg detect the heat of the burn [1 mark].
They send a sensory pathway (afferent) to the spinal cord [1 mark].
The spinal cord makes a quick unconscious response (does not involve the brain) and sends a motor/efferent pathway back to the leg while simultaneously sending a message to alert the brain [1 mark].
Lucas automatically jumps away from the fire [1 mark].

3 a It is hypothesised that students who are told the test is going towards their reports will experience greater stress levels when compared to students who are completing the test with no bearing of results [1 mark].
  b A limitation of using heart rate as the measure is that other external factors can impact upon heart rate [1 mark]. For example, the individual may have an increased heart rate due to the fact they are worried about having their heart rate recorded, or because the test is being sprung on them. The heart rate increase may not be due to anxiety about the actual test itself [1 mark].
  c The primary appraisal of group 1 is likely that of either threat/harm/loss or challenge [1 mark] – essentially, individuals in this group will view it as a threatening situation. This is because they have been told that the test result will be recorded on their reports, and also because most students did not have time to prepare for the test [1 mark]. This is evident in the graph: the heart rates of students in group 1 increased more significantly than heart rates for individuals in the control group [1 mark]. The graph does show an increase in the heart rates of students in the control group, though this increase is less than that of students in the experimental group [1 mark]. This highlights that while some students in the control group may have had a primary appraisal of stress, most considered it an irrelevant stressor [1 mark], and hence didn't need to move onto a secondary appraisal [1 mark].
  d This would highlight an uneven allocation as students at risk of failure are more likely to experience higher stress levels than those who are not [1 mark], as the test poses an even higher threat to these individuals [1 mark]. If the experimental group contains more students at risk of failure it may impact on the overall averages of heart rates; that is, the heart rates may be higher due to the threat of failure as opposed to the stress related to reporting of test results [1 mark]. These results cannot be generalised to the population due to the confounding variable present as we cannot determine whether the results were due to the independent variable or due to the uneven allocation of students [1 mark].

4 (Sample answer)
The transactional model of stress and coping is the psychological model to explain their stress [1 mark]. This model is split into primary appraisals and secondary appraisals. Our primary appraisals are the initial determination about whether the stimulus is stressful. In terms of this model, both Lia and Geoff both viewed this as a stressful situation, with Lia determining it as a threat of potential harm or loss due to the financial instability she will face and Geoff viewing it as a challenge to focus his attention on university [1 mark]. A secondary appraisal is where the individual assesses their resources to cope with the stressor. Geoff would have had a secondary appraisal of having the resources to cope with the stressor because he had financial support from his family to back him up after quitting [1 mark]. Lia, on the other hand, would have appraised that she didn't have the resources to cope as without the job she was struggling financially, which was impacting on her mental health

and her grades [1 mark]. A key benefit of a psychological perspective is that it allows for individual interpretation; we can see both Lia and Geoff interpreted the stressor differently based on their own circumstances [1 mark]. However, a criticism of this model is that it doesn't account for the physiological aspects of the stress response and focuses purely on the psychological [1 mark]. To overcome this limitation, the general adaptation syndrome model aims to address the physiological aspects of the stress response. In this model both Lia and Geoff would experience their stressors in the same way. The GAS model consists of three stages: alarm reaction, resistance and exhaustion [1 mark]. The alarm reaction stage is the initial stage and is split into shock and countershock. This was experienced when Lia and Geoff first quit their jobs; initially, their ability to cope would have dropped below normal and then increased during countershock [1 mark]. Both Geoff and Lia are now in the resistance stage as they are actively dealing with the stressor: Geoff is coping well due to high levels of stress resistance; however, due to the other life factors Lia was facing, her ability to cope with these was reduced [1 mark]. As such, her immune system slowed and Lia got sick, and this impacted on her ability to overcome the stressor. While this model aims to overcome the limitations of the transactional model (i.e. it ignores the physiological aspect), this model rejects any subjectivity or personal interpretation of the stressor [1 mark]. Hence, it is important to understand that both Lia and Geoff are going through a similar physiological response even though their psychological interpretations of the stressor are very different and based on their own circumstances. It is therefore important to consider both models to gain an in-depth understanding of the stress response.

CHAPTER 4

ACTIVITY 4.1

Apply it

2 Giving stickers – positive reinforcement. Adding in something desirable (stickers) will increase the likelihood that the behaviour (five minutes of focus) will be repeated again in the future.
Taking away recess time – negative punishment. Removal of something desirable (recess time) will decrease the likelihood that the behaviour (fidgeting etc.) will be repeated in the future.

Exam ready

3 B

ACTIVITY 4.2

Apply it

2 a Retention is where the learner forms a mental representation of the behaviour being learned and retains that information.
  b When you watch the video, you might need to pause and rewind to see the folding techniques a few times before you are successful at making the origami sunglasses. This is suggested in the text of the video as well.
3 a Reproduction is when the learner is physically and mentally capable of carrying out the behaviour they have seen.
  b As a VCE student you should be dextrous enough to fold the A4 paper, and have access to the A4 paper.
4 a Motivation is the incentive to repeat the behaviour being observed, and reinforcement is the desirable consequence that comes from repeating the behaviour that has been observed.
  b You must have the desire to make the origami sunglasses (to enhance your learning, to please your teacher, to make very cool origami sunglasses). Then, if you are praised for making the glasses or see someone else being praised for making the glasses, you are more likely to make the origami glasses again in the future. This is positive reinforcement (or vicarious reinforcement if you see someone else being praised).

Exam ready

3 C

END-OF-CHAPTER EXAM

Section A: Multiple-choice

1 B	5 B	9 C	13 D	17 D
2 D	6 B	10 B	14 B	18 C
3 B	7 C	11 A	15 D	19 C
4 D	8 D	12 D	16 C	20 D

Section B: Short answer

1 a [1 mark for each correctly identified element]
   NS: new soft drink
   UCS: people having a good time
   CS: new soft drink
   UCR: people experiencing positive emotions in response to having a good time
   CR: people experiencing positive emotions about the soft drink

[Total 5 marks]

b Long-term potentiation (LTP). Each time the person watches the advertisement linking the soft drink (conditioned stimulus) to people having a good time (unconditioned stimulus), the neural circuits that associate good times with the soft drink are stimulated [1 mark].

Linking repeated pairing of UCS and CS to LTP. This repeated stimulation of the neural connections at the synapses in the neural circuit results in the neurons becoming more efficient at carrying the information associating the soft drink with good times [1 mark].

Linking the learning to LTP. This strengthening of neural pathways in response to experience is long-term potentiation, which is the neural basis for learning that the soft drink is associated with good times, causing the conditioned reflex of positive emotions [1 mark].

2 a Antecedent: Ava not given a chocolate before dinner time [1 mark]
Behaviour: Ava throws a tantrum [1 mark]
Consequence: Simran ignores Ava's tantrum [1 mark]

b Dopamine is released in anticipation of rewards [1 mark]. When Ava asks for chocolate before dinnertime, she is expecting to receive the desired reward (chocolate treat) and her dopamine levels will increase [1 mark].

c [1 mark for each correctly identified cognitive process of observational learning explained with reference to Noah learning to throw tantrums.]
» Attention: Noah has watched closely when Ava has thrown tantrums when she does not get chocolate before dinnertime. [1 mark]
» Retention: Noah has made a mental representation of Ava throwing a tantrum. [1 mark]
» Reproduction: When Noah does not get what he wants he repeats the actions he has seen Ava carry out and he throws a tantrum. [1 mark]
» Motivation and reinforcement: If Noah has a strong desire for the object he wants, he is more likely to throw a tantrum. He also may have seen Ava receive the positive reinforcer of chocolate on some occasions when she throws a tantrum, and this is likely to strengthen his tantrum-throwing behaviour. [1 mark]

3 a [1 mark for explanation, with an example, for each of the ways of knowing, being and doing.
Student answers will vary.]
'Ways of knowing' refers to Australian First Nations Peoples' different ways of understanding the world. Their understanding of *Country* is a system of interrelated living entities in which humans are embedded, and that includes the landscape, waterways, skies, animals, plants, climate and the Songlines/Songspirals, Dreaming and ancestors. Knowledge is held by all living entities and they can teach people the knowledge [1 mark].
'Ways of being' refers to the ways that an individual attends to and interacts with other entities in the system. An example is the practice of *Dadirri* – a word from the peoples of the Daly River that means 'deep listening'. Only through listening deeply to *Country* can people understand the inter-relationships between entities and how they inform adaptive behaviours [1 mark].
'Ways of doing' refers to the methods for learning and communicating knowledge through cultural activities; for example, singing stories and linking them to dance and drawn symbols [1 mark].

b [1 mark for explanation of role of The Dreaming.
1 mark for explanation of role of Songlines.
1 mark for linking to knowledge holders (stores) and knowledge transmission.
Student responses will vary in the selection of details.]
The Dreaming is the body of ancestral knowledge that has been passed down orally over thousands of generations. It is encoded in the physical landscape, which can be 'read' like a book. The land and sky are like libraries in which the knowledge is stored [1 mark].
Songlines are the sung narratives that represent the body of ancestral knowledge that has been passed down over many generations. They are used to encode trade and ceremonial routes, and knowledge of plants, animals and natural resources [1 mark]. They also include knowledge of cultural laws and the kinship relationships that form the basis of knowledge transmission. The Songlines contain different layers of meaning that enable stories to be told in different ways depending on the stage of life of the audience [1 mark].

CHAPTER 5

ACTIVITY 5.1

Apply it

2 It is likely that people better recalled the second list due to the use of chunking. Chunking is a technique that can be used to increase the capacity of STM beyond the usual 5–9 items. Chunking refers to the process of taking individual pieces of information and grouping them into larger units. The items in the first list were presented individually (not in chunked form), meaning that they would quickly exceed the capacity of STM.

Exam ready

3 B

ACTIVITY 5.2

Apply it

2 The amygdala is involved in the consolidation of emotionally significant memories (particularly those related to fear). As such, given the damage to SM's amygdala, she is unable to attach information regarding fear to her memory.

Exam ready

3 D

ACTIVITY 5.3

Apply it

2 Aphantasia is a condition in which a person struggles or is unable to bring an image to mind. People with aphantasia often experience severely deficient autobiographical memory (SDAM). As such, it would be expected that a person with aphantasia would find it difficult to recall their first concert as they are unable to visualise the venue, the band, who they were with, and so on.

Exam ready

3 D

ACTIVITY 5.4

Apply it

2 The average person would remember only 5–9 items from the list. This is because the average person's short-term memory has a capacity of 5–9 items. Given that the list contains 10 items, it would be unlikely that a person could recall all items using maintenance rehearsal. Maintenance rehearsal increases the duration of short-term memory but does not increase the capacity of short-term memory.

Exam ready

3 D

END-OF-CHAPTER EXAM

Section A: Multiple-choice

1 C	5 A	9 B	13 D	17 C
2 B	6 B	10 C	14 A	18 C
3 A	7 D	11 A	15 C	19 D
4 B	8 D	12 B	16 A	20 B

Section B: Short answer

1 a The multi-store model explains the transfer of information from STM to LTM as a process of encoding via maintenance rehearsal [1 mark]. Maintenance rehearsal involves mentally repeating the information to be remembered [1 mark].

b The evidence that supports the multi-store model account of transfer from STM to LTM via maintenance rehearsal is the primacy effect [1 mark], where words at the beginning of the list are recalled better than those in the middle [1 mark]. The evidence that supports this explanation comes from studies that eliminate the primacy effect by preventing maintenance rehearsal during encoding [1 mark].

2 When Eliza imagines herself in a future experience she needs to construct the imagined scene using information from her general semantic knowledge [1 mark, note not semantic autobiographical knowledge]. Thinking about the meaning of the word 'formal' brings to mind general knowledge (schema) of the kinds of things to expect at a formal celebration [1 mark]. This allows her to mentally 'set the stage/scene' in which she can then imagine herself participating [1 mark].

3 Aboriginal peoples' use of Songlines differs from the method of loci by linking sung narratives [1 mark] and performances [1 mark] with areas of *Country*.

Unit 3 Area of Study 2 review

Multiple-choice questions

1 C	7 D	13 A	19 C	25 A
2 A	8 C	14 C	20 D	26 D
3 C	9 C	15 D	21 C	27 B
4 C	10 D	16 A	22 D	28 C
5 B	11 D	17 B	23 C	29 C
6 A	12 C	18 B	24 B	30 B

Short answer questions

1 a This is a procedural memory [1 mark] because it is a memory regarding how to do things – in this case it is Mila's memory of how to ride a bike. It is not a memory of personal experience and thus cannot be episodic [1 mark].

b The basal ganglia [1 mark] and cerebellum [1 mark] are the areas of the brain responsible for storing implicit procedural memories. The basal ganglia process and adjust voluntary movement [1 mark]; the cerebellum stores classically conditioned memories [1 mark].

2 Alzheimer's disease begins in the hippocampus [1 mark], so if Ashita is in the early stages this will be the first place that begins to degrade. This can be determined due to the symptoms seen in early onset, such as general forgetfulness and struggling to encode information from short-term to long-term memory [1 mark]. The hippocampus helps commit short-term memories to long-term memory. Damage to the hippocampus interrupts this role, leading to forgetting of new memories before pre-existing ones [1 mark].

3 a Operant conditioning [1 mark]. This is because Michael is an active participant of the conditioning of the response [1 mark], and he is consciously aware that in doing his work on time he will receive a sticker [1 mark].

b Any two of the following:
- Motivation/reinforcement [1 mark]: Lucas was motivated by the reinforcement of receiving a sticker as a reward from Ms Lewis if he did his work, as he had watched that happen to Michael [1 mark].
- Reproduction [1 mark]: Lucas successfully replicated the behaviour, which shows he had the ability to complete the task as it was within his skill set to manage his time better to complete work on time [1 mark].
- Attention [1 mark]: Lucas paid direct attention to Michael completing his work and receiving the sticker, which is how he came to the decision to replicate the action [1 mark].
- Retention [1 mark]: Lucas made a mental representation of the skills of Michael managing his time and completing all of the work so he could do the same in the next class [1 mark].

4 a Before conditioning: UCS – smell of food, UCR – hunger, NS – bowling centre [1 mark]
During conditioning: repeated associations of smell of food (UCS) with being in the bowling centre (NS) leading to hunger (UCR) [1 mark]
After conditioning: Hunger (CR) whenever in the bowling centre (CS) [1 mark]

b Antecedent: seeing the ad for the 2-for-1 deal [1 mark]
Behaviour: purchasing the 2-for-1 deal [1 mark]
Consequence: positive reinforcement through getting a free item [1 mark]

CHAPTER 6

ACTIVITY 6.1

Apply it

2 While it is clear that sleep exists (we all do it every day), the existence of sleep and the stages of REM and NREM cannot be directly observed. Instead, we gather data from devices such as an EEG, EMG or EOG to observe changes in brainwave patterns, muscle activity and eye movements as physiological evidence of the existence of these two distinct stages of sleep.

Exam ready

3 B

Try it

1

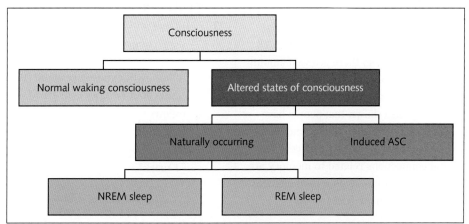

ACTIVITY 6.2

Apply it

2 An EEG detects, amplifies and records electrical activity of the brain in the form of brainwave patterns. In REM sleep, the EEG will record high-frequency and low-amplitude brain wave patterns. In NREM sleep, the brain wave patterns will have lower frequency and higher amplitude. An EMG detects, amplifies and records electrical activity of the muscles in the body. In REM sleep the muscles will be atonic and record no activity. In NREM sleep some activity will be recorded. An EOG detects, amplifies and records electrical activity of the muscles around the eyes. In REM sleep high activity will be recorded. In NREM sleep low activity will be recorded.

Exam ready

3 D

ACTIVITY 6.3

Try it

1

 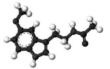

Suprachiasmatic nucleus   Melatonin molecule

The eyes respond to light (natural or artificial) and transmit signals to the SCN. Then light-induced activation of the SCN suppresses the release of melatonin from the pineal gland.

 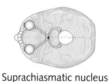 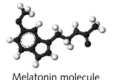

Suprachiasmatic nucleus   Melatonin molecule

Conversely, melatonin production and secretion is increased when it is dark.

Apply it

2 a Circa means 'around'. Dian means 'day'. A circadian rhythm repeats once a day.
   b Ultra means 'many'. Dian means 'day'. An ultradian rhythm repeats many times a day.

Exam ready

3 B

ACTIVITY 6.4

Apply it

2 Life outside the womb can be highly stimulating, and newborns will need a lot of sleep to cope with this. All experiences are new for a baby, so there is a vast amount to be learnt each day. It is important that their brain has time to process it all. This explains the need for so much sleep. Studies have also shown that people have trouble retaining information in their short- and long-term memory without the support of REM sleep. REM sleep is when neural connections related to memory are consolidated. With so much to remember each day, this explains a baby's need for so much REM sleep.

Exam ready

5 A

END-OF-CHAPTER EXAM

Section A: Multiple-choice

1 B	5 C	9 C	13 D	17 D
2 B	6 C	10 B	14 A	18 A
3 B	7 C	11 A	15 C	19 C
4 D	8 D	12 C	16 D	20 D

Section B: Short answer

1 » Infants will receive about 14 hours of total sleep compared to adults who require about 8 hours. [1 mark]
   » Infants will experience about 35 per cent REM and adults about 20 per cent REM. [1 mark]
   » Infants will experience about 65 per cent NREM and adults about 80 per cent. [1 mark]
   » Infant sleep cycles are shorter than those of adults [1 mark]

» A reason infants experience more total sleep is that their brain and body are still developing and require more sleep to do so, whereas adults are fully developed. [1 mark]

[There are more possible answers to this question; this is a guide only.]

2 EEG: an EEG detects, amplifies and records the electrical activity of the brain [1 mark]. It would be useful in identifying someone in REM as the recording would be beta-like waves (low amplitude, high frequency) [1 mark] and if in NREM stage 3 the recording would be mostly delta waves (high amplitude, low frequency) [1 mark] so the scientist would be able to make a judgement.
[Or:]
EOG: an EOG detects, amplifies and records the electrical activity of the muscles surrounding the eye that are involved in rotating and moving the eye [1 mark]. During REM, the EOG would detect high amounts of electrical activity and this may indicate that the participant's eyes are moving rapidly [1 mark], whereas in NREM stage 3, the electrical activity would be minimal as the eyes do not move much during slow-wave sleep [1 mark].

3 The SCN is located in the hypothalamus and it is the body's 'master clock'.
It receives signals from the optic chiasm/nerve from the eyes about the light/dark conditions in the environment. If the signal is 'dark' light conditions, then the SCN will signal to the pineal gland [1 mark] to release melatonin [1 mark], a hormone that makes us feel drowsy and assists sleep onset [1 mark].
In the morning when light conditions change to 'light', the SCN will receive this signal from the eyes and signal to the pineal gland to inhibit the release of melatonin; instead, it will signal the release of cortisol, which will make us more alert [1 mark].
The SCN helps regulate sleep by signalling the release of hormones to either wake us up or make us sleepier.

CHAPTER 7

ACTIVITY 7.1

Apply it

2 Affective: amplified emotional responses (quicker and more intense); increased irritability
Behavioural: excessive sleepiness; microsleeps; reduced reaction times
Cognitive: reduced levels of alertness and inattention; reduced decision-making ability

Exam ready

3 C

ACTIVITY 7.2

Try it

1 Some key points to include in the infographic are:
» a definition of the disorder
» who the disorder affects
» the causes of the disorder
» the symptoms of the disorder
» treatment options for the disorder.

Apply it

2 [Any two of the following would be accepted (direct comparisons must be made, and one difference must reference ASPD and DSPS)]:
» Age: ASPD tends to affect individuals who are middle-aged or elderly, whereas DSPS tends to affect teenagers.
» Timing of sleeping and waking: A person with ASPD tends to fall asleep between 6 p.m. and 9 p.m. and wake at 2 a.m. to 5 a.m.; whereas a person with DSPS tends to fall asleep at approximately 2 a.m. and finds it difficult to wake before 10 a.m.
» Treatment options: people with DSPS are encouraged to take melatonin in the evening to encourage sleep; whereas those with ASPD are encouraged to use bright light therapy in the evening to prevent falling asleep early in the evening.
» Any other appropriate differences.

Exam ready

3 D

ACTIVITY 7.3

Apply it

2 [Any three of the following issues would be accepted]:
» Jeremy has naps after work; improvement: Jeremy should resist napping after work.
» Jeremy has a coffee in the evening; improvement: Jeremy should avoid having a coffee once home from work.
» Jeremy goes to the gym too late at night for his routine; improvement: Jeremy should go to the gym earlier in the evening. Alternatively, Jeremy could also try going to the gym prior to work.
» Jeremy eats a banana and has a protein shake once home from the gym; improvement: Jeremy should not be eating so close to bedtime.

- » Jeremy checks his emails/does work too late in the evening; improvement: Jeremy should not be checking his emails when in bed. The bedroom should not be a place for work.
- » Jeremy continually checks the time on his phone throughout the night; improvement: Jeremy needs to avoid checking his phone throughout the night/potential for blue light to encourage wakefulness.
- » Jeremy engages in negative thought patterns regarding his sleep (he's a 'bad sleeper'); improvement: Jeremy needs to replace these negative thoughts with more positive thoughts regarding sleep.

Exam ready

3 A

END-OF-CHAPTER EXAM

Section A: Multiple-choice

1 C	5 D	9 A	13 C	17 B
2 D	6 C	10 A	14 A	18 A
3 C	7 C	11 C	15 C	19 B
4 D	8 A	12 B	16 A	20 C

Section B: Short answer

1

Functioning	Changes due to sleep deprivation
Cognitive	» Illogical thinking » Difficulty remembering things » Difficulty making decisions
Affective	» Amplified emotions » Irritable » Difficulty reading emotions/feeling other emotions
Behavioural	» Clumsiness » Slower reflex response » Difficulty coordinating movements » Difficulty processing information

2 Violet may be experiencing a delay in the release of melatonin (she is teenager) and therefore doesn't feel sleepy until 1–2 hours later in the night than adults would [1 mark]. She needs about nine hours of sleep to meet her requirements (as she is an adolescent) and so waking early for school means that she is shortening the amount of sleep she gets [1 mark] and hence experiences sleep deprivation [1 mark].
Hannah is younger and her body releases melatonin earlier than Violet's [1 mark], making her sleepy early in the evening [1 mark]. She is in grade 4 so may require more sleep than Violet but is able to get more sleep as she goes to sleep earlier [1 mark].

3 Similarity: both are circadian rhythm sleep disorders, so they affect the timing of sleep [1 mark].
Difference: DSPS is a delay in sleep onset and a delay in waking time, whereas ASPD is an advance in sleep onset and an advance in wake time [1 mark].
Bright light therapy could treat DSPS by exposing the individual to bright light early in the morning to make them wake earlier and extending the time awake [1 mark] (making it so they are more tired in the evening and more likely to want to sleep) [1 mark].
It could be used in the evening for ASPD to keep the individual awake in the evening and delay the onset of sleep, effectively pushing the onset of sleep later [1 mark].

Unit 4 Area of study 1 review

Multiple choice questions

1 D	7 A	13 A	19 D	25 C
2 C	8 D	14 D	20 C	26 D
3 D	9 B	15 A	21 B	27 D
4 C	10 C	16 D	22 D	28 C
5 D	11 A	17 C	23 B	29 B
6 A	12 D	18 C	24 D	30 B

Short answer questions

1 a This refers to when individuals struggle to stay awake at night and fall asleep typically earlier in the night [1 mark]. This then leads them to awaken in the early hours of the morning [1 mark].
   b As it is a shift of the body clock backwards [1 mark]; our circadian rhythms are out of sync with the external world [1 mark] highlighting a circadian phase disorder
   c Bright light therapy can be used to shift the sleep wake cycle forward to enable individuals to fall asleep later in the night [1 mark]. This can be done through the administration of bright light late at night when the individual would typically feel sleepy to induce later onset of melatonin and hence later onset of sleepiness [1 mark].

2 a Lui is most likely suffering from jetlag [1 mark], this is evident by the disruption to his sleep wake cycle caused by constant changing of time zones [1 mark].
   b Due to the constant change in time zones Lui would have a mismatch between the external world and his internal body clock [1 mark], with melatonin released during hours of daylight due to the lack of adjustment to his current time zone as he isn't spending enough time to adjust to the time zone shift [1 mark].

Due to this mismatch Lui is finding himself sleepy during the daytime where he may find it difficult to sleep due to the sunlight, then is unable to sleep at night due to lack of melatonin release, hence he is experiencing less and poorer quality sleep [1 mark].

  c Lui would experience an affective impact of lack of emotional regulation [1 mark] due to the sleep deprivation, this may lead to things such as him being overly irritable with friends and becoming easily upset or angry [1 mark].
Lui would also experience a decrease in cognitive functioning, an example of this is that he may have trouble problem solving [1 mark] and committing things to memory and as such may be more forgetful than usual [1 mark].

  d Lui could use bright light therapy to help promote the release of melatonin later in the night [1 mark]. This therapy utilises bright light exposure for individuals to delay the onset of melatonin. In Lui's case exposure to bright light therapy early in the morning when he is typically tired would help suppress melatonin production [1 mark], leading to release later in the day and as such inducing sleepiness at a later time of day aiming to align his sleep-wake cycle with the external day-night cycle [1 mark].

  e Because Lui travels for work frequently, whilst the bright light therapy will help realign him, because he will then travel again [1 mark] it will lead to his circadian rhythms becoming out of sync with the external environment again [1 mark].

3
  a » because Jessica frequently goes out on weekends, this offsets her sleep wake cycle, in which it then takes her the week to reset it only to be ruined again on the weekend [1 mark].
     » eating before bed disrupts sleep due to digestion [1 mark].
  b Jessica should have roughly 8 hours of sleep [1 mark] whereas Jane only requires 7 hours of sleep [1 mark].
  c Jessica could stop eating 2 hours prior to sleep [1 mark]; this would lead to less disruptive sleep as her sleep cycle isn't being interrupted by digestion allowing energy to be conserved and used for restoration during sleep [1 mark].
  d This would enable her pre-sleep routines to be observed [1 mark] as well as things waking her during the night which wouldn't be directly observable with EMG [1 mark].

4 Correct axes [1 mark]
  Correct shape graph [1 mark]
  a
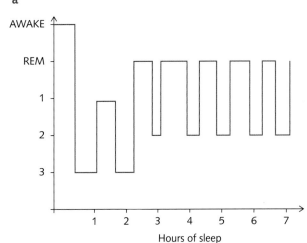

  b This is due to the blue light emitted from the IPAD. This blue light suppresses the release of melatonin which is used to induce sleepiness [1 mark]. As such Anna is falling asleep later at night due to the interruption of the blue light.

5 Any two of: [1 mark each]
  Adolescents sleep for roughly 9 hours when compared to the elderly who sleep for an average of 6.5 hours.
  Adolescents tend to fall asleep later in the night when compared to the elderly
  Adolescents experience more deep sleep than the elderly.

CHAPTER 8

ACTIVITY 8.1

Try it

1 Before AFL: Absence of mental ill-health and likely at least moderate wellbeing (located in the right middle region of the dual-continuum space). During AFL career: Presence of mental disorder and languishing mental wellbeing (located in the lower left quadrant of the dual-continuum space). After AFL career: Presence of mental ill-health (treated effectively) and flourishing (located in the left upper quadrant of the dual-continuum space).

Apply it

2 Jo: bottom right
3 Sam: top right
4 Lee: top left
5 Sia: bottom left

Exam ready
6 B

ACTIVITY 8.2

Apply it
2a [Answers will vary.]
2b [Answers will vary.]
2c [Answers will vary.]
2d [Answers will vary.]

Exam ready
3 C. Revitalisation of traditional languages is core to connecting with culture.

ACTIVITY 8.3

Try it
1 Strengths are referred to in references to the depth of culture going back more than 60 000 years, ancestral ties to the land, and sovereignty never being ceded. Impacts of government policy referenced include First Nations Australians being the most incarcerated peoples on the planet, and the 'structural nature of our problem', referring to lack of voice and powerlessness through government policies.

Apply it
2 The restorative actions include seeking a voice to parliament to be able to advise on matters that directly impact Aboriginal and Torres Strait Islander communities, seeking Makarrata – a coming together after struggle – to enable self-determination, and a commission to tell the truth of Australia's history.

Exam ready
3 A. The self-determination that is core to the Voice is most relevant to restoring community control of policies and programs that affect First Nations Australians.

ACTIVITY 8.4

Try it
1 [Answers will vary.]

Apply it
2 [Answers will vary.]

Exam ready
3 C

END-OF-CHAPTER EXAM

Section A: Multiple choice

1 A	5 D	9 D	13 D	17 A
2 B	6 A	10 A	14 B	18 D
3 C	7 C	11 B	15 A	19 B
4 D	8 D	12 B	16 D	20 C

Section B: Short answer

1 Mental wellbeing is a state of wellbeing in which the individual realises his or her own abilities, can cope with the normal stresses of life, can work productively and fruitfully, and is able to make a contribution to his or her community [1 mark]. A mental health disorder (mental disorder, mental illness) is any condition characterised by cognitive and emotional disturbances, abnormal behaviours, impaired functioning or any combination of these [1 mark].

2 The three levels of functioning related to people's self-reported mental wellbeing are:
   » flourishing – a state of optimal mental wellbeing in which a person both feels good (has emotional wellbeing) and functions effectively (has psychological and social wellbeing) [1 mark]
   » languishing – when a person experiences both low levels of positive emotions and low levels of psychological and social functioning [1 mark]
   » moderate mental wellbeing – people who self-reported a score in between languishing and flourishing [1 mark].

3 The Social and Emotional Wellbeing (SEWB) model is the core framework for understanding the determinants of physical and mental health for Aboriginal and Torres Strait Islander peoples [1 mark]. The SEWB model is a holistic approach because it looks at the whole person, not just their mental wellbeing needs [1 mark]. It considers the relationships between people, family, kin and community, including the importance of connection to land, culture, spirituality and ancestry, and how these affect the individual [1 mark].

CHAPTER 9

ACTIVITY 9.1

Try it

1

	Biological	Psychological	Social
'Even thinking about the object of my fear makes me feel anxious'		✓ (cognitive)	
'Phobias are fuelled by avoidance behaviour'		✓ (behavioural operant conditioning)	
'Lower levels of GABA increase a person's chance of developing a phobia'	✓		
'Phobias develop when a person creates an association between a stimulus and fear'		✓ (behavioural classical conditioning)	
'Phobic responses occur in the presence of the phobic stimulus'		✓ (behavioural classical conditioning)	
'I try to manage my phobia because I don't want to be seen as different'			✓
'My friends and family say I'm exaggerating, but the bird nearly killed me!'		✓ (cognitive, memory bias, catastrophic thinking)	

Apply it

2 Operant conditioning perpetuates a phobia via avoidance behaviour. Jamie has learned to walk down the streets without dogs so that he can avoid being confronted by the phobic stimulus (dogs). Jamie's avoidance behaviour is strengthened due to negative reinforcement (removal of fear/discomfort).

Exam ready

3 D

ACTIVITY 9.2

Try it & Apply it

1, 2 [A full-mark response would need to include all of the following points.]
- » The participant is taught correct breathing techniques (long, slow breaths).
- » Creation of a fear hierarchy from least fearful experience to most fearful.
- » Acknowledgement that individual does not move to the next level until they are completely in control.
- » Exposure to elements of the fear hierarchy from least fearful to most fearful, while using breathing techniques.
- » Explanation of systematic desensitisation: reciprocal inhibition; replacement of fear with calmness.

Exam ready

3 B

END-OF-CHAPTER EXAM

Section A: Multiple-choice

1 B	5 C	9 A	13 C	17 D
2 C	6 A	10 B	14 A	18 D
3 B	7 A	11 C	15 C	19 A
4 A	8 B	12 B	16 A	20 A

Section B: Short answer

1 An anxious person's breathing may consist of small, shallow breaths (hyperventilation), which may raise oxygen levels and reduce the amount of carbon dioxide in the blood [1 mark]. Carbon dioxide assists in the regulation of the body's reaction to anxiety and panic [1 mark]. Breathing retraining is the process of identifying incorrect breathing habits and replacing them with correct ones [1 mark]. A person who suffers from anxiety should learn how to breathe from their diaphragm, rather than their chest, through their nose in a slow, even and gentle way [1 mark]. Correct breathing will help to boost carbon dioxide levels in the blood and reduce the body's reaction to anxiety and panic [1 mark].

2 Psychoeducation is a psycho-social approach in which a person experiencing a mental health problem or disorder and their family are provided with information to help them understand the condition and how they can contribute to managing it [1 mark]. The goal of psychoeducation for the families and supporters of a phobic person is to help them understand the illness [1 mark]. This helps to reduce the stigma associated

with the illness [1 mark]. Additional goals are to teach them the skills needed to help support the phobic person cope and recover, and to help themselves adapt to living with a person suffering from a phobia [1 mark].
3 Individuals with a phobia may have dysfunctional levels of GABA [1 mark]. A benzodiazepine agent acts as a GABA agonist in the brain [1 mark]. This promotes GABA's inhibitory effect and reduces physiological arousal [1 mark]. The benzodiazepine agent would help reduce the extreme anxiety experienced due to a phobia [1 mark].

CHAPTER 10

ACTIVITY 10.1

Apply it

2 Biological: Dylan is not eating well. He skips breakfast and has sugar and caffeine instead of a proper lunch. Dylan needs to eat a mixture of fruit, vegetables, proteins and grains to provide his body with adequate nutrients. Dylan is also not getting enough sleep as he is going to bed late and waking up early to go to work. His lack of sleep can have a negative impact on his overall functioning, so he should develop better sleeping habits.
Psychological: Dylan is feeling stressed, which is having an impact on his ability to unwind at the end of the day and also on his ability to sleep. Dylan could make use of mindful meditation to focus on his breathing and slow down his racing thoughts.
Social: Dylan is missing out on the social support that his friends can provide him. He might suggest that he sees his friends on the weekend in the afternoon instead of at night.

Exam ready

3 D

END-OF-CHAPTER EXAM

Section A: Multiple-choice

1 B	5 B	9 D	13 C	17 C
2 C	6 B	10 D	14 D	18 C
3 A	7 D	11 A	15 C	19 D
4 B	8 C	12 A	16 D	20 B

Section B: Short answer

1 Community support organisations include:
   » Beyond Blue
   » the Black Dog Institute
   » headspace
   » SANE Australia.
   [1 mark for each correctly identified community support organisation]

2 Nutritional intake can affect our mental wellbeing both directly and indirectly: directly, through the nutrients that keep our body and brain physically healthy so they can function at optimal level [1 mark], and indirectly, through the emotional impact of having a physical condition caused by an unhealthy diet [1 mark].
Sleep is vitally important to us for replenishing and revitalising the physiological processes that keep the mind and body functioning at an optimal level [1 mark].
3 Cultural continuity relates to the domain of spirituality and ancestors through passing ancestral knowledge across generations [1 mark] and integrating an individual (and/or community) with the deep history of their culture [1 mark]. This is protective of social and emotional wellbeing because it strengthens the individual's (and/or community's) sense of personal continuity within a continuing culture [1 mark].

Unit 4 Area of Study 2 review

Multiple-choice questions

1 C	7 D	13 A	19 D	25 D
2 B	8 B	14 B	20 A	26 B
3 A	9 B	15 C	21 D	27 C
4 B	10 A	16 D	22 B	28 D
5 C	11 C	17 C	23 C	29 B
6 B	12 A	18 C	24 B	30 A

Short-answer questions

1 a Psychological: cognitive bias as she has played it up in her head to be much more of a fearful event than was actually experienced [1 mark]
Social: stigma around seeking treatment as she may find it embarrassing as it is a fear not many people have [1 mark]
Biological: long-term potentiation strengthening the connections between fear and balloons each time she sees one [1 mark]
  b [Students must name and describe one relevant external factor. An example follows.] Psychoeducation for family and friends [1 mark]. Description: This can help by providing education to her loved ones about how to best support Lauren in overcoming her fear of balloons [1 mark].
  c [Students must name and describe one relevant biological protective factor. An example follows.] Mindfulness meditation [1 mark]; Explanation: to help ground and focus Lauren when she is near balloons to help overcome her feelings of anxiety [1 mark].

2 a Primary [1 mark]: appraisal as harm/loss as she is experiencing stress because of the loss of her relationship with her boyfriend [1 mark]. Secondary [1 mark]: insufficient ability to cope with the stressor as she is unable to function in her everyday life as usual [1 mark].

b [Students must name and describe one relevant strategy. An example follows.]
Emotion focused: any example such as confiding in friends [1 mark].

c [Answers will vary. An example follows.]
She could utilise her friends and community to energise and support her so that she does not feel isolated or alone [1 mark]. These friends could be from university or her extracurricular support groups [1 mark].

d Resilience is the ability to bounce back after stressful situations [1 mark]. If Dushani displayed resilience then she may be affected by the breakup in the short term but would be able to recover from the stressful event relatively quickly and get back to her normal activities, making it less likely for this to develop into a mental disorder [1 mark].

3 a Level of functioning (low) [1 mark] justification (because he is struggling to keep up with life demands and is impacting on his sleep which is likely to be impacting on his ability to function in day to day life) [1 mark]

b It is likely he is experiencing a mental disorder due to the persistent and constant anxiety with no known cause or trigger [1 mark] rather than a mental health problem which is usually associated with more short-term symptoms [1 mark].

c Internal factor: overthinking/catastrophic thinking leading to feelings of anxiety and rumination of thoughts [1 mark].
External factor: stigma around seeking treatment preventing him for seeking help for his constant feelings of anxiety [1 mark].

4 Internal sources of stress are caused by physiological issues within the individual [1 mark]; these can include things such as GABA dysfunction [1 mark]. External sources of stress come from the environment [1 mark]; these can include things such as stigma around seeking treatment [1 mark].

5 Cultural determinants are the protective factors that come from Aboriginal and Torres Strait Islander peoples [1 mark] being able to learn, practice and pass on their traditional ways of knowing, being and doing [1 mark]. Two of the strongest cultural determinants of SEWB are cultural continuity [1 mark] and self-determination [1 mark]. [Students must provide an example of each of cultural continuity and self-determination for 2 more marks.]

References

Ackermann, S., & Rasch, B. (2014). Differential effects of non-REM and REM sleep on memory consolidation. *Current neurology and neuroscience reports. 14*(2), 430.

Addis, D. R., Sacchetti, D. C., Ally, B. A., Budson, A. E., & Schacter, D. L. (2009). Episodic simulation of future events is impaired in mild Alzheimer's disease. *Neuropsychologia, 47*(12), 2660–2671.

Ader, R., & Cohen, N. (1993). Psychoneuroimmunology: Conditioning and stress. *Annual Review of Psychology, 44*, 53–85.

Adesope, O. O., Trevisan, D. A., & Sundararajan, N. (2017). Rethinking the use of tests: A meta-analysis of practice testing. *Review of educational research, 87*(3), 659–701.

Ahn, J. N., Hu, D., & Vega, M. (2019). Do as I do, not as I say: Using social learning theory to unpack the impact of role models on students' outcomes in education. *Social and Personality psychology compass, 14*. https://doi.org/10.1111/spc3.12517.

Alcohol and Drug Foundation (2021). *What are benzodiazepines?* https://adf.org.au/drug-facts/benzodiazepines/

Alvarez, T. A., & Fiez, J. A. (2018). Current perspectives on the cerebellum and reading development. *Neuroscience & biobehavioral reviews, 92*, 55–66.

American Academy of Sleep Medicine. (2007, June 13). Reduced sleep quality can aggravate pre-existing psychological conditions. https://aasm.org/reduced-sleep-quality-can-aggravate-preexisting-psychological-conditions

American Psychiatric Association. (2013). *Diagnostic and statistical manual of mental disorders* (5th ed.). https://doi.org/10.1176/appi.books.9780890425596

American Psychological Association (2020). *Building your resilience.* https://www.apa.org/topics/resilience/building-your-resilience

Anders, T. F., Sadeh, A., & Appareddy, V. (1995). Normal sleep in neonates and children. In Ferber R. K. M. (Ed.). *Principles and practice of sleep medicine in the child.* (pp. 7–18). Philadelphia: Saunders.

Andrews, S. (2020). Cloaked in Strength—how possum skin cloaking can support Aboriginal women's voice in family violence research. *AlterNative: An International Journal of Indigenous Peoples, 16*(2), 108–116.

AP. (2011, January 14). Exam anxiety relief found – Research. *The Australian.* https://www.news.com.au/breaking-news/examanxiety-relief-found--research/news-story/b23131a4bd1261b3bf9dc3547e6881e9

Atkinson, J., Nelson, J., Brooks, R., Atkinson, C., & Ryan, K. (2014). Addressing individual and community transgenerational trauma. *Working together: Aboriginal and Torres Strait Islander mental health and wellbeing principles and practice, 2*, 289–307.

Atkinson, R. C., & Shiffrin, R. M. (1968). Human memory: A proposed system and its control processes. In K. W. Spence & J. T. Spence (Eds.), *The psychology of learning and motivation: II.* (pp. 89–195). Academic Press. doi: 10.1016/S0079-7421(08)60422-3

Australian Human Rights Commission (AHRC). (2022). Self-determination. https://humanrights.gov.au/our-work/rights-and-freedoms/right-self-determination

Australian Institute of Family Studies (2019). Australian teens not getting enough sleep. Media release. https://aifs.gov.au/media-releases/australian-teens-not-getting-enough-sleep

Australian Institute of Health and Welfare. (2020). *Mental health snapshot.* https://www.aihw.gov.au/reports/australias-health/mental-health

Babson, K. A., & Feldner, M. T. (2010). Temporal relations between sleep problems and both traumatic event exposure and PTSD: A critical review of the empirical literature. *Journal of anxiety disorders, 24*, 1–15.

Baddeley, A. D., Hitch, G. J., & Allen, R. J. (2019). From short-term store to multicomponent working memory: The role of the modal model. *Memory & cognition, 47*(4), 575–588.

Bandura, A. (1965). Influence of models' reinforcement contingencies on the acquisition of imitative responses. *Journal of personality and social psychology, 1*(6), 589.

Barlow, D. H. (2002). *Anxiety and its disorders* (2nd ed.). New York: Guilford Press.

Barnes, J. (2013). *Essential biological psychology.* London: Sage Publications.

Bastiaanssen, T. F. S., Cussotto, S., Claesson, M. J., Clarke, G., Dinan, T. G. & Cryan, J. F. 2020. Gutted! Unraveling the role of the microbiome in major depressive disorder. *Harvard review of psychiatry, 28*. https://doi: 10.1097/HRP.0000000000000243

Bawaka Collective, including Burarrwanga, L., Ganambarr, R., Ganambarr-Stubbs, M., Ganambarr, B., Maymuru, D., Lloyd, K., Wright, S., Suchet-Pearson, S., & Daley, L. (2022). Songspirals bring Country into existence: singing more-than-human and relational creativity. *Qualitative inquiry*, 10778004211068192.

Bawaka Country including Wright, S., Suchet-Pearson, S., Lloyd, K., Burarrwanga, L., Ganambarr, R., ... & Maymuru, D. (2015). Working with and learning from Country: decentring human author-ity. Cultural geographies, *22*(2), 269–283.

Bayram, E., Caldwell, J. Z., & Banks, S. J. (2018). Current understanding of magnetic resonance imaging biomarkers and memory in Alzheimer's disease. *Alzheimer's & dementia: Translational research & clinical interventions, 4*, 395–413.

Belenky, G., Wesensten, N. J., Thorne, D. R., Thomas, M. L., Sing, H. C., Redmond, D. P., Russo, M. B., & Balkin, T. J. (2003). Patterns of performance degradation and restoration during sleep restriction and subsequent recovery: A sleep dose-response study. *Journal of sleep research, 12*(1), 1–12.

Berg, H., Söderlund, M., & Lindström, A. (2015). Spreading joy: examining the effects of smiling models on consumer joy and attitudes. *Journal of consumer marketing, 32*(6), 459–469.

Bessarab, D., & Ng'Andu, B. (2010). Yarning about yarning as a legitimate method in Indigenous research. *International journal of critical Indigenous studies, 3*(1), 37–50.

Better Health Channel (2021). *Breathing to reduce stress.* State of Victoria. https://www.betterhealth.vic.gov.au/health/healthyliving/breathing-to-reduce-stress

Better Health Channel (2022). *Healthy eating and dieting.* https://www.betterhealth.vic.gov.au/health/healthyliving/healthy-eating#australian-dietary-guidelines

Better Health Channel. (2020). *Wellbeing.* State of Victoria. https://www.betterhealth.vic.gov.au/health/healthyliving/wellbeing

Beyond Blue (2022) Specific phobias. https://www.beyondblue.org.au/the-facts/anxiety/types-of-anxiety/specific-phobias

Beyond Blue. (2015) *Benzodiazepines. (tranquillisers and sleeping pills).* Retrieved from www.beyondblue.org.au

Bilton, N., Rae, J., & Yunkaporta, T. (2020). A conversation about Indigenous pedagogy, neuroscience and material thinking. In *Teaching Aboriginal Cultural Competence* (pp. 85–97). Singapore: Springer.

Binks, P. G., Waters, W. F., & Hurry, M. (1999). Short-term total sleep deprivation does not selectively impair higher cortical functioning. *Sleep, 22*(3), 328–34.

Blazhenkova, O. (2016). Vividness of object and spatial imagery. *Perceptual and motor skills, 122*(2), 490–508.

Bliss, T. V., & Collingridge, G. L. (1993). A synaptic model of memory: Long-term potentiation in the hippocampus. *Nature, 361,* 31–39.

Bostan, A. C., & Strick, P. L. (2018). The basal ganglia and the cerebellum: nodes in an integrated network. *Nature reviews Neuroscience, 19*(6), 338–350.

Boynton, R., & Swinbourne, A. (2017). *How virtual reality spiders are helping people face their arachnophobia.* https://theconversation.com/how-virtual-reality-spiders-are-helping-people-face-their-arachnophobia-73769

Brando, S. (2020). Box B4: Marine Mammal Training. *Zoo animal learning and training,* 197–201.

Breedlove, S. M., & Watson, N.V. (2013). *Biological psychology: An introduction to behavioral, cognitive and clinical neuroscience* (7th ed.). Massachusetts: Sinauer Associates.

Breit, S., Kupferberg, A., Rogler, G., & Hasler, G. (2018). Vagus nerve as modulator of the brain-gut axis in psychiatric and inflammatory disorders. *Frontiers in psychiatry, 9,* 44. https://doi.org/10.3389/fpsyt.2018.00044

Briere, J. (2002). Treating adult survivors of severe childhood abuse and neglect: Further development of an integrative model. In J. E. B. Myers, L. Berliner, J. Briere, C. T. Hendrix, T. Reid, & C. Jenny (Eds.), *The APSAC handbook on child maltreatment* (2nd ed., pp. 175–203). Newbury Park, CA: Sage Publications.

Buddle, C. (2014). *Explainer: why are we afraid of spiders?* https://theconversation.com/explainer-why-are-we-afraid-of-spiders-26405

Cane, S. (2013). *First footprints: the epic story of the first Australians* (p. 316). Sydney, Australia: Allen & Unwin.

Cao, J., Herman, A. B., West, G. B., Poe, G., & Savage, V. M. (2020). Unraveling why we sleep: Quantitative analysis reveals abrupt transition from neural reorganization to repair in early development. *Science advances, 6*(38), eaba0398 https://doi.org/10.1126/sciadv.aba0398

Carr-Gregg, M. (2007, January 18). Sweet dreams make better teens. *Herald Sun.* Retrieved from www.heraldsun.com.au/news/opinion/sweet-dreams-make-better-teens-story-e6frfhqf-1111112848092

Carskadon, M. A. (2002). Factors influencing sleep patterns of adolescence. In M. A. Carskadon (Ed.), *Adolescent sleep patterns: Biological, social, and psychological influences.* New York: Cambridge University Press.

Carter, R., Aldridge, S., Page, M., & Parker, S. (2014). *The brain book.* London: Dorling Kindersley Limited.

Chang, P. P., Ford, D. E., Mead, L. A., Cooper-Patrick, L., & Klag, M. J. (1997). Insomnia in young men and subsequent depression. *American journal of epidemiology, 146,* 105–114.

Cheng, C. (2003). Cognitive and motivational processes underlying coping flexibility: A dual-process model. *Journal of personality and social psychology, 84,* 425–438.

Cheng, C., & Cheung, M. W. L. (2005). Cognitive processes underlying coping flexibility: Differentiation and integration. *Journal of personality, 73,* 859–886.

Clayton, R. B., Leshner, G., Bolls, P. D., & Thorson, E. (2017). Discard the smoking cues – keep the disgust: An investigation of tobacco smokers' motivated processing of anti-tobacco commercials. *Health communication, 32*(11), 1319–1330.

Colquhoun, W. P. (1976). Accidents, injuries and shift work. In Department of Health and Human Services (Ed.). *Shift Work and Health.* Washington, DC: Government Printing Office.

Colten, H. R., Altevogt, B. M., & Institute of Medicine (US) Committee on Sleep Medicine and Research (Eds.). (2006). *Sleep disorders and sleep deprivation: An unmet public health problem.* National Academies Press (US).

Conway, M. A., & Loveday, C. (2015). Remembering, imagining, false memories & personal meanings. *Consciousness and cognition, 33,* 574–581.

Conway, M. A., & Pleydell-Pearce, C. W. (2000). The construction of autobiographical memories in the self-memory system. *Psychological review, 107*(2), 261.

Copinschi, G., & Caufriez, A. (2013). Sleep and hormonal changes in aging. *Endocrinology and metabolism clinics of North America, 42*(2), 371–389. https://doi.org/10.1016/j.ecl.2013.02.009. Epub 2013 Mar 29. PMID: 23702407

Coren, S. (1996). *Sleep thieves.* New York: Free Press.

Corkin ch 5 pages 183-4 is cited, is this the right ref to add?: Corkin, S. (2013). *Permanent present tense: The unforgettable life of the amnesic patient, H. M.* Basic Books/Hachette Book Group.

Cowan, N. (2015). George Miller's magical number of immediate memory in retrospect: Observations on the faltering progression of science. *Psychological review, 122*(3), 536.

Cragg, B. G. (1975). The development of synapses in kitten visual cortex during visual deprivation. *Experimental neurology, 46,* 445–451.

Craik, F. I. M., & Tulving, E. (1975). Depth of processing and the retention of words in episodic memory. *Journal of experimental psychology: General, 104*(3), 268–294.

Crowley, S. J, Lee, C., Tseng, C. Y., Fogg, L. F., & Eastman, C. I. (2004). Complete or partial circadian re-entrainment improves performance, alertness, and mood during night-shift work. *Sleep, 27,* 1077–1078.

Cryan, J. F., O'Riordan, K. J, Cowan, C. S. M., Sandhu, K. V., Bastiaanssen, T. F. S., Boehme, M., Codagnone, M. G., Cussotto, S., Fulling, C., Golubeva, A. V., Guzzetta, K. E., Jaggar, M., Long-Smith, C. M., Lyte, J. M., Martin, J. A., Molinero-Perez, A., Moloney, G., Morelli, E., Morillas, E., ... & Dinan, T. G. (2019). The microbiota–gut–brain axis. *Physiological reviews, 99*(4), 1877–2013. https://doi.org/10.1152/physrev.00018.2018

Dahl, R. E., & Lewin, D. S. (2002). Pathways to adolescent health sleep regulation and behavior. *Journal of adolescent health, 31*(6 Suppl.), 175–184.

Davis, C. P., & Yee, E. (2019). Features, labels, space, and time: Factors supporting taxonomic relationships in the anterior temporal lobe and thematic relationships in the angular gyrus. *Language, Cognition and Neuroscience, 34*(10), 1347–1357.

Davydov, D. M., Stewart, R., Ritchie, K., & Chaudieu, I. (2010). Resilience and mental health. *Clinical psychology review, 30*(5), 479–495.

Dawes, A. J., Keogh, R., Andrillon, T., & Pearson, J. (2020). A cognitive profile of multi-sensory imagery, memory and dreaming in aphantasia. *Scientific reports, 10*(1), 1–10.

Dawson, D., & Reid, K. (1997). Fatigue, alcohol and performance impairment. *Nature, 388,* 235.

De Palma, G., Collins, S. M., Bercik, P., & Verdu, E. F. (2014). The microbiota–gut–brain axis in gastrointestinal disorders: stressed bugs, stressed brain or both? *The journal of physiology, 591*(14), 2989–2997.

Debiec, J., & Olsson, A. (2017). Social fear learning: from animal models to human function. *Trends in cognitive sciences, 21*(7), 546–555.

Delgado, M. R., Olsson, A., & Phelps, E. A. (2006). Extending animal models of fear conditioning to humans. *Biological psychology, 73*(1), 39–48.

Dementia Australia (2022) https://www.dementia.org.au/about-dementia/what-is-dementia

Dickerson, B., & Eichenbaum, H. (2010). The episodic memory system: Neurocircuitry and disorders. *Neuropsychopharmacol 35,* 86–104.

Diener, E., Suh, E. M., Lucas, R. E., & Smith, H. L. (1999). Subjective well-being: Three decades of progress. *Psychological bulletin, 125*(2), 276–302.

Dinges, D. (1995). An overview of sleepiness and accidents. *Journal of sleep research, 4,* 4–11.

Doran, S. M., Van Dongen, H. P., & Dinges, D. F. (2001). Sustained attention performance during sleep deprivation: Evidence of state instability. *Archives Italiennes de biologie, 139,* 253–267.

Dresler, M., Shirer, W. R., Konrad, B. N., Müller, N. C., Wagner, I. C., Fernández, G., ..., & Greicius, M. D. (2017). Mnemonic training reshapes brain networks to support superior memory. *Neuron, 93*(5), 1227–1235.

Dudgeon, P., Bray, A., & Walker, R. (2020). Self-determination and strengths-based Aboriginal and Torres Strait Islander suicide prevention: an emerging evidence-based approach. In *Alternatives to suicide* (pp. 237–256). Academic Press.

Duke University (2018, 22 Sep). *The gut–brain connection.* [Video]. YouTube. https://www.youtube.com/watch?v=oym87kVhqm4

Duke University (2017, July 11). Live-in grandparents helped human ancestors get a safer night's sleep: sleep changes common with age may have helped our ancestors survive the night. *ScienceDaily.* http://www.sciencedaily.com/releases/2017/07/170711215825.htm

Durmer, J. S., & Dinges, D. F. (2005). Neurocognitive consequences of sleep deprivation. *Seminars in neurology, 25*(1), 117–129.

Ecole Polytechnique Fédérale de Lausanne. (2015, February 18). How stress can lead to inequality. *ScienceDaily,* https://www.sciencedaily.com/releases/2015/02/150218141309.htm

Edalati, H., & Conrod, P. J. (2019). A review of personality-targeted interventions for prevention of substance misuse and related harm in community samples of adolescents. *Frontiers in psychiatry, 9,* 770.

El Haj, M., Gallouj, K., & Antoine, P. (2019). Mental imagery and autobiographical memory in Alzheimer's disease. *Neuropsychology, 33*(5), 609.

Feinstein, J. S., Buzza, C., Hurlemann, R., Follmer, R. L., Dahdaleh, N. S., Coryell, W. H., ..., & Wemmie, J. A. (2013). Fear and panic in humans with bilateral amygdala damage. *Nature neuroscience, 16*(3), 270–272.

Feinstein, J.S., Adolphs, R., Damasio, A., & Tranel, D. (2011). The human amygdala and the induction and experience of fear. *Current biology; 21*(1), 34–8. doi: 10.1016/j.cub.2010.11.042

First Nations National Constitutional Convention (2017). *The Uluru Statement from the Heart.* Alice Springs, Northern Territory: Central Land Council Library. https://ulurustatement.org/the-statement

Flinker, A., Korzeniewska, A., Shestyuk, A. Y., Franaszczuk, P. J., Dronkers, N. F., Knight, R. T., & Crone, N. E. (2015). Redefining the role of Broca's area in speech. *Proceedings of the National Academy of Sciences, 112*(9), 2871–2875.

Flom, R., Lee, K., & Muir, D. (Eds.). (2017). *Gaze-following: Its development and significance.* Psychology Press.

Flórez, K. R., Shih, R. A., & Martin, M. T. (2014). Nutritional Fitness and Resilience.

Foerde, K., & Shohamy, D. (2011). The role of the basal ganglia in learning and memory: insight from Parkinson's disease. *Neurobiology of learning and memory, 96*(4), 624–636.

Folkard, S., & Monk, T. H. (1979). Shift work and performance. *Human Factors, 21*(4), 483–492.

Foster, J. A., & McVey Neufeld, K-A. (2013). Gut–brain axis: how the microbiome influences anxiety and depression. *Trends in neuroscience*, 36(5). http://dx.doi.org/10.1016/j.tins.2013.01.005

Frazer, B., & Yunkaporta, T. (2021). Wik pedagogies: Adapting oral culture processes for print-based learning contexts. *The Australian journal of Indigenous education*, 50(1), 88–94.

Fresco, D. M., Williams, N. L., & Nugent, N. R. (2006). Flexibility and negative affect: Examining the associations of explanatory flexibility and coping flexibility to each other and to depression and anxiety. *Cognitive therapy and research*, 30, 201–210.

Fullana, M. A., Dunsmoor, J. E., Schruers, K. R. J, Savage, H. S., Bach, D. R., & Harrison, B. J. (2020). Human fear conditioning: From neuroscience to the clinic. *Behaviour research and therapy*, 124, 103528.

Furness, J. B. (2007) Enteric nervous system. Retrieved from http://www.scholarpedia.org/article/Enteric nervous system

Futa, K., Nash, C., Hansen, D., & Garbin, C. (2003). Adult survivors of childhood abuse: An analysis of coping mechanisms used for stressful childhood memories and current stressors. *Journal of family violence*, 18(4), 227–239.

Gan, Y., Shang, J., & Zhang, Y. (2007). Coping flexibility and locus of control as predictors of burnout among Chinese college students. *Social behavior and personality*, 35, 1087–1098.

Gazziniga, M., Heatherton, T., & Halpern, D. (2013). *Psychological science* (4th ed.). New York: W. W. Norton & Company, Inc.

Gee, G. J. (2016). Resilience and recovery from trauma among Aboriginal help seeking clients in an urban Aboriginal community controlled health organisation. (Doctoral dissertation, University of Melbourne, Melbourne School of Psychological Sciences).

Gee, G., Dudgeon, P., Schultz, C., Hart, A., & Kelly, K. (2014). Aboriginal and Torres Strait Islander Social and Emotional Wellbeing. In P. Dudgeon, H. Milroy, & R. Walker (Eds.), *Working together: Aboriginal and Torres Strait Islander mental health and wellbeing principles and practice* (2nd ed.). Telethon Kids Institute, Kulunga Aboriginal Research Development Unit, Department of the Prime Minister and Cabinet (Australia).

Geia, L. K., Hayes, B., & Usher, K. (2013). Yarning/Aboriginal storytelling: Towards an understanding of an Indigenous perspective and its implications for research practice. *Contemporary nurse*, 46(1), 13–17.

Gibbons, C., Dempster, M., & Moutray, M. (2011). Stress, coping and satisfaction in nursing students. *Journal of advanced nursing*, 67(3), 621–632.

Gillaspy, J. A. Jr., Brinegar, J. L., & Bailey, R. E. (2014). Operant psychology makes a splash – in marine mammal training (1955–1965). *Journal of the history of the behavioral sciences*, 50(3), 231–248.

Gillberg, M., & Akerstedt, T. (1998). Sleep loss performance: No 'safe' duration of a monotonous task. *Physiology & behavior*, 64(5), 599–604.

Glanzer, M., & Cunitz, A. R. (1966). Two storage mechanisms in free recall. *Journal of verbal learning and verbal behavior*, 5(4), 351–360.

Graves, L., Pack, A., & Abel, T. (2001). Sleep and memory: A molecular perspective. *Trends in neurosciences*, 24, 237–243.

Gray, P. (2011) *Psychology* (6th ed.) New York: Worth.

Grazia Magazine (n.d.). *Kendall Jenner's extremely strange phobia*. https://graziamagazine.com/articles/kendall-jenner-phobia/

Green, J., & Turpin, M. (2013). If you go down to the soak today: symbolism and structure in an Arandic children's story. *Anthropological linguistics*, 358–394.

Greenberg, D. L., & Knowlton, B. J. (2014). The role of visual imagery in autobiographical memory. *Memory & cognition*, 42(6), 922–934.

Greenough, W. T., & Juraska, J. H. (1979). Experience-induced changes in fine brain structure: Their behavioral implications. In M. E. Hahn, C. Jensen, & B. C. Dudek (Eds.), *Development and evolution of brain size: Behavioral implications* (pp. 295–320), New York: Academic Press.

Greenough, W. T., & Volkmar, F. R. (1973). Pattern of dendritic branching in occipital cortex of rats reared in complex environments. *Experimental neurology*, 40, 491–504.

Guy-Evans, O. (2021, Sept 27). *Hypothalamic–pituitary–adrenal axis*. Simply Psychology. www.simplypsychology.org/hypothalamic–pituitary–adrenal-axis.html

Hagenauer, M. H., Perryman, J. I., Lee, T. M., & Carskadon, M. A. (2009). Adolescent changes in the homeostatic and circadian regulation of sleep. *Developmental neuroscience*, 31(4), 276–284.

Hale, L., & Guan, S. (2015). Screen time and sleep among school-aged children and adolescents: a systematic literature review. *Sleep medicine reviews*, 21, 50–58. https://doi.org/10.1016/j.smrv.2014.07.007

Hanh, T. N. (1998). *The heart of the Buddha's teaching: Transforming suffering into peace, joy, and liberation*. New York: Broadway Books

Hardwick, R. M., Caspers, S., Eickhoff, S. B., & Swinnen, S. P. (2018). Neural correlates of action: Comparing meta-analyses of imagery, observation, and execution. *Neuroscience & biobehavioral reviews*, 94, 31–44.

Hawkins, J., & Ahmad, S. (2016). Why neurons have thousands of synapses, a theory of sequence memory in neocortex. *Frontiers in neural circuits*, 23.

Hawkins, R. D., & Bower, G. H. (1989). Computational models of learning in simple neural systems. In G. H. Bower (Ed.), *The psychology of learning and motivation* (vol. 23). San Diego: Academic Press.

Health Direct. (2018). Dehydration. Healthdirect.gov.au; Healthdirect Australia. https://www.healthdirect.gov.au/dehydration

Henrich, J., Heine, S. J., & Norenzayan, A. (2010). The weirdest people in the world?. *The behavioral and brain sciences*, 33(2–3), 61–135. doi: 10.1017/S0140525X0999152X

Herculano-Houzel, S. (2009). The human brain in numbers: a linearly scaled-up primate brain. *Frontiers in human neuroscience*, 31.

Hobson, J. A. (1999). Order from chaos. In R. Conlan (Ed.), *States of mind*. New York: Wiley.

Hobson, J. A. (2001). *The Oxford companion to the body*. Oxford: Oxford University Press.

Hobson, J. A., Pace-Schott, E. F., Stickgold, R., & Kahn, D. (1998). To dream or not to dream? Relevant data from new neuroimaging and electrophysiological studies. *Current Opinion in Neurobiology*, *8*(2), 239–244.

Hone, L. C., Jarden, A., Schofield, G. M., & Duncan, S. (2014). Measuring flourishing: The impact of operational definitions on the prevalence of high levels of wellbeing. *International Journal of Wellbeing*, *4*(1), 62–90.

Horne, R., & Foster, J. A. (2018). Metabolic and Microbiota Measures as Peripheral Biomarkers in Major Depressive Disorder. *Frontiers in psychiatry*, *9*. https://doi.org/10.3389/fpsyt.2018.00513

Hovane V, Dalton (Jones) T and Smith P (2014). 'Aboriginal Offender Rehabilitation Programs'. Chapter 30 in Dudgeon P, Milroy H and Walker R (eds) Working Together: Aboriginal Torres Strait Islander Mental Health and Wellbeing Principles and Practice. Department of Prime Minister and Cabine

Human Adaptation Institute ch 6 page 238 should there be a citation/ref given for the info in the introduction about the Deep Time Project?

Huth, A. G., De Heer, W. A., Griffiths, T. L., Theunissen, F. E., & Gallant, J. L. (2016). Natural speech reveals the semantic maps that tile human cerebral cortex. *Nature*, *532*(7600), 453–458.

Irish, M. (2016). Semantic memory as the essential scaffold for future-oriented mental time travel. *Seeing the future: Theoretical perspectives on future-oriented mental time travel*, 389–408.

Irish, M., & Piguet, O. (2013). The pivotal role of semantic memory in remembering the past and imagining the future. *Frontiers in behavioral neuroscience*, *7*, 27.

Irish, M., & Vatansever, D. (2020). Rethinking the episodic-semantic distinction from a gradient perspective. *Current Opinion in Behavioral Sciences*, *32*, 43–49.

James, W. (1890). Principles of psychology. New York, NY: Holt.

Jenni, O. G., & Carskadon, M. A. Sleep Research Society. *SRS Basics of Sleep Guide*. Westchester, I. L.: Sleep Research Society; 2000. Normal human sleep at different ages: Infants to adolescents; pp. 11–19.

Johns Hopkins Medicine. (2012, April 18). New medication offers hope to patients with frequent, uncontrollable seizures. *Science Daily*.

Kahn, M., Sheppes, G., & Sadeh, A. (2013). Sleep and emotions: Bidirectional links and underlying mechanisms. *International Journal of Psychophysiology*, *89*, 218–228.

Kao, P., & Craigie, P. (2013). Evaluating student interpreters' stress and coping strategies. *Social Behavior and Personality*, *41*(6), 1035–1043

Karni, A., Tanne, D., Rubenstein, B. S., Askenasy, J. J. M., & Sagi, D. (1994). Dependence on REM sleep of overnight improvement of a perceptual skill. *Science*, *265*, 679–682.

Kay, A. A. (2021, August 27). *How online mindfulness training can help students thrive during the pandemic*. https://theconversation.com/how-online-mindfulness-training-can-help-students-thrive-during-the-pandemic-166264#:~:text=This%20study%20revealed%20that%20online,their%20actions%20with%20their%20values

Kelly, L. (2015). Knowledge and power in prehistoric societies: Orality, memory, and the transmission of culture. Cambridge University Press.

Keogh, R., Pearson, J., & Zeman, A. (2021). Aphantasia: The science of visual imagery extremes. In *Handbook of clinical neurology* (vol. 178, pp. 277–296). Elsevier.

Keyes, C. L. (2002). The mental health continuum: From languishing to flourishing in life. *Journal of health and social behavior*, *43*(2), 207–222.

Keyes, C. L. (2007). Promoting and protecting mental health as flourishing: A complementary strategy for improving national mental health. *American psychologist*, *62*(2), 95.

Keyes, C. L., Wissing, M., Potgieter, J. P., Temane, M., Kruger, A., & van Rooy, S. (2008). Evaluation of the mental health continuum-short form (MHC-SF) in Setswana-speaking South Africans. *Clinical psychology and psychotherapy*, *15*(3), 181–192.

Killgore, W. D. S., Balkin, T. J., & Wesensten, N. J. (2006). Impaired decision making following 49 hours of sleep deprivation. *Journal of sleep research*, *15*, 7–13.

Kirk, M., & Berntsen, D. (2018). The life span distribution of autobiographical memory in Alzheimer's disease. *Neuropsychology*, *32*(8), 906.

Kivistö, J., Soininen, H., & Pihlajamaki, M. (2014). Functional MRI in Alzheimer's disease. In *Advanced brain neuroimaging topics in health and disease-methods and applications*. IntechOpen.

Kolb, B., & Whishaw, I. Q. (2003). *Fundamentals of human neuropsychology* (5th ed.). New York: Worth.

Kozlowska, K., Walker, P., McLean, L., & Carrive, P. (2015). Fear and the defense cascade: Clinical implications and management. *Harvard review of psychiatry*, *23*(4), 263–287. https://doi.org/10.1097/HRP.0000000000000065

Lancet, The (2014), Long-term effectiveness of dopamine agonists and monoamine oxidase B inhibitors compared with levodopa as initial treatment for Parkinson's disease (PD MED): a large, open-label, pragmatic randomised trial. vol. 384, (9949) https://www.thelancet.com/journals/lancet/article/PIIS0140-6736(14)60683-8/fulltext

Lazarus, R. S., & Folkman, S. (1984). *Stress, appraisal and coping*. New York: Springer.

LeDoux, J. E. (2020). Thoughtful feelings. *Current biology*, *30*(11), R619–R623.

LeDoux, J. E. (2022). As soon as there was life, there was danger: the deep history of survival behaviours and the shallower history of consciousness. *Philosophical transactions of the Royal Society B*, *377*(1844), 20210292.

LeMoult, J., & Joormann, J. (2012). Attention and memory biases in social anxiety disorder: The role of comorbid depression. *Cognitive therapy and research*, *36*, 47–57. https://doi.org/10.1007/s10608-010-9322-2

Levine, B., Svoboda, E., Hay, J. F., Winocur, G., & Moscovitch, M. (2002). Aging and autobiographical memory: Dissociating episodic from semantic retrieval. *Psychology and Aging, 17*(4), 677–689. https://doi.org/10.1037/0882-7974.17.4.677

Liddle, J., Langton, M., Rose, J. W., & Rice, S. (2022). New thinking about old ways: Cultural continuity for improved mental health of young Central Australian Aboriginal men. *Early Intervention in Psychiatry, 16*(4), 461–465.

Lindquist, K. A., Wager, T. D., Kober, H., Bliss-Moreau, E., & Barrett, L. F. (2012). The brain basis of emotion: A meta-analytic review. *The behavioral and brain sciences, 35*(3), 121.

Littleton, H., Horsey, S., John, S., & Nelson, D. (2007). Trauma coping strategies and psychological distress: A meta-analysis. *Journal of traumatic stress, 20*(6), 977–988.

Louie, K., & Wilson, M. A. (2001). Temporally structured replay of awake hippocampal ensemble activity during rapid eye movement sleep. *Neuron, 29*, 145–156.

Macedo, D. M., Smithers, L. G., Roberts, R. M., Haag, D. G., Paradies, Y., & Jamieson, L. M. (2019). Does ethnic-racial identity modify the effects of racism on the social and emotional wellbeing of Aboriginal Australian children? *PloS one, 14*(8), e0220744.

Madaboosi, S., Grasser, L. R., Chowdury, A., & Javanbakht, A. (2021). Neurocircuitry of contingency awareness in Pavlovian fear conditioning. *Cognitive, affective, & behavioral neuroscience, 21*(5), 1039–1053.

Makin, S. (2016, 1 September). Where words are stored: The brain's meaning map. *Scientific American Mind, 27*(5), 17. doi:10.1038/scientificamericanmind0916-17a

Malmberg, K. J., Raaijmakers, J. G., & Shiffrin, R. M. (2019). 50 years of research sparked by Atkinson and Shiffrin (1968). *Memory & cognition, 47*(4), 561–574.

Mandai, O., Guerrien, A., Sockeel, P., Dujardin, K., & Leconte, P. (1989). REM sleep modifications following a Morse code learning session in humans. *Physiology & behavior, 46*, 639–642.

Manoogian, E. N. C., Chaix, A., & Panda, S. (2019). When to eat: The importance of eating patterns in health and disease. *Journal of biological rhythms, 34*(6), 579–581. https://doi.org/10.1177/0748730419892105

Maquet, P. (2001). The role of sleep in learning and memory. *Science, 294*, 1048–1052.

Mattis, J., & Sehgal, A. Circadian rhythms, sleep, and disorders of aging. *Trends in endocrinology and metabolism: TEM, 27*(4), 192–203. https://doi.org/10.1016/j.tem.2016.02.003

Maynard, J. (2021). Yuraki – an Australian Aboriginal perspective on deep history. In *The Routledge companion to global Indigenous history* (pp. 722–735). Routledge.

Mayo Clinic (2022). Mindfulness exercises. https://www.mayoclinic.org/healthy-lifestyle/consumer-health/in-depth/mindfulness-exercises/art-20046356

Mayo Clinic (2014). Mental health: Overcoming the stigma of mental illness. https://www.mayoclinic.org/diseases-conditions/mental-illness/in-depth/mental-health/art-20046477

McCorry, L. K. (2007). Physiology of the autonomic nervous system. *American journal of pharmaceutical education, 71*(4), 78. https://www.ncbi.nlm.nih.gov/pmc/articles/PMC1959222/

McDermott, K. B. (2021). Practicing retrieval facilitates learning. *Annual review of psychology, 72*, 609–633.

McGaugh, J. L. (2013). Making lasting memories: Remembering the significant. *Proceedings of the National Academy of Sciences, 110*(supp. 2), 10402–10407.

McGaugh, J. L., & LePort, A. (2014). Remembrance of all things past. *Scientific American, 310*(2), 40–45.

McLeod, S. A. (2008). *Serial position effect.* Simply Psychology. www.simplypsychology.org/primacy-recency.html

McLeod, S. A. (2015). *Systematic desensitization as a counter conditioning process.* Simply Psychology. http://www.simplypsychology.org/Systematic-Desensitisation.html

Mental Health Foundation Australia (2022). *Fight stigma.* https://www.mhfa.org.au/CMS/FightStigma

Miller, A. H. (1998). Neuroendocrine and immune system interactions in stress and depression. *Psychiatric clinics of North America, 21*(2), 443–463.

Miller, G. A. (1956). The magical number seven, plus or minus two: Some limits on our capacity for processing information. *Psychological review, 63*(2), 81.

Min, M., Farkas, K., Minnes, S., & Singer, L. (2007). Impact of childhood abuse and neglect on substance abuse and psychological distress in adulthood. *Journal of traumatic stress, 20*(5), 833–844.

Mindful.org (2022) What is mindfulness? https://www.mindful.org/what-is-mindfulness/

Minkel, J., Htaik, O., Banks, S., & Dinges, D. (2010). Emotional expressiveness in sleep-deprived healthy adults. *Behavioral sleep medicine, 9*, 5–14.

Mollica, F., & Piantadosi, S. T. (2019). Humans store about 1.5 megabytes of information during language acquisition. *Royal Society Open Science, 6*, 181393. doi: 10.1098/rsos.181393

Muller, V. (2021, October 9). Hair samples show meditation training reduces long-term stress. Max Planck Institute. *Neuroscience News.* https://neurosciencenews.com/medication-cortisol-stress-19443/

Munawar, K., Kuhn, S. K., & Haque, S. (2018). Understanding the reminiscence bump: A systematic review. *PloS one, 13*(12), e0208595.

Murdock, B. B. (1962). The serial position effect of free recall. *Journal of experimental psychology, 64*(5), 482–488.

Nakamaru-Ogiso, E., Miyamoto, H., Hamada, K., Tsukada, K. and Takai, K. (2012). Novel biochemical manipulation of brain serotonin reveals a role of serotonin in the circadian rhythm of sleep–wake cycles. *European Journal of Neuroscience, 35*: 1762-1770. https://doi.org/10.1111/j.1460-9568.2012.08077.x

National Institute of Neurological Disorders and Stroke (NIH). (2013, February 5). Reflex control could improve walking after incomplete spinal injuries. *ScienceDaily.*

National Sleep Foundation, 2020. https://www.sleepfoundation.org/

Neale, M., & Kelly, L. (2020). *Songlines: The power and promise.* Thames & Hudson Australia.

Neuroscientifically Challenged (2017) Retrieved from https://neuroscientificallychallenged.com/posts/sorting-out-dopamines-role-in-reward

Newsom, R., & DeBanto, J. (2020). Aging and sleep. Updated 18 March, 2022. https://www.sleepfoundation.org/aging-and-sleep

Newton, N. C., Conrod, P. J., Slade, T., Carragher, N., Champion, K. E., Barrett, E. L., ..., & Teesson, M. (2016). The long-term effectiveness of a selective, personality-targeted prevention program in reducing alcohol use and related harms: A cluster randomized controlled trial. *Journal of child psychology and psychiatry, 57*(9), 1056–1065.

Nguyen, T. T., Zhang, X., Wu, T-C, Liu, J., Le, C., Tu, X. M., Knight, R., & Jeste, D. V. (2021). Association of loneliness and wisdom with gut microbial diversity and composition: An exploratory study. *Frontiers in psychiatry.* https://doi.org/10.3389/fpsyt.2021.648475

Nuss, P. (2015). Anxiety disorders and GABA neurotransmission: a disturbance of modulation. *Neuropsychiatric disease and treatment, 11*, 165–175. 10.2147/NDT.S58841

Ohio State University. (2012, February 22). If you're afraid of spiders, they seem bigger: Phobia's effect on perception of feared objects allows fear to persist. *ScienceDaily.*

Palombo, D. J., Sheldon, S., & Levine, B. (2018). Individual differences in autobiographical memory. *Trends in cognitive sciences, 22*(7), 583–597.

Partanen, E., Kujala, T., Näätänen, R., Liitola, A., Sambeth, A., & Huotilainen, M. (2013). Learning-induced neural plasticity of speech processing before birth. *Proceedings of the National Academy of Sciences, 110*(37), 15145–15150.

Patterson, K., & Ralph, M. A. L. (2016). The hub-and-spoke hypothesis of semantic memory. In *Neurobiology of language*, (pp. 765–775). Academic Press.

Pavlov, I. P. (1927). *Conditioned reflexes.* New York: Dover.

Pavlov, I. P. (1928). Scientific study of the so-called psychical processes in the higher animals. In I. P. Pavlov & W. H. Gantt (Trans.), *Lectures on conditioned reflexes: Twenty-five years of objective study of the higher nervous activity (behaviour) of animals* (pp. 81–96). Liverwright Publishing Corporation.

Pearson, J., Naselaris, T., Holmes, E. A., & Kosslyn, S. M. (2015). Mental imagery: functional mechanisms and clinical applications. *Trends in cognitive sciences, 19*(10), 590–602.

Petrican, R., Palombo, D. J., Sheldon, S., & Levine, B. (2020). The neural dynamics of individual differences in episodic autobiographical memory. *Eneuro, 7*(2).

Pew Research Center for the People & the Press. (2018). Pew Research Center: Spring 2018 *Global Attitudes Survey (Version 3)* [Dataset]. Cornell University, Ithaca, NY: Roper Center for Public Opinion Research. doi:10.25940/ROPER-31115445

Pulhman, L., Vrtička, P., Linz, R., Stalder, T., Kirschbaum, C., Engert, V., & Singer, T. (2021). Contemplative mental training reduces hair glucocorticoid levels in a randomized clinical trial. *Psychosomatic medicine, 83*, 894–905.

Queensland Health. (2020). *Finn's story.* https://mentalwellbeing.initiatives.qld.gov.au/stories/finn

Rah, J. C., & Choi, J. H. (2022). Finding needles in a haystack with light: Resolving the microcircuitry of the brain with fluorescence microscopy. *Molecules and cells, 45*(2), 84.

Ramirez, G., & Beilock, S. L. (2011). Writing about testing worries boosts exam performance in the classroom. *Science, 331*(6014), 211–213.

Ramirez, K. (2012). Marine mammal training: the history of training animals for medical behaviors and keys to their success. *Veterinary clinics: Exotic animal practice, 15*(3), 413–423.

ReachOut Australia (2022) How to challenge negative thoughts. https://au.reachout.com/articles/how-to-challenge-negative-thoughts

Renoult, L., & Rugg, M. D. (2020). An historical perspective on Endel Tulving's episodic-semantic distinction. *Neuropsychologia, 139*, 107366.

Reynolds, H. (2022). *Forgotten War.* Sydney: New South Publishing

Rice, V. H. (2012). Theories of stress and its relationship to health. In V. H. Rice (Ed.), *Handbook of stress, coping, and health: Implications for nursing research, theory, and practice* (pp. 22–42). Sage Publications, Inc.

Richardson, G. S., Miner, J. D., & Czeisler, C. A. (1989). Impaired driving performance in shift workers: The role of the circadian system in a multifactorial model. *Alcohol, drugs and driving, 5–6*(4–1), 265–273.

Riemann, D., & Voderholzer, U. (2003). Primary insomnia: A risk factor to develop depression? *Journal of affective disorders, 76*, 255–259.

Rizzolatti, G., Fogassi, L., & Gallese, V. (2001). Neurophysiological mechanisms underlying the understanding and imitation of action. *Nature reviews neuroscience, 2*(9), 661–670.

Roediger III, H. L., & Karpicke, J. D. (2006). Test-enhanced learning: Taking memory tests improves long-term retention. *Psychological science, 17*(3), 249–255.

Roediger III, H. L., & Karpicke, J. D. (2018). Reflections on the resurgence of interest in the testing effect. *Perspectives on psychological science, 13*(2), 236–241.

Roelofs, K. (2017). Freeze for action: neurobiological mechanisms in animal and human freezing. *Philosophical Transactions of the Royal Society B, 3722016020620160206.* http://doi.org/10.1098/rstb.2016.0206

Rosenzweig, M. R., Leiman, A. L., & Breedlove., S. M. (1999). *Biological psychology: An introduction to behavioral, cognitive and clinical neuroscience.* Massachusetts: Sinauer Associates.

Rubin, D. C. (2005). A basic-systems approach to autobiographical memory. *Current directions in psychological science, 14*(2), 79–83.

Ryan, R. M., & Deci, E. L. (2001). On happiness and human potentials: A review of research on hedonic and eudaimonic well-being. *Annual review of psychology, 52*(1), 141–166.

Ryff, C. D. (1989). Happiness is everything, or is it? Explorations on the meaning of psychological well-being. *Journal of personality & social psychology, 57*(6), 1069–1081.

Sakai, M. (2021). Freezing and tonic immobility: Their definitions and naming. In: M. Sakai (Ed.) *Death-feigning in insects. Entomology monographs.* Springer, Singapore. https://doi.org/10.1007/978-981-33-6598-8_1

Schacter, D.L., Benoit, R.G., Szpunar, K.K (2017). Episodic future thinking: mechanisms and functions. *Current opinion in behavioral sciences, 17,* 41–50.

Schmahmann, J. D. (2019). The cerebellum and cognition. *Neuroscience letters, 688,* 62–75.

Scoville, W. B., & Milner, B. (1957). Loss of recent memory after bilateral hippocampal lesions. *Journal of neurology, neurosurgery, and psychiatry, 20*(1), 11.

Seligman, M. E. P., & Csikszentmihalyi, M. (2000). Positive psychology: An introduction. *American Psychologist, 55*(1), 5–14.

Selye, H. (1976). *The stress of life.* New York: Freeman.

Shanahan, T. L., Kronauer, R. E., Duffy, J. F., Williams, G. H., & Czeisler, C. A. (1999). Melatonin rhythm observed throughout a three-cycle bright-light stimulus designed to reset the human circadian pacemaker. *Journal of biological rhythms, 14,* 237–253.

Sheldon, S. H. (2002). Sleep in infants and children. In T. K. Lee-Choing, M. J. Sateia, & M. A. Carskadon (Eds.), *Sleep medicine* (pp. 99–103). Philadelphia: Hanley and Belfus.

Sheldon, S., McAndrews, M. P., Pruessner, J., & Moscovitch, M. (2016). Dissociating patterns of anterior and posterior hippocampal activity and connectivity during distinct forms of category fluency. *Neuropsychologia, 90,* 148–158.

Simmons, B. L., & Nelson, D. L. (2001). Eustress at work: The relationship between hope and health in hospital nurses. *Health care management review, 26*(4), 7–18.

Sivak, L., Westhead, S., Richards, E., Atkinson, S., Richards, J., Dare, H., Zuckermann, G., Gee, G., Wright, M., Rosen, A., Walsh, M., Brown, N., & Brown, A. (2019). 'Language Breathes Life' – Barngarla Community Perspectives on the wellbeing impacts of reclaiming a dormant Australian Aboriginal Language. International journal of environmental research and public health, 16(20), 3918–1–3918–17

Skinner, B. F. (1953). *Science and human behavior.* New York, NY: The Free Press.

Smith, C. (1995). Sleep states and memory processes. *Behavioural brain research, 69,* 137–145.

Smith, C., & Lapp, L. (1986). Prolonged increases in both PS and number of REMS following a shuttle avoidance task. *Physiology & behavior, 36,* 1053–1057.

Snyder, C. R. (2001). *Coping with stress: Effective people and processes.* New York: Oxford University Press.

Soderstrom, N. C., Kerr, T. K., & Bjork, R. A. (2016). The critical importance of retrieval—and spacing—for learning. *Psychological science, 27*(2), 223–230.

Stainton, A., Chisholm, K., Kaiser, N., Rosen, M., Upthegrove, R., Ruhrmann, S., & Wood, S. J. (2019). Resilience as a multimodal dynamic process. *Early intervention in psychiatry, 13*(4), 725–732.

Steinvorth, S., Levine, B., & Corkin, S. (2005). Medial temporal lobe structures are needed to re-experience remote autobiographical memories: evidence from HM and WR. *Neuropsychologia, 43*(4), 479–496.

Stephenson, L. J., Edwards, S. G., & Bayliss, A. P. (2021). From gaze perception to social cognition: The shared-attention system. *Perspectives on psychological science, 16*(3), 553–576.

Stepnowsky, C. J., & Ancoli-Israel, S. (2008). Sleep and its disorders in seniors. *Sleep medicine clinics, 3*(2), 281–293. https://doi.org/10.1016/j.jsmc.2008.01.011

Stickgold, R. (2005). Sleep-dependent memory consolidation. *Nature, 437,* 1272–1278.

Strikwerda-Brown, C., Shaw, S. R., Hodges, J. R., Piguet, O., & Irish, M. (2022). Examining the episodic-semantic interaction during future thinking–A reanalysis of external details. *Memory & cognition, 50*(3), 617–629.

Sundaram, R. S., Gowtham, L., & Nayak, B. S. (2012). The role of excitatory neurotransmitter glutamate in brain physiology and pathology. *Asian journal of pharmaceutical and clinical research, 5*(2), 1–7.

Tang, Y.-Y., & Tang, R. (2015). Social cognitive neuroscience, cognitive neuroscience, clinical brain mapping. In A. W. Toga (Editor-in-chief), *Brain Mapping: A comprehensive reference.* Academic Press, Elsevier.

Tannenbaum, M. B., Hepler, J., Zimmerman, R. S., Saul, L., Jacobs, S., Wilson, K., & Albarracín, D. (2015). Appealing to fear: A meta-analysis of fear appeal effectiveness and theories. *Psychological bulletin, 141*(6), 1178. https://doi.org/10.1037/a0039729

Tapia-Osorio, A., Salgado-Delgado, R., Angeles-Castellanos, M., & Escobar, C. (2013). Disruption of circadian rhythms due to chronic constant light leads to depressive and anxiety-like behaviors in the rat. *Behavioural brain research, 252,* 1–9. https://doi.org/10.1016/j.bbr.2013.05.028

Tees, R. C. (1968). Effect of early visual restriction on later visual intensity discrimination in rats. *Journal of comparative and physiological psychology, 66,* 224–227.

Tempesta, D., Couyoumdjian, A., Curcio, G., Moroni, F., Marzano, C., De Gennaro, L., & Ferrara, M. (2010). Lack of sleep affects the evaluation of emotional stimuli. *Brain research bulletin, 82,* 104–108.

Thursby, E., & Juge, N. (2017). Introduction to the human gut microbiota. *The biochemical journal, 474*(11), 1823–1836. https://doi.org/10.1042/BCJ20160510

Tilley, A. J., Wilkinson, R. T., Warren, P. S. G., Watson, W. B., & Drud, M. (1982). The sleep and performance of shift workers. *Human factors, 24,* 629–641.

Trafton, A. (2019, August 26). Two studies reveal benefits of mindfulness for middle school students. MIT News. https://news.mit.edu/2019/mindfulnessmental-health-benefits-students-0826

Tranah, G., Stone, K., & Ancoli-Israel, S. (2017). Circadian rhythms in older adults. In M. Kryger, T. Roth and W. C. Dement (Eds.), *Principles and practice of sleep medicine* (6th ed., pp. 1510–1515.e4). https://doi.org/10.1016/B978-0-323-24288-2.00154-9

Transforming Indigenous Mental Health and Wellbeing. (2022). *Social and emotional wellbeing fact-sheet.* https://timhwb.org.au/wp-content/uploads/2021/04/SEWB-fact-sheet.pdf

Transport Accident Commission. (2021). Fatigue trial test a wake-up to drowsy drivers. https://urldefense.com/v3/__https://www.tac.vic.gov.au/about-the-tac/media-room/news-andevents/2021/fatigue-trial-test-a-wake-up-to-drowsydrivers

Tulving, E. (2002). Episodic memory: From mind to brain. *Annual review of psychology, 53*(1), 1–25.

Turner, A. M., & Greenough, W. T. (1983). Synapse per neuron and synaptic dimensions in occipital cortex of rats reared in complex, social or isolation housing. *Acta Stereologica, 2*(1), 239–244

Tyng, C. M., Amin, H. U., Saad, M. N. M., & Malik, A.S. (2017). The influences of emotion on learning and memory. *Frontiers in psychology, 8*, 1454. doi: 10.3389/fpsyg.2017.01454

Ungunmerr, M-R. (2017). To be listened to in her teaching: Dadirri: Inner deep listening and quiet still awareness. *EarthSong Journal: Perspectives in ecology, spirituality and education, 3*(4), 14–15.

Ungunmerr-Baumann, M. R., Groom, R. A., Schuberg, E. L., Atkinson, J., Atkinson, C., Wallace, R., & Morris, G. (2022). Dadirri: an Indigenous place-based research methodology. AlterNative: An International Journal of Indigenous Peoples, 18(1), 94–103.

Université de Montréal. (2013, October 3). Three hours is enough to help prevent mental health issues in teens. *ScienceDaily.*

University of Colorado at Boulder. (2022, January 27). Even dim light before bedtime may disrupt a preschooler's sleep: Broad range of intensity can sharply dampen sleep-promoting melatonin. *ScienceDaily.* http://www.sciencedaily.com/releases/2022/01/220127104208.htm

University of Utah 2016 ch 2 page 63 cited but ref details not provided

Valverde, F. (1971) Rate of and extent of recovery from dark-rearing in visual cortex of the mouse, *Brain research, 33*, 1–11.

Van Der Helm, E., Gujar, N., & Walker, M. P. (2010). Sleep deprivation impairs the accurate recognition of human emotions. *Sleep, 33*, 335.

Van Dongen, H. P., Maislin, G., Mullington, J. M., & Dinges, D. F. (2003). The cumulative cost of additional wakefulness: Dose-response effects on neurobehavioral functions and sleep physiology from chronic sleep restriction and total sleep deprivation. *Sleep, 26*(2), 117–126.

Vannucci, M., Pelagatti, C., Chiorri, C., & Mazzoni, G. (2016). Visual object imagery and autobiographical memory: Object Imagers are better at remembering their personal past. *Memory, 24*(4), 455–470.

Verywell Mind (2020). The different types of social support. https://www.verywellmind.com/types-of-social-support-3144960

Verywell Mind. (2021). *DSM-5 Diagnostic criteria for a specific phobia.* https://www.verywellmind.com/diagnosing-a-specific-phobia-2671981

Videbeck, S. L., & Miller, C. J. (2017). *Psychiatric-mental health nursing.* Wolters Kluwer.

Volkmar, F. R., & Greenough, W. T. (1972). Rearing complexity affects branching of dendrites in the visual cortex of the rat. *Science, 176*, 1445–1447.

Walk, R. D., & Walters, C. P. (1973). Effect of visual deprivation on depth discrimination of hooded rats. *Journal of comparative and physiological psychology, 85*, 559–563.

Wang, F. (2022). *Neuroscience of mindfulness meditation.* The Wharton School, The University of Pennsylvania. https://neuro.wharton.upenn.edu/community/winss_scholar_blog2/

Watkins, N. W. (2018). (A) phantasia and severely deficient autobiographical memory: Scientific and personal perspectives. *Cortex, 105*, 41–52.

Watson, D., Clark, L. A., & Tellegen, A. (1988). Development and validation of brief measures of positive and negative affect: the PANAS scales. *Journal of personality and social psychology, 54*, 1063–1070. doi: 10.1037//0022-3514.54.6.1063

Watson, J. B. (1913). Psychology as the behaviorist views it. *Psychological review, 20*(2), 158–177.

Watson, J. B. (1924). *Behaviorism.* New York: People's Institute Publishing Company.

Watson, J. B., & Rayner, R. (1920). Conditioned emotional reactions. *Journal of experimental psychology, 3*(1), 1–14. https://doi.org/10.1037/h0069608

Westerhof, G. J., & Keyes, C. L. (2010). Mental illness and mental health: The two continua model across the lifespan. *Journal of adult development, 17*(2), 110–119.

Westerman, T. (2021). Culture-bound syndromes in Aboriginal Australian populations. *Clinical Psychologist, 25*(1), 19–35.

Wilkie, M. (1997). *Bringing them home: Report of the National Inquiry into the Separation of Aboriginal and Torres Strait Children from their Families.* Canberra: Human Rights and Equal Opportunity Commission.

Williamson, A. M., & Feyer, A. (2000). Moderate sleep deprivation produces impairments in cognitive and motor performance equivalent to legally prescribed levels of alcohol intoxication. *Occupational and environmental medicine, 57*, 649–655.

Wissing, M. P., Schutte, L., Liversage, C., Entwisle, B., Gericke, M., & Keyes, C. (2021). Important goals, meanings, and relationships in flourishing and languishing states: Towards patterns of well-being. *Applied research in quality of life, 16*(2), 573–609.

Wolitzky-Taylor, K. B., Horowitz, J. D., Powers, M. B., & Telch, M. J. (2008). Psychological approaches in the treatment of specific phobias: A meta-analysis. *Clinical psychology review, 28*, 1021–1037.

Wolpe, J., & Plaud, J. J. (1997). Pavlov's contributions to behavior therapy: The obvious and the not so obvious. *American psychologist, 52*(9), 966–972. https://doi.org/10.1037/0003-066X.52.9.966

World Health Organization. (2019). *International statistical classification of diseases and related health problems* (11th ed.). https://icd.who.int

World Health Organization. (2022). *Mental disorders.* https://www.who.int/news-room/fact-sheets/detail/mental-disorders

Wu, W-L., Adame, M. D., Liou, C-W. Barlow, J. T., Lai, T-T., Sharon, G., Schretter, C. E., Needham, B. D., Wang, M. I., Tang, W., Ousey, J., Lin, Y-Y., Yao, T-H., Abdel-Haq, R., Beadle, K., Gradinaru, V., Ismagilov, R. F., & Mazmanian, S. K. (2021). Microbiota regulate social behaviour via stress response neurons in the brain. *Nature, 595,* 409–414 https://doi.org/10.1038/s41586-021-03669-y

Yong, G. (2014). Shift work and diabetes mellitus: a meta-analysis of observational studies. *Occupational and environmental medicine* (Online First). Retrieved 11 August 2014.

Yunkaporta, T. (2019). *Sand talk: How Indigenous thinking can save the world.* Text Publishing.

Zeman, A. Z., Dewar, M., & Della Sala, S. (2015). Lives without imagery – Congenital aphantasia.

Zeman, A., Milton, F., Della Sala, S., Dewar, M., Frayling, T., Gaddum, J., ..., & Winlove, C. (2020). Phantasia–the psychological significance of lifelong visual imagery vividness extremes. *Cortex, 130,* 426–440.

Zubrick, S. R., Shepherd, C. C., Dudgeon, P., Gee, G., Paradies, Y., Scrine, C., & Walker, R. (2014). Social determinants of social and emotional wellbeing. *Working together: Aboriginal and Torres Strait Islander mental health and wellbeing principles and practice, 2,* 93–112.

Index

ABC model of operant conditioning 152–8
   in animal training 154–5
   for behaviour modification 156–8
Aboriginal and Torres Strait Islander community-controlled health organisations 409
Aboriginal and Torres Strait Islander peoples
   approaches to learning 137, 166–76
   cultural continuity 342–3, 406–7
   culturally responsive mental health assistance 328
   Elders and ancestors 339, 340
   impact of colonisation on 167, 168, 339, 408
   intergenerational trauma 340
   language and cultural groups 168
   learning *with* and *from* 167–8
   negative impacts of government policies on 167, 168, 339–40
   resistance to invasion of lands during Frontier Wars 408
   self-determination 340, 408–9
   understandings of social and emotional wellbeing 324–5, 335–43, 406–10
Aboriginal and Torres Strait Islander peoples' ways of knowing 137, 166, 168–75
   Country 168, 169–70, 223, 339
   Dadirri (deep listening) 171–2
   embedded in relationships 168, 169–70
   kinship system 169–70
   as multimodal system 169, 170
   oral cultures 170, 218, 222–4
   sand talk symbols for ways of knowing 173–5
   similarities to Western approaches to learning 175
   situating ourselves as learners 166–8
   Songlines 170–1, 222–4
   system of knowledge patterned on *Country* 170–1
   yarning 173
accuracy 33, 34, 35, 430
acknowledgements 436
acronyms 218
acrostics 218–19
action potential 66, 69, 76
acute stress 92, 113
adequate diet, and mental wellbeing 394, 395
adequate sleep, and mental wellbeing 394–7
adolescence
   building resilience to help prevent mental health problems 333
   screen time and sleep 265–6
   sleep requirements 267–8
   sleep–wake shifts 296–7
adrenal glands 95, 96
adrenaline 56, 57, 95, 97, 364
adrenocorticotropic hormone (ACTH) 97
adults, sleep requirements 268
Advanced Sleep Phase Disorder (ASPD) 295, 297, 300
advertising campaigns, classical and operant conditioning in 158–60
affective functioning, and partial sleep deprivation 284, 286
afferent nerves 110
after conditioning 141, 142
agonists 67
agoraphobia 349
aim(s) 9, 426
   does the evidence support the aims? 38
alarm-reaction stage 101, 102
alcohol consumption
   affect on sleep patterns 306
   comparison with the effects of sleep deprivation 289–90, 291–2
allocortex 204
alpha waves 251, 252
altered states of consciousness (ASC) 243, 244
Alzheimer's disease 214–17
   autobiographical memory in 215–16
   causes 214–15
   post-mortem brain lesion study 214
amygdala 203–4, 365, 400
amyloid plaques 214
analysing and evaluating research 29–42
analysing data 13, 33–7, 432
ancestor mind (sand talk symbol) 174
ancestors (Aboriginal and Torres Strait Islander peoples) 339, 340
anecdotes 21
animal training, three-phase process of operant conditioning in 154–6
antagonists 67
Antecedent-Behaviour-Consequence (ABC) model of operant conditioning 152–8
antecedents 152, 153–4, 155, 157
anterior (front) temporal lobe hub 205
anterograde amnesia 198, 206
anti-anxiety benzodiazepine agents (GABA agonists) 373–4
anti-epileptic drugs 70
anxiety 68–9, 347, 348, 394
   exam anxiety 351–2
   and gut microbiota 112
   interaction with stress 347
   levels of 348
   medications to relieve symptoms of 373–4
   and mental wellbeing 347–8
   stress and phobia, key similarities and differences 350
anxiety disorders 68–9, 327, 347–8, 349
   harmful effects of stigma associated with 370
   and lack of sleep 395
   see also phobias
aphantasia 212, 213, 216
approach strategies 116
appropriate anxiety 348
arachnophobia 365, 371–2
   phobias effect on perception of feared objects 383
   systematic desensitisation 377, 378–80
   virtual reality spiders to help people face their 379–80
arousal 56–7
ARRM-R (cognitive processes in observational learning) 161
ascending tracts 52, 53
Atkinson and Shiffrin's multi-store model of memory 188–97
attention 161
Australian Indigenous Psychology Education Project (AIPEP) 167
authentic social support 402
autobiographical memory (ABM) 200, 209–11
   in Alzheimer's disease 215–16
   distribution over the lifespan 210
   individual differences in ABM and episodic-future thinking 210–11
   role of mental imagery in 212

where are you on the mental imagery/ABM spectrum? 213
see also episodic-autobiographical memory (EAM); semantic autobiographical memory (SAM)
autonomic nervous system (ANS) 52, 55, 61, 95, 110, 220, 221
   divisions 55–9, 95
   role 55
autonomic reflexes 140
avoidance behaviours 381, 382
avoidance strategies 118
axon terminals 66
axons 66

'bad' bacteria 111
Bandura, Albert
   learned aggression in children through observation of an adult model 162–3
   observational learning 160–2
   and social-cognitive learning 160–4
bar charts 29, 30, 431, 432
basal ganglia 207
Bawaka Collective 169
before conditioning 141, 142
before you begin your experiment: things to consider 430
behaviour 137
   ABC model of operant conditioning 152–3
behaviour modification
   ABC model of operant conditioning for 156–8
   to change social media habits 156–8
behavioural functioning, and partial sleep deprivation 285–6
behavioural models
   of phobic anxiety 366–8
   see also classical conditioning; operant conditioning
behaviourism 138
behaviourist approaches to learning 137–59
beneficence 39, 429
benzodiazepines 373–4
beta blockers 373
beta waves 251
between-subjects design 25–6
biological factors contributing to specific phobia 363, 364–5, 370
biological interventions to manage specific phobia 373–5
biological protective factors to maintain mental wellbeing 393–7

biological responses to stress 94–8, 101–4
biological rhythms 258–64
biological stressors 91
biopsychosocial approach to maintaining mental wellbeing 392–410
   mindfulness meditation 392–3, 398–401, 403–4
   protective factors 393–405
biopsychosocial model to explain specific phobias 363–71
   evidence-based interventions 373–84
blood alcohol concentration (BAC) 289, 290, 291–2
blue light, and sleep 303–4
Bobo doll experiment (Bandura) 162–4
body and behaviours (Aboriginal and Torres Strait Islander peoples) 339
body scan meditation 399
body temperature, and sleep 304–5
brain 50, 51, 52
   areas supporting long-term implicit and explicit memory 203–4
   changes in response to experience 75–6
   as master and commander 52
   and reward system 71
   video games affect on the 78–80
   see also gut–brain axis
brain imaging studies 215, 216
brain lesion study 214
brain's semantic map where words are stored 206
brainwave activity during sleep (EEG) 251–2
breathing retraining 373, 374–5
bright light therapy 299–300
broad hypothesis 9–10
Broca's area 205
building resilience 332, 333

caffeine 394, 396
calming effect 57–8
capacity (memory) 189, 190
case studies (examples)
   Australian fur seal training 155–6
   Dadirri: deep listening with Dr Miriam-Rose Ungunmerr-Baumann 172
   finding yourself (mental wellbeing) 345–6
   H.M. (patient) and memory 186–8
   how video games affect your brain 78–9
   possum-skin cloaks in social and emotional wellbeing 342–3

   sand talk symbols for five ways of knowing 174–5
   SEMA3 (smartphone app) 19–20
   Seven Sisters Songline 223–4
   Tanderrum gathering 407
   using behaviour modification strategies to change social media habits 156–8
case studies (methodology) 17, 383–4
   evaluation 145–6
catastrophising (catastrophic thinking) 368, 381
causal relationship 17, 23, 25
causation vs correlation 16–17, 18
central nervous system (CNS) 51–3, 54, 59, 68, 109, 220
   interaction with gut microbiota 110, 111, 112
cerebellum 207
cerebral cortex 189, 204, 207
challenging unrealistic or anxious thoughts 381–2
charts 29
chemical messengers
   at the synapse 66–7
   beyond the synapse 71–2
childhood amnesia 210
children, sleep requirements 260–1, 267
chronic sleep deprivation, making up for lost sleep 287
chronic stress 92–3, 113
   meditation to reduce 99
chunking 190
circadian rhythm sleep disorders 294–6
   caused by intrinsic and extrinsic factors 294
   daytime and night-time effects 295
   sleep-wake shifts in adolescence 296–7
   treatment 299–300
   see also Advanced Sleep Phase Disorder (ASPD); Delayed Sleep Phase Syndrome (DSPS)
circadian rhythms 258–61, 294
   comparison with ultradian rhythm 262
   and constant light exposure in rats 263–4
   correlational study 308–9
   and homeostatic sleep demand 259, 305
   and shift work 297–9
classical conditioning 138–47
   in advertising campaigns 158–60

classical conditioning (*continued*)
  Ivan Pavlov's discovery of 138–40
  Pavlov's three-phases of 141, 142
  and precipitation of a specific phobia 366–7
  survival value of 146–7
  terminology 140–1
  Watson' study of emotional response in human infant 144–6, 366
  what is learned? 142
classification 17
cognition, and gut microbiota 112
cognitive appraisal, Transactional Model of Stress and Coping 106–9
cognitive behavioural strategies to maintain mental wellbeing 397–8
cognitive behavioural therapy (CBT) 376–7, 398
cognitive bias 368
cognitive distortions 381
cognitive functioning
  and alcohol consumption 290
  and partial sleep deprivation 186, 284–5, 290
cognitive model of phobic anxiety 368
cognitive processes during learning 161
collecting data 11–13
colonisation, impact on Aboriginal and Torres Strait Islander peoples 167, 168, 339, 408
communication statement 435
community (Aboriginal and Torres Strait Islander peoples) 339
complex phobias 363
computational cognitive modelling 18
conclusions 13, 38
  evidence-based 433
  posters 436
conditioned fear response
  in an infant ('Little Albert' experiment) 144–6, 366
  in humans (fMRI and skin conductance response) 146–7
  observational learning 165
conditioned response (CR) 140, 141
conditioned stimulus (CS) 140, 141, 142, 145
conditioning 201
  see also classical conditioning; operant conditioning
conducting research *see* psychological research investigations; research investigations
confidence, and stress 105

confidentiality 39, 429
confounding variable 23–4, 25
conscious responses 61, 63
consciousness 243, 249
  states of 243–4
consequences
  operant conditioning 148, 152, 153–4, 157
  social-cognitive model of learning 161, 162–4
context-specific effectiveness 115
continuum 325
continuum of mental disorder, and levels of functioning 327–8
continuum of mental wellbeing, and levels of functioning 326–7
contradictory data 36
control condition 21
control group 21
controlled experiments 10, 14–15, 21–2, 428, 438
  designs 25–8
  participant allocation 24, 25–8
  variables 22–5
controlled variables 24, 25
convenience sampling 12
coping 115
coping flexibility 115–16
coping skills 115
coping strategies 115–20
  building resilience 332–3
  investigation 117
  practical ways 119–20
correlation 15–16
  direction 16
  strength 16, 17
  vs causation 16–17, 18
correlational studies 15–17, 23, 308–9
corticotropin-releasing hormone (CRH) 97
cortisol 95, 96–8, 101
  biological functions 96–7
  during the flight-or-fight-or-freeze response 97
  negative effects of excess cortisol on the body 97–8
  released by HPA axis 97
  and sleep 259, 268, 296–7
  stress and confidence 105
counterbalancing 27
counterconditioning 377
countershock 101
Country 168, 169, 223, 339
  and Dadirri (deep listening) 171
  and kinship system 169–70

  as multimodal system 169, 170
  system of knowledge patterned on 170–1
COVID-19 pandemic
  and disconnection from family, friends and regular routines 402
  influence on mental wellbeing, literature review 334
  online mindfulness training to help students during 403–4
cranial nerves 51
CRAP test 426, 427
creative activities 120
cultural continuity (Aboriginal and Torres Strait Islander peoples) 406–7
  revival and adaptation of possum-skin cloak making 342–3, 407
  and self-determination 408
  Tanderrum gathering 407
cultural determinants of social and emotional wellbeing (SEWB) (Aboriginal and Torres Strait Islander peoples) 406–10
  cultural continuity 406–7
  self-determination 340, 408–9
culture (Aboriginal and Torres Strait Islander peoples) 339

Dadirri (deep listening) 171–2
data 11, 428
data analysis 13, 33–7, 432
data collection 11–13, 431
data display 29, 30, 431–2
data evaluation 432
data interpretation 433
daylight 303
debriefing 39–40, 429
decay 190
deception, use of in research 40, 429
declarative memory 198, 365
deep listening (Dadirri) 171–2
deep sleep (stage 3 NREM) 246, 248, 251
Deep Time Project 242
deeper levels of encoding (long-term memory) 191–2
deeply relaxed 251
dehydration 394
Delayed Sleep Phase Syndrome (DSPS) 295–6, 297, 300
delta waves 251, 252, 287
dendrites 65, 66, 74
dependent variable (DV) 10, 22–3, 25, 427, 428

depression
  and disrupted sleep patterns 263–4
  and gut microbiota 112
deprived environment 74
descending tracts 52, 53
descriptive statistics 30–1
designing a research investigation 10–11, 425–30
designing and conducting a scientific experiment 8–41
designs for controlled experiments 25–8
diabetes 306
Diagnostic and Statistical Manual of Mental Disorders (*DSM-5-TR*™) 17, 79, 327, 347–8, 350
digestive system functions, enteric nervous system control 110, 111
digestive tract 59
direction of correlation 16
discussion (poster) 436
displaying data in tables, bar charts and line graphs 29, 30, 431–2
distress 91, 92, 97
distribution of data 29–31
dopamine 71–2, 78, 79, 207
drawing conclusions 13
Dreaming mind (sand talk symbol) 174
dreams 246, 254
drinking patterns, and sleep 305–6
drowsiness 262
drowsy drivers, fatigue trial test 288–9
drugs, affecting neurotransmitters 67
dual-continuum model of mental health and wellbeing 325–32, 343–53
  continuum of mental disorder and levels of functioning 327–8
  continuum of mental wellbeing and levels of functioning 326–7
  influence of internal and external factors 344–6
  interaction between mental wellbeing and mental disorder 330–1
  stress, anxiety and phobia 346–53
Dudgeon, Professor Pat 409
duration (memory) 189, 190
during conditioning 141, 142

eating patterns, and sleep 305–6
eating well 119–20
echoic memory 189
elaborative encoding 194
electroencephalograph (EEG) 250–2, 253
electromyograph (EMG) 252–3
electro-oculograph (EOG) 252, 253

embedded in relationships (Aboriginal and Torres Strait Islander peoples)
  kinship system 169–70
  learner and knowledge 168, 170
  patterned on *Country* 170
emotional control and responses, and partial sleep deprivation 284
emotional response in human infant, classically conditioned 144–6
emotional wellbeing 326
encoding and retrieval processes
  long-term memory 191–2
  short-term memory 189–90
end-of-chapter exams 84–6, 124–6, 179–82, 230–7, 275–7, 311–19, 356–7, 387–8, 412–14
endocrine system 110
energising social support 402
enriched environment 74
enteric nervous system (ENS) 52, 55, 58–9, 110
  manages functions of digestive system 110, 111
  neural connections to CNS 59
environmental triggers, and specific phobias 369
epileptic seizures, drug control 70
episodic-autobiographical memory (EAM) 198–9, 209–10
  self-report questionnaire 213
episodic-future thinking 199
  individual differences in ABM and 210–11
  role of mental imagery in 212
episodic memory 198–9, 200, 201, 365
  neural basis 204–5
  role in remembering the past and imagining the future 210–11
errors 35–6, 432
ethical concepts (for psychology research) 38, 39, 429
ethical guidelines (for psychology research) 38, 39–40, 429
ethics 38–40, 429–30
eustress 92, 97, 346
evaluating a research design 38–9
evaluating data and investigation methods 37–8
evidence-based conclusions 433
evidence-based interventions for specific phobias 373–84
  biological interventions 374–5
  psychological interventions 376–80
  social interventions 381–3

exam anxiety, writing about to boost exam performance 351–2
excitatory neurotransmitters 68
exercise 119, 396
exhaustion stage 102–3
experience, effect on neural development 74–5
experimental condition 21
experimental group 23
explanatory power 194–5
explicit memory 198–200, 201, 365
  autobiographical memory component 200, 209–11
  hippocampal-neocortical networks for 206
  roles of neocortex, hippocampus and amygdala in 203–4
external factors
  influencing mental wellbeing 344
  interactions with internal factors (affecting mental wellbeing) 344–5
  supporting resilience 332
external stressors 91
external validity 8, 35
extinction 145
extraneous variables (EV) 23, 25, 429
extrinsic factors, in circadian rhythm sleep disorders 294
Exxon Valdez oil spill 282
eye blinks 261, 262

family and friends
  lack of understanding about person with phobias 370
  social support from, to assist mental wellbeing 402
  as supporters of phobic person 381, 382
family and kinship (Aboriginal and Torres Strait Islander peoples) 339
fatigue
  affect on performance 291–2
  and drowsy drivers 288–9
fear 56, 144, 146
fear conditioning experiments 144–6, 165, 366
fear hierarchy 377–8, 380
fear of public speaking 368
fear of spiders 365, 371–2, 378–80, 383
fieldwork/field studies 17–18
First Nations peoples 168
  sung narratives 223, 224
flight-or-fight-or freeze response (FFF) 57, 94–8, 365
flight-or-fight reactions 95

flooding 377
flourishing 326–7, 330, 331
fMRI (functional magnetic resonance imaging) 147, 164, 210, 215
fonts 435
foods, and mental wellbeing 394
fortune-telling 381
free recall tests 192
freeze response 95–6
frequency table 431

gambling behaviour 153–4
gamma-amino butyric acid (GABA) 68, 364
   agonists, for treatment of specific phobia 373–4
   dysfunction, contribution to anxiety disorders/phobia 68–9, 364, 365
gastrointestinal (GI) system 59, 109, 110
Gayaa Dhuwi (Proud Spirit) Declaration 337
Gay'wu Group of Women of Bawaka 171
gaze following and shared-attention in infants 165
Gee, Dr Graham 336
General Adaptation Syndrome (GAS) 101–4
   alarm-reaction stage 101, 102
   exhaustion stage 102–3
   resistance stage 101–2
   strengths and limitations 103
generalisability 433
ghrelin 305, 306
glossary 463–72
glossophobia 368
glutamate 68, 76, 78
   vs GABA (neural transmission) 69–70
'good' bacteria 111, 112
graphs 29, 431–2
group allocation and variables 24
gut
   connection to vagus nerve 110, 111
   impact on psychological processes and behaviour 112–14
gut–brain axis (GBA) 59, 109–14
gut control 58–9
gut microbiota 109, 111
   interaction with central nervous system 110, 111, 112
   role in regulation of HPA axis 112, 113
   and stress 113–14

health and safety 40
heartbeats 261, 262
Hebb's rule 75

highly superior autobiographical memory (HSAM) 210–11
hippocampal-neocortical networks
   for consolidation of episodic memory 204–5
   for explicit memory 206
   retrieval of ABMs and episodic-future thinking 210
hippocampus 203, 204, 206, 365
histograms 29, 30, 31
H.M. (Henry Molaison) (patient), memory and the brain 186–8, 198, 201
holistic framework 336, 339
homeostasis 57, 95, 96
homeostatic sleep drive 259, 305
human nervous system 50–83
   and method of loci 220–1
   modelling interaction with the external world 59–60
hydration, and mental wellbeing 394, 395
hyperphantasia 212
hypnic jerks 248
hypnogogic state 248
hypnograms 247
hypothalamic–pituitary–adrenal (HPA) axis 97, 101, 112, 113
hypothalamus 258, 304
hypothesis 9–10, 426–8, 433

iconic memory 189
identification 17
illness, in resistance and exhaustion stages 101, 102
imaginal flooding 378
immediate serial recall 190
immune system 110
impaired actions and reactions, and partial sleep deprivation 285–6
implicit memory 201
   neural basis 203
   role of basal ganglia and cerebellum in 207
incomplete data 36
independent variable (IV) 10, 22–3, 25, 427, 428
inequality, and stress 105
infants
   conditioned fear response 144–5, 366
   gaze following and shared-attention in 165
   REM sleep 246–7, 267
   sleep requirements 267
informed consent 40, 429
inhibitory neurotransmitters 68–9

initialisms 218
insomnia 270
insulin 206, 305
integrity (ethical concept) 39, 429
interference 190, 194
intergenerational trauma 340
internal factors
   affecting mental wellbeing 344
   interactions with external factors (affecting mental wellbeing) 344–5
   supporting resilience 332
internal organs and glands
   control of 55
   parasympathetic response 57–8, 59
   sympathetic response 56–7, 58, 59
internal stressors 91
internal validity 8, 35
International classification of disease (ICD), 11th edition (ICD-11) 327
internet gaming disorder 79
interneurons 62
intrinsic factors, in circadian rhythm sleep disorders 294
introduction (poster) 436
involuntary association 142
involuntary behaviours 138
involuntary muscles 55, 61
involuntary responses 55, 61, 62

jet lag 259
justice 39, 429

K-complex 252
kinship mind (sand talk symbol) 174
kinship system (Aboriginal and Torres Strait Islander peoples) 169–70, 339

languishing 327, 330, 331
laughter 120
Lazarus and Folkman's Transactional Model of Stress and Coping 106–9
learned aggression in children through observation of an adult model (Bandura's study) 162–4
learning 73
   Aboriginal and Torres Strait Islander peoples' approaches 137, 166–76
   behaviourist approaches 137–59
   by modelling 369
   classical conditioning 138–47, 158–9
   definition 137
   operant conditioning 148–60
   process of learning based on performance on a puzzle maze (investigation) 438–9

and sleep deprivation 285
social-cognitive approach 137, 160–6
within the womb 136–7
learning and memory
neural basis of 73–81
synaptic changes as a mechanism for 75
and synaptic plasticity 75–8
learning associations 138
leptin 305, 306
level of processing (long-term memory) 191
levels of functioning 325–6
and continuum of mental disorder 327–8
and continuum of mental wellbeing 326–7
and mental ill-health 328
levodopa (L-dopa) 71–2
life span
changes in sleep over the 265–72
recommended hours of sleep by different age groups 269, 396
light, and melatonin 260–1, 262
light sleep 251, 252
limitations
of conclusions 38
of your study 13, 433
line graphs 29, 30, 431, 432
literature reviews 9, 20–1, 334
'Little Albert' experiment 144–6, 366
'lock-and-key' process at the synapse 67
logbook 8–9, 425
long-term depression (LTD) 77–8
long-term memory (LTM) 191–4, 195
components 201
retrieval and elaborative encoding as protection against forgetting from 194
retrieval of information from 189, 190
structure 198–202
transfer of information from STM to 189
see also episodic memory; explicit memory; implicit memory; semantic memory
long-term potentiation (LTP) 76–7, 78, 142, 204–5
role in contributing to phobia 364–5
long-term retention 189
practicing retrieval improves 192–3
low 'glycaemic index' foods 394

McKague, Meredith 166–7
magnetic resonance imaging (MRI) 215

maintenance rehearsal 189
maladaptive behaviours 373, 376, 382
matched groups 26
matched-participants design 26
materials 428
mean 31
measurement errors 35–6
measures of central tendency 31, 431
measures of variability 431
medial temporal lobe 203
median 31
medications, to treat specific phobia 373–4
meditation 392
mindfulness meditation 392–3, 398–401, 403–4
simple meditation practice 392–3
to reduce stress 99, 119
meditative state 251
melatonin 258–9, 260–1, 262, 267, 268, 296–7, 299, 305
memory 73
and the brain (patient H.M.) 186–8, 198, 201
and learning, neural basis 73–81
long-term 189, 190, 191–4, 198–208
multi-store model 188–97
psychobiological process of 183–229
short-term 188, 189–91
and sleep deprivation 285
what is it? 188–97
see also specifics, e.g. semantic memory
memory bias 368
memory techniques 218–25
mental disorders 324, 327
anxiety disorders 347–8, 349
continuum of and levels of functioning 327–8
diagnosis 327
interaction with mental wellbeing 330–1
and level of functioning 327, 328
myths versus facts 329
seeking advice about 328–9
self-report scales to measure functioning in six domains 327
stigma associated with 328, 329–30, 370
WHO definition 327
mental health 324
and mental wellbeing, dual-continuum model 325–32, 343–53
and stress, anxiety and phobia 346–50
WHO definition 324
Mental Health Continuum Scale 326
mental health problems 328
seeking advice about 328–9

mental ill-health 324, 329
and levels of functioning 328
mental imagery 199–200, 212
lack of 212
role in ABM and episodic future-thinking in 212
self-report questionnaires 213
mental wellbeing
Aboriginal and Torres Strait Islanders understanding of 324–5, 335–43
and anxiety 347–8
biopsychosocial approach to maintaining 392–410
continuum of, and levels of functioning 326–7
core components 326
definition 324
finding yourself (case study) 345–6
importance of sleep for 282–3
and improving sleep–wake patterns 303–9
influence of COVID-19 pandemic on 334
influence of internal and external factors on 344–5
interaction with mental disorder 330–1
maintenance of 390–410
and mental health, dual-continuum model 325–32, 343–53
mindfulness meditation for 392–3, 398–401, 403–4
as multidimensional framework 325
protective factors to maintain 393–405
and resilience 331–2
and stress 346–7
ways of considering 325–43
see also social and emotional wellbeing (SEWB)
method 428–9
method of loci 219–21
history 222
and the human nervous system 220–1
methodology see research methodology
microbiome–gut–brain axis 111
microbiota see gut microbiota
microsleeps 286
middle school students, mindfulness training 399–400
mild anxiety 348
mind and emotions (Aboriginal and Torres Strait Islander peoples) 339

mindfulness meditation 392–3, 398–401
   online 403–4
'mirror neurons' activated during observation of movement 164–5
mirror-tracing task 187, 201, 207
mixed design 25, 27–8
mnemonics 218–25
   based on written language 218–19
   method of loci 219–22
   sung narratives in oral cultures 222–4
mode 31
modelling 18, 369
moderate anxiety 348
moderate mental wellbeing 327
mood 72
motivation and reinforcement 161
motor cortex 207
motor messages 51, 52
motor neurons 54, 61, 62, 63
multi-store model of memory 188–97
   explanatory power 194–5
   strengths and weaknesses 194–5
multidimensional framework of mental wellbeing 325
multimodal system of knowledge 168, 169
myelin sheath 66

naturally occurring altered states of consciousness 244, 245
   see also sleep
negative correlation 16
negative punishment 149, 150, 151, 155
negative reinforcement 148, 149, 150, 151, 155
negatively skewed distribution 31
neocortex 203, 204, 205
nerve impulses 66
nervous system 50–1, 65
   roles and subdivisions 51–60
   see also specifics, e.g. peripheral nervous system
neural basis
   of episodic memory 204–5
   of explicit and implicit memories 203–9
   of learning and memory 73–81
   of semantic memory 205–6
neural development
   experience effect on 74–5
   and synaptic plasticity 75–8
neural transmission 65–73
neurofibrillary tangles 214, 215
neuromodulators 71–2

neurons 52, 53, 65, 66
   chemical communication between 66–7
   structure 65–6
neurotransmitters
   chemical messengers at the synapse 65, 66–7, 71
   drugs affecting 67
   dysfunction, contribution to anxiety disorder 364
   excitatory and inhibitory effects 68–70
   in synaptic plasticity 76–8
   vs neuromodulators 71
   see also specifics, e.g. gamma-amino butyric acid (GABA)
neutral stimulus (NS) 140, 141
newborns, sleep requirements 266–7
night-shift work/workers 297, 298, 307
no correlation 16
no mental ill-health 328
non-maleficence 39, 429
non-rapid eye movement (NREM) sleep see NREM sleep
non-scientific ideas 9
noradrenaline 56, 95, 97, 364
normal distribution 29
normal waking consciousness (NWC) 243, 244, 251, 252
not encouraging avoidance behaviours 382
NREM sleep 245–6, 259, 261, 394–5
   across the life span 266–70
   and body and brain temperature 305
   in Delayed Sleep Phase Syndrome 295
   and shift workers 298
   and sleep deprivation 286, 287
   stages 245, 247, 248, 251–2, 253, 261, 262, 269
nutritional intake, and mental wellbeing 394, 395

obesity 306
object imagery 212
observation of movement, mirror neurons activate during 164–5
observational learning 160–2
   conditioned fear response 165
   four cognitive processes in 161
   gaze following and shared-attention in infants 165
   learned aggression in children through observation of an adult model 162–4
   'mirror neurons' activated during 164–5
older adults

sleep patterns 269
sleep requirements 268–9
sound night's sleep more elusive in 270–1
olfactory bulb 204
online mindfulness meditation 403–4
operant conditioning 138, 148–60
   in advertising campaigns 158, 160
   and perpetuation of a specific phobia 367–8
   reinforcement and punishment 148–50
   three-phase process (ABC model) 152–8
   to improve study habits 150–1
opinions 21
oral cultures 170, 218
   sung narratives of 222–4
order effects 26, 27
outliers 30, 36–7, 433
overgeneralisation 381

PANAS wellbeing survey 10, 15, 22
paradoxical sleep 246
parasympathetic nervous system 52, 55, 57–8, 95, 109–10
   and freeze response 96
   physiological changes in internal organs and glands 58, 59
Parkinson's Disease (PD), treatment 71–2
partial sleep deprivation 283
   effects of 284–6
   recovering from 288–9
participant allocation in controlled experiments 24, 25–8
patella 'knee-jerk' reflex 62, 63, 64
pattern mind (sand talk symbol) 175
Pavlov, Ivan 144
   and discovery of classical conditioning 138–40
   salivation in dogs 138–9, 141
   three-phases of classical conditioning 141, 142
Perceived Stress Scale (PSS) 117
percentage 33
percentage change 33
peripheral nervous system (PNS) 51, 52, 53, 54–6, 220–1
   role 54
   subdivisions 54–6
   working with CNS 54
personal errors 35, 36
phobias 327, 349–50, 351
   anxiety and stress, key similarities and differences 350

top ten 363
see also specific phobias
physical activity 56
physiological responses to stress 91, 92, 94–100
    flight-or-fight-or-freeze response 57, 94–8, 365
    and gut microbiota 113
    Selye's General Adaptation Syndrome 101–4
pineal gland 258
planning your investigation 14–21
polysomnogram 253
population 7, 9, 10, 11
positive correlation 16, 17
positive punishment 149, 150, 151, 155
positive reinforcement 148, 149, 150
positively skewed distribution 31
possum-skin cloaks in social and emotional wellbeing 342–3, 407
posterior (back) temporal lobe hub 205
posters 434–8
postsynaptic neurons 66, 67, 68, 76
precision 33, 34, 35, 430
predictions (controlled experiments) 9–10
prefrontal cortex 204, 207
pre-school children, sleep 260–1
presynaptic neurons 66, 67, 76
primacy effect 194, 195
primary appraisal 106
primary data 11, 431
probiotics 112, 113
procedural memory 201
procedure 428–9
process development 18–19
processing quantitative data 29–33
product, process or system development 18–19
protective factors to maintain mental wellbeing 393–405
    biological protective factors 393–7
    psychological protective factors 397–401
    social protective factors 402
psychobiological process
    of memory 183–229
    stress as example of 88–123
psychoeducation 380–2
    challenging unrealistic or anxious thoughts 381–2
    for families and supporters of a phobic person 381, 382
    not encouraging avoidance behaviours 382

psychological constructs 6
psychological factors contributing to specific phobia 363, 366–8, 370
psychological internal stressors 91
psychological interventions to manage specific phobia 376–80
psychological models 7
psychological processes and behaviour, gut impact on 112–14
psychological protective factors that maintain mental wellbeing 397–401
psychological research investigations
    analysing and evaluating research 29–42
    controlled experiment in detail 21–9
    methodologies 14–21, 428
    process 8–14
    see also research investigations
psychological responses to stress 91–2, 104–9
    Lazarus and Folkman's Transactional Model of Stress and Coping 106–9
psychological theory 7
psychological wellbeing 326
psychotherapies 373, 376–80
punishing consequences 148
punishment 148, 149, 151
    measuring the effects of 149–50

qualitative data 11
quality of data, and sample size 37
quantitative data 11, 428, 431
    processing 29–33
quasi-experimental designs 26
questionnaires 11, 429

R U OK?Day 324, 325, 329
random allocation 26
random errors 35–6
random sampling 12
randomised controlled trial (RCT) 21
rapid eye movement (REM) sleep
    see REM sleep
recency effect 194, 195
receptor cells 54
receptor sites 66
reciprocal inhibition 377
recognition memory 192
recommendations 13–14
recording your data 13
references 436
reflex control 64
reflexes 62–3, 64, 138, 140

reinforcement 148, 149, 150–1
    measuring the effects of 149–50
relaxation response 95
relaxation techniques 373, 373–5
relaxation training (CBT) 376
REM rebound 288
REM sleep 245, 246–7, 252, 259, 261, 262, 395
    across the life span 266–70
    and body temperature 305
    in Delayed Sleep Phase Syndrome 295
    and shift workers 298
    and sleep deprivation 285, 286, 287, 288
reminiscence bump 210
repeatability 34, 430
repeated-measures design 27
representative sample 12
reproducibility 34, 430
reproduction 161
rerouting 77, 78
Research Australia 424
research design, evaluating 38
research investigations 438–9
    conducting 11–14, 431–4
    designing 10–11, 425–30
    see also psychological research investigations
research methodology 14–21, 428
    posters 436
research process 8–41
research question 9, 425–6
resilience 331–2, 393
    building 332, 333
    internal and external factors supporting 332
    resistance stage 101–2
respect 39, 429
response, definition 138
restoration theory (of sleep) 265
results (poster) 436
retention 161
retention interval 190
retrieval 189, 192–3
retrieval cues 192
retrieval failure 194
retrograde amnesia 198
re-uptake 67
reward system 71
rewarding consequences 148, 153–4
risk assessment form 40
risk assessments 40
risk factors of suffering mental health problems or mental disorder 331, 338, 339, 344, 393
rotating shift schedule 298–9

safety 40, 429
safety data sheets (SDS) 40
salivation response in dogs (Pavlov) 138–9, 140–1
sample 7, 9, 10, 11, 428
sample size 13
  and quality of data 37
sampling 11–13
sampling methods 12–13, 428
sand talk symbols for five ways of knowing 173–5
schemas 199
school-aged children, screen time and sleep 265–6
school counsellor 329
schools, mindfulness training in 399–400
science communication 434–8
scientific investigations *see* psychological research investigations; research investigations
scientific posters 434–8
  example 437
  sections 435–6
screen time and sleep among school-aged children and adolescents 265–6
seal training 155–6
secondary appraisal 106–7
secondary data 11
seek help to reduce stress 120
self-determination
  Aboriginal and Torres Strait Islander peoples 340, 408–9
  definition 408
self-reporting bias 429
self-reports (sleep) 253–4
Selye's General Adaptation Syndrome 101–4
SEMA3: an example of a product, process or system development (case study) 19–20
semantic autobiographical memory (SAM) 209, 210
semantic hubs 205–6
semantic memory 198, 199–200, 201
  neural basis of 205–6
  role in remembering the past and imagining the future 210–11
  where words are stored 206
semantic networks 199
semantic scaffolding theory 210
sensations 54
sensory information 51, 52, 61
sensory memory 188, 189

sensory neurons 54, 62, 63
sensory stimuli, responses to 61–5
serial position curve 194, 195
serial position effect 194, 196
serotonin 72, 364
Seven Sisters Songline 223–4
severe anxiety 348
severely deficient autobiographical memory (SDAM) 210, 211
shift work 259, 297
  and circadian rhythms 297–9
  effects of 298–9
  reducing the effects of 299
  and weight gain 307
shift work disorder 297–8
shock 101
short-term memory (STM) 188, 189–91, 194, 195
  capacity 190
  duration 190
  encoding and retrieval processes 189–90
simulation 18
sitting meditation 399
skewed distribution 30, 31
skin conductance response (SCR), in socially conditioned fear studies 146–7, 165
Skinner, B. F., operant conditioning 148–55
Skinner box 149–50, 153
sleep 72, 120, 244
  adequate, and mental wellbeing 394–7
  categories of 244–7
  changes in over the life span 265–72
  and circadian rhythms 258–61, 263–4
  demand for: how much is enough? 266–70, 396
  how to get better sleep 396
  importance of to mental wellbeing 282–3
  as naturally occurring altered state of consciousness 244, 245
  objective methods to measure physiological changes 250–3
  in pre-school children 260–1
  as a psychological construct 244–50
  purpose of 245, 265
  recommendations across the life span 269, 396
  reduced quality aggravates pre-existing psychological conditions 397
  and screen time among school-aged children and adolescents 265–6

self-reports 253–4
  subjective methods for recording 253–5
  and ultradian rhythms 261–2
  video monitoring 255–6
sleep cycles 247–8, 261–2, 293
sleep demand 258
sleep deprivation 282, 283–4, 305
  in adolescence 296
  effects of 284–8
  effects of, comparison with alcohol consumption 289–90, 291–2
  see also partial sleep deprivation; total sleep deprivation
sleep-deprivation psychosis 288
sleep diaries 254–5
sleep disorders 293–4, 301
  circadian rhythm 294–6, 297, 299–300
  and shift work 297–9
  sleep–wake shifts in adolescence 296–7
sleep hygiene 303
  improving 303–6, 396
sleep laboratory 250
sleep schedule 396
sleep spindles 252
sleep–wake cycles 72, 258, 267, 294, 296, 297, 299
sleep–wake homeostasis 258, 259
sleep–wake patterns
  improving, and mental wellbeing 303–9
  regulation 258–64, 300
sleep–wake shifts in adolescence 296–7
  timing of melatonin and cortisol release 296–7
sleepiness scale 291
slow-wave sleep (SWS) 245, 248
smartphone apps 18–20
social and emotional wellbeing (SEWB) 336
  as holistic framework 336, 339
social and emotional wellbeing (SEWB), Aboriginal and Torres Strait Islanders understanding of 324–5, 335–43, 406
  collectivist understanding 337
  cultural determinants 406–10
  as culturally responsive 337
  experiences and expressions of individuals, families and communities 338, 341–2
  historical, political, cultural and social contexts/determinants 337, 338, 339–41, 406

nine guiding principles 337
revitalising traditional possum-skin cloak making 342–3, 407
seven domains of connections 337, 338–9, 406
visual representation 337–8
social anxiety disorder 349
social behaviour, and gut microbiota 112
social-cognitive approaches to learning 137, 160–6
social-cognitive model of learning
Bandura's studies 160–4
current applications 164–6
four cognitive processes 161
social connections 119
social factors contributing to specific phobia 363, 369–70
social interventions to manage specific phobias 381–3
social media habits, behaviour modification to change 156–8
social networks 402
social protective factors to maintain mental wellbeing 402
social support, and mental wellbeing 402
social wellbeing 326
soma 65, 66
somatic nervous system (SNS) 52, 54, 61, 62, 220, 221
somatic reflexes 140
Songlines (Songspirals) 170–1, 222–4
spatial imagery 212
specific hypothesis 10
specific phobias 327, 349–50, 362–3
biological factors contributing to 363, 364–5, 370
biopsychosocial model to explain 363–72
development 363–73
DSM-V criteria 350
evidence-based interventions 373–84
investigation (case study) 383–4
psychological factors contributing to 363, 366–8, 370
social factors contributing to 363, 369–70
sub-categories 349
spinal cord 51, 52
cross-section 52
role 52–3, 61
and spinal reflex 62
spinal nerves 51, 52
spinal reflexes 62–3

spine 52
spontaneous recovery 145
sprouting 76, 78
stage 1 NREM sleep 247, 248, 252, 253, 261, 262, 269
stage 2 NREM sleep 247, 248, 252, 261, 262, 269
stage 3 NREM sleep 245, 246, 247, 248, 261, 262, 267, 269
standard deviation (SD) 32
states of consciousness 243–4
stigma 369
around seeking treatment for mental illness contributing to phobia 369–70
associated with mental disorders 328, 329–30, 370
stimuli (stimulus)
definition 138
responses to 61–5
stimulus–response associations 138
stimulus–stimulus associations 138
Stolen Generations 168, 339–40, 409
story mind (sand talk symbol) 174
stratified sampling 12
strength of correlation 16, 17
stress 90, 91, 247
acute or chronic 92–3, 99, 113
as an example of psychobiological process 91–123
anxiety and phobia, key similarities and differences 350
building resilience 332
can lead to inequality 105
and confidence 105
coping strategies 115–20
and gut microbiota 113–14
interaction with anxiety 347
manifestations of 91–2
meditation to reduce chronic stress 99
and mental wellbeing 346–7
relationship with stressors and stress reaction 93–4
stress hormones 56, 57–8, 95
and confidence 105
physiological effects 95, 96–8
see also adrenaline; cortisol; noradrenaline
stress responses 93–4, 98
as biological processes 94–8, 101–4
and dysbiosis in gut microbiota 113
flight-or-fight-or-freeze response 94–8
Lazarus and Folkman's Transactional Model of Stress and Coping 106–9

physiological responses 91, 92, 94–8, 101–4
as psychobiological response 93
as psychological process 91–2, 104–9, 113
Selye's General Adaptation Syndrome 101–4
stressors 91–2
cognitive appraisal 106–9
relationship with stress and stress reaction 93–4
and release of cortisol 97
stretch reflex 62, 63
sung narratives of oral cultures 222–4
suprachiasmatic nucleus (SCN) 258, 259, 294, 303
sympathetic nervous system 52, 55, 56–7, 109
and flight-or-fight reactions 95
and freeze reaction 96
physiological changes in internal organs and glands 56–7, 58, 59
synaesthesia 212
synapse 66
'lock-and-key' process 67
and neurotransmitters 66–7, 71
synapse formation 74
synaptic changes
as a mechanism for learning and memory 75
process of long-term depression (LTD) 77–8
process of long-term potentiation (LTP) 76–7, 78
synaptic cleft 66, 71
synaptic plasticity 75–8
synaptic pruning 77, 78
synaptic vesicles 66
system development 18–19
systematic desensitisation 377–80
systematic errors 35, 36

tables 29, 30
Tanderrum gathering (of five tribes of the Kulin Nation of Victoria) 407
tau protein neurofibrillary tangles 214, 215
temperature, and sleep 304–5
theta waves 251, 252
thinking processes disruption, and partial sleep deprivation 284–5
threatening anxiety 348
three-phase process of classical conditioning 141, 142
in advertising campaigns 158–60

three-phase process of operant conditioning 152–4
  in advertising campaigns 160
  in animal training 154–6
  for behaviour modification 156–8
title (poster) 435
tonic immobility 96
tonically active systems 95
Torres Strait Islanders 168
  see also Aboriginal and Torres Strait Islander peoples
total sleep deprivation 283, 288
transfer of information (from STM to LTM) 189
Transactional Model of Stress and Coping 106–9
  primary appraisal: how significant or threatening is the event? 106
  secondary appraisal: how can I cope with the event? 106–7
  strengths and limitations 108
treatment group 21
true value 33
trypophobia 362

ultradian rhythms 261–2
  comparison with circadian rhythm 262
Uluru Statement from the Heart 341
uncertainty (investigations) 36, 432–3

unconditioned response (UCR) 140, 141
unconditioned stimulus (UCS) 140, 141, 142, 145
unconscious responses 61–5
Ungunmerr-Baumann, Dr Miriam-Rose, on Dadirri (deep listening) 171–2
Unit 3, Area of Study 1 review 127–31
Unit 4
  Area of Study 2 review 415–19
  Outcome 3 422, 425, 431, 434
unrealistic or anxious thoughts, challenging 381–2

vagus nerve 110
  link to gastrointestinal (GI) system 110, 111
validity 7–8, 34–5, 430
variability 32
variables 7, 10, 16, 17, 22–5, 428
  summary 24–5
vertebrae 52
video games affect on the brain 78–80
video monitoring (sleep) 255–6
visual deprivation 74
visual imagery 212
voluntary behaviours 138
voluntary movements 54
voluntary participation 40, 429
voluntary responses 61

walking meditation 399
water 394
Watson, John B.
  behaviourism 138
  classically conditioned emotional response in human infant 144–6, 366
  'Little Albert' experiment 144–6, 366
ways of knowing (Aboriginal and Torres Strait Islander peoples) 137, 166, 167, 168–76
WEIRD samples 13
Wernicke's area 205
wholegrain cereals 394
winding down (before sleep) 396
withdrawal reflex 62, 140
withdrawal rights 40, 429
within-subjects design 25, 27
'working' memory 189

yarning 173–5
Yunkaporta, Dr Tyson 173
  sand talk symbols for ways of knowing 173–5
  on yarning 173

zeitgebers 258, 259, 303, 304, 306